Imperial College London

AF606587

HIV Protocols

METHODS IN MOLECULAR BIOLOGY™

John M. Walker, SERIES EDITOR

489. Dynamic Brain Imaging: Methods and Protocols, edited by Fahmeed Hyder, 2009
485. HIV Protocols: Methods and Protocols, edited by Vinayaka R. Prasad and Ganjam V. Kalpana, 2009
484. Functional Proteomics: Methods and Protocols, edited by Julie D. Thompson, Christine Schaeffer-Reiss, and Marius Ueffing, 2008
483. Recombinant Proteins From Plants: Methods and Protocols, edited by Loïc Faye and Veronique Gomord, 2008
482. Stem Cells in Regenerative Medicine: Methods and Protocols, edited by Julie Audet and William L. Stanford, 2008
481. Hepatocyte Transplantation: Methods and Protocols, edited by Anil Dhawan and Robin D. Hughes, 2008
480. Macromolecular Drug Delivery: Methods and Protocols, edited by Mattias Belting, 2008
479. Plant Signal Transduction: Methods and Protocols, edited by Thomas Pfannschmidt, 2008
478. Transgenic Wheat, Barley and Oats: Production and Characterization Protocols, edited by Huw D. Jones and Peter R. Shewry, 2008
477. Advanced Protocols in Oxidative Stress I, edited by Donald Armstrong, 2008
476. Redox-Mediated Signal Transduction: Methods and Protocols, edited by John T. Hancock, 2008
475. Cell Fusion: Overviews and Methods, edited by Elizabeth H. Chen, 2008
474. Nanostructure Design: Methods and Protocols, edited by Ehud Gazit and Ruth Nussinov, 2008
473. Clinical Epidemiology: Practice and Methods, edited by Patrick Parfrey and Brendon Barrett, 2008
472. Cancer Epidemiology, Volume 2: Modifiable Factors, edited by Mukesh Verma, 2008
471. Cancer Epidemiology, Volume 1: Host Susceptibility Factors, edited by Mukesh Verma, 2008
470. Host-Pathogen Interactions: Methods and Protocols, edited by Steffen Rupp and Kai Sohn, 2008
469. Wnt Signaling, Volume 2: Pathway Models, edited by Elizabeth Vincan, 2008
468. Wnt Signaling, Volume 1: Pathway Methods and Mammalian Models, edited by Elizabeth Vincan, 2008
467. Angiogenesis Protocols: Second Edition, edited by Stewart Martin and Cliff Murray, 2008
466. Kidney Research: Experimental Protocols, edited by Tim D. Hewitson and Gavin J. Becker, 2008.
465. Mycobacteria, *Second Edition*, edited by Tanya Parish and Amanda Claire Brown, 2008
464. The Nucleus, Volume 2: Physical Properties and Imaging Methods, edited by Ronald Hancock, 2008
463. The Nucleus, Volume 1: Nuclei and Subnuclear Components, edited by Ronald Hancock, 2008
462. Lipid Signaling Protocols, edited by Banafshe Larijani, Rudiger Woscholski, and Colin A. Rosser, 2008
461. Molecular Embryology: Methods and Protocols, Second Edition, edited by Paul Sharpe and Ivor Mason, 2008
460. Essential Concepts in Toxicogenomics, edited by Donna L. Mendrick and William B. Mattes, 2008
459. Prion Protein Protocols, edited by Andrew F. Hill, 2008
458. Artificial Neural Networks: Methods and Applications, edited by David S. Livingstone, 2008
457. Membrane Trafficking, edited by Ales Vancura, 2008
456. Adipose Tissue Protocols, Second Edition, edited by Kaiping Yang, 2008
455. Osteoporosis, edited by Jennifer J. Westendorf, 2008
454. SARS- and Other Coronaviruses: Laboratory Protocols, edited by Dave Cavanagh, 2008
453. Bioinformatics, Volume 2: Structure, Function, and Applications, edited by Jonathan M. Keith, 2008
452. Bioinformatics, Volume 1: Data, Sequence Analysis, and Evolution, edited by Jonathan M. Keith, 2008
451. Plant Virology Protocols: From Viral Sequence to Protein Function, edited by Gary Foster, Elisabeth Johansen, Yiguo Hong, and Peter Nagy, 2008
450. Germline Stem Cells, edited by Steven X. Hou and Shree Ram Singh, 2008
449. Mesenchymal Stem Cells: Methods and Protocols, edited by Darwin J. Prockop, Douglas G. Phinney, and Bruce A. Brunnell, 2008
448. Pharmacogenomics in Drug Discovery and Development, edited by Qing Yan, 2008.
447. Alcohol: Methods and Protocols, edited by Laura E. Nagy, 2008
446. Post-translational Modifications of Proteins: Tools for Functional Proteomics, *Second Edition*, edited by Christoph Kannicht, 2008.
445. Autophagosome and Phagosome, edited by Vojo Deretic, 2008
444. Prenatal Diagnosis, edited by Sinhue Hahn and Laird G. Jackson, 2008.
443. Molecular Modeling of Proteins, edited by Andreas Kukol, 2008.
442. RNAi: Design and Application, edited by Sailen Barik, 2008.
441. Tissue Proteomics: Pathways, Biomarkers, and Drug Discovery, edited by Brian Liu, 2008
440. Exocytosis and Endocytosis, edited by Andrei I. Ivanov, 2008
439. Genomics Protocols, Second Edition, edited by Mike Starkey and Ramnanth Elaswarapu, 2008
438. Neural Stem Cells: Methods and Protocols, Second Edition, edited by Leslie P. Weiner, 2008
437. Drug Delivery Systems, edited by Kewal K. Jain, 2008
436. Avian Influenza Virus, edited by Erica Spackman, 2008
435. Chromosomal Mutagenesis, edited by Greg Davis and Kevin J. Kayser, 2008
434. Gene Therapy Protocols: Volume 2: Design and Characterization of Gene Transfer Vectors, edited by Joseph M. Le Doux, 2008
433. Gene Therapy Protocols: Volume 1: Production and In Vivo Applications of Gene Transfer Vectors, edited by Joseph M. Le Doux, 2008
432. Organelle Proteomics, edited by Delphine Pflieger and Jean Rossier, 2008
431. Bacterial Pathogenesis: Methods and Protocols, edited by Frank DeLeo and Michael Otto, 2008
430. Hematopoietic Stem Cell Protocols, edited by Kevin D. Bunting, 2008
429. Molecular Beacons: Signalling Nucleic Acid Probes, Methods and Protocols, edited by Andreas Marx and Oliver Seitz, 2008
428. Clinical Proteomics: Methods and Protocols, edited by Antonia Vlahou, 2008
427. Plant Embryogenesis, edited by Maria Fernanda Suarez and Peter Bozhkov, 2008
426. Structural Proteomics: High-Throughput Methods, edited by Bostjan Kobe, Mitchell Guss, and Thomas Huber, 2008
425. 2D PAGE: Sample Preparation and Fractionation, Volume 2, edited by Anton Posch, 2008

METHODS IN MOLECULAR BIOLOGY™

HIV Protocols

Second Edition

Edited By

Vinayaka R. Prasad, PhD

and

Ganjam V. Kalpana, PhD

Albert Einstein College of Medicine, Bronx, NY, USA

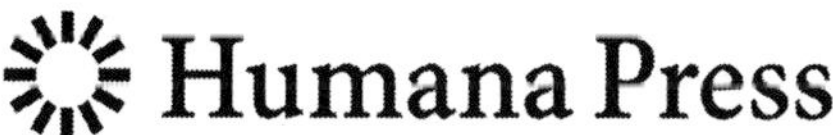

Editor
Vinayaka R. Prasad
Albert Einstein College of Medicine
Bronx, NY
USA
prasad@aecom.yu.edu

Ganjam V. Kalpana
Albert Einstein College of Medicine
Bronx, NY
USA
kalpana@aecom.yu.edu

Series Editor
John M. Walker
University of Hertfordshire
Hatfield Herts
UK

ISBN: 978-1-58829-859-1 e-ISBN: 978-1-59745-170-3
ISSN: 1064-3745 e-ISSN: 1940-6029
DOI 10.1007/978-1-59745-170-3

Library of Congress Control Number: 2008938341

Printed on acid-free paper

springer.com

Preface

Why another book of HIV protocols? The question is sure to arise in the minds of the readers. The AIDS epidemic continues unabated despite strong advances in therapeutics and an unprecedented level of efforts in vaccine development. Although some of the reasons for the failure in AIDS control can be attributed to poor prevention measures or paucity of antiretrovirals, a major drawback is the absence of efficacious, potent antiretrovirals that can both suppress viral load completely and that do not have toxic side effects. Toxicity can lead to nonadherence, which in turn results in poor virus control, emergence of drug resistance and the eventual clinical drug failure. Development of novel drugs and vaccines require definition of new targets, better definition of already known viral targets and understanding the interplay between viral and host factors. Similarly, in order to develop an AIDS vaccine, we need to be equipped with effective methods to measure immune response. More importantly, these studies require the development of efficient and powerful in vitro and in vivo systems to study viral replication and pathogenesis. Therefore, HIV researchers have a real need for access to well-described, state-of-the-art methods to study HIV.

Approaches in HIV/AIDS investigation have continuously advanced in tune with the evolution of modern experimental science. Development of new technologies in investigating familiar aspects of HIV replication or immune response to it have led to new insights that have improved our understanding of the biology of HIV. In compiling this collection, our objective is threefold. First, we aim to document up-to-date protocols available for select aspects of HIV biology. Second, we bring together both virological and immunological approaches in a single volume. Third, we provide a comprehensive account of techniques that are not already part of an existing HIV protocol book.

HIV Protocols, Second Edition, is organized into five sections. Section I delineates the methods to isolate full-length DNA clones of HIV-1 from patient samples, isolation of HIV-1 particles free of contaminating cellular proteins and a method to titer these virus particles. Section II delineates the study of early and late events. Early events include entry, reverse transcription, nuclear transport, integration as well as recombination, a process that occurs during reverse transcription. A thorough study of early events would be incomplete without the analysis of complexes formed during reverse transcription and prior to integration as well as the interaction between viral and host proteins within these complexes. In the subsection on late events, we take you through methods to study assembly and particle production within the producer cells and the use of cell free systems to study the interaction of viral proteins and nucleic acids including the cognate $tRNA^{Lys,3}$.

No HIV-1 investigation is complete without the analysis of the dynamics of host-virus interactions. HIV-1, being an intra-cellular parasite, not only invades the host, but also subverts cellular antiviral mechanisms and hijacks host proteins for its own purposes. Section III explores approaches to investigate the interplay between the host and the virus by employing genetic, molecular and cellular techniques including novel small animal models. Methods to investigate specific, immunological techniques to understanding host-HIV-1 interplay are discussed in Section IV. Chapters in this section delineate methods to study mucosal immunity, T-cell responses and antiviral responses in cell culture and in Rhesus monkey models.

The last section of the book delves into the intense battle between the host and the HIV-1. The virus continues to evade antiretrovirals owing to its ability to develop drug resistance and its baffling ability to evolve and escape the immune system. The chapters included should facilitate investigations of drug resistant viruses and virus evolution.

We would like to draw the readers' attention to the Notes sections in each chapter. These notes come from the experts who have used these methods successfully many times and contain many 'tricks' and little details that are rarely mentioned in standard protocols. We find them to be a unique aspect of the Methods in Molecular Biology series.

We would like to thank Humana Press for the opportunity to edit this book, the series editor, Dr. John Walker, for his continuous support and guidance and David Casey of Humana Press for his patience and support. We extend our sincerest gratitude to all contributors for their submission of critical contributions to this collection. The advice of Dr. Barbara Shacklett of the University of California, Davis in the selection of immunological topics is specially acknowledged. The administrative assistance of Ms. Emilia Ortiz in the production of the book went beyond the call of duty and we are thankful to her. We also wish to express our gratitude to our graduate students, post-doctoral fellows and other close colleagues at Einstein who helped with editing the book for scientific content (Dhivya Ramalingam, Elizabeth Hanna Luke, Sonald Duclair and Vasudev Rao) or for style (Andrea Provost, Aviva Joseph) and in the beta testing and improving of the subject index (Chisanga Lwatula, James Gaudette, Jennifer Cano, Melissa Smith, SeungJae Lee, Sheeba Mathew, Sohrab Khan, Supratik Das).

We would like to thank Humana Press for the opportunity to edit this book as well as the series editor, Dr. John Walker, for his continuous support and guidance. We acknowledge the advice of Dr. Barbara Shacklett of the University of California, Davis in the selection of immunological topics. The administrative assistance of Ms. Emilia Ortiz often went beyond the call of duty. We also wish to express our gratitude to colleagues at Einstein who helped with editing for scientific content (Dhivya Ramalingam, Elizabeth Hanna Luke, Sonald Duclair and Vasudev Rao) or for style (Andrea Provost). Finally, we extend our sincerest gratitude to all contributors for their submission of critical contributions to this collection.

Vinayaka R. Prasad, PhD
Ganjam V. Kalpana, PhD

Contents

Contributors

AWET ABRAHA, BA • Division of Infectious Diseases, Department of Medicine, Case Western Reserve University, Cleveland, OH, USA

CHRISTOPHER AIKEN, PHD • Department of Microbiology and Immunology, Vanderbilt University School of Medicine, Nashville, TN, USA

RAMA RAO AMARA, PHD • Department of Microbiology and Immunology, Emory Vaccine Center, Yerkes National Primate Research Center, Emory University, Atlanta, GA, USA

NATHALIE J. ARHEL, PHD • Institut Pasteur, Molecular Virology and Vectorology Group, Virology Department, Paris, France

ERIC J ARTS, PHD • Division of Infectious Diseases, Department of Medicine; Department of Molecular Biology and Microbiology, Case Western Reserve University, Cleveland, OH, USA

BEN BERKHOUT, PHD • Laboratory of Experimental Virology, Academic Medical Center, Amsterdam, The Netherlands

JOAN W. BERMAN, PHD • Department of Pathology, Albert Einstein College of Medicine, Bronx, NY, USA

BLUMA G. BRENNER, PHD • McGill University AIDS Centre, Lady Davis Institute for Medical Research, Jewish General Hospital, Montreal, Quebec, Canada

MICHAEL I. BUKRINSKY, MD, PHD • The George Washington University, Washington, DC, USA

HAROLD BURGER, MD, PHD • Wadsworth Center, New York State Department of Health, Albany, New York & Albany Medical College, Albany, NY, USA

SHAN CEN, PHD • Department of Medicine, McGill University, Lady Davis Institute for Medical Research, Jewish General Hospital, Montreal, QC, Canada

PIERRE CHARNEAU, PHD • Institut Pasteur, Molecular Virology and Vectorology Group, Virology Department, Paris, France

MARIO P.S. CHIN, PHD • HIV Drug Resistance Program, National Cancer Institute, Frederick, MD, USA

J. WILLIAM CRITCHFIELD, PHD • Department of Medical Microbiology and Immunology and School of Medicine, University of California, Davis, CA, USA

ATZE T. DAS, PHD • Laboratory of Experimental Virology, Academic Medical Center, Amsterdam, The Netherlands

SUPRATIK DAS, PHD • Department of Molecular Genetics, Albert Einstein College of Medicine, Bronx, NY, USA

SIDDHARTHA A.K. DATTA, PHD • HIV Drug Resistance Program, National Cancer Institute, Frederick, MD, USA

KRISTA A. DELVIKS-FRANKENBERRY, PHD • HIV-Drug Resistance Program, NCI-Frederick, Frederick, MD, USA

ALAN ENGELMAN, PHD • Department of Cancer Immunology and AIDS, Dana-Farber Cancer Institute and Division of AIDS, Harvard Medical School, Boston, MA, USA
ELISEO A. EUGENIN, PHD • Department of Pathology, Albert Einstein College of Medicine, Bronx, NY, USA
ARIBERTO FASSATI, MD, PHD • Wohl Virion Centre and MRC Centre for Medical Molecular Virology, Division of Infection and Immunity, University College London, London, UK
ROBERT J. FISHER, PHD • Protein Chemistry Laboratory, Research Technology Program, SAIC-Frederick, Inc., NCI-Frederick, Frederick, MD, USA
ERIC O. FREED, PHD • Virus-Cell Interaction Section, HIV Drug Resistance Program, NCI-Frederick, National Institutes of Health, Frederick, MD, USA
YONG GAO, MD, PHD • Division of Infectious Diseases, Department of Medicine, Case Western Reserve University, Cleveland, OH, USA
NATASSIA GAZNICK, BS • Molecular Medicine Program, Mayo Clinic College of Medicine, Rochester, MN, USA
STEPHEN P. GOFF, PHD • Howard Hughes Medical Institute, Department of Biochemistry and Molecular Biophysics, Columbia University, College of Physicians and Surgeons, New York, NY, USA
HARRIS GOLDSTEIN, MD • Departments of Microbiology and Immunology and Pediatrics, Albert Einstein College of Medicine, Bronx, NY, USA
ROBERT J. GORELICK, PHD • AIDS Vaccine Program, SAIC-Frederick, Inc., NCI-Frederick, Frederick, MD, USA
FEI GUO, PHD • Lady Davis Institute for Medical Research, Jewish General Hospital Montreal, QC, Canada
WEI-SHAU HU, PHD • HIV Drug Resistance Program, National Cancer Institute, Frederick, MD, USA
SERGEY N. IORDANSKIY, PHD • The D.I. Ivanovsky Institute of Virology, Moscow, Russia
AVIVA JOSEPH, PHD • Department of Microbiology and Immunology, Albert Einstein College of Medicine, Bronx, NY, USA
GANJAM V. KALPANA, PHD • Department of Molecular Genetics, Albert Einstein College of Medicine, Bronx, NY, USA
KIMDAR SHEREFA KEMAL, PHD • Wadsworth Center, New York State Department of Health, Albany, NY, USA
LAWRENCE KLEIMAN, PHD • Department of Medicine, McGill University, Lady Davis Institute for Medical Research, Jewish General Hospital, Montreal, QC, Canada
DONNA LEMONGELLO, BS • Department of Medical Microbiology and Immunology and School of Medicine, University of California, Davis, CA, USA
JAISRI R. LINGAPPA, MD, PHD • Departments of Pathobiology and Medicine, University of Washington, Seattle, WA, USA
MANUEL LLANO, MD, PHD • Molecular Medicine Program, Mayo Clinic College of Medicine, Rochester, MN, USA
HOLDEN T. MAECKER, PHD • Palo Alto, CA, USA

DIBYAKANTI MANDAL, PHD • Department of Microbiology and Immunology, Albert Einstein College of Medicine, Bronx, NY, USA
JEAN L. MBISA, PHD • HIV-Drug Resistance Program, NCI-Frederick, Frederick, MD, USA
JIRI MESTECKY, MD, PHD • University of Alabama, Birmingham, AL, USA
ZINA MOLDOVEANU, PHD • University of Alabama, Birmingham, AL, USA
DAVID C. MONTEFIORI, PHD • Department of Surgery, Laboratory for AIDS Vaccine Research and Development, Duke University Medical Center, Durham, NC, USA
MICHAEL D. MOORE, PHD • HIV Drug Resistance Program, National Cancer Institute, Frederick, MD, USA
IMMACULATE NANKYA, MBCHB, PHD • Division of Infectious Diseases, Department of Medicine, Department of Molecular Biology and Microbiology, Case Western Reserve University, Cleveland, OH, USA
KENNETH N. NELSON, BA • Division of Infectious Diseases, Department of Medicine, Case Western Reserve University, Cleveland, OH, USA
MAUREEN OLIVEIRA, BSC • McGill University AIDS Centre, Lady Davis Institute for Medical Research, Jewish General Hospital, Montreal, Quebec, Canada
AKIRA ONO, PHD • Department of Microbiology and Immunology, University of Michigan Medical School, Ann Arbor, MI, USA
DAVID E. OTT, PHD • AIDS Vaccine Program, SAIC-Frederick, Inc., NCI-Frederick, Frederick, MD, USA
VINAY K. PATHAK, PHD • HIV-Drug Resistance Program and AIDS Vaccine Program, SAIC-Frederick, Inc., NCI-Frederick, Frederick, MD, USA
ERIC M. POESCHLA, MD • Molecular Medicine Program, Mayo Clinic College of Medicine, Rochester, MN, USA
VINAYAKA R. PRASAD, PHD • Department of Microbiology and Immunology, Albert Einstein College of Medicine, Bronx, NY, USA
VASUDEV R. RAO, MBBS, MS • Department of Microbiology and Immunology, Albert Einstein College of Medicine, Bronx, NY, USA
ALAN REIN, PHD • HIV Drug Resistance Program, National Cancer Institute, Frederick, MD, USA
MILAN REINIS, PHD • Institute of Molecular Genetics, Academy of Sciences of the Czech Republic, Prague
ANDREA RUBIO, BSC • National Reference Center for AIDS, Department of Microbiology, School of Medicine, University of Buenos Aires, Buenos Aires, Argentina
KAORI SANGO, MS • Department of Microbiology and Immunology, Albert Einstein College of Medicine, Bronx, NY, USA
BARBARA L. SHACKLETT, PHD • Department of Medical Microbiology and Immunology and Division of Infectious Diseases, Department of Internal Medicine, School of Medicine, University of California, Davis, CA, USA
ANDREW G. STEPHEN, PHD • Protein Chemistry Laboratory, Research Technology Program, SAIC-Frederick, Inc., NCI-Frederick, Frederick, MD, USA
BETH K. THIELEN, BS • Departments of Pathobiology and Medicine, University of Washington, Seattle, WA, USA

JAMES A. THOMAS, PHD • AIDS Vaccine Program, SAIC-Frederick, Inc., NCI-Frederick, Frederick, MD, USA
RYAN M. TROYER, PHD • Department of Microbiology, Immunology and Pathology, Colorado State University, Fort Collins, CO, USA
SUSANA T. VALENTE, PHD • Howard Hughes Medical Institute, Department of Biochemistry and Molecular Biophysics, Columbia University, College of Physicians and Surgeons, New York, NY, USA
ABDUL A. WAHEED, PHD • Virus-Cell Interaction Section, HIV Drug Resistance Program, NCI-Frederick, National Institutes of Health, Frederick, MD, USA
MARK A. WAINBERG, PHD • McGill University AIDS Centre, Lady Davis Institute for Medical Research, Jewish General Hospital, Montreal, Quebec, Canada
BARBARA WEISER, MD • Wadsworth Center, New York State Department of Health, and Albany Medical College, Albany, NY, USA
OTTO O. YANG, MD • Division of Infectious Diseases, Department of Medicine, Department of Microbiology, Immunology, and Molecular Genetics and AIDS Institute, Geffen School of Medicine, UCLA Medical Center, Los Angeles, CA, USA

Section I

Preparation of Virus Particles and Their Analysis

Chapter 1

Methods for Viral RNA Isolation and PCR Amplification for Sequencing of Near Full-Length HIV-1 Genomes

Kimdar Sherefa Kemal, Milan Reinis, Barbara Weiser, and Harold Burger

Abstract

HIV-1 in plasma represents the viral quasispecies replicating in the patient at any given time. Studies of HIV-1 viral RNA from plasma or other body fluids therefore reflect the virus present in real time. To obtain near full-length genomic sequences derived from virion RNA it is first necessary to carefully isolate and amplify the RNA.

The procedure described below, involves viral RNA extraction, reverse transcription (RT) of the extracted RNA to produce cDNA copies, and PCR amplification of long HIV-1 gene fragments using site-specific, overlapping primers. The primers are based on subtype B HIV-1 strains, and plasma specimens are used in the procedures. However, the protocol can easily be adapted to other HIV-1 subtypes by modifying the primers to match the subtype of interest.

Key words: HIV-1, HIV-1 viral RNA, HIV-1 primers, Long RT-PCR amplification of HIV-1.

1. Introduction

HIV-1 infection is characterized by continuous replication of $\sim$ 9 kb RNA genomes resulting in a viral swarm of closely related molecules called quasispecies *(1, 2)*. Sequence variation is a hallmark of lentivirus infection; surviving viral species reflect replication and selection *(3)*. When studying the relationship between HIV-1 sequences and pathogenesis, it is highly desirable to analyze the complete HIV-1 genome because variability in multiple regions of the genome may play a role in pathogenesis and virus–host interactions *(4–6)*. Full-length HIV-1 sequencing provides essential data needed to address vaccine design and molecular epidemiology *(7, 8)*. Full-length sequence analysis also helps to identify the presence of dual infections and recombination in vivo

Vinayaka R. Prasad, Ganjam V. Kalpana (eds.), *HIV Protocols: Second Edition, vol. 485*

DOI 10.1007/978-1-59745-170-3_1 Springerprotocols.com

(9–11). Furthermore, knowledge of the complete genomic HIV-1 RNA sequence and HLA type of the infected individual makes it possible, with the use of an immunologic database (http://hiv-web.lanl.gov/content/immunology/index), to predict cytotoxic T-cell epitopes encoded by the virus *(4, 5)*.

HIV-1 in plasma represents the replicating virus population at any given time; by contrast, proviral DNA, integrated in cellular genomes, represents a repository of older sequences, the majority of which have been shown to lack replication competence *(12, 13)*.

Amplification of a full-length HIV-1 genome requires careful step-by-step procedures involving viral RNA extraction, synthesis of cDNA copies of the viral RNA (RT-PCR), PCR amplification of target gene fragments using specific primers, and the purification of the amplified PCR products for further analyses. Successful amplification of a near full-length HIV-1 viral RNA depends on several factors, including viral RNA load in the specimen, quality of specimen, viral RNA isolation methods used, cDNA synthesis from RNA using RT-PCR, primer selection, and PCR amplification conditions *(6, 9, 14–17)*. The quality of the specimen, such as plasma, is directly affected by factors such as storage temperature, repeated freezing and thawing, and general handling of the specimen from initial processing to the RNA extraction steps. Although we cannot rule out the possibility of amplifying full-length HIV-1 genomes from specimens with viral loads as low as 5,000 copies/mL, we recommend using samples with viral loads of at least 10,000 copies/mL. Specimens should be aliquoted into 1 mL volume and stored at temperatures below −80 °C. Repeated freeze and thaw steps need to be avoided (*see* **Note 1**).

2. Materials

2.1. HIV-1 Viral RNA Isolation

1. RNAgents® Total RNA Isolation System kit (Promega). The kit includes combined guanidine thiocyanate crystals and the Citrate/Sarcosine/β-Mercaptoethanol (CSB) buffer in a single bottle of denaturing solution. Phenol/chloroform/isoamyl alcohol solution, isopropanol, 2 M sodium acetate (pH 4.0), and nuclease free water are also included.
2. Yeast t-RNA (Sigma).
3. Glycogen (Sigma).
4. Siliconized Flat Top microfuge tubes (Fisher Scientific).

2.2. Reverse Transcription and cDNA Synthesis

SuperScript™ First-strand Synthesis System for RT-PCR kit (Invitrogen).

2.3. PCR Amplification

1. GeneAmp XL PCR Kit (Applied Biosystems).
2. AmpliWax, PCR Gem 50 (Applied Biosystems).
3. 10 mM dNTP Mix with dTTP (Applied Biosystems).
4. TE buffer (pH 8.0).

2.4. PCR Primers

All the PCR primers and PCR amplification programs specific for each fragment are listed in **Table 1.1**.

2.5. Agarose Gel Analysis

1. Agarose (Fisher Scientific).
2. 6X gel-loading buffer: 0.25% bromophenol blue, 0.25% xylene cyanol, 30% glycerol, water up to desired volume. A stock solution can be prepared and aliquots can be made in 1.5 mL microfuge tubes and stored in a refrigerator. Once a tube is taken out of the refrigerator, it can be stored at room temperature and used.
3. Tris-Borate-EDTA (TBE) buffer: 0.45 M Tris-borate, 0.01 M EDTA, pH 8.3.
4. Ethidium bromide tablets, 100 mg/tablet (Sigma). To have a 10 mg/mL stock solution, dissolve one tablet in 10 mL deionized water and keep the solution at room temperature away from direct light (*see* **Note 2**).
5. DNA ladder, 1 kb plus (Invitrogen).
6. Disposable scalpels.
7. QIAquick Gel Extraction Kit (Qiagen).

3. Methods

HIV-1 viral RNA isolation is one of the critical steps for the success of the rest of the procedures. RNA should always be handled with care; gloves should be worn at all times to help eliminate the introduction of endonucleases. It is important to work in an RNase-free environment and use RNase-free reagents. Work should be done in a Bio-safety level 2 (BSL-2) laminar flow hood; RNA isolation areas should be separate from DNA or PCR amplification areas in the lab. It is also important to have dedicated pipettes for RNA isolation. If this is not possible, always clean the hood and the pipettes with RNase and DNA contaminant removing reagents and turn on the UV light, in the laminar flow hood, for 20–30 min after use. To increase viral recovery, we suggest pelleting the plasma virions in siliconized microfuge tubes. We also recommend using 0.5–1.0 mL sample volumes, particularly if the viral load is low ($<$ 10,000 copies/mL). When available, 1 mL is preferred. Using this method, RNA for full-length amplification can be recovered from plasma samples with viral loads as little as 5,000 RNA copies/mL (*see* **Note 3**).

Table 1.1
PCR primers and amplification conditions used to amplify a 9-kb HIV-1 subtype B genome as four overlapping fragments

Primer	Primer sequences (5′ − 3′)	HXB2 location
Fragment 1/partial 5′ LTR and gag		
517F[A]	CTT-AAG-CCT-CAA-TAA-AGC-TTG-CCT-TGA	517–543[B]
2348R[C]	TAC-TGT-ATC-ATC-TGC-TCC-TGT-ATC	2348–2325
548F	TCA-AGT-AGT-GTG-TGC-CCG-TCT-G	549–570
2314R	TCC-TTT-AGT-TGC-CCC-CCT-ATC	2314–2294
PCR conditions:	1X 94 °C: 5 min	
	5X 94 °C: 15 s, 45 °C: 45 s, 72 °C: 3 min	
	30X 94 °C: 15 s, 55 °C: 45 s, 72 °C: 3 min	
	1X 72 °C: 10 min	
	hold at 4 °C: until next step	
Fragment 2/pol		
1999F	AAT-TGC-AGG-GCC-CCT-AGA-AAA-AAG-GGC-TGT	1999–2028
5304R	TTC-TAT-GGA-GAC-TCC-CTG-ACC-CAA-ATG-CCA	5304–5275
2138F	AGA-GCA-GAC-CAG-AGC-CAA-CAG	2138–2158
5202R	TCC-CCT-AGT-GGG-ATG-TGT-ACT	5222–5202
PCR conditions:	1X 94 °C: 5 min	
	5X 94 °C: 15 s, 45 °C: 45 s, 72 °C: 5 min;	
	30X 94 °C: 15 s, 55 °C: 60 s, 72 °C: 4 min;	
	1X 72 °C: 10 min	
	hold at 4 °C: until next step	
Fragment 3/env and accessory genes		
4650F	ATT-CCC-TAC-AAT-CCC-CAA-AGT-CAA-G	4650–4674
9626R	CTT-GAA-GCA-CTC-AAG-GCA-AGC-TTT-ATT-G	9626–9599
4956F	TGG-AAA-GGT-GAA-GGG-GCA-GTA-GTA-ATA-CAA-G	4956–5014
9173R	TGG-TGT-GTA-GTT-CTG-CCA-ATC-AGG-GAA-G	9173–9146
PCR conditions:	1X 94 °C: 5 min	
	5X 94 °C: 15 s, 55 °C: 45 s, 72 °C: 5 min;	
	30X 94 °C: 15 s, 60 °C: 5 min	
	1X 72 °C: 10 min	
	hold at 4 °C: until next step	

(continued)

Table 1.1 (continued)

Fragment 4/*nef and LTR*

8081F	TGG-AAT-AAC-ATG-ACC-TGG-AGG-G	8091–8112
9626R	CTT-GAA-GCA-CTC-AAG-GCA-AGC-TTT-ATT-G	9626–9599
8296F	CAT-AAT-GAT-AGT-AGG-AGG-CTT-GG	8279–8301
9608R	TGA-AGC-ACT-CAA-GGC-AAG-C	9608–9590
PCR conditions:	1X 94 °C: 5 min	
	35X 94 °C: 15 s; 55 °C: 45 s; 72 °C: 3 min	
	1X 72 °C: 10 min	
	hold at 4 °C: until next step	

[A]F: stands for forward primers, the numbers indicate the nucleotide positions of the forward or reverse genome fragment.
[B]Indicates nucleotide positions in HIV-1 HXB2 strain.
[C]R: stands for reverse primers.

3.1. RNA Extractions

3.1.1. Beginning

1. Clarify plasma by spinning at 400 × *g*, at room temperature, for 10 min (*see* **Note 4**).
2. Transfer the plasma to sterile, siliconized, 1.5 mL microfuge tubes.
3. To pellet virions, centrifuge the plasma 18, 500 × *g*, at 10 °C, for 90 min.
4. Remove most of the now virion-free plasma without disturbing the pellet (*see* **Note 5**).
5. To extract virion-associated RNA, re-suspend the pellet in 300 μL pre-chilled denaturing solution.
6. Vortex the pellet until it is completely dissolved. This usually takes 5–10 min (*see* **Note 6**).
7. Mix in order: 1 μL yeast tRNA (1 μg/μL), 1 μL glycogen (2 μg/μL) and 30 μL of 2 M sodium acetate, pH 4.0 (*see* **Note 7**).
8. Add 32 μL of the above mixture into each tube, and then add 0.5 mL phenol/chloroform/isoamyl alcohol solution, being careful to remove only from the lower organic phase (*see* **Note 8**).
9. Vortex for 10 s, then chill on ice for 20 min.
10. Centrifuge at 18, 500 × *g* for 20 min, at 4 °C.
11. Transfer the top phase, which contains the RNA, to fresh microfuge tubes (*see* **Note 9**).
12. To precipitate the RNA, add an equal volume of isopropanol (0.3–0.5 mL) and keep the tubes at −20 °C for at least 2 h (*see* **Note 10**).

3.1.2. After 2 h or the Next Day

1. Pellet the RNA by centrifugation at 18,500 × *g* for 30 min at 4 °C. Discard the supernatant.
2. Re-suspend the RNA pellet in 200 μL denaturing solution. Vortex until RNA is dissolved; usually 1 min is enough.
3. Add 300 μL isopropanol and precipitate the RNA at −20 °C for at least 2 h.
4. After 2 h, pellet the RNA by centrifugation at 18,500 × *g* for 25 min at 4 °C.
5. Wash the pellet, once with 1 mL and then with 0.5 mL of 75% ice-cold ethanol, by centrifugation at 18,500 × *g* for 20 min at 4 °C each time. During each wash break up the pellet by using RNase-free pipette tips before each centrifugation; do not vortex (*see* **Note 11**).
6. Air-dry the pellet for 5–20 min, depending on the amount of ethanol left in the tubes or the size of the pellet.
7. Dissolve the RNA in 25 μL nuclease-free water by using the pipette tips. Do not vortex (*see* **Note 12**).
8. Store the RNA at −20 °C until use. For long-term use store at −80 °C.

3.2. Reverse Transcription PCR (RT-PCR)

The procedure used for RT-PCR is essential for synthesis of a long cDNA. The procedure is given below, with details of the reagents given in **Tables 1.2 and 1.3**. There are a number of commercial kits available for RT-PCR and each laboratory has its own preferences. The method described here uses the SuperScript^{TM} III First-Strand Synthesis System for RT-PCR kit (Invitrogen). SuperScript^{TM} III Reverse Transcriptase enzyme is used to synthesize cDNA at the temperature range of 42–55 °C, providing increased specificity, higher yields of cDNA, and more full-length product than other reverse transcriptase enzymes (*see* **Note 13**). We recommend using oligo(dT) primer, as it is more specific and allows the synthesis of full-length cDNA fragments.

1. Prepare RNA/Primer master mix which includes oligo$(dT)_{20}$ and the dNTP mix according to **Table 1.2**.
2. Dispense 10 μL of the master mix into each tube.

Table 1.2
RNA/primer master mix

Component	Volume	Final concentration
50 μM oligo$(dT)_{20}$	1 μL	2.5 μM
10 mM dNTP mix	1 μL	500 μM
Nuclease free water	5 μL	–

Table 1.3
cDNA synthesis master mix

Component	Volume	Final concentration
10X RT buffer	2 μL	1X
25 mM $MgCl_2$	4 μL	5 mM
0.1 M DTT	2 μL	10 mM
RNaseOUT™ (40 U/μL)	1 μL	2 U
SuperScript™ III RT enzyme (200 U/μL)	1 μL	10 U

3. Add 3 μL RNA into the tube containing the master mix (*see* **Note 14**).
4. Incubate the mixture at 65 °C for 5 min in a thermocycler.
5. Place the tubes on ice for at least 1 min.
6. Prepare cDNA synthesis master mix which includes the RT buffer, salts, and enzymes according to **Table** 1.3.
7. Add 10 μL of the cDNA synthesis mix to each of the RNA/primer mixture from **step 5**.
8. Synthesize the cDNA by incubating the mixture at 45 °C for 60 min, and terminate the reaction by heating the mixture at 70 °C for 15 min (*see* **Note 15**).
9. To remove the RNA template from the cDNA:RNA hybrid, add 1 μL RNase H to the reaction and incubate for 20 min at 37 °C, then inactivate the RNase H by heating the mixture at 70 °C for 15 min.
10. Transfer the cDNA to microfuge tubes and store at −80 °C until use. The cDNA is stable for several months when stored at −80 °C. Always thaw cDNA on ice and return the remaining cDNA to the freezer immediately after each use.

3.3. PCR Amplification

The near full-length HIV-1 genome is amplified as three, large, overlapping fragments and one optional, short, fragment to complete the 3′-LTR fragment (*see* **Note 16**). Nested PCR methods, using outer and inner primer pairs, are employed in all steps. Although there are a number of companies providing qualified kits for PCR amplification, this protocol describes the procedures using the GeneAmp® XL PCR Kit (Applied Biosystems) and in-house primers listed in **Table** 1.1. PCR is performed in thin-walled tubes in a final volume of 100 μL. Hot Start methods are used to separate the primers from the enzyme, by using AmpliWax PCR gem beads. In order to avoid cross contamination, there should be a separate reagent preparation (clean room), template hood, and electrophoresis areas.

3.3.1. PCR Set up

1. Bring all reagents (except enzymes) and primers to room temperature in the clean reagent hood or bench. Enzymes should be removed from the freezer just prior to use. After thawing, vortex all reagents, except the enzyme, briefly centrifuge in a microcentrifuge to ensure that all liquid is brought to the bottom. Keep all reagents on ice while in use, and immediately return to the freezer when finished.
2. Prepare the lower master mix (master mix A) in 1.5 mL microfuge tube, on ice according to **Table 1.4**. Prepare enough master mix to accommodate the number of reactions including negative and positive controls (*see* **Note 17**).
3. Add 40 µL of master mix A into each PCR tube.
4. Add one AmpliWax® PCR Gem 50 into each tube.
5. To melt the wax, incubate the mixture in a thermocycler at 80 °C for 3 min followed by 25 °C incubation for 1 min to solidify the wax.
6. Prepare the upper layer mix (master mix B) according to **Table 1.4**.
7. Add 55 µL of master mix B on top of the solidified wax.
8. Take the tubes into the template area to add 3–5 µL of cDNA (*see* **Note 18**).

Table 1.4
PCR master mixes

Master mix A

Component	Volume	Final concentration
3.3X XL Buffer II	12 µL	1X
GeneAmp 10 mM dNTP	8 µL	800 µM (200 µM each)
$Mg(OAc)_2$, 25 mM	5 µL	1.25 µM
blend		
Forward Primer, 20 µM	1 µL	0.2 µM
Reverse Primer, 20 µM	1 µL	0.2 µM
Water	13 µL	–

Master mix B

Component	Volume	Final concentration
3.3X XL Buffer II	18 µL	1X
Water	35–37 µL	–
rTth DNA polymerase	2 µL	0.04 U
XL, 2 U/µl		

9. Change gloves and bring the reaction mixture to the PCR amplification room and run the PCR by using the PCR program specified in **Table** 1.1.
10. Perform the secondary PCR (nested PCR) using the primers and the PCR conditions listed in **Table** 1.1, and using 5 μL first round (primary) PCR products as templates. After the secondary PCR procedure is complete, the products need to be analyzed by agarose gel electrophoresis for the presence of amplified PCR products, and appropriate DNA ladder should be used to determinant the size of the amplified PCR products, based on the primers used. The amplified PCR product can be stored at 4 °C until electrophoresis is performed; for long-term storage use −20 °C freezer.

3.4. Agarose Gel Electrophoresis

3.4.1. Agarose Gel Preparation

1. Dilute the TBE buffer to 0.5X with distilled/deionized water. Add ethidium bromide to a final concentration of 0. 5 μg/mL.
2. Prepare a 1% (w/v) agarose solution by melting the agarose in a microwave.
3. Cool the agarose to 55 °C, under cold running water, before pouring on the tray.
4. Set up the gel tray with appropriate comb size and pour the agarose into the gel tray.
5. Remove any bubbles or debris and allow the agarose to solidify at room temperature for at least 45 min.
6. Once agarose has solidified, gently remove the comb and prepare gel apparatus to receive running buffer.
7. Fill the gel-running apparatus with 0.5X TBE buffer containing 0. 5 μg/mL ethidium bromide.
8. Mix 5 μL of the secondary PCR product with 2–3 μL gel-loading buffer and load into each well. Always load 10 μL of the 1 kb DNA ladder into the first well to check the size of the amplified PCR product.
9. Run the gel at 100–120 V for about 45–60 min, occasionally checking with UV transilluminator for proper separation of bands.
10. If there is no band seen or the size of the fragment is not correct, the PCR procedure has to be repeated (*see* **Note 19**). If the size is correct, proceed with DNA extractions.

3.4.2. DNA Extractions

1. Prepare 1% agarose gel using thicker comb sizes.
2. Mix the PCR product with 20 μL loading buffer and load the whole mixture into each well.
3. Run the electrophoresis as described above, and gel extract the PCR product following the QIAquick gel extraction protocol (Qiagen), or your preferred gel extraction protocol.

4. The gel extracted PCR product is ready for sequence analyses using specific sequencing primers or it can be stored at −20 °C for several months.

4. Notes

1. When aliquoting plasma or other specimens, always thaw on ice and return the specimen to the freezer immediately. We do not recommend thawing and freezing more than two times.
2. Ethidium bromide (EtBr) is carcinogenic and it has to be handled with care. Use a dedicated pipette for EtBr and dispose the tips in a clearly labeled biohazard waste container.
3. For samples with HIV-1 RNA loads of $< 5,000$ copies/mL, sufficient viral RNA can be obtained by using 2–3 mL of plasma. Each 1 mL sample will be processed independently until it is pooled immediately after the phenol/chloroform extraction step.
4. This step is not always necessary, but it is particularly helpful for "dirty" or old samples.
5. The supernatant can be saved and used for studies involving antibody, chemokine, or other soluble factors.
6. The vortexing can be done by placing the tubes in an eppendorf mixer for 10 min at room temperature. After each vortex give each tube a quick pulse in a centrifuge to ensure no aerosolization/carry-over of sample.
7. This mixture facilitates the co-precipitation of RNA and is important in recovery of small amounts of RNA. For samples with high viral loads ($\geq 10,000$ copies/mL of plasma), 30 μL sodium acetate without the t-RNA and glycogen is sufficient.
8. Bring the phenol/chloroform solution to room temperature (15–20 min in the hood) before use.
9. Avoid taking material from the interface, which contains genomic DNA; use ART 200 aerosol resistant tips to take a maximum of 200 μL volume a time. Do not use tips with 1,000 μL capacity, these might cause disturbance at the interface.
10. To maximize the success of the RNA extraction, and increase the RNA yield from samples with relatively low amounts of RNA, we recommend at least 2-h incubation at −20 °C at this step (overnight incubation is preferred).
11. During the ethanol wash, a small white precipitate could be seen on the inside wall of the tube. However, the absence of visible precipitate does not indicate failure of the RNA extraction. Whether or not there is a visible precipitate, always avoid

touching the bottom and the side of the tubes, and leave a small amount of fluid with the RNA precipitate.

12. Do not overdry the pellet; RNA will be difficult to re-suspend. Moreover, a trace amount of ethanol will not interfere with RT-PCR.
13. SuperScript™ III Reverse Transcriptase enzyme is used to synthesize cDNA at a temperature range of 42–55 °C.
14. For samples with low viral RNA copies, or if the PCR fails to amplify with 3 μL RNA, repeat the PCR using 5 μL RNA.
15. Problems with RT-PCR due to intrinsic secondary structures can be overcome by increasing the temperature to 50–55 °C.
16. Fragment 4 can be amplified by using primary PCR templates from fragment 3 and secondary PCR primers for fragment 4. When sequencing fragment 4, use primers specific for genomic regions 8800–9626. There is no need to sequence the whole fragment 4, as most of the 5′-end overlaps with fragment 3.
17. Use both positive and negative controls during the initial stages of these procedures. Once the procedure is mastered, we strongly recommend including negative controls, either the negative control from the RT-PCR and/or using water as a template, with each run. Positive controls, especially lab isolates or patient samples with high viral load are sources of contamination, especially when working with samples with low viral RNA copy numbers. Therefore, positive control use should be avoided or minimized. If there is a band in the negative control, the results have to be rejected and the PCR be repeated using fresh cDNA samples.
18. For samples with low viral RNA copies ($< 10,000$ copies/mL of plasma) use 5 μL cDNA.
19. It is important that for every PCR the optimum magnesium concentration is determined empirically by doing titration in increments of 0.1 mM from 0.8 to 1.5 mM and testing with a known PCR-positive template. The primer concentration can also be adjusted between 0.1 and 0.25 μM.

References

1. Coffin, J. M. (1995) HIV population dynamics in vivo: implications for genetic variation, pathogenesis, and therapy. *Science* 267, 483–489.
2. Leitner, T., Foley, B., Hahn, B., et al. (Eds.) (2005) HIV Sequence Compendium 2005; Published by Theoretical Biology and Biophysics Group, Los Alamos National Laboratory, NM, LA UR 06 0680.
3. Malim, M., Emerman, M. (2001) HIV-1 sequence variation: drift, shift, and attenuation. *Cell* 104, 469–472.
4. Addo, M. M., Yu, X. G., Rathod, A., et al. (2003) Comprehensive epitope analysis of human immunodeficiency virus type 1 (HIV-1)-specific T-cell responses directed against the entire expressed HIV-1 genome demonstrate broadly directed responses, but no correlation to viral load. *J Virol* 77, 2081–2092.
5. Altfeld, M., Addo, M. M., Shankarappa, R., et al. (2003) Enhanced detection of human immunodeficiency virus type 1-specific T-cell responses to highly variable regions by using

peptides based on autologous virus sequences. *J Virol* 77, 7330–7340.

6. Fang, G., Burger, H., Chappey, C., et al. (2001) Analysis of transition from long-term nonprogressive to progressive infection identifies sequences that may attenuate HIV type 1. *AIDS Res Hum Retroviruses* 17, 1395–1404.
7. Gaschen, B., Taylor, J., Yusim, K., et al. (2002) Diversity considerations in HIV-1 vaccine selection. *Science* 296, 2354–2360.
8. Kesturu, G. S., Colleton, B. A., Liu, Y., et al. (2006) Minimization of genetic distance by the consensus, ancestral, and center-of-tree (COT) sequences for HIV-1 variants within an infected individual and the design of reagents to test immune reactivity. *Virology* 348, 437–448.
9. Fang, G., Weiser, B., Kuiken, C., et al. (2004) Recombination following superinfection by HIV-1. *AIDS* 18, 153–159.
10. Philpott, S., Burger, H., Tsoukas, C., et al. (2005) Human immunodeficiency virus type 1 genomic RNA sequences in the female genital tract and blood: compartmentalization and intrapatient recombination. *J Virol* 79, 353–363.
11. Dowling, W. E., Kim, B., Mason, C. J., et al. (2002) Forty-one near full-length HIV-1 sequences from Kenya reveal an epidemic of subtype A and A-containing recombinants. *AIDS* 16, 1809–1820.
12. Li, Y., Kappes, J. C., Conway, J. A., et al. (1991) Molecular characterization of human immunodeficiency virus type 1 cloned directly from uncultured human brain tissue: identification of replication-competent and – defective viral genomes. *J Virol* 65, 3973–3985.
13. Li, Y., Hui, H., Burgess, C. J., et al. (1992) Complete nucleotide sequence, genomic organization, and biological properties of human immunodeficiency virus type 1 in vivo: evidence for limited defectiveness and complementation. *J Virol* 66, 6587–6600.
14. Fang, G., Weiser, B., Visosky, A. A., et al. (1996) Molecular cloning of full-length HIV-1 genomes directly from plasma viral RNA. *J Acquir Immune Def Syndr* 12, 352–357.
15. Rousseau, C. M., Birditt, B. A., McKay, A. R., et al. (2006) Large-scale amplification, cloning and sequencing of near full-length HIV-1 subtype C genomes. *J Virol Meth* 136, 118–125.
16. Fang, G., Zhu, G., Burger, H., et al. (1998) Minimizing DNA recombination during long RT-PCR. *J Virol Meth* 76,139–148.
17. Salminen, M. (1999) PCR amplification and cloning of virtually full-length HIV-1 provirus, in *Methods in Mol Medicine: HIV protocols* (Michael, N.L., and Kim, J.H., Eds.) Humana, Totowa, NJ, pp. 89–98.

Chapter 2

Purification of HIV-1 Virions by Subtilisin Digestion or CD45 Immunoaffinity Depletion for Biochemical Studies

David E. Ott

Abstract

The presence of cellular proteins outside and inside retroviruses can indicate the roles they play in viral biology. However, experiments examining retroviruses can be complicated by the contamination of even highly purified virion preparations with nonviral particles (either microvesicles or exosomes). Two useful methods have been developed that can remove contaminating particles from virus stocks to produce highly pure virus preparations. One approach, the subtilisin digestion procedure, enzymatically removes the proteins outside the virions. While this method is well suited for the analysis of the interior proteins in the virions, it removes the extracellular domains of the integral membrane proteins on the virion. To preserve the proteins on the exterior of the virion for biochemical studies, a CD45 immunoaffinity depletion procedure that removes vesicles by capture with antibody-linked microbeads is employed. These methods allow for the isolation of highly purified virion preparations that are suitable for a wide variety of experiments, including the biochemical characterization of cellular proteins both on and in HIV virions, examination of virion/cell interactions, and imaging of virions.

Key words: HIV, protein analysis, retroviruses, virus purification, subtilisin-digestion, CD45 immunoaffinity depletion, microvesicles, exosomes.

1. Introduction

Meaningful biochemical analyses of virions and the use of virions in a variety of biological experiments require relatively

This project has been funded in whole or in part with federal funds from the National Cancer Institute, National Institutes of Health, under contract N01-CO-12400. The content of this publication does not necessarily reflect the views or policies of the Department of Health and Human Services, nor does mention of trade names, commercial products, or organizations imply endorsement by the U.S. Government.

Vinayaka R. Prasad, Ganjam V. Kalpana (eds.), *HIV Protocols: Second Edition, vol. 485*

DOI 10.1007/978-1-59745-170-3_2 Springerprotocols.com

pure preparations. Unlike many viruses, retroviruses, which are released continuously from infected cells, can contain a significant amount of contaminating cellular proteins *(1)*. While rapid harvests and density centrifugation can eliminate some of the proteins in cell culture supernatants that are secreted from cells or released from dying cells, there are particles (i.e., both exosomes and microvesicles) that can contaminate even the most carefully prepared virion preparation *(2, 3)*. While these nonviral particles have a wide range of densities and sizes, a fraction of them have the same size and density as retroviruses *(2, 3)*. Thus, these vesicular contaminants are impossible to remove by physical methods *(4)*. The amount of contaminating particles varies widely depending on the cell type producing the virions. Generally, epithelial cell lines produce limited amounts of these contaminating proteins. Macrophages appear to produce an intermediate level of vesicular contamination *(5)*. In contrast, lymphoid cells, especially T cells produce a significant amount of vesicle-associated protein, up to the same level as the virion proteins *(2–4)*. Since T cell lines are typically used for producing large-scale HIV stocks *(6)*, these preparations can be heavily contaminated.

To overcome this problem, we have developed two methods that are not based on particle density or size to remove these vesicles. For the study of proteins inside the virions, an enzymatic method that removes the proteins in these vesicles using the nonspecific serine protease subtilisin can be used. We have observed that proteins in contaminating vesicles are digested by subtilisin digestion while those inside the virions are protected by the virion membrane. Therefore, after digestion vesicles are less dense due to the removal of all or most of their protein while treated virions, which still contain the interior proteins, retain their density. This allows for the removal of these once protein-laden particles from the preparation by density centrifugation. The subtilisin treatment also removes the proteins that are merely adhered to the outside of the virions, including the extracellular domain of the transmembrane proteins on the virus. This process removes >95% of the proteins associated with microvesicles *(7–9)*, allowing the cellular proteins in the treated virions to be analyzed.

One drawback for the protease digestion approach is that potentially important proteins on the exterior of the virus are also digested and therefore lost. To study all the proteins on and in the virion, we have developed a secondary approach using CD45 depletion of the virion preparation *(4)*. CD45 is a surface protein found in the plasma membrane of hematopoeitic cells. CD45 is incorporated into many vesicles released from hematopoeitic cells that co-purify with density-purified virion preparations. The key attribute exploited here is that HIV appears to exclude CD45 from the surface of the virus *(4, 10)*, probably due to its relatively bulky cytoplasmic domain. Thus, removing particles with CD45 removes most of the contamination from density-purified virion preparations. One important aspect of this technique is that virions must be prepared from CD45-expressing cell lines, i.e., hematopoietic cells such as T cells and macrophages (*see* **Note 1**).

2. Materials

2.1. Density Centrifugation of Virions

1. Dulbecco's phosphate-buffered saline (PBS) without calcium or magnesium. Store at 4°C.
2. 20% *w/v* sucrose in PBS: 100 g sucrose in PBS, 500 mL final volume. Store at 4°C.
3. TNE: 10 mM Tris–HCl, pH 8.0, 100 mM NaCl, 2 mM Na_2EDTA. Store at room temperature.
4. 5 μm low protein-binding filter.

2.2. Subtilisin Digestion of Virions

1. 2X digestion buffer: 40 mM Tris–HCl, pH 8.0 and 2 mM $CaCl_2$. Store at room temperature.
2. 2X subtilisin digestion buffer: 2 mg Subtilisin (*see* **Note 2**) dissolved in 1 mL of 2X digestion buffer. Store at −20°C or lower.
3. 2000X phenylmethyl sulfonyl fluoride (PMSF) solution: 10 mg/mL PMSF in ethanol, store at room temperature (*see* **Note 3**).
4. PBS, store at 4°C.
5. Mouse anti-gp41^{TM} monoclonal antibody: Perkin Elmer Life and Analytical Science, Shelton, CT, catalog number NEA-9303001EA, aliquot and store at −20°C or lower.
6. Rabbit anti-gp41^{TM} serum: Fitzgerald Industries International, Concord, MA, catalog number 20-HR92. Store at −20°C.

2.3. CD45 Immunoaffinity Depletion of Virions

1. PBS, store at 4°C.
2. Anti-CD45 microbeads: Miltenyi Biotec, Inc., Auburn, CA, catalog number 130-045-801, Store at 4°C.
3. Magnetic separator: Dynal MPC-S or MPC-L, Invitrogen Corp., Carlsbad, CA (*see* **Note 4**).
4. Mouse CD45 monoclonal antibody: B-D Biosciences Pharmingen, San Jose, CA, catalog number 610265, Store at −20°C.

3. Methods (*see* Note 4)

3.1. Virus Isolation and Preparation

To produce the starting material for subtilisin digestion or CD45 immunoaffinity depletion, virions are isolated from clarified tissue culture supernatants by density centrifugation. This separates most of the soluble proteins from the virions. Repeating this process can increase the purity of the virions with respect to the soluble proteins, but can result in a loss of sample, especially with smaller samples. For comparative purposes, a vesicular or 'mock virus' preparation can be prepared using supernatants from uninfected control cell culture maintained under the same conditions as the virus-producing cells.

1. Remove virus containing supernatant from infected or transfected cells by decanting and clarify to remove cells by centrifugation at 2,000 × *g* for 10 min or by filtering through a 5 μm, low protein-binding filter (*see* **Note 6**).
2. Isolate virions by ultracentrifugation through a 20% sucrose pad at 125,000 × *g* for 1 h at 4°C. Fill the ultracentrifuge tube to half of the tube volume with clarified supernatant, then underlay with one-tenth of the total tube volume using 20% sucrose. Fill the remainder of the tube with either more clarified supernatant or with PBS.
3. After ultracentrifugation, remove the supernatant and sucrose with a pipette and invert the tube to dry.
4. Resuspend the pellet in TNE (digestion) or PBS (depletion) at a ratio of between 1-to-80 and 1-to-1000 (PBS-to-initial volume).

3.2. Subtilisin Digestion of Virions

This procedure was developed to study the proteins on the interior of HIV by removing the proteins on the exterior of the virion, including those proteins that are simply adhered to the surface and those in nonviral particles *(7--9)*. It is thought that removing most of the protein from these vesicules makes them less dense. Therefore, they no longer co-fractionate with virions postdigestion. Studies with purified microvesicles have found that protease digestion alone removes nearly all of the protein (> 95%) in vesicles *(7)*, so perhaps they are permeable to the protease while virions are not. This technique has been used in our laboratory successfully on a variety of retroviruses including, murine leukemia virus *(11)*, Rous sarcoma virus, equine infectious anemia virus *(12)*, mouse mammary tumor virus, simian immunodeficiency virus *(11)*, human T cell leukemia virus, and Mason–Pfizer monkey virus, so this technique likely works for all retroviruses and possibly any enveloped virus.

1. Add concentrated virus preparation (either prepared as above or from another source) to an equal volume of subtilisin

digestion buffer at a ratio of 1-to-1 and incubate at 37 °C overnight (18 h; *see* **Note 7**). For a negative control, include a mock-digested control that consists of a matched virion preparation in 1X digestion buffer that contains no protease. Add 2X digestion buffer without subtilisin to the virion sample at a 1-to-1 ratio.

2. After digestion, add PMSF from the 2000X stock to a final concentration of 5 μg/mL and incubate for 15 min at room temperature to inhibit the protease (*see* **Note 3**).
3. Re-isolate the digested virions from the subtilisin, protease digestion products, and digested vesicles. Centrifuge the preparations in PBS through a 20% sucrose pad as described above in **Section 3.1**. Choose an ultracentrifuge tube that is appropriate for the amount of virions to be analyzed (*see* **Note 8**). Fill the ultracentrifuge tube to approximately one-fourth of the tube volume with PBS. Gently underlay, at the bottom of the tube, half of the tube volume with 20% sucrose. Add the digestion or mock digestion mix to the top of the PBS solution, and top of the tube with PBS if needed. Ultracentrifuge at > 100,000× *g* for 1 h at 4°C.
4. After ultracentrifugation, carefully draw off the PBS layer and a small amount (~1/10) of the sucrose cushion first. Take the supernatant from the very top of the solution. It is very important not to contaminate the lower fractions with protease-containing solution, so change pipettes and then remove all but ~0.1 mL of sucrose solution, thereby leaving only a small layer of sucrose just above the pellet. Rapidly draw off the remainder of the sucrose with a finer (i.e., smaller bore) pipette or micropipetter (e.g., Pipetman) and invert the tube (*see* **Note 9**).
5. Wrap a small wipe, e.g., Kimwipe, around a pair of forceps and wipe out any remaining liquid that might contain subtilisin from the sides of the tube. Keep the tube inverted and take care to stay away from the pellet when wiping.
6. Turn the tube upright, and resuspend the pellet with the desired volume of PBS or other solution that is appropriate for the next intended procedure (*see* **Note 10**).
7. To perform a quality control of the virion digestion/reisolation, examine equal amounts of the mock-digested and digested samples by an appropriate protein analysis, e.g., SDS-PAGE of the total proteins or immunoblotting for proteins on the outside of HIV. For SDS-PAGE analysis, compare the mock-treated with digested virions for a decrease in actin (a prominent band at 42 kDa) or several of the cellular proteins present on the surface. The mature Gag protein bands, i.e., p24 or p17, should be approximately equal between the two preparations while the amounts of cellular proteins should diminish in the digested samples. Even those cellular proteins that are present inside the virion

(actin) should decrease as many of the contaminating vesicles carry the same proteins *(1)*. Additional evidence for effective removal of proteins on the outside of the virion by subtilisin digestion can be provided by an immunoblot to examine the removal of $gp120^{SU}$ or the truncation of $gp41^{TM}$ from a 41-kDa molecule down to a 22-kDa species (a truncated form with only the transmembrane and cytoplasmic tail sequences) (*see* **Note 11**). Likewise, blotting for an interior viral protein, commonly a Gag protein, confirms the integrity of the interior proteins (*see* **Note 12**).

3.3. CD45 Immunodepletion of HIV Preparations

CD45 does not appear to be on the surface of virions yet is associated with most vesicles produced from T cells and other hematopoetic cells *(4, 10)*. This phenomenon has been exploited to remove vesicles from HIV-1 T cell and macrophage preparations by CD45 immunodepletion with anti-CD45 paramagnetic microbeads *(4, 5)*. It is important to consider that most non-hematopoietic cells do not express CD45 so this procedure will not work for preparations from epithelial cells such as HeLa or HEK293. Also, while this process can remove most of the contaminating particles, it may not remove all of them as some vesicles might not contain CD45 *(4)*. Therefore, it is also important to determine if your protein of interest can be removed from a vesicle-only, 'mock virus', preparation from uninfected cells with this procedure.

1. Prepare the microbeads for use by first separating them from the stock material provided by the manufacturer by placing the suspension in a microcentrifuge tube that is then placed in the magnetic separator (the MPC-S holds standard microcentrifuge tubes) at 4 °C to attract the beads to the side of the tube (*see* **Note 13**). Attraction of the beads to the magnet can take hours to overnight as the beads are very small and readily remain in suspension. The beads are reddish-brown so it is apparent when the beads are captured on the magnet side of the tube. The amount of bead-containing supernatant will vary with the amount of virus preparation to be treated. For purified virions, use 2 μL of beads (initial stock solution volume) for each microgram of total protein of the density-purified preparation. If the amount of total protein is not known, an estimate of 1–2 μg total protein per milliliter of initial cell culture supernatant before density centrifugation (*see* **Note 14**). Use at least 10 μL of beads so that you can manipulate them without loss, as smaller amounts of beads are difficult to observe (*see* **Note 15**).
2. Remove the stock solution which contains unbound antibody by carefully removing the supernatant with a pipette, taking

care not to disturb the beads on the side of the tube. Wash twice by first removing the tube from the separator, resuspending the beads in PBS and then separating the beads with the magnet. After the second wash, resuspend the beads in a volume that is convenient for adding to the virus samples (*see* **Note 16**).

3. Add the beads to the virus solution and mix gently. Incubate for 1 h at room temperature before placing the tube in the magnetic separator and beginning the separation at 4 °C until the beads are captured on the side of the tube usually ~20 h. Carefully remove the virus-containing supernatant from the tube, taking great care not to contaminate the supernatant with vesicle-containing microbeads (*see* **Note 17**).
4. For quality control purposes, the removal of CD45 from the preparation can be monitored by immunoblotting samples of both treated and untreated preparations with a CD45 antibody at a dilution of 1:1000. CD45 should be present in the starting material and absent from the treated material (*see* **Note 18**). The general removal of cellular proteins also can be examined by SDS-PAGE.
5. Use the preparation for experiments or analyses (*see* **Note 19**). For treated cell culture fluids, the virus can be pelleted at this point. Losses of the virions can be up to ~30% of the starting material when depleting pelleted virus.

4. Notes

1. The subtilisin digestion and CD45 immunoaffinity depletion method are complementary methods. Both have their advantages, so the choice of method is dictated by the desired study. Subtilisin works best for experiments that focus on the proteins inside the virion. The removal of vesicular proteins is complete or nearly so and there is essentially no loss of the proteins inside the virion. CD45 depletion is most useful for studies that require the exterior of the virion to be intact. As noted above, this approach is restricted to virions produced from cell lines that express CD45. Unlike the digestion procedure, CD45-depleted virions maintain their infectivity *(4)* so they can be used for virological studies. Also, virion-mediated cell signaling or attachment studies can use depleted virions. A drawback of this method is the possibility that not all of the vesicles might contain CD45 and remain in the depleted virus preparation. Our experience has been that this varies with the cell-type used for virus production *(4, 5)*. In either case, the appropriate controls outlined below should help determine

the level of vesicular contamination, in any. This complication has not been observed with the digestion procedure.

2. There are several preparations of subtilisin commercially available from a variety of sources. Any high quality preparation from a reputable supplier should be suitable as this class of enzyme is relatively nonspecific. Two sources that we have used are subtilisin A (Sigma-Aldrich Chemical Company, catalog number P 5380) or subtilisin Carlsberg (Fluka Chemical Company, catalog number 82490).
3. Stocks of PMSF can be made with either acetone or ethanol but be careful to maintain a relatively fresh stock (<2 months old) especially when using ethanol solutions as they tend to degrade faster. The main complication with acetone is that it is more volatile, thus making the stock solution harder to accurately pipette. Several companies provide serine protease inhibitor cocktails which are also suitable.
4. Any high strength magnetic separator/concentrator can be used. There are other suppliers for tube-based magnetic separation available. One can also simply put the tube next to a strong magnet, though this is likely to be less efficient.
5. Working with HIV entails a certain risk that when managed correctly is minimal. When this type of work is carried out with poor training and practices, there is a significant risk of contacting AIDS, a serious and incurable deadly disease. Do not attempt the following procedures without the proper training, containment, engineering controls, and work practices.
6. Smaller pore filters, e.g. 0.22 or 0.45 μm, can be used but they should be the 'low protein binding' type.
7. The digestion time is not very critical. The protease is in gross excess so the digestion time could be reduced to a couple of hours. Likewise, digestion times can be extended because the virion membrane prevents the protease from attacking the interior of the virus. A 3-day digestion of virions produced no discernable reduction of the interior viral Gag proteins.
8. The size of tube to be used is important to maintain a good separation between the protease digestion and the virions, yet minimize the loss of protein on the tube. The suggested tubes for virion amounts based on p24 or total protein values are: <30 μg of p24 or <300 μg of total protein, use an Beckman SW60Ti tube and rotor (or equivalent); 30–100 μg of p24 or 300–1,000 μg of total protein, use an Beckman SW41Ti tube and rotor (or equivalent); and >100 μg of p24 or $>1{,}000$ μg of total protein, use an Beckman SW28Ti tube and rotor (or equivalent). For tube dimensions and other specifications, check the Beckman-Coulter website (www.beckmancoulter.com) or contact your local Beckman sales office.

9. The virion pellet after digestion is quite loose and will disintegrate if treated roughly. Using a micropipetter allows for more precision in removing the last bit of sucrose solution. It is better to leave a small amount of sucrose if the pellet starts to disintegrate. If a considerable amount of the pellet does break loose, then the whole pellet can be resuspended in the remaining sucrose solution, removed from the tube, diluted with PBS and then respun in the same sized tube or a smaller tube. A smaller surface area helps prevent the pellet from breaking up.
10. If you are lysing the virions to run a gel, then you can add 1 × PMSF (5 μg/mL) to the lysing buffer you typically use if you want to be cautious. Keep in mind that subtilisin can partially digest proteins even in SDS. However, do not add PMSF to a sample if the next analysis is sensitive to hydrophobic compounds (e.g., reversed-phase chromatography) as this interferes with the column or if a downstream step requires the use of a serine protease.
11. Perkin Elmer Inc. (http://las.perkinelmer.com/) manufactures a good gp41^{TM} monoclonal antibody (cat# NEA-9303001EA) and Fitzgerald II Inc. (www.fitzgerald-fii.com/) has a gp41^{TM} polyclonal antiserum [cat# 20-HR92] that is not as specific as the Perkin-Elmer antibody but it is still good. Internal proteins are best monitored by antibodies or antiserum against Gag, which are available from many sources (www.linscott.com, www.abcam.com, www.biocompare.com). If you have a good patient serum that detects both the interior Gag proteins and the exterior Env proteins, both assays can be accomplished in one blot.
12. If you experience problems, usually seen as a digestion of Gag detected either by SDS-PAGE or immunoblot, then either the PMSF is no longer effective and should be replaced or some of the subtilisin in the PBS layer has contaminated the preparation after the ultracentrifugation step. If this is the case, then be more careful removing the subtilisin-containing material. Also note that nonlysed sample can be centrifuged through sucrose again.
13. Our experimentation has found that the Miltenyi Biotec microbeads are especially effective while those of other suppliers have not been able to remove all of the CD45 from preparations. It is unclear whether it is the small (50 nm) size of the Miltenyi beads or the antibody used on the bead that provides this efficacy.
14. Virus containing cell culture supernatants can also be treated directly. In this case, add ~1–2 μL of beads/ml of clarified supernatant (cells removed). This takes larger magnets (the MPC-L handles 15 mL conical tubes and ~7 mL of medium)

and more time for the microbeads to migrate through the bigger volume. However, this approach appears to generally give a better yield than ultracentrifuged preparations.

15. The speed at which the beads are captured by the magnet is inversely proportional to volume. Therefore, if time is critical it is better to distribute smaller volumes of supernatant in more tubes rather than a larger volume in one tube.
16. Consider that the larger the volume the longer it takes to collect the beads. However, working with very small volumes of virus-bead mixture can result in a loss of sample that would be avoidable with a larger volume.
17. It is better to leave a little supernatant in the tube rather than contaminate your sample with some beads. This small bit can be recovered by placing the tube back on the separator and re-separating the beads. In the same way, if there is bead contamination of the sample during removal, then put the supernatant back on the separator, collect the beads on the side of the tube, and try again to remove the supernatant. As a precaution, the treated supernatant can be put through another cycle of separation to ensure that any stray microbeads are removed from the preparation.
18. If you note a ~30-kDa band [in the SDS-PAGE], then there are beads in the treated preparation, as this comes from the light chain of the antibody-bound beads that is released by the reducing SDS-PAGE gel. A slight signal seems unavoidable and harmless. A heavy signal indicates that another cycle of separation is needed.
19. Since the CD45 immunoaffinity depletion may not remove all of the vesicles that contain your protein of interest, it is helpful to compare depleted preparations produced from infected and uninfected cells, i.e., look at virus vs. vesicle protein levels. Immunoblotting for the protein of interest should show that the specific protein in the vesicle or 'mock virus' samples should be greatly reduced or eliminated by the process.

Acknowledgment

This project has been funded in whole or in part with Federal funds from the National Cancer Institute, National Institutes of Health, under Contract No. N01-CO-12400. The content of this publication does not necessarily reflect the views or policies of the Department of Health and Human Services, nor does mention of trade names, commercial products, or organization imply endorsement by the U.S. Government.

References

1. Ott, D. E. (2002) Potential roles of cellular proteins in HIV-1. *Rev Med Virol* 12, 359–374.
2. Bess, J. W., Jr., Gorelick, R. J., Bosche, W. J., et al. (1997) Microvesicles are a source of contaminating cellular proteins found in purified HIV-1 preparations. *Virology* 230, 134–144.
3. Gluschankof, P., Mondor, I., Gelderblom, H. R., et al. (1997) Cell membrane vesicles are a major contaminant of gradient-enriched human immunodeficiency virus type-1 preparations. *Virology* 230, 125–133.
4. Trubey, C. M., Chertova, E., Coren, L. V., et al. (2003) Quantitation of HLA class II protein incorporated into human immunodeficiency type 1 virions purified by anti-CD45 immunoaffinity depletion of microvesicles. *J Virol* 77, 12699–12709.
5. Chertova, E., Chertov, O., Coren, L. V., et al. (2006) Proteomic and biochemical analysis of purified human immunodeficiency virus type 1 produced from infected monocyte-derived macrophages. *J Virol* 80, 9039–9052.
6. Bess, J. W., Jr., Powell, P. J., Issaq, H. J., et al. (1992) Tightly bound zinc in human immunodeficiency virus type 1, human T-cell leukemia virus type 1, and other retroviruses. *J Virol* 66, 840–847.
7. Ott, D. E., Coren, L. V., Johnson, D. G., et al. (2000) Actin-binding cellular proteins inside human immunodeficiency virus type 1. *Virology* 266, 42–51.
8. Ott, D. E., Coren, L. V., Johnson, D. G., et al. (1995) Analysis and localization of cyclophilin A found in the virions of human immunodeficiency virus type 1 MN strain. *AIDS Res Hum Retroviruses* 11, 1003–1006.
9. Ott, D. E., Coren, L. V., Kane, B. P., et al. (1996) Cytoskeletal proteins inside human immunodeficiency virus type 1 virions. *J Virol* 70, 7734–7743.
10. Esser, M. T., Graham, D. R., Coren, L. V., et al. (2001) Differential incorporation of CD45, CD80 (B7-1), CD86 (B7-2), and major histocompatibility complex class I and II molecules into human immunodeficiency virus type 1 virions and microvesicles: implications for viral pathogenesis and immune regulation. *J Virol* 75, 6173–6182.
11. Ott, D. E., Coren, L. V., Copeland, T. D., et al. (1998) Ubiquitin is covalently attached to the p6Gag proteins of human immunodeficiency virus type-1 and simian immunodeficiency virus and the p12Gag protein of Moloney murine leukemia virus. *J Virol* 72, 2962–2968.
12. Ott, D. E., Coren, L. V., Sowder, R. C., II, et al. (2002) Equine infectious anemia virus and the ubiquitin-proteasome system. *J Virol* 76, 3038–3044.

Chapter 3

Calculating HIV-1 Infectious Titre Using a Virtual $TCID_{50}$ Method

Yong Gao, Immaculate Nankya, Awet Abraha, Ryan M. Troyer, Kenneth N. Nelson, Andrea Rubio, and Eric J. Arts

Abstract

Studies of HIV-1 replication kinetics and fitness require an accurate determination of the level of infectious HIV-1 present in virus stocks. The standard technique for measuring the level of replication-competent infectious virus in culture supernatants or patient samples is the tissue culture dose for 50% infectivity ($TCID_{50}$), which provides an accurate assessment of the level of infectious HIV-1. However, it is a time-consuming technique which typically takes two or more weeks to complete and requires PHA-stimulated PBMC from HIV-1 seronegative donors or an appropriate cell line. Thus rapid, cell-free surrogate measures for $TCID_{50}$ are desirable. Here, we introduce the virtual $TCID_{50}$ technique: a new cell-free method estimating a surrogate of infectious titer by comparing the reverse transcriptase activity in virus stock to that of reference viruses with a known $TCID_{50}$ value. We have demonstrated that the virtual $TCID_{50}$ obtained through this technique is comparable to the actual infectious $TCID_{50}$. This method greatly simplifies the process of accurate HIV-1 titration and is particularly beneficial for studies which require titration of large number of HIV-1 isolates.

Key words: HIV-1, $TCID_{50}$, Virtual $TCID_{50}$

1. Introduction

Accurate determination of HIV-1 infectious titer is critically important for studies on replication kinetics or competitive fitness of different HIV-1 isolates. Endpoint dilution of HIV-1 to determine tissue culture dose for 50% infectivity ($TCID_{50}$) is considered as the standard method for measuring the level of replication-competent, infectious HIV-1 in culture supernatants or patient samples *(1–3)*. This assay is highly informative because it involves direct determination of virus infectivity in the host cells

Vinayaka R. Prasad, Ganjam V. Kalpana (eds.), *HIV Protocols: Second Edition, vol. 485*

DOI 10.1007/978-1-59745-170-3_3 Springerprotocols.com

of interest. PHA-stimulated PBMCs from HIV-1 seronegative donors are frequently used as host cells for $TCID_{50}$ measurement because these cells are more representative of the in vivo target cells than cell lines. However, determination of $TCID_{50}$ by infection of PBMC ex vivo is a very time-consuming, laborious process and the values derived are often donor dependent. Thus, many laboratories utilize surrogate assays, such as capsid p24 antigen measurement in supernatant, to estimate infectious titer. These assays typically quantify a viral component protein but do not directly determine virus infectivity. We and others have found that antigen capture assays cannot differentiate between noninfectious virus particles and virions, and is frequently a very poor estimate of infectious titer, with results differing by up to 350-fold from the actual infectious $TCID_{50}$ *(4, 5)*. In contrast, we have found that a simple well-standardized reverse transcriptase (RT) activity assay provides a very good estimate of infectious HIV-1 titer *(4)*.

Here we describe a simple cell-free method for estimating $TCID_{50}$ (or infectious titer) by RT activity. This method provides rapid determination of HIV-1 titer (within 1 day) without requiring laborious and time-consuming multi-week replication in cell culture. It also provides improved accuracy over other surrogate measures of $TCID_{50}$ *(4)*. This is achieved by (1) comparing RT activity of samples to that of reference viruses whose $TCID_{50}$ is determined on the human PBMCs; (2) utilizing virus dilution series so that $TCID_{50}$ is estimated within the linear range of the assay; (3) separately analyzing viruses with different co-receptor usage (CCR5 or CXCR4). This technique greatly simplifies the process of infectious HIV-1 determinations for studies in which many HIV-1 isolates must be titrated.

2. Materials

2.1. Viruses

1. Reference viruses: each experiment needs —three to five reference viruses as controls. These $TCID_{50}$s of the viruses have been measured by real $TCID_{50}$ assay (*see* **Section 3.2.**). The reference viruses should have the same tropism (CCR5, CXCR4 or dual) as your samples *(6)* and are preferably propagated in the same PBMCs (or cell line) as your sample viruses (*see* **Notes 1** and **2**).
2. Sample viruses: propagated in PBMCs (or cell line) as your reference viruses (*see* **Note 3**).

2.2. PBMCs Preparation

1. Biohazard level 2 safety facility.
2. HIV-seronegative whole blood donor (preferably same donor for the entire experiment, *see* **Note 3**).
3. Ficoll-Hypaque.

4. Sterile, phosphate-buffered saline for tissue culture.
5. RPMI-1640–2 mM L-glutamine supplemented with 10% fetal bovine serum (FBS), 10 mM HEPES buffer, 100 IU of penicillin/mL, 100 μg of streptomycin/mL, 1 ng of recombinant interleukin-2 (Invitrogen)/ml, and 2 μg of phytohemagglutinin (PHA; Invitrogen)/mL (complete, medium II).
6. 37 °C incubator supplemented with 5% CO_2.

2.3. Infectious $TCID_{50}$

1. Biohazard level 2 safety facility.
2. Facility for radioactive isotope usage.
3. Virus supernatant to be used for titration.
4. PHA-stimulated PBMCs (need approximately 2.4×10^6 cells per virus titration).
5. RPMI-1640–2 mM L-glutamine (medium I).
6. Medium II (as above).
7. Phosphate-buffered saline sterile for tissue culture.
8. 96-well flat-bottom microtiter plates.
9. 96-well round-bottom microtiter plates.
10. Multichannel pipet.
11. RT master mix: 50 mM Tris-HCl (pH 7.8), 75 mM KCl, 2 mM dithiothreitol (DTT), 5 mM $MgCl_2$, 5 μg/mL of poly (rA), 6.25 μg/mL oligo (dT), 0.5%(v/v) NP40; make 1-mL aliquots and store at −20 to −70 °C.
12. 10 mCi/mL α-32P dTTP – Perkin Elmer, cat. no. NEG-505A (>400 Ci/mmol).
13. DEAE Filtermat for use with 1450 MicroBeta (96-well format) (Perkin Elmer, cat. no. 1450-522).
14. 1× saline-sodium citrate (SSC).
15. 85% Ethanol.

2.4. Virtual $TCID_{50}$

1. Biohazard level 2 safety facility.
2. Facility for radioactive isotope usage.
3. Virus supernatant to be used for titration.
4. Phosphate-buffered saline.
5. 96-well round-bottom microtiter plates.
6. Multichannel pipet.
7. RT master mix: 50 mM Tris-HCl (pH 7.8), 75 mM KCl, 2 mM DTT, 5 mM MgCl2, 5 μg/mL of poly (rA), 6.25 μg/mL oligo (dT), 0.5%(v/v) NP40; make 1-mL aliquots and store at −20 to −70 °C.
8. 10 mCi/mL α-32P dTTP – Perkin Elmer, cat. no. NEG-505A (>400 Ci/mmol) (For those who do not want to use radioactive material, *EnzChek® Reverse Transcriptase Assay Kit* will be another option. *see* http://probes.invitrogen.com/media/pis/mp22064.pdf?id=mp22064).

9. DEAE Filtermat for use with 1450 MicroBeta (96-well format) (Perkin Elmer, cat. no. 1450-522).
10. 1× SSC.
11. 85% Ethanol.

3. Methods

3.1. PBMC Preparation

Isolate PMBC from heparin-treated venous blood of HIV-seronegative donor by Ficoll–Hypaque density gradient centrifugation. Expected yield of PBMCs is 1–2 × 10^6 cells/mL of whole blood.

1. Resuspend pelleted PBMC in the appropriate volume of complete medium II to yield a final cell concentration of approx 1 × 10^6 cells/mL and incubate for 48 h.
2. 48 h post-stimulation, evaluate cell number and viability of cells by trypan blue exclusion.
3. Centrifuge the cells for 5 min at 1,200 × *g* at room temperature. Discard the supernatant and save the pellet.
4. Resuspend the cell pellet with freeze medium (10% DMSO, 70% FBS, 20% RPMI-1640–2 mM L-glutamine) to achieve a final cell concentration of 2 × 10^6 cells/mL.
5. Make a 1-ml aliquot of resuspended cells in cryovials.
6. Freeze cells at −80 °C for short-term storage and in liquid nitrogen freezer for long-term storage (*see* **Note 4**).

3.2. Infectious $TCID_{50}$

The following steps should be performed in a biohazard level 2 safety facility.

1. Retrieve the cryovials containing frozen cells and thaw by immersing the bottom half of the vials in a 37 °C water bath and swirl the tube until cells are thawed.
2. In a tissue culture hood, spray the outside of the vials with 70% ethanol. Open vial and resuspend cells with complete medium II by using a sterile pipette to an approximate cell concentration of 1 × 10^6 cells/mL. Transfer the cells to a sterile conical tube; you may consolidate multiple vials.
3. Centrifuge resuspended cells for 5 min at 1,200 × *g* at room temperature.
4. Discard the supernatant and resuspend pelleted cells in media to an approximate cell concentration of 1 × 10^6 cells/mL. Transfer to a tissue culture flask and incubate for 2–3 days.
5. On day 2 or 3 after stimulation of cells, while the cells are still in a logarithmic phase of growth, count viable cells by trypan blue exclusion.
6. Using a multichannel pipettor, plate 96-well flat-bottom plates with 100,000 cells per well in a total volume of 150 μL

of media II. Twenty-four wells are required for each virus to be titrated (eight dilutions in triplicate; *see* **Fig. 3.1**).

7. Once cells are plated, place into incubator until infection.
8. Thaw and make 10-fold serial dilutions of virus supernatants in medium I (*see* **Fig. 3.1** and **Notes 5**, **6**).
9. In triplicate: add 100 μL of each serial dilution to the appropriate wells in the 96-well plate containing PBMCs (*see* **Fig. 3.1**). Leave some wells uninfected for negative controls and incubate in 37 °C incubator supplemented with 5% CO_2.
10. 72 h post-infection, spin down plates and carefully harvest 150 μL of the cell-free supernatant and discard appropriately.
11. Add 150 μL of complete medium II and incubate in 37 °C incubator supplemented with 5% CO_2.
12. Using a multi-channel pipettor, harvest 25 μL of cell-free supernatant on days 6, 8, 10, and 12 post-infection and store in −80 °C freezer to be used for RT activity assay (*see* **Note 7**).

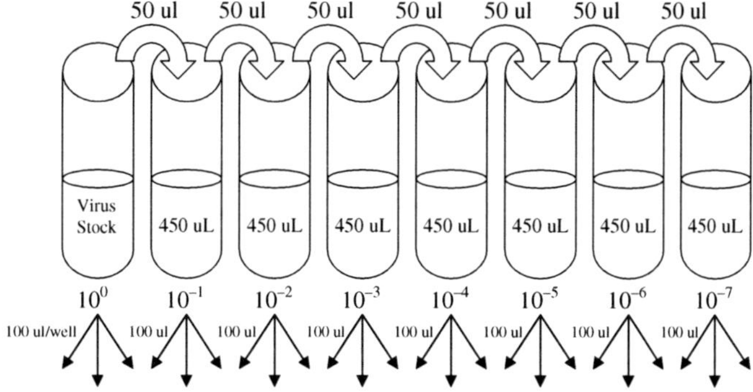

Fig. 3.1. Preparation of virus for real $TCID_{50}$ assay. (**A**) Schematic illustration of the 10-fold serial dilution step for virus titration. (**B**) 100,000 cells/well are plated as illustrated in the figure. Fill the wells marked X with 250 μL of PBS (to reduce rate of evaporation). Infect cells by adding 100 μL of the serially diluted virus (as well as undiluted) as depicted in the figure.

3.3. RT Assay

Reverse transcriptase assay measures the activity of the RT enzyme present in HIV in the culture supernatant. The RT master mixture contains the necessary substrate to drive HIV RT activity. This step requires the use of both radioactivity and biological agent (HIV) facility at the same time. You must have the proper training and certification to work with radioactive and biological agents.

1. Thaw viruses (harvested supernatant) at room temperature.
2. Aliquot 10 μL of the 25 μL supernatant into 96-well round-bottom microtiter plates.
3. Aliquot 10 μL of positive control (supernatant previously known to have positive RT activity) and negative control (supernatant from uninfected PBMC) in duplicate.
4. Add 1 μL fresh 10 mCi/mL [α-^{32}P]-dTTP into 1 mL RT master mixture.
5. Using a multi-channel pipettor, add 25 μL of the above RT mixture into each well containing 10 μL cell-free supernatant and incubate at 37 °C incubator overnight.

   Caution: Tips used to load radioactive RT master mixture into wells containing supernatant may possess two biohazards (HIV and ^{32}P radionucleotide) and must be discarded properly. Generally it is best to drop the mixture into the wells and completely avoid touching the virus. The tips can then be discarded as solid radioactive waste. If contaminations by both biohazard agents occur, you must aspirate bleach into the tips and discard the bleach as liquid and the tips as solid radioactive waste *(1)*.

   After more than 2 h incubation with the RT master mix, any HIV will be rendered noninfectious by the NP40 detergent in the mix. The remaining steps can be performed on the bench top employing appropriate radiation safety measures.
6. After the overnight incubation, using the multi-channel pipettor, blot 10 μL of the reaction mixture from each well onto the 96-well format DEAE filtermat. It is best to avoid touching the paper with the pipette tips to prevent tearing of the filtermat.
7. Allow the filtermat to dry at room temperature (~ 10 min). Wash the filtermat five times with 1× SSC solution and twice with 85% ethanol by rocking in a shaker platform for 5 min each. The SSC wash liquid waste must be discarded as radioactive waste, but the ethanol washes can be discarded as regular waste.
8. Allow filters to dry (on heat block or at room temperature), wrap with Saran wrap and expose overnight onto autoradiography film (Kodak BioMax MR).
9. Develop the film and use along with quantitative data to calculate $TCID_{50}$ (*see* **Note 8**).

10. To generate quantitative data, count the filters with a Matrix 96 β-counter (Packard, Meriden, CT) (*see* **Note 9**).
11. The 50% tissue culture infective dose values can then be calculated for each virus using the Reed–Muench technique *(1)*. The Reed–Muench accumulative calculation method yields the $TCID_{50}$ value as infectious units per milliliter (IU/mL).

3.4. Virtual TCID50

1. Pick three to five viruses of known $TCID_{50}$ value as controls. They should have been treated the same way as your viruses (e.g., propagated on same PBMCs), and should not have been freeze/thawed subsequent to titration.
2. Starting with undiluted viruses and controls, make seven additional fourfold dilutions for each virus in medium I (this can be done in rows across a 48-well plate).
3. Perform an RT assay on the serially diluted samples and controls at the same time. The dilutions for each virus should be set up so that 10 μL of each of the eight dilutions (including undilute) are added in triplicate down the round bottom 96-well plate from row A through row H.
4. Calculate the IU at each dilution for the controls, i.e., $TCID_{50}$ (IU/mL) × dilution × volume of virus added.
5. Plot log IU vs. log RT value for the triplicate values of each of the control viruses in the linear range. For each control, find the equation of the best fit line (*see* **Fig. 3.2**). It may be neces-

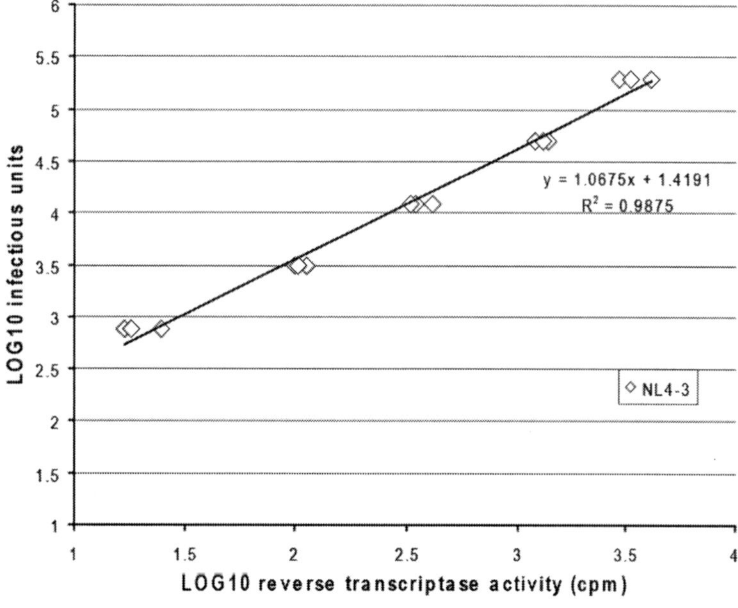

Fig. 3.2. Example of equation for virtual $TCID_{50}$ calculation. Plot log IU vs. log RT Value for the triplicate values of each of the controls, in the linear range. The equation of this line is what will be used to calculate the $TCID_{50}$.

Table 3.1
Example of calculation for virtual $TCID_{50}$

Virus	Dilution factor	Mean log RT value	log IU	IU	$TCID_{50}$	Base 10
C25	0.0625	2.00172	3.54128	3477.673	55642.777	4.745408
	0.015625	1.25996	2.72994	536.963	34365.662	4.536124
	0.00390625	1.07768	2.53057	339.291	86858.678	4.938813
					Average	4.740115

Note: For all experimental samples, find the mean RT value for each dilution value in the linear range. Calculate the LOG of the mean RT value. Plug this value into the equation achieved in Fig. 3.2 to determine log IU and convert into IU. Divide this number by (dilution of virus × volume of virus added) to get the $TCID_{50}$ value. Take the log of this number, to give a base 10 value. Average the results of a sample from within a linear range and this is your $TCID_{50}$.

sary to remove points outside the linear range for each virus in order to deal with poor sensitivity or saturation problems and replot the graph, and find a best fit line using all these points. The equation of this line is what will be used to calculate the $TCID_{50}$.

6. For all experimental samples, find the mean RT value for each dilution value in the linear range (do not use RT values which approach those found for virus-negative controls). Calculate the log of the mean RT value. Insert this value into the equation achieved in step 5 to determine a corresponding log IU.
7. Use the value from the equation to determine IU. Divide this number by (dilution of virus × volume of virus added) to get the $TCID_{50}$ value (*see* **Fig. 3.2** and **Table 3.1**). Take the log of this number, to give a base 10 value (i.e., a meaningful number). Average the results of a sample from within a linear range – this is your virtual $TCID_{50}$.

4. Notes

1. It is critically important that reference viruses have the same co-receptor usage (CXCR4, CCR5 or dual tropic) as the viruses which are being assayed. For determining virtual $TCID_{50}$ of both CXCR4 and CCR5 viruses, you must use separate sets of reference viruses and construct separate standard curves for the two tropisms. The authors recommend that co-receptor usage be determined by infection of U87.CD4.CCR5 and U87.CD4.CXCR4 cells.
2. The best results for dual-tropic (R5/X4) viruses are found when dual-tropic reference viruses are used. However, results for dual-tropic viruses may have greater variability than

exclusive CCR5 or CXCR4 viruses since dual-tropic viruses may vary in the extent of CCR5 versus CXCR4 usage.

3. It is important to note that the infectious $TCID_{50}$ assay and the following experiments should be performed with PBMCs from the same donor. Cell lines, such as U87 CD4 CCR5/CXCR4, can also be used for this assay. In this case, all the assays and the following experiments should be carried out in the same cell line at a comparable passage.
4. For successful cryopreservation and recovery of cells, it is best to freeze the cells by utilizing freezing containers (Fisher cat. no. 15-350-50) that cool cells at −1 °C/min.
5. Prior to performing virus titration, make multiple aliquots of the virus stocks and freeze at −80 °C or in liquid nitrogen. It is generally best to make small volume aliquots (200–500 μL) to avoid wasting.
6. For titrating multiple viruses, you may use 48-well plates for making serial dilutions.
7. Harvesting directly into 96-well plate makes it possible to use a multi-channel pipettor when setting up RT activity assay.
8. Visualizing the filter on an X-ray film helps identify any potential irregularities that may have occurred due to spotting or washing (i.e., background level) as well as verify the numerical data that accompanies it.
9. Alternatively, you can cut out the radioactive areas that represent each spot from the filters and count in a liquid scintillation counter. It is also possible to read it on a phosphor imager screen to generate numerical data.

References

1. Reed, L. J., Muench, H. (1938) A simple method of estimating fifty per cent endpoints. *Am J Hyg* 27, 493–497.
2. Ball, S. C., Abraha, A., Collins, K. R., et al. (2003) Comparing the ex vivo fitness of CCR5-tropic human immunodeficiency virus type 1 isolates of subtypes B and C. *J Virol* 77, 1021–1038.
3. Quinones-Mateu, M. E., Ball, S. C., Marozsan, A. J., et al. (2000) A dual infection/competition assay shows a correlation between ex vivo human immunodeficiency virus type 1 fitness and disease progression. *J Virol* 74, 9222–9233.
4. Marozsan, A. J., Fraundorf, E., Abraha, A., et al. (2004) Relationships between infectious titer, capsid protein levels, and reverse transcriptase activities of diverse human immunodeficiency virus type 1 isolates. *J Virol* Oct, 78(20), 11130–11141
5. Heneine, W., Yamamoto, S., Switzer, W. M., et al. (1995) Detection of reverse transcriptase by a highly sensitive assay in sera from persons infected with human immunodeficiency virus type 1. *J Infect Dis* 171, 1210–1216.
6. Campbell, T. B., Schneider, K., Wrin, T., et al. (2003) Relationship between in vitro human immunodeficiency virus type 1 replication rate and virus load in plasma. *J Virol* 77, 12105–12112.

Section II

Methods to Study HIV-1 Replication

Subsection A

Early Events

Chapter 4

Cell-Free Assays for HIV-1 Uncoating

Christopher Aiken

Abstract

Uncoating is an essential step in the retrovirus life cycle about which little is known. Uncoating is defined as the specific dissociation of the capsid shell from the viral core in the host cell cytoplasm. In this chapter, biochemical assays for studying HIV-1 uncoating in vitro are described. These techniques have proven useful for characterizing HIV-1 mutants that exhibit defects in the uncoating step of infection.

Key words: Uncoating, retrovirus, HIV-1, core, capsid, disassembly, reverse transcription.

1. Introduction

The term *uncoating* has been used to refer to the early post-entry steps in virus infection immediately following fusion of the viral and cellular membranes and delivery of the viral core into the cytoplasm. For most enveloped viruses, particularly retroviruses, the details of uncoating are essentially unknown. Uncoating can be defined more specifically as the shedding of the viral capsid from the retroviral core (reviewed in Ref. *(1)*). For HIV-1, this appears to involve disassembly of the conical capsid and the release of the CA protein into a soluble form (**Fig. 4.1**). The intracellular location and timing of HIV-1 uncoating during infection are not known.

Retroviruses assemble as immature particles that undergo maturation to become infectious. Maturation requires cleavage of the viral Gag and Pol polyproteins by the viral protease into individual proteins; these rearrange to form the mature viral core. For lentiviruses, such as HIV-1, the shell of the core is formed by a conical capsid composed of CA protein molecules arranged as a

Vinayaka R. Prasad, Ganjam V. Kalpana (eds.), *HIV Protocols: Second Edition, vol. 485*

DOI 10.1007/978-1-59745-170-3_4 Springerprotocols.com

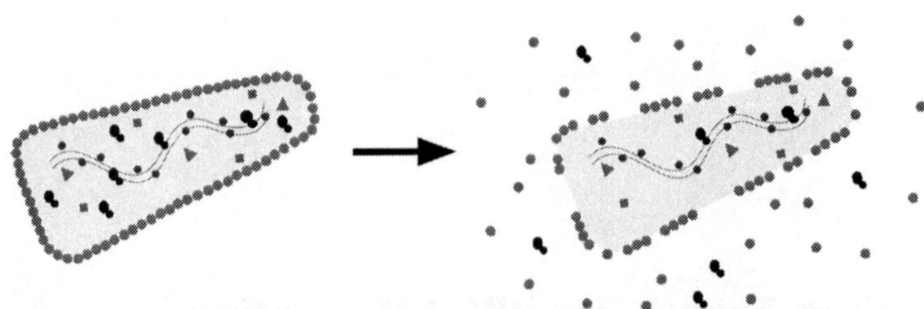

Fig. 4.1. Schematic of HIV-1 uncoating. During incubation at 37 °C, purified HIV-1 cores spontaneously release the CA and RT proteins into a soluble form. The extent of uncoating is determined by p24 ELISA after separating free from core-associated CA by ultracentifugation.

lattice of hexagons. The conical shape of lentiviral cores suggests that some unique aspect of their biology, such as the ability to infect nondividing cells, may be dependent on a specific uncoating mechanism. Studies of HIV-1 mutants containing substitutions in CA have revealed that infection is critically dependent on the proper stability of the HIV-1 capsid *(2)*. Species-specific restriction factors potently inhibit HIV-1 infection by targeting the viral capsid. These studies indicate that uncoating is a key step in HIV-1 infection that may be attractive for targeted antiviral therapy.

Historically, retroviral uncoating has been difficult to study due to the lack of sensitive and specific assays for this process. The perception that a high percentage of HIV-1 particles are defective may also have intimidated researchers from developing assays for uncoating due to potential difficulties in interpreting experimental outcomes. Here we describe a quantitative method for assaying HIV-1 uncoating in vitro by using purified viral cores. The approach involves purifiying HIV-1 cores by sedimentation of intact virions through a detergent. Samples of the cores are then incubated at 37 °C, and the extent of CA release is quantified by p24 ELISA after the cores have been pelleted in an ultracentrifuge. This approach has been employed successfully to analyze the effects of viral mutations on HIV-1 capsid stability *(2–4)* and may also prove useful for identifying cellular activities that influence HIV-1 uncoating *(5)*. The procedures are based on original methods for isolating intact cores from virions of retroviruses other than HIV-1 *(6--13)*.

2. Materials

2.1. Cells and Media

1. Dulbecco's Modified Eagle's Medium (DMEM), supplemented with 10% fetal bovine serum and penicillin/ streptomycin.
2. 293T cells (American Type Culture Collection).

2.2. Transfection of 293T Cells

1. 2X Bes-buffered saline solution (2X BBS): 50 mM BES (pH 6.95), 280 mM NaCl, and 1.5 mM Na_2HPO_4. Adjust pH to 6.95 at room temperature with NaOH. Filter sterilize and store in 10-mL aliquots at −20 °C.
2. $CaCl_2$ solution (2.5 M): Dissolve 36.8 g of $CaCl_2 2H_2O$ in purified water (MilliQ or equivalent) and adjust the volume to 100 mL. Sterilize the solution by vacuum filtration through a 0. 22 μm pore-size filter, and store it in aliquots at −20 °C.
3. Dulbecco's phosphate-buffered saline (PBS), sterile.
4. HIV-1 proviral plasmid DNA (pNL4-3 or equivalent, purified by Qiagen maxi-prep or cesium chloride density gradient centrifugation).

2.3. Density Gradient Ultracentrifugation

1. STE buffer: 10 mM Tris-HCl [pH 7.4], 100 mM NaCl, 1 mM EDTA, diluted from a 10X concentrated stock solution. Adjust the pH to 7.4 with HCl after diluting to the final (1X) concentration.
2. STE buffer containing 30% (w/vol) sucrose: prepare by adding 30 g sucrose to 10 mL of 10X STE buffer. Add purified water to final volume (100 mL). Adjust pH to 7.4. Store at room temperature.
3. STE containing 70% (w/vol) sucrose: Add 70 g sucrose to 10 mL of 10X STE buffer. Add purified water to final volume (100 mL). Adjust pH to 7.4. Store at room temperature.
4. 1% Triton X-100 in STE buffer. Add 1 mL of 10% Triton X-100 (in water) to 9 mL of 1X STE buffer. Store at 4 °C.

2.4. p24 ELISA

1. Phosphate-buffered saline, non-sterile: Prepare a 10X concentrated stock by dissolving 80 g NaCl, 2 g KCl, 6.1 g Na_2HPO_4, 2 g KH_2PO_4 in 900 mL of deionized water. Adjust the volume to 1000 mL. Make 1:10 dilution in water prior to use.
2. ELISA sample diluent: mix 10 mL of newborn calf serum and 5 mL of 10% Triton X-100 with 85 mL of PBS. This solution should be prepared fresh. Alternatively, a larger quantity can be prepared and the solution filter sterilized and stored at 4 °C.
3. Blocking buffer: 5% calf serum in PBS. Mix 5 mL of newborn calf serum with 95 mL sterile PBS (sufficient for blocking four plates). This solution should be prepared fresh.
4. ELISA wash solution: 0.2% Tween 20 in PBS. Add 2 mL Tween 20 to 1 L of PBS
5. Coating antibody: monoclonal antibody to p24 (183-H12-5C; NIH AIDS Research and Reference Reagent Program) produced in Fibercell hollow fiber bioreactor (Fibercell

Systems, Inc.). Dilute to 4 μg/mL in PBS immediately prior to use (10 mL per plate required).

6. Primary antibody: Add 5 μL of a 1:10 dilution of HIV-Ig (hyperimmune human patient serum, NIH AIDS Research and Reference Reagent Program) to 10 mL of ELISA sample diluent.
7. Secondary antibody: add 2 μL ImmunoPure Goat Anti-Human IgG (H + L), peroxidase conjugated (0.4 mg/mL; Pierce) to 10 mL of ELISA sample diluent.
8. Recombinant p24 protein or p24 standard from a commercial ELISA kit.
9. HRP substrate (TMB Microwell Peroxidase Substrate System; KPL, Inc.)
10. 4N H_2SO_4 solution in water. Slowly add 100 mL of 18 N H_2SO_4 to 350 mL water with stirring. Allow to cool to room temperature.
11. Immulon 2HB 96 well plates (Thermo Electron Corp.).

2.5. Specialized Equipment Needed

1. SW28Ti and SW41Ti rotors and compatible ultracentrifuge (Beckman Instruments, Inc.)
2. TLA45 rotor and tabletop ultracentrifuge (Beckman Instuments, Inc.).
3. Highspeed microfuge tubes (Beckman Instruments, Inc.).
4. Auto Densi-Flow density gradient fractionator (Labconco Corp., Kansas City, MO) or other peristaltic pump.
5. 20 ml linear gradient former (GM-20; CBS Scientific Co., Inc., Solana Beach, CA)
6. Refractometer (optional; Leica Abbe Mark II Plus or equivalent).

3. Methods

The procedure for isolating HIV-1 cores involves transfection of 293T cells to produce HIV-1 particles, concentration of the virus by ultracentrifugation, and ultracentrifugation of the concentrated virus particles through a layer of Triton X-100 into a linear density gradient. Exposure of the enveloped virus particles to detergent results in release of the viral cores which sediment to an equilibrium density of 1.24–1.27 g/mL. Because the density of intact HIV-1 particles is significantly lower (~1.16 g/mL), the presence of CA protein in the denser fractions is a good indication that the viral membrane has been disrupted. A significant fraction of the viral CA protein is not associated with the viral core, and separation of the free and core-associated CA ultracentrifugation allows quantification of the percentage of CA protein associated with the cores by p24 ELISA. In our experience, the yield of p24 in the core fractions is a highly sensitive measure of the stability of the viral capsid.

3.1. Isolation of HIV-1 Cores

The first step in studying HIV-1 uncoating in vitro is to isolate HIV-1 cores by brief detergent treatment of virions followed by ultracentrifugation (*see* **Note 1**). This involves the following steps:

3.1.1. Production of HIV-1 Particles

This is typically performed by transient transfection of 293T cells with HIV-1 proviral DNA {e.g. pNL4-3 *(14)*, available through the NIH AIDS Research and Reference Reagent Program}. Most molecular virology labs are familiar with this technique. Because the procedure varies considerably among laboratories, we describe our standard calcium phosphate-based approach here *(15)*.

1. Culture 293T cells in DMEM containing 10% FBS and antibiotics (penicillin and streptomycin) in an incubator calibrated at 37 °C and 5% CO_2. Cells are detached from nearly confluent dishes with trypsin-EDTA. Two million cells are seeded in 9 mL medium in 100 mm plastic culture dishes and cultured for 1 day prior to transfection. The next day, cultures will exhibit approximately 25% confluence. Because 30 mL of virus is typically used to isolate cores, we typically transfect six dishes per virus to be analyzed.
2. For each dish of cells to be transfected, mix 20 μg of proviral plasmid DNA with sterile water to a total volume of 450 μL. Add 50 μL 2.5 $CaCl_2$ solution and mix well. Add 0.5 mL 2X BBS solution. Mix well by pipetting up and down. Incubate transfection mixture for 10–20 min at room temperature.
3. Pipet transfection mixture directly onto one dish of cultured 293T cells. Add the mixture dropwise over most of the culture area. The color of the medium will change when the drops contact the media. After adding the entire 1 mL volume, rock the plate back and forth and side-to-side and place the dish overnight (~16 h) in an incubator calibrated to 35 °C and 3% CO_2.
4. Aspirate medium and rinse gently with 4 mL PBS.
5. Aspirate PBS and add 5 mL fresh medium.
6. Culture the cells in an incubator calibrated at 37 °C and 5% CO_2 for 24–48 h.
7. Withdraw supernatant by pipetting; transfer to conical centrifuge tube and centrifuge for 5 min at $1,500 \times g$ to pellet cells and debris. Clarify supernatant by passing through 0.45 μm pore-size syringe filter {for volumes > 30 mL, a vacuum filtration unit (Nalgene or equivalent) may be employed}.

3.1.2. Concentration of HIV-1 Particles by Ultracentrifugation

1. Place 30 mL virus suspension in a 38.5 mL polyallomar centrifuge tube (for SW28 rotor). Using a 5-mL pipet, carefully underlay the virus with 5 mL of a solution of 20% sucrose in PBS.
2. Centrifuge for 2.5 h at 28,000 rpm ($141,000 \times g$ at r_{max}) to pellet virus particles.

3. Aspirate supernatant; resuspend pellet in 0.5 mL STE buffer. Pipet gently to resuspend, taking care to avoid foaming as much as possible. Transfer to 1.5 screw-cap eppendorf tube and place at 4 °C for 1–3 h to allow small clumps of virus to disperse.
4. Gently pipet the suspension up and down several more times. Centrifuge concentrated virus suspension for 1 min at 8000 rpm (6000 × *g*) in an Eppendorf centrifuge to remove residual clumps.

3.1.3. "Spin-thru" Detergent Treatment of Virions

1. Prepare a ~12-mL linear gradient of 30–70% sucrose in STE buffer for the SW41 Ti rotor. We use a 20-mL gradient former and place 6 mL of 70% sucrose on the near side (closest to outlet port) and 6 mL of 30% sucrose solution on the far side. It is important to prime the channel between the two chambers prior to filling the second one to permit free flow from one chamber to the other. Using the Auto Densi-Flow gradient former, pump the gradient from the bottom to the top of the tube. It is recommended that the procedure be practiced a few times to ensure consistency. Balance tubes by adding 30% sucrose in STE buffer until the masses of the tubes are equivalent. Place gradients at 4 °C until cooled (2–4 h).
2. Overlay the gradient with 0.25 mL of 1% Triton X-100 dissolved in 15% sucrose/STE buffer. This step must be performed carefully to maintain distinct layers (*see* **Note 2**).
3. Overlay with 0.25 mL of a solution of 7.5% sucrose/STE buffer. This will serve as a "barrier" layer to minimize mixing of the virus suspension and the detergent-containing layer until centrifugation.
4. Gently overlay with up to 0.5 mL of virus suspension. This step is critical; too much disturbance will result in mixing of the layers and a low yield of cores. A useful method is to widen the bore of a 1-mL pipet tip by trimming with scissors. This will reduce the velocity of the solution as it is pipeted, thus minimizing the mixing of the barrier and detergent layers. *See* **Fig. 4.2** for a diagram depicting the gradient prior to ultracentrifugation.
5. Place the tubes in precooled SW41Ti buckets and ultracentrifuge at 35,000 rpm (210,000 × *g* at r_{max}) for 16–20 h at 4 °C (*see* **Note 3**).

3.1.4. Recovery of Cores from the Dense Fractions of the Gradient

1. Following ultracentrifugation, the gradients are fractionated from the top of the gradient with the Auto-Densi-Flow.
2. Fractions of 1 mL are taken from the top of the gradient, and the tubes are placed in an ice bucket immediately upon collection. The purified viral cores are typically present in the bottom half of the gradient whereas soluble CA protein is present at the top fractions.

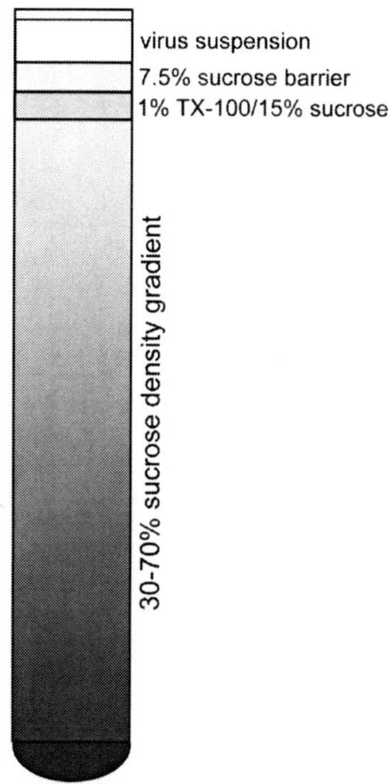

Fig. 4.2. Construction of density gradients for isolation of HIV-1 cores. A 30–70% sucrose density gradient is prepared and successively overlaid with a layer of detergent, a barrier layer to prevent premature mixing of virus and detergent, and the concentrated virus suspension. Upon ultracentrifugation, virions pass through the detergent layer, releasing HIV-1 cores that then sediment to their equilibrium density.

3. Withdraw 50 μL of each fraction and set aside for p24 ELISA (*see* **Section 3.3**) and/or reverse transcriptase activity assay.
4. After collecting fractions, add 0.5 mL of cold STE buffer to each fraction and mix several times by inversion. This procedure reduces the viscosity of the solution, minimizing sampling errors.

3.1.5. Localization of HIV-1 Cores and Storage of Samples

The remainder of the cores fraction can be used immediately for uncoating assays or flash frozen in liquid nitrogen for future use.

1. The fractions containing intact cores must be identified prior to further analysis. This can be performed either by assaying the fractions for p24 by ELISA or by measuring reverse transcriptase activity. Both the methods should reveal a peak near fraction 10, but it is best to determine this empirically until reproducibility has been established. Once an investigator has become proficient with the method, a rapid approach to predict the fractions containing the cores is to determine the density of each fraction by refractometry or gravimetric

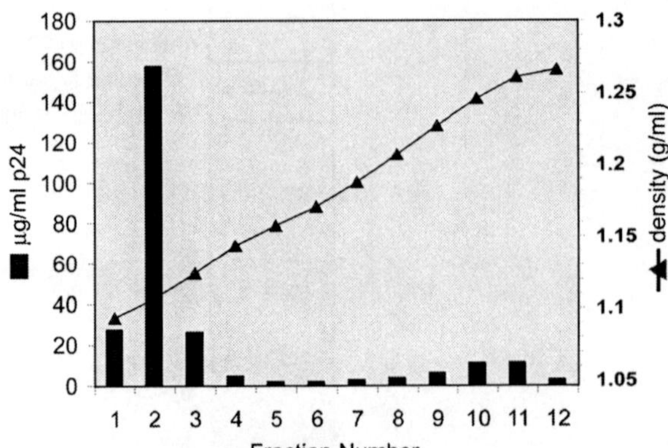

Fig. 4.3. Distribution of CA in the density gradient after ultracentifugation. Fractions from a gradient were collected from top to bottom and analyzed for p24 by ELISA and density by refractometry. In this experiment, a quantity of HIV-1 corresponding to approximately 250 μg of p24 was applied to the gradient.

analysis. Lentiviral cores typically have a density of 1.24–1.27 g/mL (*see* **Fig. 4.3**).

2. Quantification of CA in the gradient fractions. The recovery of the CA protein in the cores has been linked to core stability. By assaying the percentage of CA in the gradient that is present in fractions of HIV-1 cores, one may glean information about the particular virus (*see* **Note 4**). For this purpose, assay a sample of each fraction for p24 by ELISA *(16)*. This procedure is described under **Section 3.3**, and an example is shown in **Fig. 4.3**.
3. If uncoating assays are not to be performed the same day, the fractions containing the HIV-1 cores can be pooled and aliquots flash-frozen in liquid nitrogen and stored at −80 °C. The yield of cores is typically 1–2 μg of p24 per milliliter depending on the quantity of virus loaded on the gradient. The uncoating assay requires approximately 50 ng of p24. The purified cores may be frozen in 0.2- to 0.3-mL aliquots each of which is sufficient for performing several uncoating reactions.

3.1.6. Analysis of Pelleted Cores by Immunoblotting Using HIV-1-specific Antibodies

1. To confirm that HIV-1 cores have been isolated, proteins present in the dense fractions are characterized by immunoblotting.
2. Following dilution of the fractions by addition of STE buffer, the tubes are subjected to ultracentrifugation for 30 min at 100,000 × *g* (45,000 rpm in a Beckman TLA-45 rotor).
3. Supernatants are removed by aspiration, and the pellets are solubilized in SDS-PAGE loading buffer and are subjected to SDS-PAGE and immunoblotting with monoclonal antibodies specific for HIV-1 CA, gp120, and gp41 {183-H12-5C, 902, and Chessie 8, respectively (NIH AIDS Research and

Reference Reagent Program)}. HIV-1 cores are substantially free of gp41 and gp120 relative to intact virions.

4. The fractions identified as HIV-1 cores should also contain viral RNA. RNA can be extracted from the fractions and the HIV-1 RNA quantified by RT-PCR.

3.2. Kinetic Assay of HIV-1 Uncoating In Vitro

Once HIV-1 cores have been purified and characterized, the stability of the capsid may be determined by studying the rate of CA dissociation upon warming to 37 °C. The extent of CA release is quantified by p24 ELISA after separating the cores from the soluble p24 by centrifugation (*see* **Note 5**).

1. Predilute cores into cold STE buffer to reduce viscosity, thus avoiding sampling errors.
2. Further dilute samples by adding 50 μL into 0.5–1.0 mL cold STE buffer. Mix gently but thoroughly by inversion several times. Do not vortex. Flicking of the inverted tubes may be necessary to achieve thorough mixing. Place tubes in a 37 °C water bath for various time periods (typically 20–30 min). Be sure to immerse the tubes to their internal liquid level to ensure thorough warming. Invert the tubes periodically throughout the incubation period (every 5–10 min). As a time zero control, dilute cores into cold buffer and keep on ice for the same time period (typically 20–30 min).
3. Following the incubations, rapidly chill the samples by placing in an ice-water bath for 10 min.
4. Pellet the cores by ultracentrifugation for 20 min at 125,000 × *g* (RCF_{max} for Beckman TLA-55 rotor at 45,000 rpm). The rotor should be precooled to 4 °C and the tubes wiped thoroughly before placing into the rotor. In our experience, it is also helpful to wipe down the rotor before placing in the centrifuge. This removes the condensation on the outside of the rotor, reducing the time necessary for the centrifuge chamber to attain a sufficient vacuum.
5. Remove the supernatant and transfer to a clean microfuge tube. Resuspend the pellet in 200 μL of ELISA sample diluent.
6. Quantify p24 in both pellet and supernatant fractions by ELISA. Calculate the percentage of the total p24 present in the supernatant (*see* **Note 6**). The extent of uncoating should increase with increasing incubation time (*see* **Fig. 4.4**).

3.3. Assay for CA protein by p24 ELISA

Many commercial p24 assay kits are available and can be used for studies described herein. For in-depth studies, use of the homemade p24 sandwich ELISA described here will result in significant cost savings. The procedure has been adapted from a previous report *(16)* and can be performed using capture and

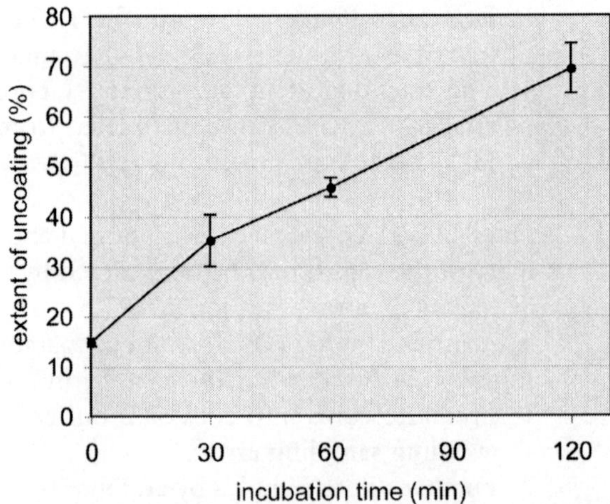

Fig. 4.4. Uncoating of HIV-1 cores occurs spontaneously during incubation at 37 °C. Samples of purified HIV-1 cores were diluted in buffer and incubated for the indicated times. Reactions were chilled and the cores were pelleted by ultracentrifugation. p24 concentrations in the pellets and supernatants were determined by ELISA, and the extent of uncoating was calculated. Error bars correspond to standard deviations from the mean values from triplicate reactions.

detector antibodies available from the NIH AIDS Research and Reference Reagent Program.

1. Coat 96-well Immulon 2HB plates with monoclonal antibody to p24 (183-H12-5C) diluted to 4 μg/mL in PBS (100 μL per well). Plates are sealed with adhesive film and incubated at 37 °C overnight to allow efficient binding.
2. Rinse wells twice with PBS and add 0.25 mL of blocking solution. Seal plate and incubate for 1 h at 37 °C to block.
3. Rinse wells three times with ELISA wash buffer. This can be performed using an automated plate washer; we use a plastic wash bottle and slap the plate on a stack of paper towels after each wash to remove most of the liquid.
4. Add standards and samples (100 μL per well) diluted in ELISA sample diluent. The assay has a dynamic range of 0.06–1 ng/mL of p24. We typically employ a series of standards containing 0, 0.06, 0.12, 0.25, 0.5, and 1.0 ng/mL p24. Because transfected 293T cells typically yield between 200 and 2000 ng/mL, dilutions ranging from 1:100 to 1:10,000 are normally sufficient. For concentrated samples of cores, dilutions of up to 1:100,000 may be necessary.
5. Seal plate and incubate at 37 °C for 2 h.
6. Wash plate three times with ELISA wash buffer.
7. Add 100 μL primary antibody solution.
8. Seal plate and incubate 1 h at 37 °C.
9. Wash plate three times with ELISA wash buffer.
10. Add 100 μL secondary antibody.

11. Seal plate and incubate 1 h at 37 °C.
12. Wash plate three times with ELISA wash buffer. After the final wash, slap the plate several times on paper towels to remove residual wash solution.
13. Add 100 μL substrate solution (made by mixing equal volumes of TMB Peroxidase Substrate and Peroxidase Substrate Solution B just prior to use).
14. Incubate at room temperature until blue color is medium strong in the well corresponding to the 1 ng/mL standard (usually 10–20 min). Do not overincubate – this will result in extreme nonlinearity in the assay.
15. Terminate the color development reaction by adding 100 μL of 4 N H_2SO_4 solution. The color will change from blue to yellow.
16. Read absorbance of each well at 450 nm with 650 nm reference in an ELISA microplate reader (e.g., Molecular Devices Emax using Softmax Pro software). Plot absorbance vs. p24 concentration for the standards to obtain the standard curve. Calculate unknown concentrations by interpolation (*see* **Note 7**).

4. Notes

1. Although our laboratory prefers to use the equilibrium ultracentrifugation to purify HIV-1 cores, other investigators have employed more rapid methods involving direct pelleting of cores *(5, 17–19)*.
2. Because HIV-1 uncoating is accelerated at elevated temperature, it is important to precool all solutions and the centrifuge buckets used for isolation of the cores. We store the ultracentrifuge buckets in the refrigerator, but usually leave the rotor itself at room temperature to minimize condensation on the rotor that can increase the time required for the centrifuge to achieve vacuum. Likewise, it is important to fractionate the gradients while they are still cold and to place the fractions in an ice-water bath immediately upon collection. We also cool the TLA45 rotor prior to use.
3. Before starting centrifugation, it is important to allow the centrifuge to attain sufficient vacuum to avoid the automatic low speed pause to establish the vacuum that occurs with some ultracentrifuges. Therefore, it is recommended that the rotor be placed in the centrifuge and the vacuum established prior to initiating the run in order to minimize the contact time between the virus and the detergent.
4. The yield of cores is subject to experimental variation. While we have not identified all of the relevant variables, it appears

that the care taken in application of the layers to the gradients, and in keeping the samples cold, are significant factors. It is often difficult for novices to avoid mixing the virus and detergent layers during application of the virus suspension onto the gradient. Members of the laboratory who have practiced the technique were able to master it, but some individuals obtained higher yields of cores than others. With repetition, an individual can obtain reproducible yields of cores. We have not observed a significant dependence of the recovery of HIV-1 cores on the quantity of virus nor of the rate of uncoating in vitro on the amount of cores in the reaction, but we cannot exclude possible effects of these variables.

5. When assaying uncoating, reactions are typically performed in duplicate. If the values are not within 15% of one another, the experiment should be performed in triplicate until the consistence is improved.
6. Although the extent of uncoating is reflected in the calculated value (percentage of total CA in the supernatant), it is important to note both the supernatant and pellet values. In some cases, an apparent increase in uncoating may result from a loss of CA from the pellet if the samples are handled improperly. This outcome could also reflect degradation of CA in the reactions.
7. In our experience, the p24 standard values fit best to a quadratic formula, which is an option in the Softmax Pro software. It is important not to let the color reaction proceed too long or the reaction will become saturated. For most uncoating studies, the sensitivity and dynamic range of the ELISA is ideal. However, the sensitivity of the assay can be enhanced by substituting additional standards (16 and 32 pg/mL) for the high standards (0.5 and 1.0 ng/mL) and extending the color reaction time.

Acknowledgments

The author thanks the former members of his laboratory Alex Kotov and Brett Forshey who contributed to the development of the HIV-1 uncoating assay, along with the present members who assisted in compiling the methods. The NIH AIDS Research and Reference Program is gratefully acknowledged for distributing reagents. Work in the author's laboratory was supported by NIH grants AI040364 and AI050423.

References

1. Aiken, C. (2006) Viral and cellular factors that regulate HIV-1 uncoating. *Curr Opin HIV AIDS* 1, 194–199.
2. Forshey, B. M., von Schwedler, U., Sundquist, W. I., et al. (2002) Formation of a human immunodeficiency virus type 1 core of optimal stability is crucial for viral replication. *J Virol* 76, 5667–5677.
3. Wacharapornin, P., Lauhakirti, D., Auewarakul, P. (2007) The effect of capsid mutations on HIV-1 uncoating. *Virology* 358, 48–54.
4. Forshey, B. M., Aiken, C. (2003) Disassembly of human immunodeficiency virus type 1 cores in vitro reveals association of nef with the subviral ribonucleoprotein complex. *J Virol* 77, 4409–4414.
5. Auewarakul, P., Wacharapornin, P., Srichatrapimuk, S., et al. (2005) Uncoating of HIV-1 requires cellular activation. *Virology* 337, 93–101.
6. Menendez-Arias, L., Risco, C., Pinto da Silva, P., et al. (1992) Purification of immature cores of mouse mammary tumor virus and immunolocalization of protein domains. *J Virol* 66, 5615–5620.
7. Stewart, L., Schatz, G., Vogt, V. M. (1990) Properties of avian retrovirus particles defective in viral protease. *J Virol* 64, 5076–5092.
8. Roberts, M. M., Oroszlan, S. (1989) The preparation and biochemical characterization of intact capsids of equine infectious anemia virus. *Biochem Biophys Res Commun* 160, 486–494.
9. Chrystie, I. L., Almeida, J. D. (1989) The recovery of antigenically reactive HIV-2 cores. *J Med Virol* 27, 188–195.
10. Benzair, A. B., Rhodes-Feuillette, A., Lasneret, J., et al. (1986) Purification and characterization of simian foamy virus type I structural core polypeptides. *Arch Virol* 87, 87–96.
11. Yoshinaka, Y., Luftig, R. B. (1977) Murine leukemia virus morphogenesis: cleavage of P70 in vitro can be accompanied by a shift from a concentrically coiled internal strand ("immature") to a collapsed ("mature") form of the virus core. *Proc Natl Acad Sci USA* 74, 3446–3450.
12. Furmanski, P., Loeckner, C. P., Longley, C., et al. (1976) Identification and isolation of the major core protein from the oncornavirus-like particle in human milk. *Cancer Res* 36, 4001–4007.
13. Bader, J. P., Brown, N. R., Bader, A. V. (1970) Characteristics of cores of avian leukosarcoma viruses. Virology 41, 718–728.
14. Adachi, A., Gendelman, H. E., Koenig, S., et al. (1986) Production of acquired immunodeficiency syndrome-associated retrovirus in human and nonhuman cells transfected with an infectious molecular clone. *J Virol* 59, 284–291.
15. Chen, C., Okayama, H. (1987) High-efficiency tranformation of mammalian cells by plasmid DNA. *Mol Cell Biol* 7, 2745–2752.
16. Wehrly, K., Chesebro, B. (1997) p24 antigen capture assay for quantification of human immunodeficiency virus using readily available inexpensive reagents. *Methods* 12, 288–293.
17. Welker, R., Hohenberg, H., Tessmer, U., et al. (2000) Biochemical and structural analysis of isolated mature cores of human immunodeficiency virus type 1. J Virol 74, 1168–1177.
18. Briggs, J. A., Wilk, T., Welker, R., et al. (2003) Structural organization of authentic, mature HIV-1 virions and cores. *EMBO J* 22, 1707–1715.
19. Accola, M. A., Ohagen, A., Gottlinger, H. G. (2000) Isolation of human immunodeficiency virus type 1 cores: retention of vpr in the absence of p6(gag). *J Virol* 74, 6198–6202.

Chapter 5

Real-Time PCR Analysis of HIV-1 Replication Post-entry Events

Jean L. Mbisa*, Krista A. Delviks-Frankenberry*, James A. Thomas, Robert J. Gorelick, Vinay K. Pathak

Abstract

The reverse transcriptase enzyme plays an essential role in the HIV-1 life cycle by converting a single-stranded viral RNA genome into a double-stranded viral DNA through a complex process known as reverse transcription. The resulting double-stranded DNA is integrated into the host chromosome to form a provirus. A small proportion of the viral DNAs form dead-end circular products, which nevertheless can serve as useful surrogate markers for monitoring viral replication. Utilizing real-time PCR technology, it is possible to track and quantify different stages of the reverse transcription process, the proviruses, and the nonintegrated dead-end reverse transcription products.

Key words: HIV-1, reverse transcription, real-time PCR, 1-LTR circle, 2-LTR circle, Alu-LTR real-time PCR.

1. Introduction

Real-time PCR uses conventional PCR methods; however, the addition of a quenched probe that becomes fluorescent upon each PCR amplification step allows for direct quantification of the number of DNA copies initially present in the sample *(1–4)*. The probe that is added to the PCR reaction (commonly referred to as the TaqMan probe), is an oligonucleotide that contains a quencher dye at the 3′ end and a fluorescent reporter dye at the 5′ end of the probe *(4)*. Different fluorescent reporter dyes can be added to the 5′ end, such as

* These two authors contributed equally to this work.

Vinayaka R. Prasad, Ganjam V. Kalpana (eds.), *HIV Protocols: Second Edition, vol. 485*

DOI 10.1007/978-1-59745-170-3_5 Springerprotocols.com

6-carboxy-fluorescein (FAM), hexachloro-6-carboxy-fluorescein, or tetrachloro-6-carboxy-fluorescein, each of which emit light at different wavelengths. The quencher at the 3′ end of the probe is typically 6-carboxy-tetramethyl-rhodamine (TAMRA). As shown in **Fig. 5.1**, a typical reaction starts when the probe binds to the target template during the cool-down phase after the melting step followed by binding of the forward and reverse primers (**Step 1**). At this stage, the quencher is in close proximity to the fluorescent reporter dye and therefore no fluorescence is emitted. The Taq polymerase that catalyzes the PCR reaction not only has a 5′-3′ polymerase activity, but also a 5′-3′exonuclease activity (**Step 2**). Therefore, as the Taq polymerase synthesizes DNA, it cleaves and displaces the probe, releasing the quencher, and allowing fluorescence to be emitted (**Step 3**) before finishing each amplification cycle (**Step 4**). As a result, for each amplification cycle, the emitted fluorescence is recorded and the cycle number at which the fluorescence becomes detectable (cycle threshold; C_t) is directly correlated to the number of templates in the reaction at that cycle. This correlation only holds true during the exponential phase of the PCR reaction. Utilizing a known copy number of input templates as a standard, a standard curve can be generated to determine the exact number of templates initially present in any unknown sample. Therefore, this technology can monitor, in real time, the progress of any PCR reaction and determine the exact DNA copy numbers per sample, with a high sensitivity that allows detection of 10 or fewer copies per sample *(5)*. As an

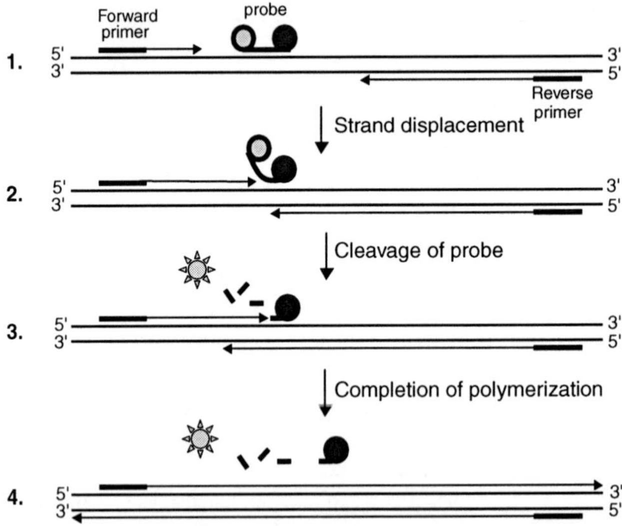

Fig. 5.1. PCR amplification using TaqMan probe chemistry. After probe and primers bind to the template (**Step 1**), Taq polymerase initiates synthesis, displaces and cleaves the probe (**Step 2** and **3**), which allows fluorescence to be emitted. Polymerization is completed after one cycle of PCR amplification (**Step 4**).

alternative to the TaqMan system, one can use SYBR Green I dye which intercalates into double-stranded DNA as it accumulates during the PCR reaction thereby producing a fluorescent signal that can be quantified *(6–8)*. The benefits of the SYBR Green dye intercalation technique are that it is less expensive and can be universally applied to all PCR reactions. However, TaqMan probe chemistry is generally preferred over SYBR Green I dye chemistry because it provides a greater level of specificity. The TaqMan probe is specific for a given sequence whereas SYBR Green I dye binds all double-stranded DNA, including any PCR artifacts (primer-dimers or nonspecific amplification products), and therefore requires a final dissociation curve analysis to verify that the reactions are optimized so that the final amplified signal is specific to the target sequence *(6–8)*.

The HIV-1 genome is initially plus-stranded RNA that must be converted into double-stranded DNA for the virus to complete its life cycle *(9)*. The unique process by which the HIV-1 reverse transcriptase carries out DNA synthesis enables one to utilize specific primer and probe sets to track and quantify early, intermediate, and late events in reverse transcription (**Fig. 5.2**) *(10–19)*. Early reverse transcription products can be monitored by using a primer/probe set that specifically amplifies minus-strand strong-stop DNA (**Fig. 5.2A**); minus-strand DNA transfer can be monitored by using a primer/probe set that flanks the U3-R-U5 junction (**Fig. 5.2B**); intermediate stages of reverse transcription during minus-strand DNA synthesis and elongation can be monitored with a primer/probe set specific to regions of the viral genome that are between the two LTRs, e.g., *gag* (**Fig. 5.2C**). Finally, plus-strand DNA transfer and late reverse transcription products can be monitored by a primer/probe set specific to the U5 and 5′ untranslated regions of the viral DNA (**Fig. 5.2D**). Quantifying the final number of integrated proviruses into the host cell chromosome relies upon the presence of natural Alu repeats found throughout all human chromosomal DNA *(11, 20, 21)*. A U5 specific forward primer and probe are used with a reverse primer that binds to the nearest Alu repeat, amplifying the integrated provirus (**Fig. 5.2G**). Every reverse transcription reaction does not always result in an integration event as some viral DNAs remain linear or circularize to form dead-end 1-LTR- and 2-LTR-circle products (**Fig. 5.2E,F**, respectively). The 2-LTR circular products can be quantified using primer/probe sets specific to U5 and U3, and the 1-LTR circle products can also be quantified using a specific primer/probe set which binds *gag* and *env* during real-time PCR *(11, 22–25)*.

The advent of real-time PCR has been instrumental to the field of retrovirology. Not only can real-time PCR be used for basic laboratory research, but it can also be used clinically to

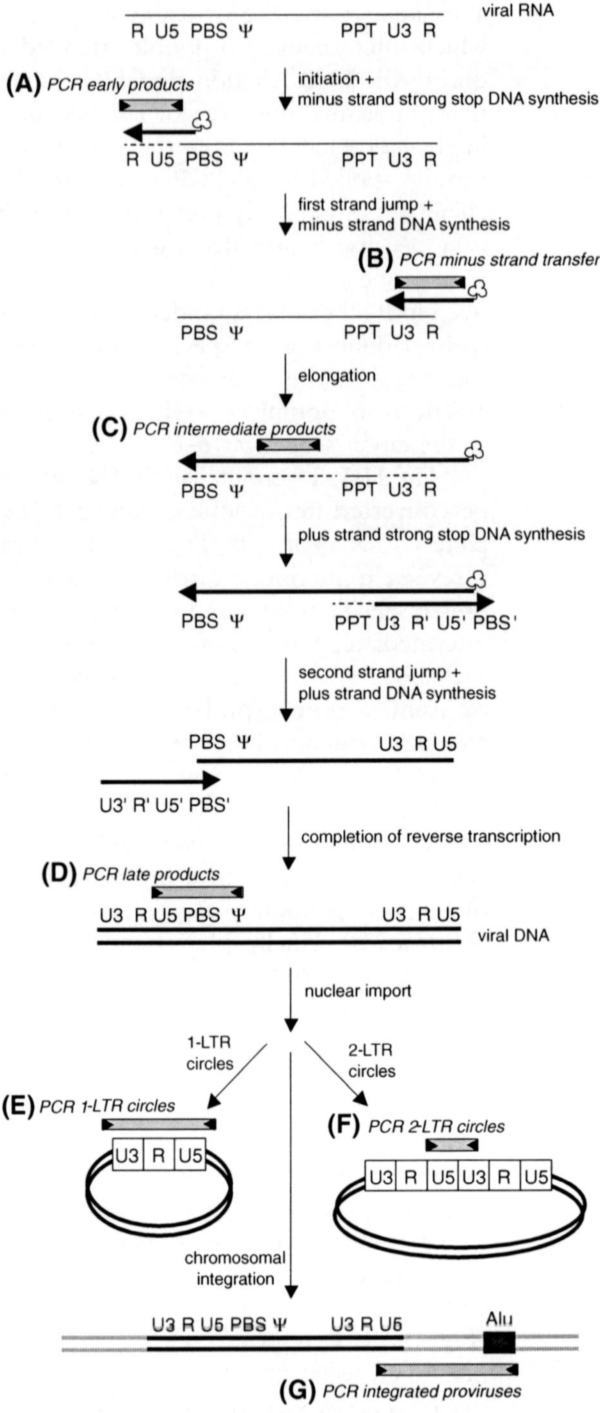

Fig. 5.2. Schematic representation of the HIV-1 reverse transcription and integration. Real-time PCR (shaded gray boxes containing arrows) can be used to analyze and quantify early, intermediate, and late reverse transcription products as well as integrated and unintegrated (1- and 2-LTR circles) HIV-1 viral DNA (Stages A–G). Thin black lines, RNA; thick black lines, DNA; thick gray lines, chromosomal DNA; dashed lines, RNase H degradation; cloverleaf, tRNA.

quantify HIV-1 viral loads and mutation frequencies in patients *(5, 26–30)*. Real-time PCR can be used to quantify the efficiency of reverse transcription and overall amount of viral DNA synthesis, quantify the levels of nonintegrated vs. integrated viral DNA for any antiviral drug treatment, and determine the effects of an HIV-1 mutant or change in experimental conditions on viral DNA synthesis. Overall, real-time PCR technology has greatly enhanced our basic knowledge of HIV-1 post-entry events.

2. Materials

2.1. Cell Culture, Virus Production, Infections

1. 293T cells and/or a cell line transduced with an HIV-1-based vector. For example, a 293T cell line has been made that stably expresses the plasmid pHDV-eGFP *(31)* (named 293T-HIV-GFP cell line for remainder of text) (*see* **Note 1**).
2. An integration standard cell line generated by transducing cells with an HIV-1-based vector capable of a single round of replication *(11, 19)* (*see* **Note 2**).
3. Dulbecco's Modified Eagle's Medium (DMEM) (Gibco/BRL, Bethesda, MD) supplemented with 10% fetal bovine serum (FBS; Hyclone, Ogden, UT), penicillin (50 U/mL; Gibco/BRL) and streptomycin (50 μg/mL; Gibco/ BRL).
4. Plasmids: an HIV-1-based vector, e.g., pHDV-eGFP *(31)*; pHCMV-G, a vector expressing the G glycoprotein of vesticular stomatis virus (VSV-G) *(32)*.
5. CalPhos Mammalian Transfection Kit (Clontech, Mountain View, CA).
6. Phosphate-buffered saline (PBS) without Ca^{2+} or Mg^{2+} (Cambrex, Walkersville, MD).
7. Millex GS 0.45-μm pore size filter (Nalgene, Rochester, NY).
8. HIV-1 p24 ELISA Kit (Perkin Elmer, Wellesley, MA).
9. DNase I (Roche, Indianapolis, IN).
10. Heat block set at 65 °C.
11. Hexadimethrine bromide (Polybrene), 1 mg/mL stock (Sigma, St. Louis, MO).

2.2. DNA Isolation

1. QIAamp DNA Blood Mini Kit (Qiagen, Germantown, MD).
2. 2 mL Collection tubes (Qiagen, Germantown, MD).
3. Dpn I restriction enzyme (NEB, Beverly, MA).

2.3. Real-Time PCR

1. A dedicated set of pipettes and filter tips for real-time PCR use only.
2. AmpliTaq Gold® DNA Polymerase, 5 U/μL (PE-Applied Biosystems, Foster City, CA).
3. 100 mM dNTP Set, PCR grade, 4 × 25 μmol (Invitrogen, Carlsbad, CA).

Table 5.1
Primer/probe sets for analysis of different stages of reverse transcription, 1- and 2-LTR circle formation, integrated proviruses, and cellular genes

Stage of reverse transcription[a]	Target sequence	Forward primer	Reverse primer	Probe	Standard template
A	RU5	hRU5-F2	hRU5-R	hRU5-P	pHDV-eGFP
B	U3U5	FST-F1	SS-R4	P-HUS-SS1	pHDV-eGFP
C	Gag	GagF1	GagR1	P-HUS-103	pHDV-eGFP
D	U5Ψ	MH531	MH532	LRT-P	pHDV-eGFP
E	1-LTR	LA1	LA15	1-LTR-P	pCR-1LTR
F	2-LTR FJ	SS-F4	LTR-R5	P-HUS-SS1	pJB1041
F	2-LTR TOT	SS-F4	LTR-R3	P-HUS-SS1	pJB1041
G	Alu-LTR	MH535	SB704	P-HUS-SS1	Special[b]
control	PBGD	nPBGD-For	nPBGD-Rev	nPBGD-Pr	pJB1057
control	CCR5	CCR5-For	CCR5-Rev	P-CCR5-01	pRB2569 *(19)*

[a]Stages labeled in **Fig. 5.2**.
[b]Standard for the Alu-LTR assay is the Integration Standard cell line, prepared as described in **Section 3.8**.

4. 10X TaqMan® Buffer II (PE-Applied Biosystems).
5. 10X TaqMan® Buffer A (containing FAM-10-ROX internal reference) (PE-Applied Biosystems).
6. 25 mM $MgCl_2$ (PE-Applied Biosystems).
7. Forward and reverse primers (100 μM stocks).
8. Dual-labeled probe (100 μM stocks).
9. Plasmid DNA to be used as standard template as listed in **Table 5.1** (*see* **Note 3**).
10. DNA from uninfected 293T cells.
11. PCR grade nuclease-free water (Ambion, Austin, TX).
12. MicroAmp® Optical 96-well Reaction Plates (PE-Applied Biosystems).
13. MicroAmp® Optical Caps for 96-well Reaction Plates (PE-Applied Biosystems).
14. Microfuge tubes (DNase/RNase free)
15. ABI7900HT (PE-Applied Biosystems) or equivalent Real-time PCR instrument.

3. Methods

The protocol outlined here describes the analysis of HIV-1 post-entry events during infection of cultured cells utilizing the Taq-Man probe chemistry. In addition, this protocol can be easily modified to analyze other types of samples, for example clinical samples. To analyze post-entry steps during HIV-1 infection of cultured cells by real-time PCR, virus is produced by transfection of 293T cells with an HIV-1-based vector and an envelope protein expression construct, target cells are infected, and DNA is isolated from the infected cells.

3.1. Virus Production from Transfected Cell Lines

1. Plate 293T target cells in DMEM at density of 2.5×10^6 cells per 100-mm-diameter dish (*see* **Note 1**).
2. Twenty four hours later, replace medium with fresh DMEM. Transfect with 10 μg of pHDV-eGFP and 2 μg of pHCMV-G using CalPhos transfection kit following manufacturer's instructions (*see* **Note 4**).
3. Incubate the plates in 3% CO_2 incubator at 37 °C for 6 h.
4. Gently rinse the plates three times with 10 mL of PBS at room temperature supplemented with 1% FBS (v/v) (*see* **Note 5**). Add 10 mL of fresh DMEM media and incubate overnight in 5% CO_2 incubator at 37 °C.
5. Next morning, gently rinse the plates again three times with 10 mL PBS and add 10 mL of fresh DMEM media. Return to incubator.
6. Gently rinse the plates again 5–6 h later once with 10 mL PBS and add 10 mL of fresh DMEM media. Incubate overnight.
7. Harvest the virus containing supernatant the next morning and clarify the samples by centrifuging at $400 \times g$ for 4 min at 4 °C. Pass the supernatant through Millex GS 0.45 μm pore size filter.
8. Take an aliquot and perform p24 ELISA analysis following manufacturer's instructions.
9. Store the remaining sample at −80 °C until ready to use for infection of target cells.

3.2. Virus Production from Stable Cell Lines

1. Plate 293T-HIV-GFP cells at density of 2.5×10^6 cells per 100-mm-diameter dish (*see* **Note 1**).
2. Twenty four hours later, replace medium with fresh DMEM. Transfect with 2 μg of pHCMV-G using CalPhos transfection kit following manufacturer's instructions (*see* **Note 4**).
3. Incubate the plates in 3% CO_2 incubator at 37 °C for 6 h.
4. Remove the media and add 10 mL of fresh DMEM media.
5. Incubate the plates in 5% CO_2 incubator at 37 °C for 36–48 h.
6. Harvest the virus containing supernatant and clarify the samples by centrifuging at $400 \times g$ for 4 min at 4 °C. Pass the supernatant through Millex GS 0.45-μm pore size filter.
7. Take an aliquot and perform p24 ELISA analysis following manufacturer's instructions.
8. Store the remaining sample at −80 °C until ready to use for infection of target cells.

3.3. Infection

1. Plate 293T target cells at a density of 1×10^5 cells per 60-mm-diameter dish (*see* **Note 1**).
2. Twenty four hours later, infect target cells with p24-normalized virus as determined by HIV-1 p24 ELISA (*see* **Note 6**). Before infection, treat the virus with 10 U/mL of DNase I plus 4 mM $MgCl_2$ at 37 °C for 1 h to reduce transfected plasmid DNA carryover, which can contribute to the signal during real-time PCR.
3. For each virus, prepare in parallel a heat-inactivated control sample by incubating the virus at 65 °C for an hour to be used for determining carryover contamination from plasmid DNA.
4. Make the inoculum, for both live and heat-inactivated virus, by diluting so that p24 values are comparable, adjust final volume to 2 mL per plate with DMEM, and add 5 μg/mL of Polybrene.
5. Infect for 3 h with gentle rocking every 30–60 min.
6. Remove inoculum, gently rinse twice with PBS plus 1% FBS and add 4 mL of fresh DMEM.
7. Harvest total cellular DNA from both live virus- and heat-inactivated virus-infected cells at desired time points by removing the media and adding 1 mL of PBS. Scrape the plates or pipet vigorously to dislodge cells, and transfer cell suspension to 1.5 mL centrifuge tubes and spin at $1000 \times g$ for 4 min at 4 °C.
8. Discard supernatant and proceed to isolate DNA (*see* **Section 3.4**) or store cell pellets at −80 °C until ready to isolate DNA.

3.4. DNA Isolation

1. Resuspend cell pellets in 200 μL of PBS.
2. Proceed to extract the total cellular DNA following the protocol as stated in the Qiagen QIAamp DNA Blood Mini Kit.
3. Elute the DNA in 200 μL of elution buffer EB provided in the kit or nuclease-free water.
4. Digest DNA sample with *Dpn*I restriction enzyme, which digests only plasmid DNA and therefore helps further to reduce plasmid DNA detection during real-time PCR. Use 0.1 U/μL in NEB Buffer 2 for 2–4 h at 37 °C.

3.5. Real-Time PCR Analysis of Postentry Reverse Transcription Steps

1. Analysis of the different stages of reverse transcription is normally carried out by completing a time course from 0 to 6 or 0 to 12 h, and subsequently comparing the different levels of reverse transcription products as time progresses. For example, collect and process samples at 0, 0.5, 1, 3, 6 and 12 h postinfection.
2. If possible, use a UV-decontaminated hood when setting up all PCR reactions to avoid any cross contamination. Only use pipettes, RNase and DNase-free Eppendorf tubes and sterile

PCR racks that are dedicated for real-time PCR use. All prepared reaction mixtures should be kept on ice.

3. Select the primer/probe set to use from the list in **Table 5.1** depending on the stage of reverse transcription to be analyzed, i.e., RU5 for initiation of reverse transcription or early RT products, U3U5 products for minus-strand DNA transfer, *gag* for late minus-strand DNA synthesis, and U5Ψ for plus-strand DNA transfer and late RT products (*see* **Note 7**). PBGD or CCR5 primer/probe sets target cellular genes and are run for each sample to normalize for DNA recovery and the number of cells loaded in each reaction. The sequences of the primers and probes are listed in **Table 5.2**.
4. Prepare a standard curve from the appropriate standard template listed in **Table 5.1** to be used to compute the copy numbers in each sample. Make a stock of 10^9 DNA copies/μL and dilute the working stock solution to 10^8 copies/μL. Prepare 11 dilutions for the standard curve: the first dilution should be made at a final concentration of 10^7 copies per 10 μL. Starting with the first dilution, make 10 additional threefold serial dilutions down to 169 copies per 10 μL.
5. In a nuclease-free tube on ice, prepare a PCR master mix with all of the PCR reagents except the DNA as outlined in **Table 5.3** (*see* **Note 8**). The reaction setup shown in **Table 5.3** is for a 50-μL reaction, but prepare enough PCR master mix for all the samples and standards in the experiment as well as a "no template" control (NTC). Include a 5% overage to compensate for pipetting error. Mix well and centrifuge briefly. It is recommended that all samples should be run in duplicate or triplicate to increase the accuracy of results.
6. Dispense 40 μL of PCR master mix into each well of a MicroAmp® Optical 96-well Reaction Plate and add 10 μL of DNA isolated from step 3.4 (100–500 ng), or 10 μL of the serial dilution of standard DNA template or 10 μL of water for the NTC. You may also run a blank containing 50 μL of water. Securely cover all the wells (including unused ones) with the MicroAmp® Optical Caps for 96-well Reaction Plates. Mix well and centrifuge briefly to ensure that the entire sample is at the bottom of the well.
7. Run the PCR reaction using the thermocycling parameters outlined in **Table 5.4** for analysis of different stages of reverse transcription or for the cellular gene controls, using an ABI Prism 7900HT sequence detection system (*see* **Note 9**).
8. After the reaction is complete, analyze the results according to the manufacturer's instructions, especially paying great attention to the baseline and threshold settings (*see* http://www3.appliedbiosystems.com/cms/groups/mcb-marketing/documents/generaldocuments/cms-042502.pdf and https:// www2.appliedbiosystems.com/support/apptech/

Table 5.2
Sequences of oligonucleotide primers and probes

Name of oligonucleotide[a]	Sequence
hRU5-F2[b]	5′-GCCTCAATAAAGCTTGCCTTGA-3′
hRU5-R[b]	5′-TGACTAAAAGGGTCTGAGGGATCT-3′
hRU5-P[b]	5′-FAM-AGAGTCACACAACAGACGGGCACACACTA-TAMRA-3′
FST-F1[b]	5′-GAGCCCTCAGATGCTGCATAT-3′
SS-R4[b]	5′-CCACACTGACTAAAAGGGTCTGAG-3′
P-HUS-SS1[b]	5′- FAM-TAGTGTGTGCCCGTCTGTTGTGTGAC-TAMRA-3′
GagF1[b]	5′-CTAGAACGATTCGCAGTTAATCCT-3′
GagR1[b]	5′-CTATCCTTTGATGCACACAATAGAG-3′
P-HUS-103[b]	5′-FAM-CATCAGAAGGCTGTAGACAAATACTGGGA-TAMRA-3′
MH531[b]	5′-TGTGTGCCCGTCTGTTGTGT-3′
MH532[b]	5′-GAGTCCTGCGTCGAGAGATC-3′
LRT-P[b]	5′-FAM-CAGTGGCGCCCGAACAGGGA-TAMRA-3′
SS-F4[b]	5′- TGGTTAGACCAGATCTGAGCCT-3′
LTR-R3[b]	5′-AGGTAGCCTTGTGTGTGGTAGATCC-3′
LTR-R5[b]	5′- GTGAATTAGCCCTTCCAGTACTGC-3′
LA1[b]	5′-GCGCTTCAGCAAGCCGAGTCCT -3′
LA15[b]	5′-CACACCTCAGGTACCTTTAAGA -3′
1-LTR-P[b]	5′-FAM-AGTGGCGAGCCCTCAGATGCTGC-TAMRA-3′
MH535[b]	5′-AACTAGGGAACCCACTGCTTAAG-3′
SB704[c]	5′-TGCTGGGATTACAGGCGTGAG-3′
nPBGD-For[c]	5′-AGGGATTCACTCAGGCTCTTTCT-3′
nPBGD-Rev[c]	5′-GCATGTTCAAGCTCCTTGGTAA-3′
nPBGD-Pr[c]	5′-FAM-TCCGGCAGATTGGAGAGAAAAGCCTG-TAMRA-3′
CCR5-For[c]	5′-CCAGAAGAGCTGAGACATCCG-3′
CCR5-Rev[c]	5′-GCCAAGCAGCTGAGAGGTTACT-3′
P-CCR5-01[c]	5′-FAM-TCCCCTACAAGAAACTCTCCCCGG-TAMRA-3′

[a] *See* **Table 5.1** for the target sequence detected by each primer/probe set.

[b] Primer sequences were designed to amplify NL4-3 viral DNA, or vectors containing these sequences. Because of sequence variation between HIV-1 isolates, it may be necessary to modify or redesign the sequences.

[c] Primer sequences are designed for human genes. The use of cells from other species will require modification or redesign of these sequences.

Table 5.3
Setup of real-time PCR reactions

Amount	Component	Final concentration
3 μL	10X Buffer A	
2 μL	10X PCR Buffer II	1X
8 μL	$MgCl_2$ (25 mM)	4 mM
0.4 μL	dNTPs (25 mM)	0.2 mM each
0.3 μL[a,b]	Forward primer (100 μM)	600 nM[a,b]
0.3 μL[a,b]	Reverse primer (100 μM)	600 nM[a,b]
0.04 μL[a,b]	Probe (100 μM)	80 nM[a,b]
0.25 μL	AmpliTaq Gold® (5 U/μL)	1.25 U
10 μL	DNA (100 to 500 ng)	
to 50 μL	PCR Grade Nuclease-free Water	

[a]For the reaction to analyze 1-LTR circle formation, the forward and reverse primers are used at 400 nM, the probe is used at 200 nM *(25)*.
[b]For Alu-LTR reactions, the forward primer is used at 50 nM, the reverse primer is used at 900 nM, the probe is used at 100 nM (*see (19)*, Bushman website, http://microb230.med.upenn.edu/).

#rt_pcr). The blank and NTC samples should not give any signal. The viral standard curve will be used to calculate the copy number for each unknown sample. A graph plot showing a typical standard curve with values from live and heat-inactivated virus samples is presented in **Fig. 5.3**. The cellular gene copy numbers can be used to normalize for the number of the cells loaded in each reaction. A good standard curve should have a y-intercept of approximately 39–41 cycles and a slope between -3.321 and -3.5, which corresponds to an efficiency of 100–93%, respectively.

3.6. Real-Time PCR Analysis of 2-LTR-Circle Formation

1. Analysis of 2-LTR-circle formation is normally done using DNA collected 24 h after infection.
2. Select a primer/probe set from the list in **Table 5.1** depending on whether you want to analyze the total number of 2-LTR circles (2-LTR TOT) or 2-LTR circles with complete junctions only (2LTR FJ). Also run a duplicate plate with PBGD or CCR5 primer/probe set for each sample to normalize for the number of cells loaded in each reaction.
3. Prepare a standard curve from the appropriate standard template listed in **Table 5.1**, to be used to compute the copy numbers in each sample as described in point 3 of **Section 3.5**.

Table 5.4
Thermocycling parameters for real-time PCR reactions

Target of amplification	Cycling parameters		
	Step	**Temperature**	**Duration**
RU5, *gag* or U5Ψ	Denature[a]	95 °C	10 min
	45 PCR cycles		
	Denature	95 °C	15 s
	Anneal and extend	60 °C	1 min
PBGD or CCR5	Denature	95 °C	10 min
	45 PCR cycles		
	Denature	95 °C	15 s
	Anneal	55 °C	30 s
	Extend	60 °C	30 s
Alu-LTR	Denature	95 °C	10 min
	45 PCR cycles		
	Denature	95 °C	15 s
	Anneal	60 °C	1 min
	Extend	72 °C	1 min
2-LTR circles	Denature	95 °C	10 min
	45 PCR cycles		
	Denature	95 °C	15 s
	Anneal and extend	66 °C	1 min
1-LTR circles	Denature	95 °C	10 min
	45 PCR cycles		
	Denature	95 °C	30 s
	Anneal	60 °C	30 s
	Extend	72 °C	1 min
	Final extension	72 °C	5 min

[a]AmpliTaq Gold (ABI) is acetylated to prevent DNA synthesis; 10 min at 95 °C activates the polymerase. Platinum Taq (Invitrogen) is bound to an antibody to prevent DNA synthesis; 2 min at 95 °C is enough to activate the polymerase.

4. Prepare the PCR master mix as described in point 4 of **Section 3.5**.
5. Dispense the PCR master mix and DNA into each well of a MicroAmp® Optical 96-well Reaction Plate as described in point 5 of **Section 3.5**.

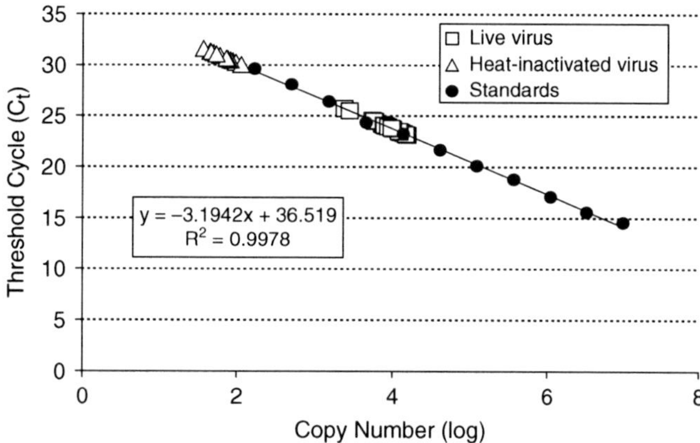

Fig. 5.3. Typical real-time PCR standard curve with values from live and heat-inactivated virus samples.

6. Run the PCR reaction using the thermocycling parameters outlined in **Table 5.4** for analysis of 2-LTR circles or for the cellular gene controls, using an ABI Prism 7900HT sequence detection system (*see* **Note 9**).
7. Analyze the data as described in point 8 of **Section 3.5**. The number of 2-LTR circles formed is normally expressed as a percentage of late RT products at 6 h after infection in a parallel infection. On average, during wild-type virus infection, 1–5% of late RT products at 6 h ultimately form 2-LTR circles, depending on the infection target cell line *(11, 19)*.

3.7. Real-Time PCR Analysis of 1-LTR-Circle Formation

1. Analysis of 1-LTR-circle formation is normally done using DNA collected 24 h after infection.
2. Prepare a standard curve from the appropriate standard template listed in **Table 5.1**, to be used to compute the copy numbers in each sample as stated in point 3 of **Section 3.5**.
3. Prepare the PCR master mix as described in point 4 of **Section 3.5** using the 1-LTR primer/probe set (**Table 5.1**). Note the difference in the 1-LTR primer and probe concentrations. Also run a duplicate plate with PBGD or CCR5 primer/probe set to normalize for the number of cells loaded for each reaction.
4. Dispense the PCR master mix and DNA into each well of a MicroAmp® Optical 96-well Reaction Plate as described in point 5 of **Section 3.5**.
5. Run the PCR reaction using the thermocycling parameters outlined in **Table 5.4** for analysis of 1-LTR circle-formation or for the cellular gene controls, using an ABI Prism 7900HT sequence detection system (*see* **Note 9**).
6. Analyze the data as described in point 8 of **Section 3.5**. The number of 1-LTR circles formed is normally expressed as a percentage of late RT product at 6 h after infection determined

from a parallel infection. On average, the proportion of 1-LTR:2-LTR circles formed during wild-type virus infection is 9:1 *(11)*.

3.8. Alu-LTR Real-Time PCR Analysis of Viral Integration

1. Analysis of integrated proviruses is normally done using DNA collected 24–48 h after infection.
2. Prepare integration standard DNA from a 293T cell line transduced with an HIV-1-based vector as previously described *(11, 19)* (*see* **Note 2**).
3. Determine the equivalent provirus copy number by correlating the Alu-LTR signal with late RT product using U5Ψ primer/probe set as previously described *(11,20,21)* (*see* **Note 10**). It is important to make this determination in the presence of the amount of cell-extract that will be present when measuring the unknowns (*see* **Note 11**). Make single use aliquots of 10^5–10^6 provirus/μL and keep at −80 °C until needed. Keep the cellular DNA as concentrated as is practical.
4. Measure the quantities of either PBGD or CCR5 to determine the number of cells present in each reaction, and dilute if necessary to insure that 100–500 ng of DNA is present when quantifying provirus. Otherwise, the signal obtained cannot be equated with quantities on the standard curve.
5. Prepare dilutions for the standard curve: the first dilution should be made at a final concentration of approximately 10^6 copies per 10 μL. Starting with this dilution, prepare threefold serial dilutions down to approximately 1,000 copies per 10 μL. Each dilution should also contain 100–500 ng of DNA from uninfected 293T cells per 10 μL, as a control for cellular DNA in the sample (*see* **Note 11**).
6. Prepare the PCR master mix as described in point 4 of **Section 3.5** using the Alu-LTR primer/probe set (**Table 5.1**). Note the difference in the Alu-LTR primer and probe concentrations.
7. Dispense the PCR master mix and DNA into each well of a MicroAmp® Optical 96-well Reaction Plate as described in point 5 of **Section 3.5**.
8. Run the PCR reaction using the thermocycling parameters outlined in **Table 5.4** for analysis of integrated proviruses or for the cellular gene controls, using an ABI Prism 7900HT sequence detection system (*see* **Note 9**).
9. Analyze the data as described in point 8 of **Section 3.5**. The number of integrated proviruses is normally expressed as a percentage of late RT product at 6 h after infection determined from a parallel infection. On average, during wild-type virus infection, 5–20% of late RT products at 6 h eventually integrate into the host genome, depending on the infection target cell line *(11)*.

4. Notes

1. Virus can be produced by transfection of other types of cells or from others type of cells transduced with an HIV-1-based vector. Similarly, infections can be carried out using other types of cells that can support infection of the HIV-1 virus.
2. It is critical to prepare integration standards in the cell line that will be infected.
3. Precise quantification of DNA standards is critical for competent determination of copy numbers. First, it is important that DNA plasmids used to measure reverse transcription progression have been linearized. Likewise, the standards used to measure LTR circles should be circular. Prior to dilution of the standard curve, make sure that the plasmid is not contaminated with other DNA forms (gel purify if necessary), then measure $A_{260/280}$ to determine concentration and purity. Pure DNA in H_2O should have a ratio of ~ 1.85. Measure the optical density (O.D.) of several dilutions to insure that the quantities determined using A_{260} correlate with the dilutions. Extra care taken at this step is never a waste of time.
4. The amount of pHDV-eGFP plasmid that can be used for transfection of a 100-mm-diameter dish can range from 2.5 to 20 μg with little effect on virus titer. The use of less plasmid DNA might help reduce plasmid DNA carryover. Do not use more than 2 μg of pCMV-G because it is cytotoxic.
5. The rinsing of transfected or infected 293T cells should be done gently because the cells can peel off the plate easily. Use pipette dispenser on slow mode, mark a single area on plate for aspirating and dispensing fluid, wash maximum of six plates at a time using a pipette capable of holding 60 mL, gently rock back and forth, and side to side 10 times each. The number of washes can be reduced if there is concern about loss of cells.
6. Normally, 150–300 ng of unconcentrated virus inoculum is sufficient to give an infection rate of approximately 40–80% GFP-positive cells by FACS analysis.
7. To analyze minus-strand DNA transfer, compare the proportion of U3U5 products to RU5 products. For plus-strand DNA transfer compare the proportion of U5Ψ products to *gag* products. Once conditions have been worked out for a particular system, it is helpful to reduce the number of freeze–thaws between analyses, i.e., try to quantify vDNA species all at the same time rather than looking at RU5, then freeze, then look at U3U5, then freeze, then look at *gag*, etc. It is very important to measure quantities of DNA that will be directly compared (late/early) with the same type of instrument. For instance, if the lab has a Stratagene and an ABI instrument, it

may cause problems to look at the early product on the Stratagene and the late product on the ABI machine.

8. There are commercially available PCR master mixes, e.g., TaqMan Universal PCR Master Mix (PE-Applied Biosystems), that contain all the PCR reagents in a single tube except for the primers, probe and DNA. Other Taq polymerases can also be substituted for the AmpliTaq Gold polymerase. However, the use of other reagents might require optimization, especially for the Alu-LTR assay.
9. The thermocycling parameters given in this protocol are optimized for the ABI Prism 7700 sequence detection system. If the ABI Prism 7900HT is being used, the reactions can be run using the ABI Prism 7700 emulation mode. The thermocycling parameters may need to be adjusted to give optimal results when using other instruments. For example, the block temperature does not always correlate well with the sample temperature so that the machine will indicate that the block is at the right temperature, but the samples exhibit a lag. In addition, different blocks heat and cool at different rates. As an example of this type of optimization, when looking at CCR5, the ABI 7700 parameters are 95 °C 15″, 55 °C 30″, 60 °C 30″. When using the Stratagene 4000×, the parameters are changed to 94 °C 30″, 55 °C 37″, 60 °C 30″. In addition, the ramp rate for heating is altered to 1.5°C/s from the default of 2.0 °C/s.
10. Verify that the HIV-1-based vector used to transduce the cells to make the integration standard has the Ψ region and sequence complementary to the reverse primer (MH532) of the U5Ψ primer/probe set. Otherwise design an alternate primer or use another primer/probe set.
11. Addition of DNA from uninfected cells is crucial for accurate quantification using Alu-LTR assay. For LTR circles and reverse transcription intermediate, extra cellular DNA does not have much effect because the primers and probes are specific for vDNA. But, because the Alu primer binds cellular DNA targets, the background DNA will sequester the Alu primer thereby decreasing the sensitivity and accuracy of the Alu-LTR reaction.

Acknowledgements

This research was supported in part by the Intramural Research Program of the NIH, National Cancer Institute, Center for Cancer Research. This project has been funded in whole or in part with federal funds from the National Cancer Institute, National Institutes of Health, under contract N01-CO-12400. The content of this publication does not necessarily reflect the views or policies of the Department of Health and Human Services, nor does mention of trade names, commercial products, or organizations imply endorsement by the U.S. Government.

References

1. Heid, C. A., Stevens, J., Livak, K. J., et al. (1996) Real time quantitative PCR. *Genome Res* 6, 986–994.
2. Lie, Y. S., Petropoulos, C. J. (1998) Advances in quantitative PCR technology: 5′ nuclease assays. *Curr Opin Biotechnol* 9, 43–48.
3. Holland, P. M., Abramson, R. D., Watson, R., et al. (1991) Detection of specific polymerase chain reaction product by utilizing the 5′-3′ exonuclease activity of Thermus aquaticus DNA polymerase. *Proc Natl Acad Sci USA* 88, 7276–7280.
4. Livak, K. J., Flood, S. J., Marmaro, J., et al. (1995) Oligonucleotides with fluorescent dyes at opposite ends provide a quenched probe system useful for detecting PCR product and nucleic acid hybridization. *PCR Methods Appl* 4, 357–362.
5. Palmer, S., Wiegand, A. P., Maldarelli, F., et al. (2003) New real-time reverse transcriptase-initiated PCR assay with single-copy sensitivity for human immunodeficiency virus type 1 RNA in plasma. *J Clin Microbiol* 41, 4531–4536.
6. Gibellini, D., Vitone, F., Schiavone, P., et al. (2004) Quantitative detection of human immunodeficiency virus type 1 (HIV-1) proviral DNA in peripheral blood mononuclear cells by SYBR green real-time PCR technique. *J Clin Virol* 29, 282–289.
7. Gibellini, D., Vitone, F., Gori, E., et al. (2004) Quantitative detection of human immunodeficiency virus type 1 (HIV-1) viral load by SYBR green real-time RT-PCR technique in HIV-1 seropositive patients. *J Virol Methods* 115, 183–189.
8. Morrison, T. B., Weis, J. J., Wittwer, C. T. (1998) Quantification of low-copy transcripts by continuous SYBR Green I monitoring during amplification. *Biotechniques* 24, 954–958, 960, 962.
9. Coffin, J. M., Hughes, S. H., Varmus, H. (1997) *Retroviruses*, Cold Spring Harbor Laboratory Press, Plainview, N.Y.
10. Voronin, Y. A., Pathak, V. K. (2004) Frequent dual initiation in human immunodeficiency virus-based vectors containing two primer-binding sites: a quantitative in vivo assay for function of initiation complexes. *J Virol* 78, 5402–5413.
11. Butler, S. L., Hansen, M. S., Bushman, F. D. (2001) A quantitative assay for HIV DNA integration in vivo. *Nat Med* 7, 631–634.
12. Karageorgos, L., Li, P., Burrell, C. J. (1995) Stepwise analysis of reverse transcription in a cell-to-cell human immunodeficiency virus infection model: kinetics and implications. *J Gen Virol* 76 (Pt 7), 1675–1686.
13. Huber, H. E., McCoy, J. M., Seehra, J. S., et al. (1989) Human immunodeficiency virus 1 reverse transcriptase. Template binding, processivity, strand displacement synthesis, and template switching. *J Biol Chem* 264, 4669–4678.
14. Majumdar, C., Abbotts, J., Broder, S., et al. (1988) Studies on the mechanism of human immunodeficiency virus reverse transcriptase. Steady-state kinetics, processivity, and polynucleotide inhibition. *J Biol Chem* 263, 15657–15665.
15. Victoria, J. G., Lee, D. J., McDougall, B. R., et al. (2003) Replication kinetics for divergent type 1 human immunodeficiency viruses using quantitative SYBR green I real-time polymerase chain reaction. *AIDS Res Hum Retroviruses* 19, 865–874.
16. Munk, C., Brandt, S. M., Lucero, G., et al. (2002) A dominant block to HIV-1 replication at reverse transcription in simian cells. *Proc Natl Acad Sci USA* 99, 13843–13848.
17. Reardon, J. E. (1993) Human immunodeficiency virus reverse transcriptase. A kinetic analysis of RNA-dependent and DNA-dependent DNA polymerization. *J Biol Chem* 268, 8743–8751.
18. Buckman, J. S., Bosche, W. J., Gorelick, R. J. (2003) Human immunodeficiency virus type 1 nucleocapsid zn(2+) fingers are required for efficient reverse transcription, initial integration processes, and protection of newly synthesized viral DNA. *J Virol* 77, 1469–1480.
19. Thomas, J. A., Gagliardi, T. D., Alvord, W. G., et al. (2006) Human immunodeficiency virus type 1 nucleocapsid zinc-finger mutations cause defects in reverse transcription and integration. *Virology* 353, 41–51.
20. Brussel, A., Sonigo, P. (2003) Analysis of early human immunodeficiency virus type 1 DNA synthesis by use of a new sensitive assay for

quantifying integrated provirus. *J Virol* 77, 10119–10124.
21. O'Doherty, U., Swiggard, W. J., Jeyakumar, D., et al. (2002) A sensitive, quantitative assay for human immunodeficiency virus type 1 integration. *J Virol* 76, 10942–10950.
22. Svarovskaia, E. S., Barr, R., Zhang, X., et al. (2004) Azido-containing diketo acid derivatives inhibit human immunodeficiency virus type 1 integrase in vivo and influence the frequency of deletions at two-long-terminal-repeat-circle junctions. *J Virol* 78, 3210–3222.
23. Lusi, E. A., Guarascio, P., Presutti, C., et al. (2005) One-step nested PCR for detection of 2 LTR circles in PBMCs of HIV-1 infected patients with no detectable plasma HIV RNA. *J Virol Meth* 125, 11–13.
24. Gillim-Ross, L., Cara, A., Klotman, M. E. (2005) HIV-1 extrachromosomal 2-LTR circular DNA is long-lived in human macrophages. *Viral Immunol* 18, 190–196.
25. Jacque, J. M., Stevenson, M. (2006) The inner-nuclear-envelope protein emerin regulates HIV-1 infectivity. *Nature* 441, 641–645.
26. Lewin, S. R., Vesanen, M., Kostrikis, L., et al. (1999) Use of real-time PCR and molecular beacons to detect virus replication in human immunodeficiency virus type 1-infected individuals on prolonged effective antiretroviral therapy. *J Virol* 73, 6099–6103.
27. Halvas, E. K., Aldrovandi, G. M., Balfe, P., et al. (2006) Blinded, multicenter comparison of methods to detect a drug-resistant mutant of human immunodeficiency virus type 1 at low frequency. *J Clin Microbiol* 44, 2612–2614.
28. Mulder, J., McKinney, N., Christopherson, C., et al. (1994) Rapid and simple PCR assay for quantitation of human immunodeficiency virus type 1 RNA in plasma: application to acute retroviral infection. *J Clin Microbiol* 32, 292–300.
29. Piatak, M., Jr., Luk, K. C., Williams, B., et al. (1993) Quantitative competitive polymerase chain reaction for accurate quantitation of HIV DNA and RNA species. *Biotechniques* 14, 70–81.
30. Piatak, M., Jr., Saag, M. S., Yang, L. C., et al. (1993) High levels of HIV-1 in plasma during all stages of infection determined by competitive PCR. *Science* 259, 1749–1754.
31. Unutmaz, D., KewalRamani, V. N., Marmon, S., et al. (1999) Cytokine signals are sufficient for HIV-1 infection of resting human T lymphocytes. *J Exp Med* 189, 1735–1746.
32. Yee, J. K., Miyanohara, A., LaPorte, P., et al. (1994) A general method for the generation of high-titer, pantropic retroviral vectors: highly efficient infection of primary hepatocytes. *Proc Natl Acad Sci USA* 91, 9564–9568.

Chapter 6

Analysis of 2-LTR Circle Junctions of Viral DNA in Infected Cells

Dibyakanti Mandal and Vinayaka R. Prasad

Abstract

The unintegrated viral DNA synthesized during human immunodeficiency virus type 1 infection includes linear and circular forms. Circular forms of viral DNA are surrogate markers for nuclear import of viral DNA during virus replication as well as events surrounding the completion of reverse transcription. Analysis of 2-LTR circles is convenient and the quantity of 2-LTR circle formed is directly proportional to the amount of viral DNA imported into the cell nucleus. In addition, correct synthesis of 2-LTR circles is an outcome of HIV-1 *Gag-Pol* function. Thus, quantitation and sequence analysis of 2-LTR circles have been very important in studying the structure and function relationship of key viral proteins. In this chapter, we describe the methods of quantitation and analysis of 2-LTR circle junctions isolated from HIV-1 infected cells.

Key words: 2-LTR circle junction, Southern blot, Real-time PCR, Sequence analysis.

1. Introduction

Following infection, HIV-1 reverse transcriptase (RT) converts viral RNA into double-stranded DNA by exploiting its multifunctional enzyme activities *(1)*. During productive infection, the viral cDNA then enters into the nucleus of infected cell accompanied by integrase (IN) and other viral proteins, in the form of preintegration complexes (PICs) *(1–3)*. In the host cell nucleus, viral cDNA integrates into the host chromosomal DNA in a reaction mediated by IN *(4–6)*. Integrated DNA then serves as the template for viral RNA production. A fraction of cDNA which remains unintegrated become circularized, a reaction that is thought to be a dead end product since circular cDNA is not the substrate for the viral IN. Circularization of viral DNA is

Vinayaka R. Prasad, Ganjam V. Kalpana (eds.), *HIV Protocols: Second Edition, vol. 485*

DOI 10.1007/978-1-59745-170-3_6 Springerprotocols.com

believed to be mediated by cellular nonhomologous end joining pathway which includes several host proteins namely Ku, DNA-PK, XRCC4 or DNA ligase IV *(7)*. These factors bind to the PICs and modulate the course of infection.

Circular DNA formation in the cell nucleus is accomplished by: (1) joining of two ends of full-length cDNA having complete LTRs to form 2-LTR circle *(7–9)*; (2) formation of 1-LTR circle as a result of recombination between 2-LTRs *(10)*; (3) circularization of stalled RT complexes that lead to 1-LTR circle; (4) integration of the linear cDNA into itself yielding an internally arranged form *(9)*. The amount of 2-LTR circles generated is always less than that of 1-LTR circles and independent of cell line used. The ratio of 2-LTR circles to 1-LTR circles generally range from 0.16 to 0.43 *(11)*. Using HIV-1-based vectors, it has been shown that 2-LTR circles accumulate in cells more slowly compared to full-length cDNA and reach a maximum abundance after 24 h, consistent with the expected precursor-product relationship. The 2-LTR circles were also found to be notably stable in the host cell nucleus.

Though 2-LTR circle junctions are only a fraction of total viral cDNA in the infected cells, they contain enormous critical information regarding the fidelity of viral enzyme functions and or key steps during viral DNA synthesis. First, the amount of 2-LTR circle junctions accumulated in the nucleus is proportional to the amount of nuclear import of viral DNA since this form of viral cDNA is found only in the host nucleus. Secondly, 2-LTR circle DNA is a byproduct rather than an intermediate of the integration event. In addition, sequence analysis of 2-LTR circle junctions gives us insight into a number of crucial processes mediated by RT including completion of strand displacement synthesis, removal of $tRNA^{Lys,3}$ primer, removal of plus strand PPT primer that were used during plus and minus strand DNA synthesis respectively.

2. Materials

2.1. Tissue Culture, Virus Generation and Isolation of Viral DNA from Infected Cell

1. Dulbecco's Modified Eagle's Medium (DMEM) (Cellgro media, Kansas city, MO) supplemented with 10% fetal bovine serum (FBS, Gibco/BRL).
2. RPMI1640 medium (Gibco/BRL) supplemented with 10% FBS. CaPO4 mediated mammalian cell transfection reagent (Specialty media).
3. Filtration unit (Costar).
4. RNase free DNase I (10,000 U/mL, Roche diagnostics).
5. Phosphate-buffered saline (Without Ca/Mg, Cellgro medium).
6. Viral DNA isolation kit (DNeasy tissue kit, Qiagen).

2.2. Southern Blot Hybridization of Viral DNA

1. Hydrochloric acid 0.25 M.
2. Denaturation solution: 1.5 M NaCl/0.5 M NaOH.
3. Neutralization solution: 1.5 M NaCl/0.5 M Tris-HCl, pH 7.0.
4. 20x and 2x solutions of Saline Sodium Citrate, 3 M Sodium Chloride, 0.3 M Sodium Citrate, pH 7.0 (SSC).
5. Whatman 3 mm filter paper sheets.
6. Nylon *or* nitrocellulose membrane.
7. UV-transparent plastic wrap (Saran Wrap) for nylon membranes.
8. UV trans-illuminator for nylon membranes.

2.3. Hybridization of Transferred DNA

1. DNA to be used as probe.
2. Aqueous prehybridization/hybridization solution, room temperature and 68 °C (Ambion).
3. 2 × SSC/0. 1% (w/v) SDS.
4. 0. 2 × SSC/0. 1% (w/v) SDS, room temperature and 42 °C.
5. 0. 1 × SSC/0. 1% (w/v) SDS, 68 °C.
6. 2x and 6x SSC.
7. Hybridization oven *or* 68 °C water bath or incubator.
8. Hybridization tube *or* sealable bag and heat sealer.
9. Additional reagents and equipment for DNA labeling by nick translation or random oligonucleotide priming, measuring the specific activity of labeled DNA and separating unincorporated nucleotides from labeled DNA and autoradiography.

2.4. Real-Time PCR

1. Primers and Taqman Probes Specific for 2-LTR circle junction *(12)* Primer Sequences: Forward – MH535: 5′AACTAGGGAACCCACTGCTTAAG-3′;
   Reverse – MH536: 5′-TCCACAGATCAAGGATATCTTGTC-3′;
   Probe:
   MH603:5′-(FAM)ACACTACTTGAAGCACTCAAGGCAA GCTTT(TAMRA)-3′.
2. PCR mix (Applied Biosystems):
   The TaqMan Universal PCR Master Mix is 2X in concentration. The mix is optimized for TaqMan reactions and contains AmpliTaq Gold DNA Polymerase, AmpErase Uracil N-Glycosylase, dNTPs with dUTP and optimized buffer components.
3. 384 well plate for PCR reaction (Applied Biosystems).
4. Plate sealer (Applied Biosystems).

2.5. Analysis of 2-LTR Circle Junction

1. Primers specific for 2-LTR circle junctions:
   2-LTR forward – 5′ GCC TGG AGC TCT GGC TAA 3′ and
   2-LTR reverse – 5′ GCC TTG TGT GGT AGA TCC A 3′

(13) for first round PCR and MH535 and MH536 (point 1 of **Section 2.4**) for nested second Round PCR.
2. TA cloning kit (Invitrogen).
3. Plasmid DNA isolation kit (Qiagen).
4. DNA alignment software BioEdit *(14)* (BioEdit v7.0.7: www.mbio.ncsu.edu/BioEdit).

3. Methods

3.1. Transfection of Proviral DNA

1. Plate 293T cells in 100 mm tissue culture dish. Plate should be 40% confluent within 24 h (*see* **Note 1**).
2. Remove media from the plate 3–5 h before transfection and replace with fresh media.
3. Transfect 293T cells with 10 μg of proviral DNA by Ca-PO4 method using mammalian cell transfection kit (Specialty Media) following manufacturers' protocol.
4. Approximately 10^6 cells were plated 24 h before transfection.
5. Add fresh DMEM media containing 10% FBS to cells 3 h before transfection.
6. In a polystyrene tube, place 0.5 mL 1x HBS (per transfection) and add 10 μL phosphate solution. In a separate tube, mix 10 μg DNA in a total volume of 430 μL and add 60 μL $CaCl_2$ solution to the DNA solution.
7. Add DNA and $CaCl_2$ mixture drop-wise to HBS solution and gently tap the tube and incubate for 20 min.
8. Add calcium phosphate-precipitated DNA to the cells.
9. 12 to 16-h post-transfection change the media and supplement with fresh DMEM containing 10% FBS.
10. Replace media after 16–18 h post-tranfection.

3.2. Determination of Virus Production

1. Collect supernatant from transfected cells (*see* **Note 2**).
2. Preserve at −70 °C in 50 mM HEPES (pH 7.4) buffer until use.
3. Quantitate p24 production using HIV-1 p24 antigen ELISA kit (Perkin Elmer). HIV-1 p24 antigen assay is to be done by following method (Summarized from perkin Elmer manual).
4. Bring all reagents to room temperature (15–30 °C) before use.
5. Add 20 μL Triton X-100 to all wells except substrate blank.
6. Add 200 μL of the 100 pg/mL stock and make twofold serial dilutions (e.g., 50, 25, 12.5 pg/mL). Add positive control and cell-free culture supernatant to the designated wells (*see* **Note 3**). Mix well with pipette.
7. Seal the plate and incubate for 2 h at 37 °C.
8. Wash plate six times with 300 μL diluted (1X) wash buffer. Blot well before addition of next reagent.

9. Add 100 μL detector antibody to all wells except substrate blank.
10. Seal plate and incubate for 60 min at 37 °C.
11. Wash plate six times with 1x wash buffer as in step '8' in Section 3.1.
12. Within 15 min of use, dilute sufficient Streptavidin-HRP (SA-HRP) Concentrate to the 1:100 working concentration with Streptavidin–HRP diluent. Mix thoroughly.
13. Add 100 μL diluted SA-HRP to all wells except substrate blank.
14. Seal the plate and incubate 30 min at room temperature (15–30 °C).
15. Wash plate six times with 1x wash buffer as in step '8' in Section 3.1.
16. Prepare sufficient OPD Substrate (O-phenyl diamine) solution by dissolving one tablet in 11 mL substrate within 15 min of use. Vortex vigorously to assure complete dissolution and protect from light. The OPD substrate solution should be colorless to pale yellow. A yellow-orange color indicates that the reagent is contaminated and must be discarded.
17. Add 100 μL OPD substrate solution to all wells including substrate blank.
18. Seal plate and incubate 30 min at room temperature (15–30 °C) in the dark.
19. Stop the reaction by adding 100 μL of Stop Solution to all wells.
20. Read the plate at 490 or 492 nm, blanking the plate reader on air (consult plate reader Instruction Manual).

3.3. Virus Infection and the Isolation of HIV cDNA from Infected Cells

1. Infect Jurkat cells (1×10^5 cells) with 60 ng p 24 (*see* **Note 5**).
2. Incubate 2 h at 37 °C.
3. Wash with PBS and supplement with fresh media.
4. Harvest cells at 2, 24 and 48 h postinfection. Isolate DNA using DNeasy tissue kit (Qiagen) following manufacturer's protocol as described below.
5. Resuspend cells in 200 μL PBS and add 20 μL Proteinase K to the cells.
6. Lyse the cells with 200 μL lysis buffer supplied with the kit.
7. Incubate at 56 °C for 10 min.
8. Add 200 μL 95% Ethanol to the tubes and mix by vortexing.
9. Apply mixture to spin column.
10. Centrifuge for 1 min at 8000 rpm ($6,000 \times g$).
11. Place the cartridge to a fresh 2-ml collection tube and add 0.5-mL wash buffer AW1 supplied with the kit.
12. Centrifuge for 1 min at 8000 rpm ($6,000 \times g$) as above.

13. Put the spin column on a fresh collection tube and add 0.5-mL wash buffer AW2 supplied with the kit.
14. Centrifuge at 14,000 rpm (20,000 × g) for 3 min.
15. Place the spin column inside a fresh 1.5 mL Eppendorf tube.
16. Add 100–200 μL elution buffer and incubate at room temperature for 1 min.
17. Centrifuge for 2 min at 14,000 rpm (20,000 × g).
18. Store DNA at 4 °C.
19. Analyze 2-LTR circle junction DNAs formed in the infected cells by restriction digestion of the cellular DNA isolated followed by Southern blot and the absolute DNA quantitation is performed by real-time PCR. A portion of the remaining DNA is used to do PCR amplification, cloning and sequencing.

3.4. Analysis of 2-LTR Circle by Southern Blot Analysis

3.4.1. Preparation of Viral DNA

Digest the viral DNA (10 μg) isolated from infected cells with MscI and XhoI overnight. An XhoI site is present at position 8897 and two MscI sites at position 2619 and 4551 are present of R3B provirus. Double digestion of DNA with XhoI and MscI will produce fragments depending on whether viral DNA is present as integrated or un-integrated forms. Restriction digestion of viral DNA will also generate different fragments corresponding to the circular DNA, either 1-LTR circle or 2-LTR circle.

3.4.2. Agarose gel Electrophoresis of Viral DNA

Analyze Viral DNA digested with XhoI and MscI enzymes on 0.7% agarose gel to separate the different sized fragments generated. Carry out the electrophoresis in 1X TAE buffer for 5 h at a constant 50 V. Photograph the gel along with appropriate molecular size markers.

3.4.3. Transfer of DNA to Nylon Membrane

Transfer viral DNA separated by agarose gel electrophoresis to a Hybond N+ membrane as described earlier *(15)*.

3.4.4. Hybridization of Transferred Viral DNA

3.4.4.1. Preparation of the Probe

Probe to be used for the hybridization is a 1.6 kb fragment corresponding to nucleotide positions 1818–3498 of $HXB2_{R3B}$. This fragment must be PCR amplified using the primers 5′ AGA AGA AAT GAT GAC AGC 3′ and 5 TGC CAG TTC TAG CTC TG 3′ *(16)*. PCR-amplified DNA is labeled with radioactive ^{32}P by random-primed DNA labeling kit following manufacturers' protocol (Roche Diagnostics).

1. Add to 25 ng template DNA (linear) sterile double distilled water to a final volume of 9 μL in a microfuge tube.
2. Denature the DNA by heating in a boiling water bath for 10 min at 95 °C and chilling quickly on ice.
3. To the DNA, add 1 μL each of dATP, dGTP and dTTP from the stock, 2 μL reaction mixture (hexanucleotide), 5 μL (50 μCi) of [α-^{32}P] dCTP, 3,000 Ci/mmol (aqueous solution) and the Klenow enzyme 1 μL (2 U/μL).

4. Mix and centrifuge briefly. Incubate at 37 °C for 30 min (*Note*: Longer incubation can increase the yield of labeled DNA).
5. Stop the reaction by adding 2 μL 0.2 M EDTA (pH 8.0) and/or by heating to 65 °C for 10 min.
6. Unincorporated deoxyribonucleoside triphosphates are removed by QIAGEN nucleotide purification kit. In brief, to the reaction mixture, add 5x volume of binding buffer and apply to the spin column. The spin column is centrifuged at 8000 rpm ($6,000 \times g$) for 1 min. Add 0.5 mL of wash buffer to the spin column and centrifuge for 1 min. Another 0.5 mL of wash buffer is added to the column and centrifuged at 14,000 rpm ($20,000 \times g$) for 2 min. Finally probe was eluted at 20 μl volume of elution buffer (incorporation is 2×10^9 dpm/μg of DNA).

3.4.4.2. Pre-hybridization

Place the Nylon membrane (Hybond N+, Amersham Pharmacia) with transferred viral DNA into the hybridization bottle and add 10 mL pre-hybridization buffer. Carry out pre-hybridization for 1 h at 42 °C.

3.4.4.3. Hybridization

1. Denature radioactive probe at 95 °C for 2 min. Add single stranded probe to hybridization bottle. Hybridize viral DNA with the radioactive probe at 42 °C overnight (16–18 h).
2. Remove the hybridization solution (into radioactive waste), and add an equal volume of 2x SSC/0.1% SDS. Incubate for 15 min at room temperature with shaking, changing the wash solution after 5 min.
3. Wash the membrane with 0.2×SSC/0.1% SDS and incubate with rotation for 15 min at room temperature, changing the wash solution after 5 min This is a low-stringency wash.
4. Pour off the final wash solution, rinse the membrane in 2× SSC at room temperature and pour off the wash buffer.

3.4.4.4. Detection of Viral DNA

After washing, blot the membrane with a filter paper (3 mm Whatman) to remove moisture. Air-dry the membrane for 10 min. Wrap the blot with saran-wrap prior to autoradiography. Expose the blot to X-ray film with an intensifying screen at −70 °C for 1–2 days.

3.4.5. Interpretation of Southern Hybridization Result

MscI and XhoI digestion of un-integrated viral DNA will generate fragments corresponding to linear, 1-LTR or 2-LTR circular DNA (*see* **Fig. 6.1** and **Note 4**). Fragments specific for 1-LTR or 2-LTR circular DNA could be easily detected by probe described above. 1-LTR circular DNA will produce 1.9- and 2.8-kb fragments whereas 2-LTR circle DNA will form 1.9- and 3.4-kb fragments. By Southern blot of viral DNA isolated from infected cells, a 2.6-kb fragment can also be detected which correspond to

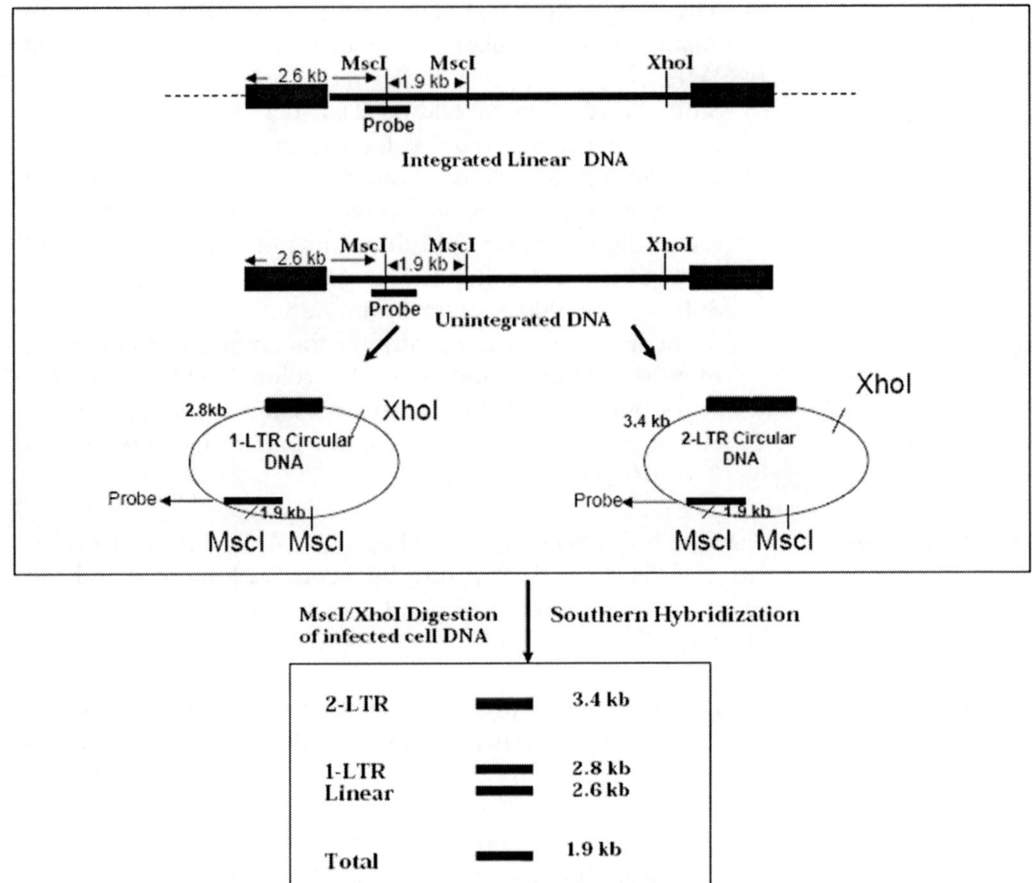

Fig. 6.1. Graphical outline of restriction digestion and Southern blot hybridization of HIV-1$_{R3B}$ infected cell DNA to analyze 2-LTR circle junction. The top panel shows the positions of the restriction sites on the linear and circular forms. The bottom panel shows the relative expected positions of the various bands on a Southern blot. Total refers to a band that is common to all restriction fragments.

integrated or un-integrated linear DNA. Intensity of each band detected is directly proportional to the amount of viral DNA formed in the infected cell. Amounts of 1-LTR and 2-LTR circular DNA species increase over time after postinfection. Quantitative measurement of 1-LTR and 2-LTR circular DNA and the amount of linear DNA formed indicates that 55% of the total viral DNA imported into the nucleus are fully processed and integrated into the host nucleus whereas 35% and 5% DNA are circularized to form 1-LTR and 2-LTR circle junctions respectively. Remaining 10% DNA remains linear *(16)*.

3.5. Real-Time PCR Quantitation of 2-LTR Circle Junctions

Quantify DNA copy numbers of RT intermediates synthesized within the HIV-1 infected cells by real-time PCR using TaqMan technology (Applied Biosystems). The primers and probes specific for 2-LTR circles were as described previously *(12, 13)*. Perform

Real time PCR on an ABI7700 apparatus (Applied Biosystems). Use approximately 500 ng of infected cell DNA for each PCR reaction.

3.5.1. PCR Amplification of 2-LTR Circle Junctions and Cloning

1. Construct a plasmid DNA p2LTR from HIV-1-infected cell DNA. PCR amplify 2-LTR junction sequence from wild-type HIV-infected Jurkat cell DNA and clone PCR product into the TA cloning vector pCR 2.1 (Invitrogen) to generate p2-LTR (see Section 3.6 for details PCR protocol and cloning). Verify clones by sequencing.
2. Quantitate the pLTR DNA by taking the $Absorbance_{260}$ of the purified DNA.
3. Calculate DNA copy number of p2-LTR from the O.D. value. Use this DNA as positive control for the real-time PCR quantitation of 2-LTR circle junctions present in the HIV-1 infected cell DNA. Also use this DNA to generate the standard curve to calculate the copy number of unknown DNA.

3.5.2. Real Time PCR to Quantify 2-LTR Circle Junction

1. Prepare real-time reaction cocktail by mixing 2x PCR mix (Applied Biosystems PE), 2-LTR specific probe and primers.
2. Dispense 30 μL of reaction mixture in a 200-μL PCR tube.
3. Add 5 μL (500 ng) of virus-infected cell DNA to the mixture.
4. Add 10 μL of the mix into 384 wells plate in triplicate.
5. Make a serial dilution of p2-LTR DNA to obtain 10^8, 10^6, 10^4 and 10^2 copies/μL of this DNA.
6. Add 5 uL of the diluted p2-LTR plasmid DNA of known copy numbers to 30 μL PCR mix.
7. Add 10 μL of the above mix to 384 well plate.
8. Cover the plate with plate sealer carefully avoiding any mixing.
9. Spin the plate for a short time at 4 °C to bring the PCR reaction mix at the bottom.
10. Place the plate into the real time PCR machine.
11. Run Real-time PCR reactions for the quantification of 2-LTR circles under the following conditions:

    1 cycle
    50 °C, 2 min
    1 cycle
    95 °C, 10 min
    40 cycle
    95 °C, 15 s
    60 °C, 1 min

3.5.3. Determination of 2-LTR Circle Junction Copy Number in the Infected Cells

1. Generate a standard curve by plotting the Ct values (cycle threshold) obtained from real time PCR reaction of p2-LTR plasmid against the DNA copy number.

2. Calculate the copy numbers of 2-LTR circle DNA in the viral infected cellular DNA from the standard curve.

3.5.4. Interpretation of 2-LTR Circle Junction Quantification Results

The standard curve generated is always linear when log copy of number (ranging from 2 to 8) of control DNA is plotted against the Ct values. 2-LTR copy number from experimental sample is obtained from the standard curve. It is possible by real-time PCR to quantify full-length viral DNA synthesized during infection. It is generally found that copy number of 2-LTR circle junction formed is 10^2 times less than full-length DNA (**Fig. 6.2**).

3.6. Sequence Analysis of 2-LTR Circle Junctions

3.6.1. PCR Amplification of 2-LTR Circle Junctions

Viral DNA isolated from 48 h infected cells is used as template to amplify 2-LTR circle junctions. A nested two-step PCR is required to amplify sufficient amounts of 2-LTR circle junction DNA for the purpose of cloning. Primers used for the first round PCR are 2-LTR forward and 2-LTR reverse and primers for the second round are MH535 and MH536 (described in the Materials section). A 200-bp DNA fragment can be amplified using primers specific for U3 and U5 sequences abutting the 2-LTR circle junction by this approach. Conditions to be employed for the PCR amplification is as follows:

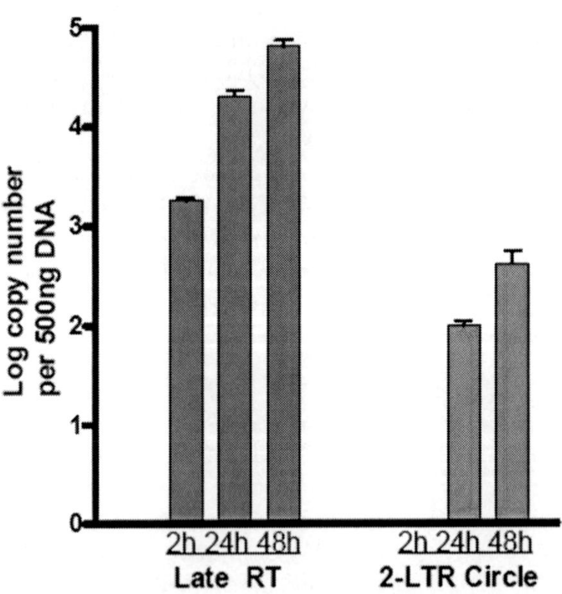

Fig. 6.2. Quantitation of 2-LTR circle Junction DNA by Real-time PCR. Jurkat cells were infected with WT R3B virus and infected cells DNA was isolated at 2, 24 and 48 h postinfection. Virus-infected cell DNA was analyzed for amount of 2-LTR circle junction DNA (*green bar*) and compared with the amount of late reverse transcription (RT) products (*Blue bars*). [Adapted from Mandal *et al.* (2006) *Nucleic Acids Res* May 24, 34(10), 2853–2863.]

Conditions for the first round PCR:

1 cycle of
95 °C, 5 min
35 Cycle of
95 °C, 15 s
50 °C, 1 min
72 °C, 30 s
1 Cycle of
72 °C, 5 min

Use 2 μL of products from the first round of amplification as template for second round PCR.

Conditions for the second round PCR:

1 cycle of
95 °C, 5 min
35 cycles of
95 °C, 15 s
55 °C, 1 min
72 °C, 30 s
1 cycle of
72 °C, 5 min

3.6.2. Agarose Gel Analysis

1. Five microliters of the PCR product obtained after the second step PCR are employed for analysis on 2% agarose gel for 1 h at 100 V.
2. Stain the gel with ethidium bromide (EtBr) and monitor under UV trans-illuminator. Check for the DNA product of 189 bp.

3.6.3. TA Cloning of PCR Products

1. Ligate PCR products to TA cloning vector (pCR 2.1, Invitrogen): Mix 1 μL PCR product, 1 μL vector, 1 μL 10x buffer and 1 μL ligase in a total volume of 10 μL volume. Incubate the mixture at 16 °C overnight.
2. Transform DH5α cells with the ligation product.
3. Clones obtained are characterized by *Eco*RI restriction digestion. Clones having insertion of ~ 188 bp PCR product are confirmed by the release of the corresponding fragment.

3.6.4. Sequencing of 2-LTR Circle Junction

Determine the nucleotide sequence of about 50 clones to get enough information about the 2-LTR circle junctions. Sequence the clones using primers specific for TA cloning vector, adjacent to the cloning site.

3.6.5. Alignment of the 2-LTR Circle Junction Sequences

Do a multiple alignment of all the sequence obtained by ClustalX program using BioEdit Software.

3.6.6. Statistical Analysis

Perform statistical analysis with the 2-LTR circle junction sequences to determine the significance in the difference between sequence distributions of different mutant viruses. Analyze the data by contingency table analysis. Subsets of pertinent groupings

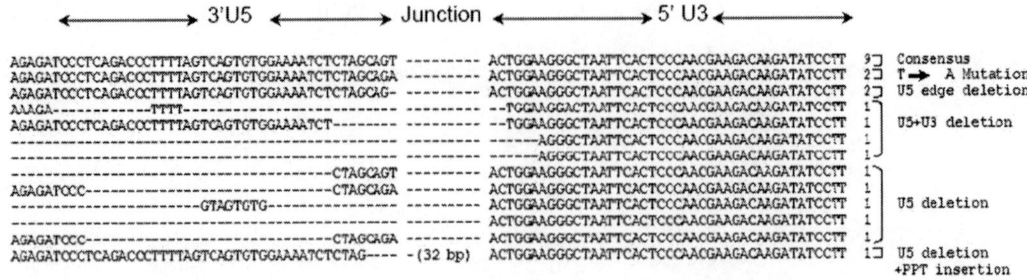

Fig. 6.3. Sequence analysis of HIV-1 2-LTR circle junction DNA. 2-LTR circle junction sequences obtained from HIV-1 R3B infected cell DNA were aligned by multiple alignment. The 3′ U5, 5′ U3 and circle junction sequence is indicated. Deletions within the 3′U5 or 5′U3 region are indicated by dashes. Insertion within the junction sequence is mentioned by number of bases. [Adapted from Mandal *et al.* (2006) *Nucleic Acids Res* May 24, 34(10), 2853–2863.]

are followed up with the traditional two-by-two Fisher's exact tests.

3.6.7. Interpretation of 2-LTR Circle Junction Sequence Analysis Result

2-LTR circle junctions are formed when full-length linear viral DNA fails to integrate into the host genome. In most cases, viral DNA fails to integrate mostly because of erroneous terminal sequences due to the error proneness of the processes involved. Sequence analysis of 2-LTR circle obtained from wild type HIV-1_{R3B} infected cell DNA showed that almost 50% circle junctions were of 'consensus' sequences (**Fig. 6.3**) without revealing any specific defect – confirming an inefficient nature of the integration reaction. Some of the sequences were found to have terminal 'GA' in the 3′ U5 boundary. Majority of the remaining sequences have deletions at the U5 edge, the U3 edge or contain a deletion that spans the U5–U3 junction. Fewer sequences display the insertion of fragments or full PPT sequence between the U5–U3 junction. Insertion of fragments of tRNA primer are also often observed. The internal insertions and deletions within the 2-LTR circle junction, which tend to involve single nucleotides are a result of error prone RT.

4. Notes

1. The number of producer cells to be plated for transfection is important depending on the time of supernatant collection. Seed $\sim 1 \times 10^6$ 293T cells on a 100-mm tissue culture dish so that, on the day transfection, the confluency of the cells reaches about 30–40%.
2. Viral supernatant can be collected as early as 6 h post-transfection or as late as 48 h post-transfection. Supernatants collected at some of the early time points (24 h) may have higher infectivity per nanogram of virus produced, but the

total amount of virus produced will be less than at peak p24 production. On the other hand, if virus supernatant is collected at very late time points, the total virus production may be increased, but the proportion of infectious virus particles decreases.

3. For p24 antigen quantitation, note that 200 μL cell-free culture supernatant generally contains p24 above the range of measurement. Serial dilutions of up to 1,000-fold would be helpful.
4. For Southern blot analysis of 2-LTR circle junctions, restriction sites are specific for the R3B molecular clone. These restriction sites may vary for different molecular clones, isolates or subtypes of HIV-1.
5. For quantitation of 2-LTR circle junction by real time PCR, infect 1×10^5 Jurkat cells with 50–60 ng of p24 equivalent of virus supernatant which gives m.o.i of $\sim$ 1.

References

1. Goff, S. P. (2001) Intracellular trafficking of retroviral genomes during the early phase of infection: viral exploitation of cellular pathways. *J Gene Med* 3, 517–552.
2. Bukrinsky, M. I., Haffar, O. K. (1997) HIV-1 nuclear import: in search of a leader. *Front Biosci* 2, d578–d587.
3. Cullen, B. R. (2001) Journey to the center of the cell. *Cell* 105, 697–700.
4. Coffin, J. M., Hughes, S. H., Varmus, H. E. (1997) *Retroviruses*. Cold Harbour laboratory press, Cold Spring Harbour, NY.
5. Hansen, M. S. T., Carteau, S., Hoffman, C., et al. (1998) Retroviral cDNA integration: mechanism, applications and inhibition. In: Setlow, J.K. (ed.), *Genetic Engineering. Principles and Methods*, Vol. 20. Plenum Press, New York, NY, pp. 41–62.
6. Bukrinsky, M. I., Sharova, N., McDonald, T. L., et al. (1993) Association of integrase, matrix, andreverse transcriptase antigens of human immunodeficiency virus type 1 withviral nucleic acids following acute infection. *Proc Natl Acad Sci USA*90, 6125–6129.
7. Li, L., Olvera, J. M., Yoder, K. E., et al. (2001) Role of the non-homologous DNA end joining pathway in the early steps of retroviral infection. *EMBO J* 20(12), 3272–3281.
8. Brown, P. O. (1997) Integration. In: Coffin, J.M., Hughes, S.H., Vermus H.E. (eds.), *Retroviruses*. Cold Spring harbor laboratory press, New York, NY.
9. Shoemaker, C., Goff, S., Gilboa, E., et al. (1980) Structure of a cloned circular Moloney murine leukemia virus DNA molecule containing an inverted segment: implications for retrovirus integration. *Proc Natl Acad Sci USA* 77(7), 3932–3936.
10. Farnet, C. M., Haseltine, W. A. (1991) Circularization of human immunodeficiency virus type 1 DNA in vitro. *J Virol* 65(12), 6942–6952.
11. Kilzer, J. M., Stracker, T., Beitzel, B., et al. (2003) Roles of host cell factors in circularization of retroviral DNA. *Virology* 314(1), 460–467.
12. Butler, S. L., Hansen, M. S., Bushman, F. D. (2001) A quantitative assay for HIV DNA integration in vivo. *Nat Med* 7(5), 631–634
13. Julias, J. G., Ferris, A. L., Boyer, P. L., et al. (2001) Replication of phenotypically mixed human immunodeficiency virus type 1 virions containing catalytically active and catalytically inactive reverse transcriptase. *J Virol* 75(14), 6537–6546.
14. Hall, T. A. (1999) BioEdit: a user-friendly biological sequence alignment editor and analysis program for windows 95/98/NT. *Nucl Acids Symp Ser* 41, 95–98.
15. Brown, T. (1999) Southern Blotting. p 2.9.2–2.9.11. In: Current protocols in molecular biology. Ausubel, F.M (Ed). John Wiley and Sons, Inc.
16. Zennou, V., Petit, C., Guetard, D., et al. (2000) HIV-1 genome nuclear import is mediated by a central DNA flap. *Cell* 101(2), 173–185

Chapter 7

HIV-1 Recombination: An Experimental Assay and a Phylogenetic Approach

Michael D. Moore, Mario P. S. Chin, and Wei-Shau Hu

Abstract

The generation of genetic diversity is a fundamental characteristic of HIV-1 replication, allowing the virus to successfully evade the immune response and antiviral therapies. Although mutations are the first step towards diversity, mixing of the mutations through the process of recombination increases the variation and allows for the faster establishment of advantageous strains within the viral population. Therefore, studying recombination of HIV-1 provides insights into not only the mechanisms of HIV-1 replication but also into the potential for spread of antiviral drug resistance mutations within and across viral subtypes. This chapter describes, in detail, a highly sensitive recombination assay designed to measure the frequency of recombination between two viruses. This assay allows us to investigate the requirements, mechanisms, and final products of recombination. Additionally, software-based phylogenetic tools are described in this chapter, which allow for the identification of specific recombination events within patient samples or viral progeny from the recombination assay.

Key words: Recombination, flow cytometry analysis, HIV-1, immunostaining, template switching, phylogenetic analysis.

1. Introduction

Recombination between two HIV-1 viruses promotes the assortment of mutations within the gene pool, thereby increasing the genetic diversity within the viral population. The resulting variation enhances the virus's ability to adapt and evolve around selective pressures, such as the host immune system and antiviral drug interventions. Currently, many circulating recombinant forms of HIV-1 have been identified and these viral strains are an integral part of the HIV-1 pandemic. Therefore, it is important to analyze the ability of HIV-1 to recombine in both experimental

Vinayaka R. Prasad, Ganjam V. Kalpana (eds.), *HIV Protocols: Second Edition, vol. 485*

DOI 10.1007/978-1-59745-170-3_7 Springerprotocols.com

systems and clinical settings in order to obtain a greater understanding of its recombination potential and the impact of these recombination events.

The recombination assay described in this chapter is based on the reconstitution of functional green fluorescence protein (GFP) through recombination between two viruses, both encoding for mutated, inactive *gfp* genes. In this system, one virus (HIV-H0) contains a reporter cassette in the Nef reading frame encoding the *heat stable antigen* gene (*hsa*), followed by an internal ribosomal entry site (IRES), and a *gfp* gene containing a nonsense mutation at its 5′ end. The other virus (HIV-T6) also contains a reporter cassette in Nef; this cassette includes a mouse thy 1.2 gene (*thy*), an IRES, and a *gfp* gene mutated at its 3′ end. Therefore, if recombination occurs between these two viruses, more specifically at a site between the two mutations in *gfp*, the resulting provirus could contain a functional *gfp* gene. As both of the parental viruses encode for a nonfunctional *gfp*, the recovery of a GFP signal can be used to score recombination events. In addition to *gfp*, the two parental viruses and the recombinants all contain another marker gene, *hsa* or *thy*; hence, the expression of HSA or Thy can be used to score all infection events. The number or cells that express HSA, Thy, or GFP can be measured using immunostaining and flow cytometry.

Using this recombination assay and by varying the distance between the two inactivating mutations in *gfp* we have demonstrated that HIV-1 recombines at a very high frequency *(1)*. We have also used this system to compare the recombination rates of HIV-1 in various target cells *(2)*. Additionally, we have measured the HIV-1 intersubtype recombination rates between subtype B and subtype C and found them to be ninefold lower than the intrasubtype recombination rates *(3)*. We further determined that a 3-nucleotide difference within the dimer initiation site of the stem loop 1 was responsible for the low recombination rate observed between these two subtypes.

As recombination has been shown to occur in experimental systems at high frequencies, it is important to evaluate its impact in clinical settings within HIV-1-infected patients. Additionally, phylogenetic analyses of various HIV-1 strains are complicated by the presence of recombinant sequences; therefore identification of such sequences is a critical part of evolutionary studies into HIV-1. Identification of recombinants involves the detailed examination of the viral sequences using computer software. The software highlighted in this chapter uses representative examples of each viral subtype to which a sample sequence is compared. Then statistical methods are employed to calculate the phylogeny and parental makeup of the potential recombinant. These techniques can also be used to identify specific recombination events within the in vitro system outlined above, providing a more detailed picture of the reverse transcriptase template switching pattern and the identification of recombination hotspots.

2. Materials

2.1. Viral Production

1. 293T cells.
2. Complete medium for 293T cells (293T-CM), consisting of Dulbecco's Modified Eagle's Medium (DMEM) (Mediatech, Inc.), 10% fetal calf serum (FCS, HyClone), 1% penicillin/streptomycin (P/S) (Gibco).
3. Trypsin (0.25%) and ethylenediamine tetraacetic acid (EDTA) (1 mM) (Gibco).
4. Transfection agent for mammalian cells. For this purpose 25 kDa poly(ethyleneimine) (PEI, Sigma) is used but can be substituted at all stages by other agents, most commonly the calcium phosphate method (Promega).
5. Plasmids HIV-H0 and HIV-T6 (**Fig. 7.1**) or derivatives thereof that are being studied.
6. Plasmid pHCMV-G, encoding the vesicular stomatitis virus glycoprotein (VSV-G) from the CMV promoter *(4)*.
7. Syringe filters (0. 45 μm pore size).

2.2. Dual Infected Cell Line Production

1. Polybrene.
2. Phosphate-buffered saline (PBS) supplemented with 2.5% FCS (PBS+FCS).
3. R-phycoerythrin (R-PE) conjugated rat anti-mouse CD24 (heat stable antigen, HSA) antibody (PE-α-HSA, BD Pharmingen).
4. Allophycocyanin (APC) conjugated anti-mouse CD90.2 (Thy1.2) antibody (APC-α-Thy, eBioscience).

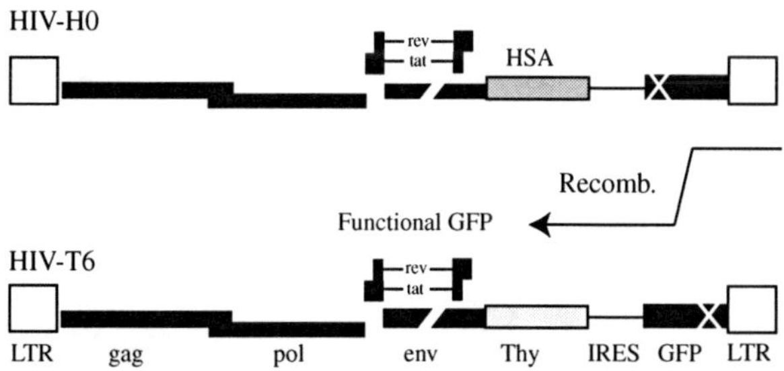

Fig. 7.1. Schematic representation of the viral genomes used in the recombination assay. The two viruses used in this study are depicted. Both are based on pNL4.3 sequence but with inactivating deletions in *vif, vpu, vpr* and *env*. Within the *nef* reading frame a double marker gene cassette has been inserted using an IRES to express the downstream gene. HIV-H0 encodes *hsa* and a *gfp* with an inactivating mutation near its 5′ end. HIV-T6 encodes *thy* and a *gfp* with an inactivating mutation near its 3′ end.

5. 1% paraformaldehyde in PBS.
6. Fluorescence-activated cell sorting apparatus with 488 and 635 nm excitation lasers and at least three channels for detection of GFP (520 nm emission), PE (590 nm emission) and APC (675 nm emission), e.g., FACSCantage SE system with the FACSDiVa digital option (BD Biosciences).

2.3. Assay of Virus Released by the Cell Lines

1. Plasmid pIIINL(AD8)env encoding the CCR5 tropic HIV-1 envelope gene *(5)*.
2. Hut/CCR5 cell line, derived from Hut78 cells modified to constitutively express the CCR5 co-receptor.
3. Complete media for Hut/CCR5 cells (Hut-CM) consisting of Roswell Park Memorial Institute-1640 (RPMI, Mediatech, Inc.), 10% FCS, 1% P/S, 1% glutamine, 500 μg/mL Gentamicin and 1 μg/mL Puromycin.
4. Flow cytometry apparatus with 488 nm and 635 nm excitation lasers and at least three channels (520, 590 and 675 nm), e.g., FACSCalibur (BD Biosciences) or CyFlow® ML (Partec).

2.4. Calculation of Virus Recombination

1. FlowJo FACS analysis software or equivalent.

2.5. Software Analysis of HIV-1 Recombination

1. Recombination identification program (RIP) (at http://www.hiv.lanl.gov/content/sequence/RIP/RIP.html).
2. SimPlot (from http://sray.med.som.jhmi.edu/SCRoftware/simplot/).

3. Methods

3.1. The Recombination Assay

The recombination assay involves the establishment of a dual-infected virus producer cell line and the subsequent phenotypic analysis of its viral progeny through target cell infection (**Fig. 7.2**). The accuracy of this assay relies on generating cells containing two proviruses, one from each virus under study, resulting in similar expression of the two viral genomic RNAs. This is achieved by infecting the cells with the two viruses separately, each at low multiplicities of infection (MOI), thereby reducing the probability of a cell containing multiple copies of a provirus derived from the same virus. Although generating a cell line is more time-consuming than cotransfection of two plasmids, viral production from cell lines yields far more accurate and consistent data than those from cotransfection. There are at least two issues associated with generating viruses used for recombination studies via cotransfection: first, DNA recombination between the two plasmids encoding the viral genome under study can occur, which generates unnecessary background signal. Secondly, during transient transfection, one cannot control the distributions of the two plasmids in the producer cells, which introduces

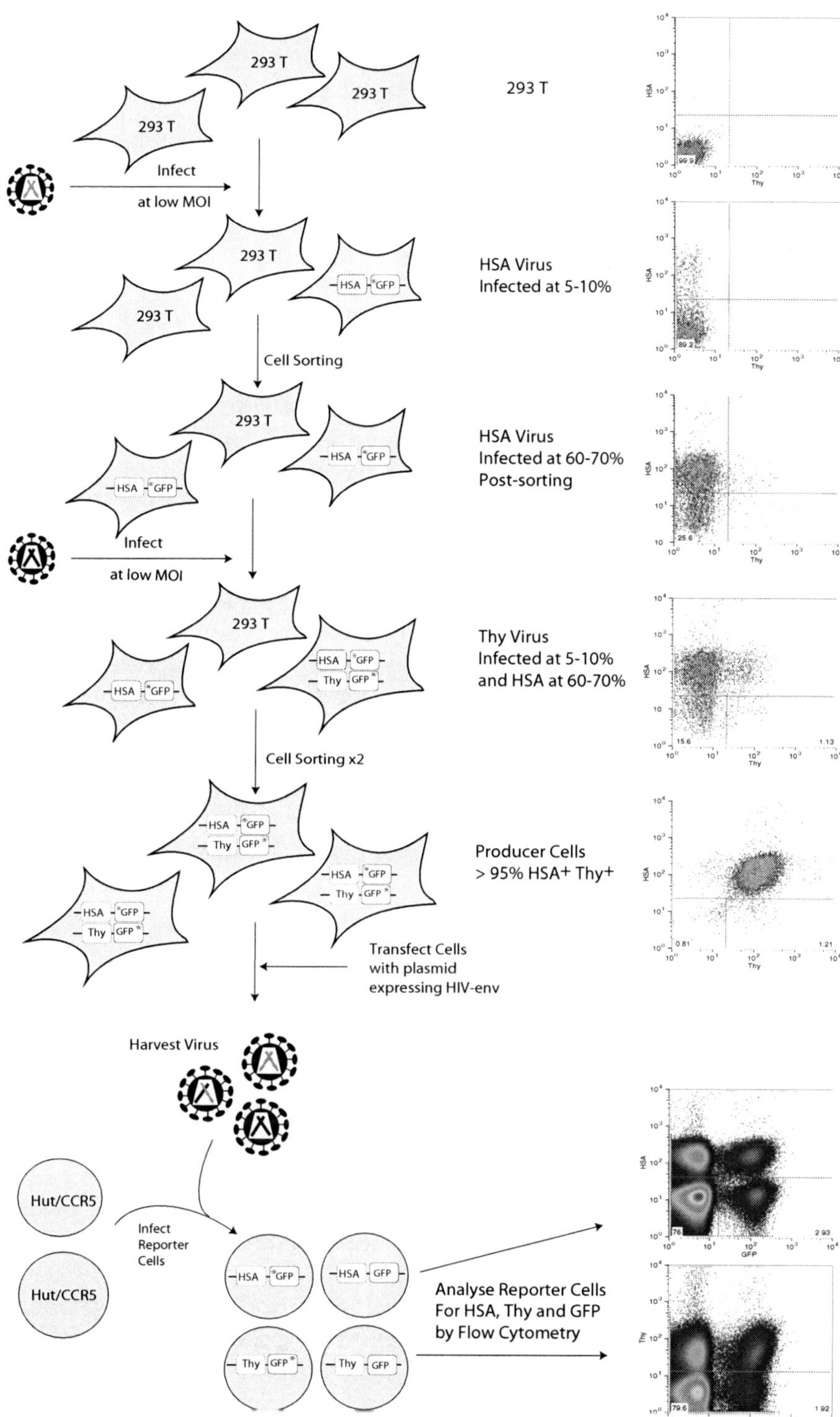

Fig. 7.2. Cell sorting schedule for production of dual infected cell lines. The cell lines used to analyze the recombination potential between two viruses were produced by two successive rounds of infection at low MOIs, each infection followed by a cell sorting event. A final cell sort for double positive cells results in more than 95% of cells being double positive. These cells are then transfected with an envelope encoding plasmid and the resulting virus is harvested and used to infect Hut/CCR5 cells. Flow cytometry is performed on the infected target cells and the data obtained are used to calculate the frequency with which GFP is reconstituted by the two parental viruses through recombination. Also shown are representative flow cytometry data for the cells at each stage of the procedure.

inaccuracy and variations in the results. The confidence in the results obtained with the dual-infected cell lines relies on infecting cells at low MOIs and collecting a large pool of independently infected cells, each containing two proviruses but integrated into different regions of the genome. Thus the result obtained from a single cell line is actually an average of multiple viruses from many different cells. Additionally, we generally prefer to generate at least two cell lines for each viral pairing and analyze more than one viral harvest from each cell line to create even greater confidence in the measurement.

3.1.1. Viral Production

1. The 293T cells are plated at 4×10^6 cells per 100-mm-diameter tissue culture dish in 10 mL of 293T-CM.
2. Twenty-four hours later, the cells are transfected with 10 μg of a plasmid encoding the viral vector (based on HIV-H0 or HIV-T6) and 4 μg of a plasmid encoding the VSV-G protein (pHCMV-G). Transfection can be accomplished by a variety of methods including calcium phosphate or PEI (*see* **Note 1**). Both techniques require an incubation of the cells in serum-free media containing the transfection mixture for 4 h, after which the transfection mixture is replaced with 10 mL of fresh 293-CM. Cells are then incubated at 37 °C in 5% CO_2 for 48 h for optimal virus production.
3. Virus is released from transfected cells into the supernatant, which is harvested using a 10 mL syringe and passed through a 0.45-μm filter to remove cellular debris. Viral supernatants are then ready to use or are frozen and stored at −70 °C.

3.1.2. Dual Infected Cell Line Production

1. Low passage 293T cells are seeded at 1×10^6 cells per 100-mm-diameter tissue culture dish in 10 mL of 293T-CM (*see* **Note 2**).
2. Twenty-four hours later the medium is replaced with 2 mL of transduction medium. Transduction medium consists of between 5×10^4 and 1×10^5 transducing units of HSA-encoding virus (equal to an MOI of between 0.05 and 0.1) generated in **Section 3.1.1**, diluted in 293T-CM and supplemented with 50 μg/mL polybrene (*see* **Note 3**).
3. The cells are incubated with the transduction medium for 2 h, which is then removed and replaced with 10 mL of fresh 293T-CM.
4. When the transduced cells reach confluency (about 72 h postinfection), the cells are trypsinized (remove medium, add 1 mL trypsin/EDTA, incubate for 1 min, remove trypsin and pipette the cells off the plate using 10 mL of 293T-CM) and divided into five 100-mm-diameter tissue culture dishes.
5. When nearing confluency again (another 48–72 h) the cells are re-trypsinized and the resulting single cell suspension is stained for the presence of the HSA marker gene.

6. The cells are pelleted by centrifugation at $300 \times g$ for 3 min in a 15-mL falcon tube and resuspended in 10 mL of cold (4 °C) PBS+FCS (1× wash).
7. The cells are washed again and resuspended at 1×10^7 cells/mL in staining solution. Staining solution consists of PBS+FCS supplemented with 0. 2 μg/mL of PE-conjugated anti-HSA antibody (*see* **Note 4**).
8. Cells are incubated on ice for 1 h to allow binding of the antibody to the cell surface marker HSA.
9. Once stained, the cells are washed twice in PBS+FCS to remove excess antibody and finally resuspended at 1×10^7 cells/mL in cold PBS+FCS ready for cell sorting.
10. For sorting, the cells are gated for live cells first, then sorted based on HSA expression by gating on PE-positive cells.
11. Successfully sorted cells (about 1×10^6 cells), collected from the machine in 10 mL of 293T-CM, are pelleted and resuspended in fresh 293T-CM and plated into a 100-mm-diameter tissue culture dish (*see* **Note 5**).
12. Once the cells near confluency they are expanded. When sufficient cells have been produced they are re-plated into a 100-mm-diameter tissue culture dish at 1×10^6 cells in 10 mL of 293T-CM.
13. The cells are then infected for the second time (repeat **Sections 1–6**) using the *thy* encoding virus at an estimated MOI of 0.05–0.1 (i.e., infection will result in 5–10% of cells positive for Thy).
14. For immunostaining of the dual infected cells, cells are washed once with PBS+FCS as described in **step 6** and resuspended at 1×10^7 cells/mL in PBS+FCS containing 0. 2 μg/mL PE-α-HSA and 2 μg/mL APC-α-Thy antibodies (*see* **Note 6**).
15. The cells are stained on ice for 1 h followed by two washes in 10 mL PBS+FCS and resuspension in PBS+FCS at 1×10^7 cells/mL.
16. The cells are sorted a second time but this time the live cells are gated for HSA^+, Thy^+, GFP^- (*see* **Note 7**).
17. Again the cells (about 1×10^6) are collected from the cell sorter in 10 mL of 293-CM, pelleted at $300 \times g$, resuspended in 10 mL fresh 293T-CM and plated in a 100-mm-diameter tissue culture dish. Once near confluency the dual infected cells are split into four 100-mm-diameter tissue culture dishes.
18. Forty-eight hours later, one plate of cells is frozen as a backup, the others are harvested as before (steps 5 + 6) and readied for a second round of HSA^+, Thy^+, GFP^- sorting (repeat **steps 14–17**).
19. The cells are then expanded and a sample can be stained, as before, for the two marker genes. The dual infected cell line

at this stage should be more than 95% double positive, if not it can be resorted a third time. A sample of the resulting cell line should be frozen for backup and the rest used promptly (minimum number of passages) to evaluate the recombination frequency between the two viruses under study.

3.1.3. Assay of Virus Released by the Producer Cell Lines

1. The newly created dually infected virus producer cell line is plated at 4×10^6 cells per 100-mm-diameter tissue culture dish in 10 mL 293T-CM.
2. Twenty-four hours later the cells are transfected with 4 μg of pIII(AD8)env, a plasmid encoding for the HIV-1 CCR5 tropic envelope glycoprotein (*see* **Note 8**).
3. Forty-eight hours post-transfection the cell supernatant, containing the released virus, is harvested with a 10-mL syringe, clarified through a 0.45-μm filter, and used to transduce the target cell line Hut/CCR5.
4. The Hut/CCR5 cells are plated at 1×10^6 cells per well of a six-well plate in 1 mL of Hut-CM, to which 1 mL of virus is added (*see* **Note 9** and **10**).
5. After four hours incubation at 37 °C the cells are harvested from the six-well plate, pelleted by centrifugation at 300 × *g* for 3 min, resuspended in 10 mL Hut-CM and seeded into a 25-cm^2 tissue culture flask.
6. Seventy-two hours post-transduction the Hut/CCR5 cells are recovered by centrifugation at 300 × *g* for 3 min and resuspended in 10 mL of cold PBS + FCS.
7. The cells are then re-pelleted at 300 × *g* for 3 min, and resuspended in cold PBS + FCS supplemented with antibodies against HSA and Thy, as before, and are incubated on ice for 1 h.
8. The immunostained cells are washed twice with PBS + FCS, as described in **Section 3.1.2 step 6**, resuspended in 1 mL of PBS supplemented with 1% paraformaldehyde fixative (fixative is optional), and incubated on ice for 20 min prior to analysis.
9. The cells are then analyzed by flow cytometry for the presence of PE (HSA$^+$), APC (Thy$^+$) and GFP. The cells are gated for live cells and correct compensation between PE and GFP is performed (**Fig. 7.3**) (*see* **Note 11**). To ensure accurate results are obtained, at least 1000 GFP positive cells or 1×10^6 live cells should be scored for each infection.

3.1.4. Calculation of Virus Recombination

1. The flow cytometry results are then analyzed using a software package, such as FlowJo.
2. The flow cytometry machine data file is first opened and the live cell population gated.
3. The cells are then divided up as either GFP$^+$ or GFP$^-$ by gating on the GFP histogram plot. These subpopulations are then displayed separately as a dotplot of PE (HSA$^+$) on

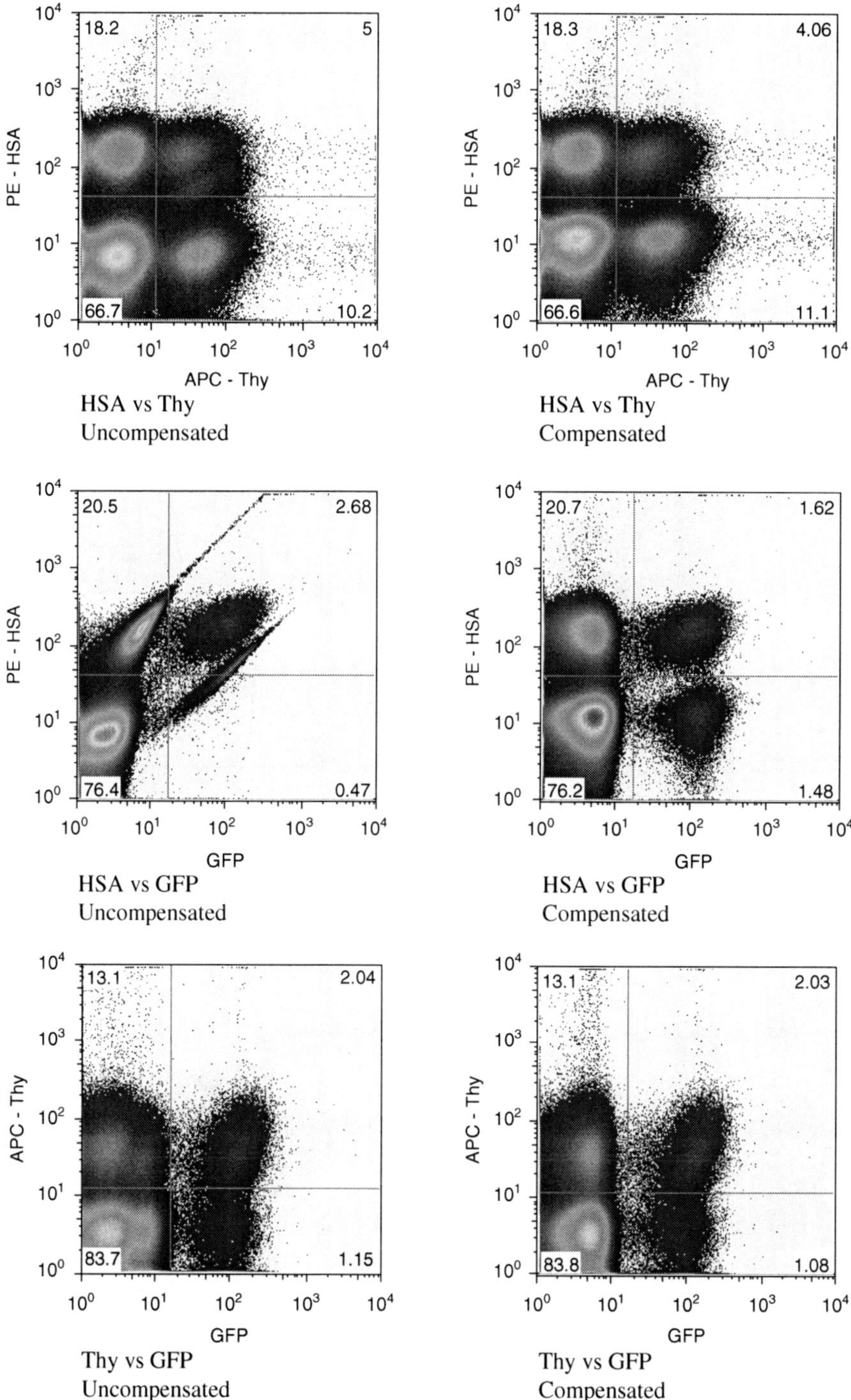

Fig. 7.3. Effects of compensation on the flow cytometry data. Due to the bleed-through of fluorescence between the PE (HSA) and GFP signals, compensation is required. A representative Hut/CCR5 infection shows the effect of compensation on the data, allowing accurate gating of the positive cells.

the x-axis and APC (Thy^+) on the y-axis. A quadrant gate is used to subdivide the $GFP^{+/-}$ populations according to HSA^+/Thy^-, HSA^+/Thy^+, HSA^-/Thy^+ and HSA^-/Thy^- (**Fig. 7.4**). The results of each subdivision in actual cell numbers are then used to calculate the MOI of each viral phenotype.

4. The total MOI is calculated by the formula below:

$$\textit{Infection MOI} = \frac{\log\left(1 - \frac{Zi}{\Upsilon}\right)}{\Upsilon \times \log\left(\frac{\Upsilon-1}{\Upsilon}\right)}$$

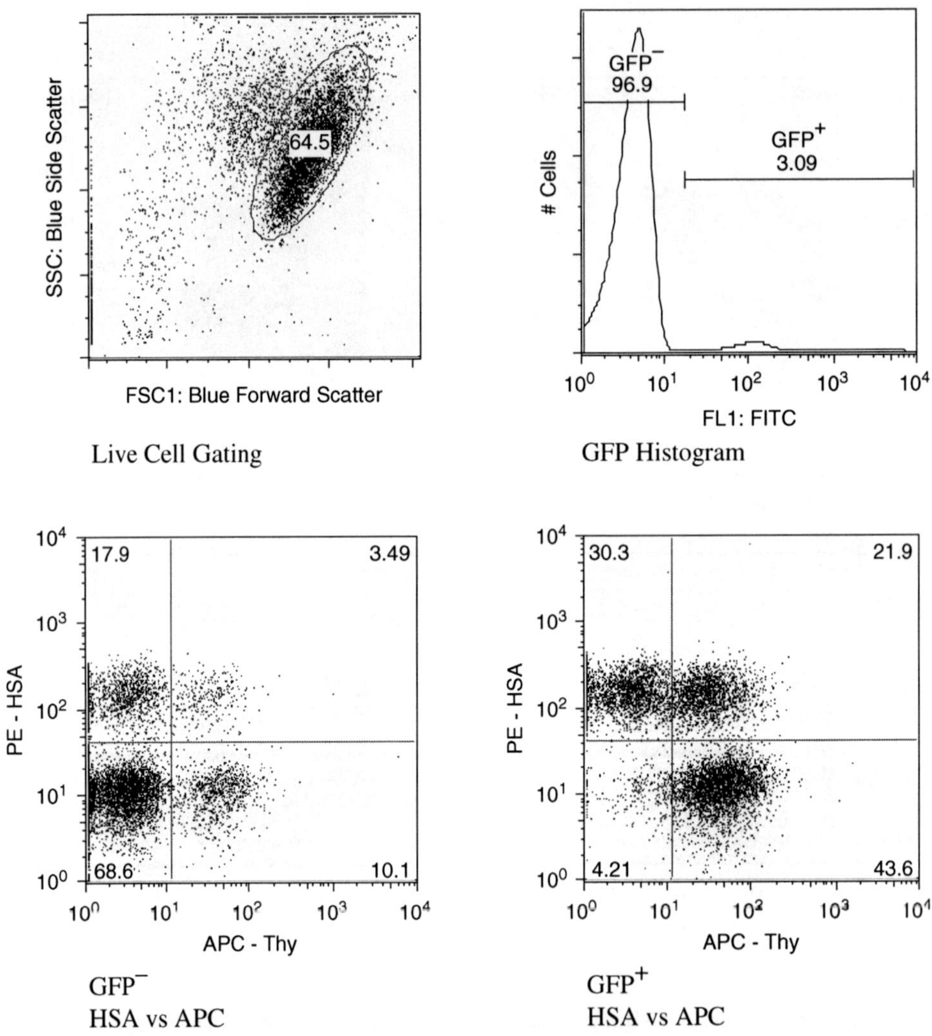

Fig. 7.4. Software analysis of flow cytometry data to obtain the recombination rate. In order to calculate the recombination rate the number of cells positive for each marker gene is required. To achieve this, the live cells are first gated upon and plotted as a histogram of GFP signal. The live cells are then divided into positive and negative according to their GFP signal. The subdivided cells are then plotted on a dot plot of HSA and Thy expression, and gated accordingly.

Where Υ represents the number of total live cells analyzed and Zi the number of infected cells (i.e., the total number of cells positive for any of the three markers: HSA, Thy or GFP).

5. The MOI of GFP is calculated by a similar formula:

$$GFP\ MOI = \frac{\log\left(1 - \frac{Zg}{\Upsilon}\right)}{\Upsilon \times \log\left(\frac{\Upsilon - 1}{\Upsilon}\right)}$$

Where Υ represents the number of total live cells analyzed and Zg the number of GFP positive cells detected.

6. Finally, the recombination rate between the two viruses is calculated using the formula below with the previously estimated MOI results (**Table 7.1**):

$$Rate\ of\ recombination = \frac{GFP\ MOI}{Infection\ MOI} \times 100\%$$

3.2. Software Analysis of HIV-1 Recombination

To identify an HIV-1 recombinant, the virus sequence is first analyzed using the RIP; if recombination is detected it can be confirmed and further analyzed using a bootscanning program called SimPlot. RIP is a web-based program accessible through the HIV Sequence Database at the Los Alamos National Laboratory *(6)*, which provides likely parental sequences for a query sequence. SimPlot is a freeware program available for download at the author's homepage *(7)*, and allows for detailed statistical comparisons between the query sequence and the parental sequences.

The RIP uses a nucleotide window of defined size to scan over a sequence alignment containing the query sequence and several background representatives. The relationship between the sequences is calculated according to similarity, Hamming, and Juke-Cantor distance measurements. The window is then advanced one nucleotide (step) and the same calculations are

Table 7.1
Calculation of the Recombination Rate. The number of live cells, GFP, HSA and Thy positive cells are all used to produce a final result for the recombination rate between two viruses. GFP MOI and Infection MOI are calculated as described in Section 3.1.4. Recombination rate is calculated as GFP MOI/Infection MOI × 100%

	No. live cells	No. GFP$^+$ cells	No. of infected cells (HSA$^+$, Thy$^+$, or GFP$^+$)	GFP MOI	Infection MOI	Recomb. rate
Virus infection	1,648,452	30,406	373,215	0.0186	0.2567	7.25%

repeated until the end of the query sequence is reached and the likely parental sequences are identified.

The SimPlot program again scans over a sequence alignment with a defined nucleotide window size and step size. For each window, the alignment is subjected to a phylogenetic analysis and a bootstrapped phylogenetic tree is built *(8)*. Bootstrapping is a statistical method to test the reliability of a phylogenetic analysis by resampling the alignment many times (default in SimPlot is set at 100). The result of the resampling is a bootstrap value, which is the percentage of times the query sequence is placed together with each of the reference sequences within an internal branch of the constructed phylogenetic tree. The bootstrap values of each window are plotted along the query sequence after the scan; the bootstrap values can be used to identify from which parental sequence each window of the query originated. Recombination breakpoints within the query sequence are detected as the point at which an increase is observed in bootstrap values for the query to one parent along with a simultaneous decrease in bootstrap values to a different parent.

3.2.1. RIP

1. Prepare the query sequence in Fasta/Pearson format and upload the sequence as a file or paste it from the clipboard.
2. Choose one of the options to define the background alignment (*see* **Note 12**).
3. Select a window size and a significance threshold (*see* **Note 13**).
4. Choose one of the options for handling nucleotide gaps within the sequence alignment (*see* **Note 14**).
5. If multistate character code is present in the sequence, select "True" to count the characters as a partial match or "False" to count the characters as mismatches (*see* **Note 15**).
6. With all the parameters defined, press "Run" to analyze the query sequence.
7. The results are presented in three graphs each presenting a different distance measurement (similarity, Hamming, and Juke-Cantor) between the query and the background sequences (*see* **Note 16**).
8. Press "Rerun" to re-analyze the query using only the best match sequence if necessary (**Fig. 7.5**) (*see* **Note 17**).
9. If the output of the RIP suggests the query is a recombinant, it is necessary to confirm the analysis using the SimPlot program (*see* **Note 18**).

3.2.2. SimPlot

1. Align the query sequence with reference sequences of your choice (*see* **Note 19**).
2. In the SimPlot program, use the "File" menu and open the aligned sequence file.

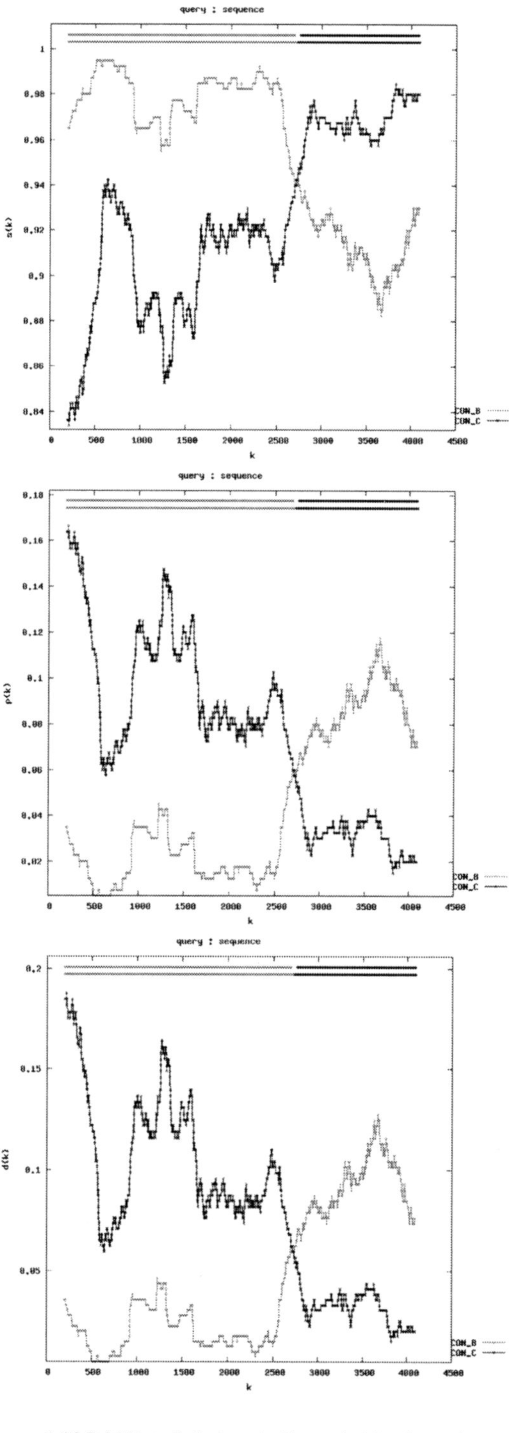

Fig. 7.5. RIP analysis of an HIV-1 recombinant. A 4.2-kb sequence of HIV-1 *gag* and *pol* is subjected to RIP analysis. Plots show the similarity, Hamming, and Juke-Cantor distances (from *top* to *bottom*) between the query and the reference sequences after "Rerun". The distance values are shown on the y-axis and the position along the query sequence is shown on the x-axis. Potential parents will have the highest similarity distance but lowest Hamming and Juke-Cantor distances. The first 2.7 kb of the query sequence most closely resembles the consensus sequence of subtype B HIV-1 (*gray*) whereas the rest of the query sequence more closely resembles the subtype C HIV-1 consensus sequence (*black*).

3. The sequences are now assigned into different groups under the "SeqPage" tab, reassign the groups if necessary and select those intended for comparison (*see* **Note 20**).
4. Select the "BootScan" tab and select the query sequence under the "Commands" menu.
5. Change the window size and step size by selecting the relevant settings at the bottom of the window if needed (*see* **Note 21**).
6. Start bootscanning by selecting "Do BootScan" under the "Command" menu.
7. The percentages of permuted trees (bootstrap value) of the scanning windows are plotted after the bootscanning is complete (**Fig. 7.6**). Zoom in on a plot area by selecting and enclosing the area of interest. Pan around the plot by dragging with the right mouse button and zoom out by dragging up to the left.
8. Reference sequences that are placed together with the query sequence in a high percentage of permuted trees are the

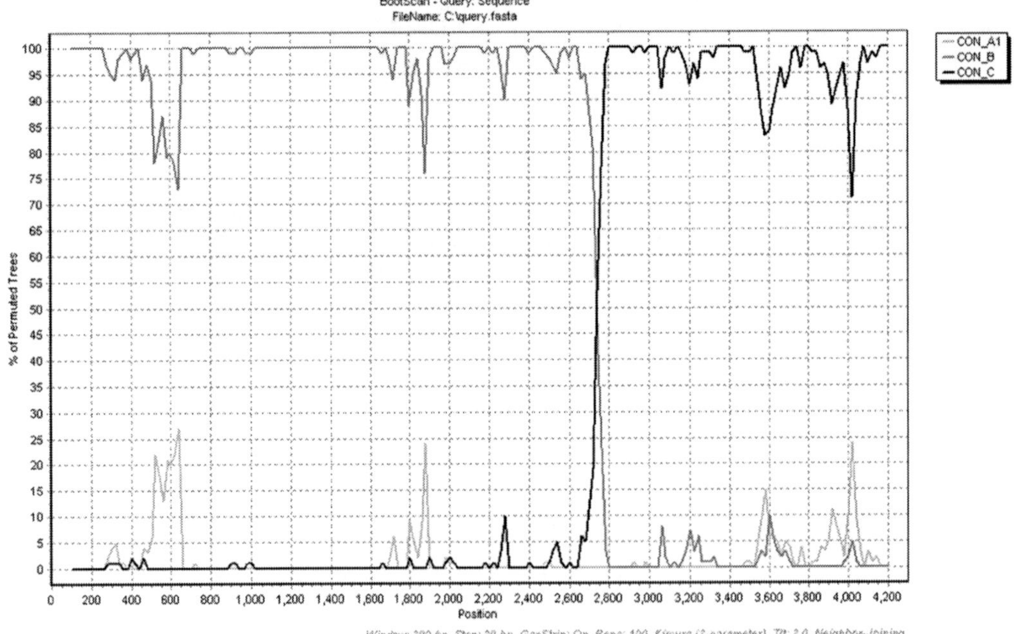

Hu W.S. Fig.6 HIV-1 recombination: An experimental assay and a phylogenetic approach

Fig. 7.6. SimPlot analysis to confirm recombination within a query sequence. The same 4.2 kb sequence of HIV-1 *gag* and *pol* is subjected to SimPlot analysis. The percentage of permuted trees is shown on the *y*-axis and the position along the query sequence is shown on the *x*-axis. SimPlot analysis indicates that the query is a recombinant of subtype B (*dark gray*) and subtype C (*black*) HIV-1. The recombination breakpoint is located approximately 2.7 kb along the recombinant sequence. HIV-1 subtype A1 (*light gray*) is used as background in this analysis.

potential parents. Repeat the analysis with the potential parental sequences until all the irrelevant reference sequences are omitted (*see* **Note 22**).

4. Notes

1. The transfection agent PEI provides efficient, consistent transfection of 293T cells. The 25-kDa branched PEI is diluted to 10 mM in water (stock, store in dark at 4 °C) then further diluted to 10 μM in DMEM prior to use (make fresh on day of transfection). Plasmid DNA is diluted in 2.5 mL DMEM and added drop wise, with constant vortex agitation, to 2.5 mL DMEM containing 100 μL of 10 μM PEI. The PEI-DNA complex is allowed to form at room temperature for 20 min before being added to the 4×10^6 293T cells seeded into a 100-mm-diameter tissue culture dish the previous day. After 4 h incubation at 37 °C, the transfection mix is removed and replaced with fresh 293T-CM.
2. The transfection efficiency of 293T cells decreases with age, particularly at over passage 40–50. As production of the dual infected cell lines can take about 20 passages and transfection of the envelope plasmid is required to obtain infectious virus from the cell lines, it is essential to start with very early passage 293T cells (preferably less than passage 10).
3. In order to estimate the correct amount of virus required to obtain a 5–10% infection rate it is important to perform a "dry-run" infection for each virus harvest. Moreover, when generating a cell line, multiple groups of cells should be infected, each with a varied virus dilution, so that at least one group meets the required infection rate of around 5–10%. To test each group for the level of infection, the medium is removed 48 h postinfection and a small area of cells (about 1 cm^2) is harvested with a pipette using the residual medium on the plate (about 500 μL). These cells are then immunostained and analyzed by flow cytometry; the cells with the correct level of infection progress to the next stage of dual infected cell line production.
4. The antibodies used to stain HSA and Thy markers are fluorescently labeled, and as such are photosensitive. It is therefore important to limit their exposure to light during the immunostaining procedure. All manipulations of the antibodies should be performed with the tissue culture cabinet overhead light switched off and the incubation on ice should be performed in the dark (i.e., covered by aluminum foil).
5. After each cell sort only around 1×10^6 cells are recovered. However, there is a degree of cell death associated with cell

sorting and the surviving cells adhere to the tissue culture plates slowly and exhibit a delay in cell division. This results in extended cell expansion periods of about 2 weeks between each cell sorting event in order to produce enough cells to progress to the next step. It is possible to reduce the lag time between sorts to about 8 days by plating the sorted cells in conditioned medium. Conditioned medium is 293T-CM that has previously been used to culture 293T cells; it requires clarification through a 0.45-μm filter prior to use and should only be used for the first two days post-sorting, after which fresh 293T-CM should be sufficient for normal cell growth.

6. To ensure efficient staining has been achieved it is useful to transduce a sample of cells with the functional GFP version of the two viruses. After staining, almost all of these cells should be double positive for the marker gene and GFP.
7. During the production of the cells lines, functional *gfp* can arise through cellular mitotic DNA recombination of the proviruses. This is a rare event, but these cells are removed during sorting of the dual infected cell lines by gating on the GFP^- population.
8. To improve viral titers and to achieve higher levels of target cell transduction, alternative transfection agents that demonstrate greater transfection efficiency with low levels of DNA can be used (e.g., FuGene6). Alternatively, transfection efficiency with PEI can be increased by supplementing the 4 μg of HIV-1 envelope plasmid with 10 μg of filler DNA (e.g., salmon sperm DNA).
9. If the level of target cell infection is too low and insufficient GFP^+ cells are detected for accurate quantification of the recombination rate, then improvements in the infection efficiency can be made by modifying the infection procedure. A spinoculation procedure can be used (centrifugation at 1200 × *g* for 45 min at 15 °C), and 10 μg/mL polybrene can be added during the transduction. If polybrene is used, it must be removed 4 h postinfection due to cytotoxic effects on Hut/CCR5 cells.
10. The calculation of recombination rates uses the MOIs of each virus and not the percentages of infections. MOI calculation is based on the Poisson distribution, thus multiple infections are taken into consideration. However, the accuracy of the MOI calculation is compromised when the infection level is too high and the system is saturated or too low and insufficient numbers of cells are scored. We generally limit our total transduction level to $< 50\%$.
11. For flow cytometry of the three different colors (PE, Thy and GFP) it is important that any cross-talk between the fluorescent channels is accounted for by using compensation. This is most evident between the GFP and PE signals, both

of which demonstrate a small amount of fluorescent bleed-through (about 1% of PE into the GFP channel and 31% of GFP into the PE channel, depending on the filter sets used). Using the FACScalibur machine such compensation must be performed prior to analyzing the samples using control cells only fluorescing in one channel. Alternatively, the PARTEC machine uses software based compensation that can be performed after acquisition of the data.

12. The default background alignment consists of a comparison to near full length consensus sequences of subtypes A1, B, C, D, F1, F2, G, H and CRF01_AE, and one sequence each from subtypes A2, J and K. It is not necessary to omit any reference sequences from the analysis at this point. A "Rerun" function in the output will perform a re-analysis of the data with only the most relevant reference sequences included.
13. Larger window sizes result in a smoother output but will reduce the likelihood of observing small recombinant segments. Smaller window sizes give better resolution but generate a more noisy output. A *z*-test is applied to the calculation of distances with confidence levels at 90%, 95% or 99%. The default window size of 400 nucleotides and significance threshold of 90% is a good starting point for detecting most of the recombination breakpoints.
14. The user can choose to analyse the sequence with nucleotide gaps present in the alignment or to remove (strip) them. Removing the gaps may result in the loss of some informative nucleotides, thus for the initial analysis, it is best to run the program with the gaps present.
15. Multistate character code may be present in the query or reference sequences, select "True" to perform a more accurate comparison.
16. Reference sequences of different subtypes are shown in different colors on the graph and each data point represents the distance value of each window analyzed. The lower colored bar of the two horizontal bars near the top of the graph represents the background sequence with the highest similarity to the query (the best match). Above it is another colored bar which shows whether the best match is significantly better than the second best match using the significance threshold selected in **Section 3.2.1 Step 3**. Following the graphs is an alignment of the query sequence to the background sequence. It is very important to check the quality of the alignment to ensure the program has been accurate.
17. If the default consensus sequences are used it will be apparent that only a few subtypes in the background are closely related to the query. By performing the "Rerun" option, the RIP automatically runs the analysis using only the best match sequences, making the results easier to read.

18. If recombination is not detected, try a smaller window but keep the threshold at 90%.
19. The program requires a minimum of four sequences in the alignment for each analysis; two or more potential parental sequences, one outgroup sequence for comparison and the query sequence. The outgroup sequence can be selected from the HIV-1 subtypes identified by the RIP to have low similarity to the query. To repeat the RIP analysis using SimPlot, the earlier alignment performed by the RIP can be downloaded and used as its input alignment. SimPlot version 3.5.1 will accept most sequence formats, such as MSF, Clustal, Fasta/Pearson, PHYLIP and NEXUS.
20. Each group should contain sequences of the same subtype. The SimPlot default setting uses the first character of the sequence name to assign groups. If the program is not assigning groups correctly, change the setting using the "Use first character to identify groups" button on the right to correct it.
21. The default window size of 200 nucleotides and step size of 20 is a good starting point for bootscanning. The default distance model and tree model for the phylogenetic analysis are Kimura-2-parameter and Neighbor-Joining, respectively, which are fast and quite accurate. There are other models to choose from using "Preferences" within the "Options" menu.
22. When the parental sequences are determined, repeat the analysis but reduce the window size and step size. Again, larger window sizes will overlook recombination breakpoints that are close together whereas smaller window sizes generate noisy results. Smaller step sizes result in a more detailed plot by adding more data points, which helps to pinpoint the exact recombination breakpoints.

Acknowledgments

We thank Dr. Jianbo Chen for reading of this chapter; Drs. Vinay K. Pathak, Jianbo Chen, and Olga Nikolaitchik for intellectual input and discussions.

References

1. Rhodes, T. D., Nikolaitchik, O., Chen, J., et al. (2005) Genetic recombination of human immunodeficiency virus type 1 in one round of viral replication: effects of genetic distance, target cells, accessory genes, and lack of high negative interference in crossover events. *J Virol* 79, 1666–1677.
2. Chen, J., Rhodes, T. D., and Hu, W. S. (2005) Comparison of the genetic recombination rates of human immunodeficiency virus type 1 in macrophages and T cells. *J. Virol* 79, 9337–9340.
3. Chin, M. P., Rhodes, T. D., Chen, J., et al. (2005) Identification of a major

restriction in HIV-1 intersubtype recombination. *Proc Natl Acad Sci USA* 102, 9002–9007.

4. Yee, J. K., Friedmann, T., Burns, J. C. (1994) Generation of high-titer pseudotyped retroviral vectors with very broad host range. *Meth Cell Biol* 43 (Pt A), 99–112.
5. Dang, Q., Chen, J., Unutmaz, D., et al. (2004) Nonrandom HIV-1 infection and double infection via direct and cell-mediated pathways. *Proc Natl Acad Sci USA* 101, 632–637.
6. Siepel, A. C., Halpern, A. L., Macken, C., et al. (1995) A computer program designed to screen rapidly for HIV type 1 intersubtype recombinant sequences. *AIDS Res Hum Retroviruses* 11, 1413–1416.
7. Lole, K. S., Bollinger, R. C., Paranjape, R. S., et al. (1999) Full-length human immunodeficiency virus type 1 genomes from subtype C-infected seroconverters in India, with evidence of intersubtype recombination. *J Virol* 73, 152–160.
8. Salminen, M. O., Carr, J. K., Burke, D. S., et al. (1995) Identification of breakpoints in intergenotypic recombinants of HIV type 1 by bootscanning. *AIDS Res Hum Retroviruses* 11, 1423–1425.

Chapter 8

Methods of Preparation and Analysis of Intracellular Reverse Transcription Complexes

Ariberto Fassati

Abstract

Shortly after penetration into the cell, HIV-1 must reverse transcribe its genome into a double-stranded DNA molecule and must gain access to the nucleus of nondividing cells for productive infection. There is limited knowledge of these early events in HIV-1 life cycle. We have developed methods, which allow the study of postpenetration events and in particular allow the isolation and purification of intracellular complexes mediating reverse transcription. Such complexes can be examined for their structure, function, protein composition and ability to access the nucleus.

Key words: HIV-1, reverse transcription complex, linear gradient, purification, reverse transcription assay.

1. Introduction

All retroviruses copy the RNA genome to form a double-stranded DNA molecule, which is subsequently integrated into the host chromosomal DNA. The process is mainly carried out in the cytoplasm soon after viral penetration into the cell and is catalyzed by reverse transcriptase (RT) *(1)*. The various steps of reverse transcription are known in some detail. It starts at the primer-binding site, where a partially unfolded tRNA is annealed to the viral genomic RNA. Then RT synthesizes approximately 150 bases of the first minus (−) DNA strand (the so called minus strong stop DNA) complementary to 5′ U5 and R elements while RNase H activity of RT degrades the RNA template. At this point two complementary R sequences are generated, which form a bridge by base pairing. The "first jump" of RT takes place and subsequently

Vinayaka R. Prasad, Ganjam V. Kalpana (eds.), *HIV Protocols: Second Edition, vol. 485*

DOI 10.1007/978-1-59745-170-3_8 Springerprotocols.com

the synthesis of the (−) DNA strand is completed. The synthesis of the (+) DNA strand starts at the polypurine tract, a short stretch of RNA resistant to RNase H activity, which acts as the primer. Synthesis of the (+) strand DNA continues until the end of the (−) DNA strand, where a second "jump" takes place, allowing RT to complete reverse transcription *(1)*.

The process is carried out in a nucleoprotein complex called the reverse transcription complex or "RTC". There is limited knowledge of the structure and composition of this complex. Electron microscopy and sedimentation velocity analyses suggested that HIV-1 RTC might be rather large (200–700 nm across) and composed of packed nucleic acids filaments with a ~ 6 nm diameter *(2, 3)*. The RTC contains viral and cellular proteins and its density is 1.32–1.34 g/mL (*2–5*).

Intracellular RTCs can be purified and studied *(4, 5)*. For example, purified RTCs obtained at different times after infection can help examine early events like uncoating, reverse transcription *(4–6)* and the activity of some restriction factors *(7)*. They can also be used to monitor their protein composition, and can be labelled fluorescently and used in the nuclear import assay to study their nuclear transport *(8, 9)*.

Here, methods to extract, purify, analyse and label HIV-1 RTCs will be described.

2. Materials

2.1. Cell Culture

1. Dulbecco's Modified Eagle's Medium (Invitrogen) with GlutaMAXTM-I and 110 mg/L sodium pyruvate supplemented with 10% heat-inactivated fetal calf serum.
2. Versene (PBSA + 0.5 mM EDTA).
3. Trypsin (Gibco/BRL, Life Technologies): 0.25% Trypsin, 0.05 mM EDTA.
4. Opti-Mem medium (Invitrogen).
5. Fugene-6 transfection reagent (Roche).
6. 293T and HeLa cells.

2.2. Plasmids and Virus Preparation

1. HIV-1 packaging construct pCMVΔ8. 2, HIV-1 vector plasmid pCSGW and VSV-G encoding plasmid pMD.G *(10, 11)*.
2. 50 mM Na Phosphate buffer pH 7.4 (filter sterilized). Make it by mixing 38.7 mL 1 M Na_2HPO_4 with 11.3 mL 1 M NaH_2PO_4 and bring to 1 L final volume with dH_2O.
3. 1 M HEPES pH 7.2 (filter sterilized) in dH_2O.
4. TE: 10 mM Tris–HCl pH 8 and 1 mM EDTA.
5. DNase I (stock 70,000 U/mL in PBS), filter sterilised. Store in aliquots at −20 °C.
6. 1 M $MgCl_2$ (filter sterlilized).

7. 25% (w/w) and 45% (w/w) sucrose (Sigma) in 50 mM Na Phosphate buffer pH 7.4 (filter sterilized) (*see* **Note 1**). Store at 4 °C.

2.3. Purification and Labeling of RTCs

1. Hypotonic buffer: 10 mM HEPES [pH 7.9], 1.5 mM $MgCl_2$, 10 mM KCl. Store at 4 °C.
2. Isotonic buffer: 10 mM Tris–HCl [pH 7.4], 160 mM KCl, 5 mM $MgCl_2$.
3. 1 M Dithiothreitol (DTT) (Fermentas) diluted in H_2O. Store at −20 °C.
4. 20 mg/mL Leupeptin and Aprotinin (Sigma) diluted in dH_2O. Store in aliquots at −20 °C.
5. Polybrene (Hexadimethrine bromide) (Sigma).
6. 5% and 20% (w/w) sucrose in 50 mM Na Phosphate buffer pH 7.4. Filter-sterilize and store at 4 °C.
7. 70% (w/w) sucrose in D_2O (deuterium oxide or “heavy water”). Make fresh.
8. Dounce homogenizer.
9. 0.4% Trypan blue solution (GIBCO).
10. YOYO-1 (small molecule that becomes highly fluorescent upon binding to nucleic acids) (1 mM in DMSO) (Molecular Probes). Store at −20 °C in the dark.
11. Import buffer: 20 mM HEPES pH 7.3, 110 mM KAc, 2 mM MgAc, 0.5 mM EGTA. Filter sterilize and store at 4 °C.
12. PBSA: PBS without $MgCl_2$ and $CaCl_2$.
13. 1 mL Float-A-lyzer dialysis membranes with 100,000 MW cut off (Spectrum Labs).
14. Glass slides, glass coverslips (Fisher) and Vectashield mounting medium (Vector Labs).

2.4. PCR

1. 10X PCR buffer (Promega): 500 mM KCl, 100 mM Tris–HCl pH 9.0, 1% Triton X-100.
2. dNTPs mix (25X stock at 2.5 mM each in dH_2O). Store in aliquots at −20 °C.
3. 25 mM $MgCl_2$.
4. Primers: Strong stop forward: 5′ GGCTAACTAGGGAACC-CACTG- 3′. Strong stop reverse complementary: 5′-CTGCTAGAGATTTTCCACACTGAC-3′. (+) strand forward: 5′-AGGGCTAATTCACTCCCAACGAAG-3′(+) strand reverse complementary: 5′- GCCGTGCGCG CTTC-AGCAAGC -3′ (*see* **Note 2**).

2.5. Endogenous Reverse Transcription Assay

1. RT Buffer: 20 mM Tris–HCl pH 8.1, 15 mM NaCl, 6 mM $MgCl_2$ and 1 mM DTT.
2. dNTP stock solution of 20 mM each. Aliquot and store at −20 °C.

2.6. Western Blot

1. Separating buffer for polyacrylamide gels (4X): 1.5 M Tris–HCl, pH 8.8, 0.4% sodium dodecyl sulfate (SDS).
2. Stacking buffer for polyacrylamide gels (4X): 0.5 M Tris–HCl pH 6.8, 0.4% SDS.
3. Western blot loading buffer (2X): 2% SDS, 125 mM Tris pH 6.8, 20% glycerol, 0.1% bromophenol blue, 2 mM EDTA and 2 mg/mL PMSF in distilled water.
4. Western blot running buffer (10X): 30 g Tris, 144 g glycine +10 g SDS in 1 L distilled water.
5. Western blot transfer buffer (10X): 30 g Tris, 144 glycine +5% MeOH in 1 L distilled water.
6. Bovine serum albumin (BSA) (Sigma) lyophilized.
7. 100% w/v Trichloroacetic acid (Sigma).
8. 10% Ammonium persulfate in H_2O. Freeze at −20 °C in single use aliquots (200 μL).
9. 30% Acrylamide/Bis solution (37.5:1) (Bio-Rad). This is a neurotoxin when unpolymerized, avoid any direct skin contact.
10. N, N, N′, N′-Tetramethylethylenediamine (TEMED) (Sigma).
11. Water saturated butanol. Mix equal volumes of water and buthanol in a glass bottle and let separate. Use upper phase.
12. Polyvinylidene Difluoride (PVDF) membrane (Hybond-P, Amersham).
13. Tris-buffered saline with Tween (TBS-T). 10X stock: 1.37 M NaCl, 27 mM KCl, 250 mM Tris–HCl pH 7.4, 1% Tween-20.
14. Blocking solution: use 5 g nonfat dry milk diluted in 50 mL TBS-T. Make fresh and mix well by vortexing to avoid particulate material.
15. Primary antibodies: anti-p24 CA antibodies AG3.0 and 24-2 (NIH AIDS Repository catalogue numbers 4121 and 6457).
16. ECL Plus detection reagent and ECL films (Amersham).
17. Hoefer dual chamber casting apparatus and wet transfer apparatus

3. Methods

3.1. Production of Virus

1. High-titre virus stocks are necessary to produce RTCs. Moreover, virus stocks must be treated with DNase I to digest any contaminating plasmid DNA carried over from transfection and purified through a double sucrose cushion. Seed 293T cells in 4 × 10 cm dishes and grow them until they are confluent. Change media every day. Once 293T cells are confluent, split them 1:4 into 12 × 10 cm plates making sure they are evenly spaced in the plate. The next day, proceed

with transfection. Caution: recombinant virus is infectious, always wear protective clothes and work in a level II biosafety cabinet.

2. For each plate prepare a mix containing 1.5 μg pCSGW plasmid, 1 μg pCMVΔ8.2 plasmid and 1 μg pMD.G plasmid in a final volume of 15 μl TE pH 8. Use sterile tubes.
3. For each plate, prepare a second mix by adding 18 μL Fugene 6 to 200 μL Opti-Mem without touching the side of the tube. Mix by flicking and add the DNA solution made in **step 2**. Incubate at room temperature for 15 min.
4. Change media on cells to 8 mL fresh media, add the transfection mix drop-wise and swirl plates gently, then return cells to incubator.
5. Change the media of the transfected cells after 24 h and collect viral-containing supernatant 48, 72, and 96 h after transfection. Forty-eight hours after transfection, check cells for GFP expression using an inverted fluorescence microscope. More than 70% cells should be GFP positive. If not, viral titers will probably be too low for RTC production.
6. Adjust pH of viral-containing supernatants by adding HEPES pH 7.4 to a final concentration of 10 mM and filter them through a 0.45-μm filter. Supernatant can now be stored frozen at −80 °C for several months. Save a couple of small (0.5 mL) aliquots for virus titration.
7. Titrate virus stocks. Plate 1.5×10^5/well HeLa cells in six-well plates. The next day change media and infect cells using 0, 0.1 μl, 0.5 μl, 2.5 μl, 12.5 μl per well of virus stock. Forty-eight hours after infection analyze the percentage of GFP+ cells by flow cytometry. To calculate viral titers, multiply the percentage of GFP+ cells by the original number of cells plated (1.5×10^5) and then by the dilution factor (for example if 10% GFP+ cells are obtained with 1 μL viral stock, then the viral titer is: $10\% \times 1.5 \times 10^5 = 1.5 \times 10^4$, which must be multiplied by 1,000 (dilution factor) to obtain the value of 1.5×10^7 cfu/mL).
8. If viral titers are $= 5 \times 10^6$ cfu/mL, proceed with purification step. Adjust the samples to 5 mM $MgCl_2$ and 70 U/mL DNase I (final concentration). Incubate at 37 °C for 40 min.
9. Work on ice. Load 20 mL DNase I-treated viral stock in an ethanol sterilized and dried 36 mL polypropylene ultracentrifuge tube. Gently layer underneath 5 mL of 25% sucrose solution using a thin pipette. Then gently layer 5 mL 45% sucrose solution underneath the first 25% sucrose layer. At this point three layers should be visible: the viral supernatant, then the 25% sucrose layer, then the 45% sucrose layer. Without disturbing the layers, place tubes into ultracentrifuge buckets, balance and spin at 100,000 × g (23,000 rpm in Sorvall Surespin rotor) for 2 h at 4 °C.

10. Collect the interface between the 25% and 45% sucrose layers by piercing a hole at the bottom of the tube with a heated 21G needle (Caution: recombinant virus is infectious, wear protective clothes and perform this operation in a level II safety cabinet; discard the needle in an appropriate sharp container immediately after use and disinfect all surfaces). Discard the first 3.5 mL and collect the subsequent 3 mL, which will contain concentrated virus.
11. Store in aliquots (1 mL) at −80 °C. Purified virus will be stable for up to 2 years.
12. Titrate concentrated and purified virus as described in **step 7**.

3.2. Extraction of RTCs from Acutely Infected Cells

RTC are extracted from acutely infected cells by hypotonic lysis and Dounce homogenization. The minimum amount of virus required for RTC purification is 6 mL of purified stock at 3×10^7 cfu/mL.

1. Plate $2 \times 175\,cm^2$ flasks with 4×10^6 HeLa cells or 1.5×10^7 293T cells (cells must be ∼ 60% confluent the next day).
2. The next day, change to 30 mL/flask fresh media containing 8 μg/mL polybrene and add 3 mL/flask of concentrated and purified viral stock. Incubate at 4 °C for 2 h (*see* **Note 3**).
3. Transfer infected cells to tissue culture incubator at 37 °C and incubate for 4 h (*see* **Note 4**).
4. Wash cells twice gently with PBS + 0. 5 mM EDTA, and trypsinise using 1.5 mL trypsin solution and 1.5 mL PBS + EDTA. Add 30 mL/flask fresh media and pellet by centrifugation at 500 × *g* for 8 min.
5. Wash pellet in 30 mL/flask PBS + EDTA. Decant supernatant and place pellet on ice. Then resuspend pellet in five pellet volumes of hypotonic buffer containing 20 μg/mL aprotinin, 20 μg/mL leupeptin, 1 mM DTT (buffer must be ice cold). Centrifuge cells in minicentrifuge at 4 °C for 5 min at 500–700 × *g*.
6. Resuspend pellet in four pellet volumes of hypotonic buffer +20 μg/mL aprotinin and leupeptin and 1 mM DTT and incubate on ice for 10 min. Place the Dounce homogeniser on ice while cells are incubating. Put samples from both flasks in the homogeniser on ice and break cells using the pestle. Check under the microscope to see if the cells are broken by premixing 5 μL sample with 5 μL trypan blue before viewing under the microscope (*see* **Note 5**).
7. Centrifuge sample at 3, 300 × *g* for 15 min at 4 °C; collect supernatant and centrifuge at 7, 500 × *g* for 20 min at 4 °C. The supernatant contains crude cytoplasmic RTCs. Resuspend the pellet from the first centrifugation in 400 μL isotonic buffer +20 μg/mL aprotinin and leupeptin and 1 mM DTT. Place in homogenizer and break nuclei using the pestle. Check that nuclei are broken by trypan blue staining

as before. Centrifuge samples at $7,500 \times g$ for 20 min and collect supernatant, which contains crude "nuclear" RTCs. Crude cytoplasmic and nuclear extracts can be stored at −80 °C for several weeks.

3.3. RTC Purification

At this point, RTCs can be purified by one or two rounds of centrifugation on sucrose gradients. RTCs are separated from other intracellular components based on their size/shape in velocity sedimentation gradients and based on their density in equilibrium gradients. If RTCs are to be used for studies on reverse transcription, only one round of purification by equilibrium density sedimentation [but not by velocity sedimentation] is necessary. If RTCs are to be used for Western blot analyses, one round of purification either by equilibrium density or by velocity sedimentation is necessary. Finally, if RTC are to be labeled and used in cell biology assays, like the nuclear import assay, two rounds of purification [by equilibrium density sedimentation and velocity sedimentation] are necessary.

1. Add 20 μg/mL final Aprotinin and Leupeptin and 1 mM DTT to both 5% and 20% sucrose solutions; keep solutions at room temperature but work quickly. Prepare a 5–20% linear sucrose gradient in 5 mL polypropylene ultracentrifuge tube. We use the Biocomp Gradient Master (Fredericton, NB, Canada) to make gradients. Incubate the gradient on ice for 5 min. Gently add to the top of the gradient 500 μL of cell extracts prepared as described in the previous section (*see* **Note 6**).
2. Balance tubes and centrifuge for 1 h at $62,000 \times g$ (23,000 rpm in a Sorvall AH-650 rotor) at 4 °C.
3. Collect twelve fractions of 400 μL each by piercing the bottom of the tube with a heated 21G needle (*see* **Note 7**). Mix fractions gently by flicking. Do not vortex.
4. Set up a PCR reaction in 50 μL final volume containing 1× PCR buffer ($MgCl_2$-free), 200 μM each dNTPs, 0.85 mM $MgCl_2$, 30 pmol each primer, 5U Taq polymerase. Add 1.5 μL each fraction (*see* **Note 8**) and always include one negative (no DNA) and one positive (10 pg retroviral vector DNA) control. Run PCR as follows: one cycle at 94 °C for 3 min followed by 28 cycles of 94 °C for 1 min, 55 °C for 1 min and 68 °C for 1 min and one final extension at 68 °C for 10 min. Resolve PCR products on a 2% NuSieve low melting + 1% agarose gel containing 0.2 μg/mL ethidium bromide (caution: ethidium bromide is a powerful mutagen always wear protective clothes and gloves). Visualize PCR bands by UV light and take a picture.
5. Locate the peak of viral DNA (either strong stop or + strand DNA) in the gradient. Discard fractions containing little or no viral DNA.

6. Prepare a 20–70% linear sucrose gradient in a 5-mL polypropylene ultracentrifuge tube. Keep the 20% sucrose solution at room temperature and add 2 μg/mL aprotinin and leupeptin and 1 mM DTT. The 70% gradient must be made in D_2O prior to use by heating up the solution in a microwave oven for a few seconds, mixing and re-heating until sucrose is completely dissolved. Do not overheat samples. Cool to about 50 °C and then make the gradient (we use the Biocomp Gradient Master) (*see* **Note 9**). Keep gradients on ice for 5 min and gently add to the top of this gradient, the fraction from the velocity sedimentation gradient containing the peak point of the viral DNA curve [approximately 400 μL].
7. Balance the tubes and centrifuge at 145,000 × *g* (35,000 rpm in a Sorval AH-650 rotor) for 16–18 h at 4 °C. Collect 12 fractions as described in **step 3**.
8. Perform PCR on each fraction as described in **step 4**, but use 5-μL aliquot instead of 1. 5-μl aliquot. Locate the peak of the viral DNA (**Fig. 8.1**) and freeze that fraction in 100-μL aliquots at −80 °C. Purified RTCs will be stable for several months in sucrose.
9. To label RTCs, use 200 μL of the fraction from density gradients and dilute 1:1 in import buffer containing YOYO-1 (1:2,500 final) so that the final YOYO-1 concentration in the mix will be 1:5,000. Incubate at room temperature for 1 h in the dark. Wash 1 mL float-A-lyzer 100,000 MWCO membranes in dH_2O and then in import buffer. Place YOYO-1-labelled samples in the float-A-lyser membrane and dialyse against 500 mL import buffer at 4 °C in the dark for 12–16 h with gentle stirring in a sterile glass bottle. Collect dialysed samples and freeze in 50-μL aliquots at −80 °C. Labeled RTCs will be stable for up to 6 months.
10. Check labeling by putting a drop of labeled RTCs on a glass slide and incubating at room temperature in the dark in a humidified atmosphere for 30 min. Wash once with import buffer for 5 min, add 5 μL Vectashield and mount with a glass coverslip. Analyse by confocal or epifluorescence microscopy. Small fluorescent dots should be visible using the 60x or 100x objective.

3.4. Endogenous Reverse Transcription Assay on RTCs

Reverse transcription can be examined using purified RTCs. The doubly purified RTCs give the weakest activity whereas RTCs purified by a single equilibrium density centrifugation step work best in this assay. Additional primers to those described here can be used to closely monitor the various steps of reverse transcription. This assay can be made quantitative if samples are analysed by Taqman PCR.

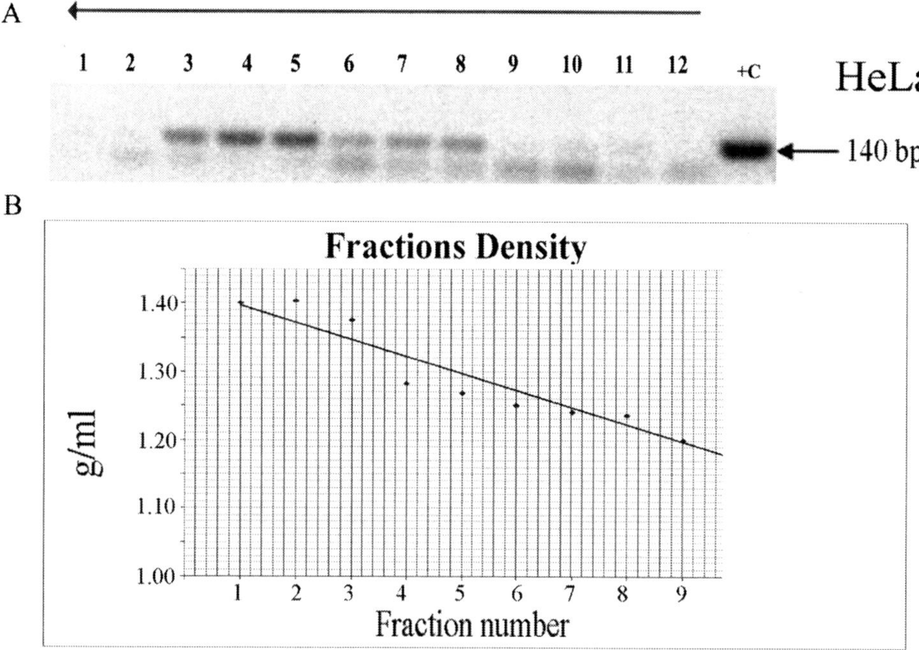

Fig. 8.1. Detection of RTCs by PCR after equilibrium density fractionation. (**A**) After equilibrium density centrifugation in a 20–70% linear sucrose gradient, 12 fractions were collected by piercing the bottom of the tube with a needle. Five microliters of each fraction were analyzed by 28 cycles PCR with primers specific for the viral strong stop DNA (140-bp size). PCR products were visualized by gel electrophoresis (2% NuSieve + 1% agarose) and ethidium bromide. The peak of the strong stop DNA is visible in fraction 5. The arrow indicates the direction of the gradient from top (low density) to bottom (high density). (**B**) The first nine fractions were weighed using a precision balance and their density calculated. The regression plot of density versus fraction number is shown.

1. Mix 60 μL RT buffer to 15 μL of the density fraction containing the peak of viral DNA and mix well with a pipette tip. Add dNTPs to the mix to a final concentration of 2 mM each and DTT to a final concentration of 1 mM. Always add controls without dNTPs. Incubate at 37 °C for 6 h. Samples can be frozen at −20 °C at this point.
2. Use 3 μL of the mix for PCR as described in **Step 4** of **Section 3.3** and run 25 cycles (*see* **Note 10**). Visualise bands by agarose gel electrophoresis and ethidium bromide staining.

3.5. Detection of Proteins in RTC Fractions by Western Blot

Fractions from both sedimentation velocity or equilibrium density centrifugation can be analyzed by Western blot to detect co-sedimentation of viral and cellular proteins with viral DNA. Due to the small amount of RTCs that can be obtained at any one time, samples must first be concentrated to be analyzed by immunoblot. Below is a method to detect p24 CA, which can be easily adapted to other proteins.

1. Dilute 300 μL of RTC fraction in 1.2 mL ice-cold 50 mM sodium phosphate buffer pH 7.4 and add 2 μg BSA and 10% (vol/vol) trichloroacetic acid. Mix and incubate at −20 °C

for 16–18 h. Centrifuge samples for 30 min at 4 °C in a benchtop minicentrifuge at maximum speed. Discard supernatant and centrifuge samples for 1 min at maximum speed. Discard any supernatant left by gently aspirating with a pipette tip. Resuspend samples in 20 μL in 2x Western blot loading buffer. Sample will turn yellow due to low pH. To adjust pH, add 1 μL 1.5 M Tris–HCl pH 8.8 and mix. Boil samples for 5 min and centrifuge for 1 min.

2. To assemble the separating gel, we use the Hoefer two chamber apparatus although alternative systems can be easily adapted. Wash the glass plate thoroughly with a detergent first, then distilled water, then ethanol and dry with a clean tissue. For each gel, prepare 10 mL solution containing 1x separating buffer, 12.5% (vol/vol) Acrylamide/Bisacrylamide solution and 100 μL of 10% ammonium persulfate stock in dH_2O. When ready to pour the gel, add 10 μL TEMED, mix gently and cast the gel. Leave approximately 1.5 cm from the top of the gel to the bottom of the comb for the stacking gel. Gel should polymerise in 30 min. Overlay the gel with 1 ml H_2O-saturated butanol.
3. To make the stacking gel, prepare 5 mL solution containing 1x stacking gel buffer, 4% Acrylamide/Bisacrylamide solution, 50 μL 10% ammonium persulfate in dH_2O. Wash out butanol from chamber with dH_2O (caution: buthanol must not touch the plastic surface of the Hoefer chamber because it will stain it) and pour the stacking gel to the top of the chamber after adding 5 μL TEMED. Insert the comb: the bottom of the wells should be 1.5 cm above the separating gel. Let gel polymerise for 20 min, then gently remove comb and rinse with water.
4. Assemble the running apparatus and place 1x running buffer in the chamber. Load 10 μL boiled samples of **step 1** in each well, including appropriate protein size standards, and run gel at 100 V constant until blue front at the bottom has come out of the gel or until size standards have sufficiently separated.
5. Prepare transfer equipment (we use the Hoefer wet transfer tank) and a tray set up large enough to lay out the cassette. Cut 4x sheets of 3 mm paper and one sheet of PVDF membrane to gel size. Prepare 1x transfer buffer and add 5% methanol. Briefly wet PVDF membrane in methanol and then submerge it in transfer buffer and equilibrate for 5 min. Wash the separating gel in transfer buffer for 5 min after cutting out the stacking gel with a razor blade.
6. Assemble transfer sandwich (from bottom to top): black electrode (anode), sponge, two sheets of 3 mm paper, gel, PVDF membrane, two sheets of 3 mm paper, sponge, white electrode. Ensure no bubbles are trapped between layers by

gently rolling a pipette on the surface of each layer. Place sandwich in tank.

7. Fill tank with transfer buffer and run in cold room for 16–18 h at 100 mA constant with gentle stirring. Then open sandwich and extract PVDF membrane. Mark position of size markers with a pencil on the PVDF membrane on one side only. This will also help in recognizing the side of the membrane facing the gel.
8. Wash membrane in TBS-T for 5 min; then incubate in 50 mL blocking solution for 1 h at room temperature with rocking.
9. Wash briefly in TBS-T and add primary anti-p24 CA antibodies AG3.0 (1:500) and 24-2 (1:5,000) in 10 mL TBS-T. Incubate for 1 h at room temperature with rocking. Wash four times with 100 mL TBS-T 15 min each wash. Add anti-mouse horseradish-conjugated secondary antibodies diluted 1:6,000 in TBS-T for 1 h at room temperature with rocking. Wash four times with TBS-T as before.
10. Incubate membrane with ECL Plus developing reagent for 5 min with shaking (expose the side of the membrane that was facing the gel), dry out excess liquid by letting it drop from the membrane in vertical position; wrap PVDF membrane in saran wrap and expose to ECL film. Never touch PVDF membrane with bare fingers.

4. Notes

1. The density of sucrose solutions should be checked by weighing 100 μL using a precision balance. Zero the balance with an empty pipette tip, then use the same tip to aspirate 100 μL and weigh again. Multiply the value obtained by 10. This will be the solution density in g/mL. The 25% sucrose solution should have a density of 1.12 g/mL and the 45% sucrose solution a density of 1.22 g/mL. Adjust the solutions if density is not right. Retroviruses have a typical density of 1.16 g/mL, thus they will sediment at the interface between the two sucrose layers.
2. Alternative primers can be used, which amplify different regions of the viral genome. Different primers may be used to monitor the progress of reverse transcription at different time points.
3. Keep the flasks at 4 °C or lower in a horizontal position and making sure they are not tilted. Often fridges are not good enough for this step and we prefer to lay flasks on the top of a metal bench in the cold room. This step is important to ensure a synchronous infection.

4. The time of incubation at 37 °C may vary according to the experimental questions. Cells can be incubated at 37 °C from a minimum of 1 h to a maximum of 18 h. Generally, we find that RTC production is maximal with incubations time of 4–6 h.
5. Breaking cells with the pestle is a critical step and should be done stepwise. Excessive lysis will result in release of genomic DNA and nuclear component in the extracts whereas too little cell lysis will result in poor RTC recovery. Start with ten strokes. Do not stand when using the pestle; your body weight will be too much. Check cells with a light microscope after mixing them 1:1 with trypan blue. Properly lysed samples should show 70–80% cells with disrupted cytoplasm and blue nuclei. If cytoplasm is not disrupted but nuclei are blue, it is best to lyse cells more with another five strokes, then check again and so on. Be careful to avoid breaking the nuclei.
6. The quality of the gradient is of paramount importance. If gradients are not linear or are badly mixed, RTCs will diffuse and will be lost. It is best to run an extra gradient in parallel and monitor its density by collecting 12 fractions and weighing them as described in Note 1. Once gradients are ready and of appropriate quality, approximately 300 μL can be gently removed from the top to make space for the cell extracts. Extracts should be layered at the top of the tube trying not to disturb the gradient and samples should be centrifuged shortly thereafter. Gradients cannot be stored for more than 10 min on ice.
7. Needles can be heated with a Bunsen flame and used while still incandescent. Use caution and wear appropriate protective clothes. The liquid will tend to fall dropwise from the gradient tube at a fairly fast rate. We prepare 13 sterile eppendorf tubes with a mark at 400 μL opened on a holder before piercing the hole in the gradient tube. We place a fingertip on the top of the tube to prevent air entering the tube, which will slow down the rate of collection. After collection, close tubes and immediately put on ice.
8. Do not use more than 1. 5 μL of velocity sedimentation fractions for PCR analysis. The Na Phosphate buffer may inhibit the reaction.
9. The 70% sucrose solution is very viscous and gradients are better mixed when this solution is still warm (50 °C). The Gradient Master does not have a program for a linear 20–70% sucrose gradient and we run the program for the 20–60% gradient twice or until there is only a small layer of 70% sucrose at the bottom of the tube. Preliminary tests should be performed to check the quality of the gradient. The quality of the gradient is critical. If the gradient is not linear or is badly mixed, RTCs will be diluted and lost. The density at the bottom

of the gradient should be 1.4 g/mL (at 4 °C) and should decrease gradually towards the top to 1.10 g/mL. Because of the viscosity of the bottom region of the gradient, following centrifugation fraction collection will take longer than with the sedimentation gradient.

10. It is important to run as few cycles of PCR as possible to maintain the reaction within its linear range. Generally 25–28 cycles should be enough.

Acknowledegments

AF is a Wellcome Trust University Fellow. I thank Steve Goff, Masha Orlova, Eran Bacharach and Guanxia Gao for their help and advise for the development of these methods.

References

1. Telesnitsky, A., Goff, S. P. (1997) Reverse transcriptase and the generation of retroviral DNA in *Retroviruses* (Coffin J.M, Hughes S. H., Varmus H.E. eds.). Cold Spring Harbor Laboratory Press, Cold Spring Harbor, NY.
2. McDonald, D., Vodicka, M. A., Lucero, G., et al. (2002) Visualization of the intracellular behavior of HIV in living cells. *J Cell Biol* 159, 441–452.
3. Nermut, M. V., Fassati, A. (2003) Structural analyses of purified human immunodeficiency virus type 1 intracellular reverse transcription complexes. *J Virol* 77, 8196–8206.
4. Fassati, A., Goff, S. P. (1999) Characterization of intracellular reverse transcription complexes of Moloney murine leukemia virus. *J Virol* 73, 8919–8925.
5. Fassati, A., Goff, S. P. (2001) Characterization of intracellular reverse transcription complexes of human immunodeficiency type-1. *J Virol* 75, 3626–3635.
6. Carr, J. M., Davis, A. J., Coolen, C., et al. (2006) Vif-deficient HIV reverse transcription complexes (RTCs) are subject to structural changes and mutation of RTC-associated reverse transcription products. *Virology* 351, 80–91.
7. Stremlau, M., Perron, M., Lee, M., et al. (2006) Specific recognition and accelerated uncoating of retroviral capsids by the TRIM5alpha restriction factor. *Proc Natl Acad Sci USA* 103, 5514–5519.
8. Fassati, A., Gorlich, D., Harrison, I., et al. (2003) Nuclear import of HIV-1 intracellular reverse transcription complexes is mediated by importin 7. *EMBO J* 22, 3675–3685.
9. Zaitseva, L., Mayers, R., Fassati, A. (2006) tRNAs promote nuclear import of HIV-1 intracellular reverse transcription complexes. *PloS Biol* 4, e332.
10. Naldini, L., Blomer, U., Gallay, P., et al. (1996) In vivo gene delivery and stable transduction of nondividing cells by a lentiviral vector. *Science* 272, 263–267.
11. Demiason, C., Parsley, K., Brouns, G., et al. (2002) High-level transduction and gene expression in hematopoietic repopulating cells using a HIV-1 based lentiviral vector containing an internal spleen focus forming virus promoter. *Hum Gene Ther* 13, 803–813.

Chapter 9

Analysis of Viral and Cellular Proteins in HIV-1 Reverse Transcription Complexes by Co-immunoprecipitation

Sergey N. Iordanskiy and Michael I. Bukrinsky

Abstract

Molecular details and temporal organization of the early (preintegration) phase of HIV life cycle remain among the least investigated and most controversial problems in the biology of HIV. To accomplish reverse transcription and intracellular transport of the viral genetic material, HIV forms multi-molecular complexes termed reverse transcription complexes (RTCs). Analysis of the kinetics of reverse transcription and nuclear import of RTCs, as well as assessment of the changes in their protein content in the course of reverse transcription and nuclear translocation is a necessary step in understanding the mechanisms of cytoplasmic maturation and nuclear import of HIV-1 RTCs. Here, we review methods that allow quantitative assessment of the dynamics of the maturation of HIV-1 RTCs and transformations of RTC protein composition associated with nuclear import of the complexes.

Key words: HIV-1, reverse transcription complex, cell fractionation, real-time PCR, RTC protein composition.

1. Introduction

The early (preintegration) phase of the HIV life cycle includes virus entry into a target cell, conversion of its genome from RNA to DNA (reverse transcription), delivery of viral DNA to the site of integration and integration itself. After fusion with the cellular plasma membrane and entry into the target cell, virus particles transform into multi-molecular complexes termed reverse transcription complexes (RTCs). They provide protection for the viral genome and reverse transcription machinery, and perform transportation of the viral genetic material toward the site of integration (reviewed in *(1, 2)*). Technical difficulties associated with observation, isolation, and analysis of RTCs at different steps of

Vinayaka R. Prasad, Ganjam V. Kalpana (eds.), *HIV Protocols: Second Edition, vol. 485*

DOI 10.1007/978-1-59745-170-3_9 Springerprotocols.com

their trafficking in the cytoplasm and especially in the nucleus explain our poor understanding of the molecular events in the early phase of the retroviral life cycle.

HIV-1 RTCs isolated from the cytoplasm of infected cells contain viral proteins, such as reverse transcriptase (RT), integrase (IN), matrix protein (MA) and Vpr *(3–5)*, as well as host cell proteins, such as nonhomologous DNA end joining proteins Ku70 and Ku80 *(6)*, lens epithelium-derived growth factor (LEDGF/p75) *(7)*, nonhistone chromosomal protein HMGA *(8)*, integrase interactor 1 (Ini1) and PML protein *(4, 5, 9)*. The capsid protein (CA) was detected in virus-specific complexes early after infection, but it was absent in cytoplasmic RTCs (cRTCs) analyzed at later time points and in nuclear RTCs (nRTCs) *(3, 10, 11)*. Nuclear import of RTCs is associated with loss of RT and CA proteins *(10, 11)*. Composition of the cytoplasmic and nuclear HIV-1 RTCs, interaction of RTC proteins with proteins of infected cell, as well as quantitative analysis of dynamics of the RTC maturation and nuclear import may be accomplished using co-immunoprecipitation of RTCs with antibodies against viral and host cell proteins and subsequent real-time PCR analysis with primers specific for HIV-1 DNA *(6, 11–17)*.

2. Materials

2.1. Plasticware and Hardware

1. Tissue culture flasks, 250 mL, 75 cm^2 with vented plug seal caps (Becton Dickinson, Franklin Lakes, NJ, Cat. No. 353110).
2. 0.45-μm-pore-size syringe filter (Fisher Scientific, Cat. No. 09-719B).
3. Ultra-clear centrifuge tubes (14 × 89 mm) (Beckman, Palo Alto, CA, Reorder. No. 344059).
4. Beckman SW-41 rotor.
5. Polypropylene cryogenic vials, 2.0 mL (Corning, Cat. No. 430489).
6. Six-well polystyrene cell culture plates (Corning, Cat. No. 3506).
7. Microplate adaptors for low-speed centrifuge suitable for six-well tissue culture plates.
8. Polyethylene cell lifters (Corning, Corning, NY, Cat. No. 3008).
9. Dounce tissue homogenizer type B, 2 mL (Fisher Scientific, Cat. No. NC9781184).
10. Safe-lock Eppendorf tubes 2.0 mL (Brinkman Instruments, Cat. No. 22 60 004-4).
11. Polyallomer centrifuge tubes (11 × 60 mm) (Beckman, Reorder. No. 328874).

12. Beckman SW-60 rotor.
13. Rotation shaker.
14. 96-Well low profile polypropylene microplates, white (MJ Research, Waltham, MA, Cat. No. MLL-9651) and Flat cap PCR tube strips (BioRad, Cat. No. TCS0803), or other plates and cap strips for real-time RCP dependent on the type of PCR thermocycler/fluorescence detector.

2.2. Reagents and Supplies

1. Dulbecco's modified Eagle's medium (DMEM) (Mediatech, Herndon, VA, Ca. No. 10-013-CV) supplemented with 2 mM glutamine (Cambrex, Walkersville, MD, Cat. No. 17-605E), 10% (v/v) fetal bovine serum (Cambrex, Cat. No. 14-472F), 100 units/mL penicillin, and 100 units/mL streptomycin (Cambrex, Cat. No. 17-602E) (complete DMEM) is filtered through 0. 22-μm-pore-size filter and stored at 4 °C.
2. Trypsin/ethylenediamine tetraacetic acid (EDTA), 0.05% Trypsin/0.53 mM EDTA in HBSS (Mediatech, Herndon, VA, Cat. No. 25-052-CI) is aliquoted and stored at −20 °C.
3. Dulbecco's phosphate-buffered saline (PBS), 1X prepared from 10X solution (Mediatech, Cat. No. 20-031-CV) is filtered through 0. 22-μm-pore-size filter and stored at 4 °C.
4. Metafectene Transfection Reagent for Mammalian Cells (Biontex, Germany, Cat. No. T020) (*see* **Note 1**) is stored at 4 °C.
5. $MgCl_2$ is dissolved as 1 M solution, filtered through 0. 22-μm-pore-size filter and stored at 4 °C.
6. DNase I, RNase-free, 10 U/μL (Roche, Indianapolis, IN, Cat. No. 10 776 785 001) is stored at −20 °C.
7. Sucrose 99 + % (Sigma, Cat. No. S-0389) is dissolved at 30% in PBS, filtered through 0. 22-μm-pore-size filter and stored at 4 °C.
8. HEPES buffer (1 M), pH 7.4 (Sigma, St. Louis, MO, Cat. No. H-0887) is diluted to 20 mM in DMEM, filtered through 0. 22-μm-pore-size filter and stored at 4 °C.
9. HIV-1 p24 ELISA Kit (PerkinElmer Life Sciences, Boston, MA, Cat. No. NEK050B) is stored at 4 °C.
10. EDTA 0.5 M, pH 8.0 (Gibco BRL, Grand Island, NY, Cat. No. 15575-020) is diluted to 0.5 mM in PBS, filtered through 0. 22-μm-pore-size filter and stored at 4 °C.
11. Hypotonic buffer (buffer H): 10 mM HEPES (pH 7.9), 1.5 mM $MgCl_2$, 10 mM KCl, 5 mM dithiothreitol (DTT), and one tablet of Complete Mini EDTA-free protease inhibitor cocktail (Roche, Cat. No. 1 836 170) per 10 mL. Filter through 0. 22-μm-pore-size filter and store at 4 °C.
12. Liquid nitrogen.
13. Triton X-100 (Sigma, Cat. No. T-9284) dissolved at 5% in buffer H, filtered through 0. 22-μm-pore-size filter and stored at room temperature.

14. Isotonic buffer A (buffer I^A): 10 mM Tris-HCl (pH 7.4), 160 mM KCl, 5 mM $MgCl_2$, 1 mM DTT, and one tablet of Complete Mini EDTA-free protease inhibitor cocktail (Roche) per 10 mL. Filter through 0.22-μm-pore-size filter and store at 4 °C.
15. Isotonic buffer B (buffer I^B): 20 mM Tris-HCl (pH 7.4), 150 mM KCl, and 5 mM $MgCl_2$. Filter through 0.22-μm-pore-size filter and store at 4 °C.
16. Iodixanol 60% (W/V) solution in water (Sigma, Cat. No. D1556) is dissolved at 25 and 30% in buffer I^B, filtered through 0.45-μm-pore-size filter and stored at room temperature.
17. EZ-Grind Kit (G Biosciences, St. Louis, MO, Cat. No. 786-139) is stored at room temperature.
18. Sucrose 99% (Sigma, Cat. No. S-0389) is diluted to 50% in buffer H or buffer I^A, filtered through 0.45-μm-pore-size filter and stored at 4 °C.
19. Buffer K: 20 mM HEPES (pH 7.3), 150 mM KCl, 5 mM $MgCl_2$, 1 mM DTT, and one tablet of Complete Mini EDTA-free protease inhibitor cocktail (Roche) per 10 mL. Filter through 0.22-μm-pore-size filter and store at 4 °C.
20. Protein G Sepharose 4 fast flow (Amersham Biosciences, Uppsala, Sweden, Cat. No. 17-0618-01) is stored at 4 °C.
21. Albumin, bovine fraction V (BSA) (Sigma, Cat. No. A-9418) is dissolved at 100 mg/mL in ddH_2O, filtered through 0.22-μm-pore-size filter, aliquoted by 0.5 mL and stored at −20 °C.
22. Sheared salmon sperm DNA (10 mg/mL) (5 Prime-3 Prime, Boulder, CO, Cat. No. IK197A), stored at −20 °C.
23. Nonimmune sera: Normal Mouse Serum (Jackson ImmonoResearch, Cat. No. 015-000-001), Normal Rabbit Serum (Jackson ImmonoResearch, Cat. No. 011-000-001), stored at 4 °C.
24. Isotype controls: ChromPure mouse IgG, whole molecule (Jackson ImmonoResearch, Cat. No. 015-000-003), and ChromPure rabbit IgG, whole molecule (Jackson ImmonoResearch, Cat. No. 011-000-003), stored at 4 °C.
25. Antibodies against HIV-1 and host-cell RTC proteins. The following antibodies react with RTC proteins in immunoprecipitation (IP) assay *(11)*: mouse monoclonal antibodies to MA, RT and IN (ABI, Columbia, MD), CA (AIDS Research and Reference Reagent Program *(18)* and PML (Santa Cruz Biotechnology, Santa Cruz, CA); rabbit antiserum to Ini1 (Santa Cruz Biotechnology).
26. Triton X-100 (Sigma, Cat. No. T-9284) dissolved at 0.1% in buffer K, filtered through 0.22-μm-pore-size filter and stored at room temperature.

27. IsoQuick Nucleic Acid Extraction Kit (ISC BioExpress, Kaysville, UT, Cat. No. G-3120-1), stored at room temperature.
28. Glycogen for molecular biology (20 mg/mL solution) (Roche, Mannheim, Germany, Cat. No. 901 393) is dissolved in IsoQuick extraction matrix (Reagent 2) to final concentration 150 μg/mL, stored at room temperature.
29. Isopropanol, DNase, RNase, protease-free (Acros Organics, NJ, Cat. No. 327270010).
30. Ethanol anhydrous, molecular biology certified (Shelton Scientific, Shelton, CT, Cat. No. IB15720) is dissolved at 70% in ddH_2O.
31. Molecular grade water (distilled, deionized), DNase, RNase and protease tested, sterile (Mediatech, Cat. No. 46-000-CM).
32. iQ SYBR Green Supermix (BioRad, Hercules, CA, Cat. No. 170-8880), stored at −20 °C.
33. Purified DNA from 8E5 cells, dissolved in ddH_2O, aliquoted and stored at −80 °C.
34. Primers specific for the negative-strand "strong-stop" HIV-1 DNA (the early reverse transcription product): M667 (5′-GGCTAACTAGGGAACCCACTG-3′) and AA55 (5′-CTGCTAGAGATTTTCCACACTGAC-3′) *(19)*, dissolved at 10 μM in ddH_2O and stored at −20 °C.
35. Primers specific for the late reverse transcription products: FOR-LATE (5′-TGTGTGCCCGTCTGTTGTGT-3′) and REV-LATE (5′-GAGTCCTGCGTCGAGAGATC-3′) *(19)*, dissolved at 10 μM in ddH_2O and stored at −20 °C.

2.3. Cell Cultures, Viruses

1. 293T/17 cell line, a derivative of the 293T (293tsA1609neo) line of embryonic kidney epithelial human cells, transformed with adenovirus 5 DNA (ATCC, Manassas, VA, Item No. CRL-11268).
2. HeLa cell line, human epithelial cervical adenocarcinoma cells (ATCC, Item No. CCL-2).
3. 8E5 cell line, a derivative of the CEM line of the lymphoblastoid human cells from peripheral blood, containing a single copy of HIV-1 LAV provirus per cell (ATCC, Item No. CRL-8993).
4. Purified DNA of HIV-1 molecular clone (plasmid encoding complete HIV-1 cDNA). We use pNLHXB, a chimera between HIV-1 NL4-3 and HXB2 molecular clones *(20)*, which is dissolved in ddH_2O and stored at −20 °C.
5. Purified plasmid encoding the Env protein of the amphotropic murine leukemia virus (MLV), for instance pcDNA-Env(MLV) vector containing sequence of MLV Env under control of CMV promoter, is dissolved in ddH_2O and stored at −20 °C.

3. Methods

The main problem in the isolation of HIV-1 RTCs is low quantity of the complexes in infected cells and their low stability. To increase the efficiency of HIV-1 infection and RTC yield, we recommend infection of HeLa cells with HIV-1 pseudotyped with Env glycoprotein of the amphotropic MLV. This approach provides a 10-fold increase in the level of infection when compared to infection of CD4-positive HeLa cells with the wild type HIV-1. The cells can be infected using spinoculation, centrifugation of the cells with virus suspension at $1,000 \times g$ for 2 h, described in O'Doherty et al. *(21)*. This method allows infection of 70–80% of the cells *(11)*. To avoid distortion of the protein composition of RTCs, they should be isolated from the cytoplasmic and nuclear fractions without detergents. Analysis of protein composition of the RTCs can be accomplished using IP followed by real-time PCR analysis of HIV-1 DNA. Sensitivity of the Western blot assay is not sufficient for adequate analysis, especially when nuclear RTC specimens are concerned. If interaction between an RTC protein and its specific antibody correlates with the content of this protein in the RTC (which is often the case), real-time PCR analysis of cDNA isolated from the immunoprecipitates can be used as a quantitative measure of this protein. This approach is handy in comparing levels of certain proteins in cytoplasmic and nuclear RTCs, or changes in protein content upon maturation of RTCs. However, this method does not allow comparison between different proteins because of differences in affinity of antibodies used for the IP.

The experimental procedures described below are based on the methods published in Refs *(11, 13, 19, 21–24)*.

3.1. Preparation of Infectious Virus

1. Collect approximately 5×10^6 293T/17 cells from an ongoing culture with trypsin/EDTA, wash with 20 mL PBS once and transfer to a 75 cm^2 tissue culture flask in 12 mL of complete DMEM to provide culture for transfection with viral molecular clone. Incubate cells overnight at 37 °C in atmosphere containing 5% CO_2 to 70–80% confluence ($\sim 1 \times 10^7$ cells).
2. Carefully remove culture medium from cell monolayer, add 10 mL of prewarmed complete DMEM and incubate cells for 1–2 h at 37 °C and 5% CO_2.
3. Prepare transfection mix containing 24 μg of HIV-1 molecular clone DNA (e.g., NLHXB *(20)*) and 6 μg of pcDNA-Env(MLV), 800 μL of serum and antibiotic-free DMEM and 90 μL of Metafectene transfection reagent according to manufacturer protocol (*see* **Note 1**). Add transfection mix drop

wise into flask with 293T/17 cells, incubate cells at 37 °C and 5% CO_2 for 6 h or overnight.

4. Carefully remove culture medium containing transfection complexes from cell monolayer, wash cells with 15 mL of pre-warmed PBS (*see* **Note 2**). Add 15 mL of complete DMEM, maintain culture at 37 °C and 5% CO_2 for 48–72 h.
5. Harvest culture medium from transfected cells, spin down floating cells by low-speed centrifugation (~ 800–1000 × *g*) at room temperature for 5 min. Filter supernatants through a 0.45-μm-pore-size filter into 15 or 50 mL plastic centrifuge tubes. To the filtered supernatant (virus suspension) add 1 M $MgCl_2$ (to final concentration of 10 mM) and RNase-free DNase I (60 U/mL), incubate on a water bath for 1 h at 37 °C.
6. Dispense virus suspension in 11-mL aliquots into Ultra-Clear centrifuge tubes (14 × 89 mm) (*see* **Note 3**). For balancing the tubes, virus suspension can be diluted by DMEM to final volume of 11 mL. Underlay 1 mL of 30% sucrose in PBS, balance tubes and spin at 24,000 RPM in a Beckman SW-41 rotor for 2 h at 4 °C. Completely remove supernatants after centrifugation and resuspend the virus pellets in 1 mL of DMEM containing 20 mM HEPES (pH 7.4). Place suspension of concentrated virus into 2 mL polypropylene cryogenic vials – 1 mL/vial – seal and keep at −70 °C.

3.2. Infection, Fractionation of Infected Cells and Isolation of HIV-1 RTCs

1. Collect 24 × 10^6 HeLa cells from maintained culture with trypsin/EDTA, wash with 40 mL of PBS twice, resuspend in 48 mL of complete DMEM and transfer into four six-well tissue culture plates (0. 5 × 10^6cells/2 mL/well) to provide culture for infection. Incubate cells overnight at 37 °C in atmosphere containing 5% CO_2.
2. Measure viral titer using p24 ELISA kit according to manufacturer's protocol (*see* **Note 4**) and normalize viral load to 0.5 pg of p24 per cell. Since virus stocks comprise high level of p24 protein, we recommend the following dilutions of virus suspension for p24 ELISA: $\times 10^{-3}$, $\times 10^{-4}$, $\times 10^{-5}$, $\times 10^{-6}$. Each sample should be duplicated. Then dilute virus stocks with complete DMEM to final p24 concentration of 0. 25 μg/mL. Virus suspension can be kept on ice at 4 °C for up to 24 h until infection.
3. Wash HeLa monolayer with cold PBS (2 mL/well) once, then add 2 mL of virus suspension per well. Cover plates by lids, seal with two layers of Parafilm and place them in microplate centrifuge adaptors for spinoculation. Centrifuge using low-speed centrifuge for 2 h at 1,000 × *g* and 18 °C (to prevent viral internalization by the cells during spinoculation).

4. Completely remove virus-containing supernatants after centrifugation, wash cell monolayers with 2 mL/well of cold PBS twice and add 2 mL per well of complete DMEM at room temperature. Incubate cells at 37 °C in atmosphere containing 5% CO_2 for different times depending on the purposes of the experiment (*see* **Note 5**).
5. Remove supernatant from cell monolayer, wash cells twice with 2 mL per well of room temperature PBS containing 0.5 mM EDTA, add trypsin/EDTA (0.5 mL per well) and incubate at 37 °C for 3 min. Then lift cells using polyethylene cell lifters. Wash cells in 50 mL centrifuge tubes with 40 mL of ice-cold PBS twice. Completely remove PBS and resuspend cellular pellet in five volumes (~ 1.2 mL) of ice-cold hypotonic buffer (buffer H). Place cell suspension into 2 mL Eppendorf microtubes and incubate on ice for 20 min.
6. Centrifuge suspension of HeLa cells in buffer H on a table-top microcentrifuge at 5,000 rpm and 4 °C for 5 min, resuspend cell pellet in three volumes (~ 0.7 mL) of buffer H and incubate on ice for 10–15 min. Prechill Dounce homogenizer type B, place cell suspension into the barrel of the homogenizer and destroy plasmamembrane of the cells by 10–15 strokes of the pestle (*see* **Note 6**). Immediately place the lysate into Eppendorf microtubes and pellet the nuclear fraction at 5,500 rpm and 4 °C for 10 min. Place the supernatant (cytoplasmic lysate) into new microtubes, keep nuclear pellet on ice.
7. Clarify cytoplasmic lysate by centrifugation at 10,000 rpm (9,500 × *g*), 4 °C for 10 min. Place clarified supernatant (cytoplasmic extract) into new microtubes. Cytoplasmic extract and nuclear pellet (*see* **step 6**) can be supplemented with sucrose solution in buffer H to a final concentration of sucrose equal 8%, snap-frozen in liquid nitrogen and stored at −80 °C.
8. For isolation of nuclear RTCs, nuclei should be washed from components of the cytoplasm to avoid cytoplasmic contamination of the nuclear complexes (*see* **Note 7**). To eliminate cytoplasmic contamination, resuspend nuclear pellet in 10 volumes of ice-cold buffer H and add 0.1 volume of 5% Triton X-100 in buffer H to a final concentration of 0.5%. Incubate suspension on ice for 10 min, shake on vortex for 10 s and spin down nuclei using table-top microcentrifuge at 6,000 rpm, 4 °C for 10 min. Wash the nuclei with 0.5–1 mL of ice-cold buffer H, pellet at the same speed and 4 °C for 5 min and completely remove the supernatant.
9. Resuspend the nuclear pellet in 0.9 mL of isotonic buffer (buffer I^B) supplemented with 25% iodixanol (Sigma) and place into 2-mL Eppendorf tubes. Underlie suspension with 30% iodixanol in buffer I^B and centrifuge at 10,500 rpm

(10,000 × g) and 4 °C for 20 min as described in Graham et al. *(25)*. Carefully remove and discard supernatant and resuspend nuclear pellet in 0.5 mL of ice-cold isotonic buffer A (buffer I^A) to wash from iodixanol.

10. Pellet nuclei again at 6,000 rpm, 4 °C for 5 min and resuspend in 50 μL of buffer I^A. Prepare grinding-resin tube from EZ-Grind kit according to manufacturer's protocol (wash resin with 200 μL of buffer I^A) and place the nuclear suspension into the tube. Destroy nuclei by 80–100 strokes of matching pestle. Place suspension into Eppendorf tube and clarify the nuclear extract by centrifugation at 13,000 rpm (16,000 × g) and 4 °C for 10 min. Collect supernatant in new microtube. Nuclear extracts can be snap-frozen with 8% sucrose in liquid nitrogen and stored at −80 °C as described for cytoplasmic extracts (**step 6**), or directly used for ultracentrifugation. RTCs can be purified from cytoplasmic and nuclear extracts using ultracentrifugation through 50% sucrose cushion. According to Stremlau et al. *(24)*, a 50% sucrose cushion is optimal for discriminating between intracellular capsids associated with entry-competent viruses and those nonspecifically associated with cells.
11. Dilute cytoplasmic and nuclear extracts to 3.5 mL with ice-cold buffers H and I^A, respectively. Place diluted suspensions into Beckman polyallomer centrifuge tubes 11 × 60 mm and underlay by 0.5 mL of 50% sucrose in buffer H for cytoplasmic and in buffer I^A for nuclear extracts. Centrifuge samples using SW-60 bucket rotor in a Beckman ultracentrifuge at 34,000 RPM (100,000 × g) for 3 h at 4 °C.
12. Remove and discard supernatants, resuspend RTC pellets in 50–100 μL of buffer K and aliquot by 10–20 μL to Eppendorf microtubes. Snap-freeze RTC suspensions in liquid nitrogen and store at −80 °C.

3.3. Purification of HIV-1 cDNA and Quantification of RTCs Using Real-Time PCR

For quantitative analysis of protein composition of HIV-1 RTCs, complexes should be normalized according to the cDNA content. Thus, real-time PCR analysis of DNA isolated from cytoplasmic and nuclear RTCs is optimal for HIV-1 DNA quantification and subsequent normalization of RTC for the IP assays.

1. Thaw aliquots of cytoplasmic and nuclear RTCs, dilute by Sample Buffer (Reagent A) from IsoQuick Nucleic Acid Extraction Kit to final volume 50 μL (*see* **Note 8**). Incubate suspensions at room temperature for 5 min and add equal volume (50 μL) of Lysis Solution (Reagent 1). Then extract DNA according to Rapid DNA extraction protocol from the manufacturer. Since the quantity of DNA in RTC samples is very low, we recommend adding glycogen solution (20 mg/mL) to IsoQuick Extraction Matrix (Reagent 2) to a final concentration of 150 μg/mL.

2. Following extraction, air dry DNA pellets and resuspend in 20 μL of DNase, RNase, and protease-free ddH_2O (RTC DNA from 4 to 4.5×10^6 cells). Use this cDNA solution for real-time PCR assay.
3. Prepare serial dilutions of cDNA from cytoplasmic and nuclear RTCs with the following coefficients of dilution: $\times 1, \times 10^{-1}, \times 10^{-2}, \times 10^{-3}, \times 10^{-4}$. Use 5 μL from each dilution for each PCR sample with total volume of 50 μL. Prepare the following serial dilutions of DNA from 8E5 cells: 1,000, 500, 250, 125, 62, 31, and 15 HIV-1 DNA copies/5 μL/sample.
4. Prepare PCR premix containing half volume of 2X iQ SYBR Green Supermix and 15 pmoles of each primer per 45 μL of mix. ddH_2O should be added to make up to the final reaction volume. Dispense 45 μL of premix into wells of 96-well low profile polypropylene white microplate and add 5 μL of standard or unknown DNA into each well.
5. Place the PCR plate in the thermal cycler and start the cycling program. For M667 and AA55 primers the cycling protocol is the following:

(1) Pre-incubation	50 °C for 2 min
(2) Initial denaturation	95 °C for 10 min
(3) Denaturation	94 °C for 10 s
(4) Annealing	62 °C for 20 s
(5) Extension	72 °C for 20 s
(6) Fluorescence data acquisition	
(7) Cycle number	45 cycles (**steps 3–6**)
(8) Final extension	72 °C for 5 min
(9) Fluorescence data acquisition	
(10) Melting curve	65–95 °C, 0.2 °C 1 s hold
(11) Re-annealing	72 °C for 5 min.

6. Analyze melting curve for each sample to control the specificity of an amplified product. Determine absolute HIV-1 RTC cDNA count using a standard curve generated from dilutions of standard DNA from 8E5 cells.
7. Serial dilutions of DNA from 8E5 cells can be used as the quantitative standards. Total DNA from nonactivated 8E5 cells can be isolated using IsoQuick Nucleic Acid Extraction Kit. Since each cell contains a single copy of HIV-1 LAV provirus *(19)*, the number of HIV-1 DNA copies is equal to the cell count.
8. Real-time PCR should be performed in triplicate for each standard and experimental sample. Since the purpose of this analysis is to assess the number of cytoplasmic and nuclear HIV-1 RTCs per cell, we recommend using primers specific for the negative-strand "strong-stop" DNA (the early reverse transcription product): M667 and AA55 (*see* **Section 2**).

3.4. Co-Immunoprecipitation of HIV-1 RTCs and Isolation of Viral cDNA

Although sensitivity of a real-time PCR method of quantitative DNA analysis is very high, for representative results the quantity of RTCs in each specimen for IP should be higher than 1×10^6.

1. Wash appropriate volume of protein G sepharose (5 μg of beads is required for each RTC sample) three times with 1–1.5 mL of buffer K. After each wash, pellet the beads by centrifugation on a table-top microcentrifuge at 5,000 rpm and 4 °C for 5 min. Resuspend protein G beads in 0.5 mL of buffer K supplemented with 10 mg/mL BSA and 1 mg/mL salmon sperm DNA. Incubate suspension of beads on rotation shaker at 10 rpm and 4 °C for 3 h or overnight.
2. Pellet protein G beads as described above, wash with 1.5 mL of ice-cold buffer K once and finally resuspend in the volume of buffer K equal to the initial one.
3. Dispense cytoplasmic and nuclear RTCs into Eppendorf microtubes; the amount of RTCs should be such as to create a large excess of the antibody to be used for IP. For a comparison of protein composition of cytoplasmic and nuclear RTCs, these RTCs should be added in equal amounts to IP reactions. Adjust volume of each aliquot of RTC suspensions to 250 μL by ice-cold buffer K. Add 4 μL of nonimmune serum and 2. 5 μg of protein G-Sepharose in buffer K to each sample. Incubate suspensions on rotation shaker for 2 h at 4 °C and 10 rpm.
4. Clarify samples by centrifugation on a table-top microcentrifuge at 5,000 rpm and 4 °C for 5 min. Place supernatants into new Eppendorf tubes and add 4 μg of appropriate antibody or purified IgG from mouse, rabbit or other species as isotype control. Incubate overnight on rotation shaker at 4 °C and 10 rpm, then add 2. 5 μg of protein G-sepharose to each sample and continue incubation for an additional 2 h.
5. Pellet immune complexes by centrifugation at 5,000 rpm and 4 °C for 10 min and completely remove the supernatants. Wash pellets three times with 1 mL of buffer K supplemented with 0.1% Triton X-100, and once with 1 mL of buffer K without Triton.
6. Pellet immune complexes as described above, resuspend in 50 μL of Sample Buffer (Reagent A) from IsoQuick Nucleic Acid Extraction Kit (*see* **Note 8**). Incubate suspension at room temperature for 10–15 min and add equal volume (50 μL) of Lysis Solution (Reagent 1). Then extract DNA as described in **Section 3.3 steps 1, 2**.

3.5. Real-Time PCR Analysis of DNA from Immune-Precipitates of HIV-1 RTCs

For examination of RTC protein composition, cDNA isolated from RTC immune-precipitates is quantitatively analyzed by real-time PCR as described in **Section 3.3, steps 3–6**. Real-time PCR should be performed in triplicate for each standard and unknown specimen dilution; DNA counts in isotype immunoprecipitates should be subtracted from the data obtained with specific antibody.

M667 and AA55 primers described in **Section 3.3** are specific for the negative-strand "strong-stop" DNA and allow analysis of protein composition of RTCs comprising both incomplete and complete reverse transcripts (immature and mature RTCs). To recognize proteins in mature RTCs (with complete cDNA), the FOR-LATE and REV-LATE primers specific for the late reverse transcripts *(19)* should be used.

1. Complete **step 3** from **Section 3.3**. Then prepare appropriate volumes of PCR premixes for primers specific for the early and late HIV-1 cDNA. Each premix should contain half the volume of 2X iQ SYBR Green Supermix and 15 pmoles of each primer (M667 and AA55 for early reverse transcription product, and FOR-LATE and REV-LATE-NL43 for late reverse transcription product detection) per 45 μL of mix. To make up to the final reaction volume, ddH_2O should be added. Dispense 45 μL of premix into wells of 96-well low profile polypropylene white microplate and add 5 μL of standard or unknown DNA dilution into each well.
2. Perform real-time PCR reaction using cycling protocol described in **step 5** of **Section 3.3**. The same protocol can be applied to PCR reaction with primers specific for early and late HIV-1 cDNA.
3. Determination of absolute cDNA counts from IP specimens allows one to calculate the values of cDNA recovery in immunoprecipitated RTCs as a percentage of total HIV-1 DNA detected in the cRTCs (*see* **Section 3.3**).

4. Notes

1. Various transfection reagents such as FuGene-6 (Roche) or Lipofectamine 2000 (Invitrogen) can be used for co-transfection of 293T/17 cells with molecular clone of HIV-1 and Env (MLV)-encoding plasmid.
2. 293T/17 cells form very unstable monolayers. Change of medium and wash of cells should be performed with 37 °C prewarmed DMEM or PBS very carefully to avoid detaching the cells.
3. Polyallomer centrifuge tubes (Beckman Coulter) for purification of viral stocks can be used as well as ultra-clear tubes.
4. Any HIV-1 p24 ELISA kits can be used for normalization of viral titers for subsequent infection.
5. For the isolation of reverse transcription complexes from infected cells, long-time incubation of cells after infection is not recommended. Since RTCs are labile and are actively

destroyed by proteasomes, duration of incubation should not exceed 24 h if cells are not treated with proteasomal inhibitors.

6. Cell lysis can be assessed using phase-contrast microscopy (use 10 μL of samples after cell homogenization).
7. Contamination of nuclear fractions with cytoplasmic components can be assessed using PCR with primers specific for mitochondrial DNA (forward primer, Mito1: 5′-GAA TGT CTG CAC AGC CAC TT-3′; reverse primer, Mito2: 5′-AGA AAG GCT AGG ACC AAA CC-3′). PCR analysis of serial dilutions of cytoplasmic and nuclear extract samples allows assessing the level of contamination.
8. We use IsoQuick Nucleic Acid Extraction Kit to extract the HIV-1 cDNA from suspensions of cytoplasmic and nuclear RTCs. However, other methods and kits for DNA extraction are acceptable.

Acknowledgements

The authors thank Larisa Dubrovsky for excellent technical assistance. This work was supported in part by NIH grants R01 AI033776 and R01 AI040386 (MB).

References

1. Nisole, S., Saib, A. (2004) Early steps of retrovirus replicative cycle. *Retrovirology* 1, 9.
2. Iordanskiy, S., Bukrinsky, M. (2007) Reverse transcription complex: the key player of the early phase of HIV replication. *Future Virol* 2, 49–64.
3. Bukrinsky, M. I., Sharova, N., McDonald, T. L., et al. (1993) Association of integrase, matrix, and reverse transcriptase antigens of human immunodeficiency virus type 1 with viral nucleic acids following acute infection. *Proc Natl Acad Sci USA* 90, 6125–6129.
4. Heinzinger, N. K., Bukrinsky, M. I., Haggerty, S. A., et al. (2004) The Vpr protein of human immunodeficiency virus type 1 influences nuclear localization of viral nucleic acids in nondividing host cells. *Proc Natl Acad Sci USA* 91, 7311–7315.
5. Miller, M. D., Farnet, C. M., Bushman, F. D. (1997) Human immunodeficiency virus type 1 preintegration complexes: studies of organization and composition. *J Virol* 71, 5382–5390.
6. Li, L., Olvera, J. M., Yoder, K. E., et al. (2001) Role of the non-homologous DNA end joining pathway in the early steps of retroviral infection. *EMBO J* 20, 3272–3281.
7. Cherepanov, P., Maertens, G., Proost, P., et al. (2003) HIV-1 integrase forms stable tetramers and associates with LEDGF/p75 protein in human cells. *J Biol Chem* 278, 372–381.
8. Farnet, C. M., Bushman, F. D. (1997) HIV-1 cDNA integration: requirement of HMG I(Y) protein for function of preintegration complexes in vitro. *Cell* 88, 483–492.
9. Turelli, P., Doucas, V., Craig, E., et al. (2001) Cytoplasmic recruitment of INI1 and PML on incoming HIV preintegration complexes: interference with early steps of viral replication. *Mol Cell* 7, 1245–1254.
10. McDonald, D., Vodicka, M. A., Lucero, G., et al. (2002) Visualization of the intracellular behavior of HIV in living cells. *J Cell Biol* 159, 441–452.
11. Iordanskiy, S., Berro, R., Altieri, M., et al. (2006) Intracytoplasmic maturation of the human immunodeficiency virus type 1 reverse transcription complexes determines their capacity to integrate into chromatin. *Retrovirology* 3, 4.
12. Bukrinsky, M. I., Sharova, N., Dempsey, M. P., et al. (1992) Active nuclear import of human immunodeficiency virus type 1 preintegra-

tion complexes. *Proc Natl Acad Sci USA* 89, 6580–6584.
13. Fletcher, T. M., Brichacek, B., Sharova, N., et al. (1996) Nuclear import and cell cycle arrest functions of the HIV-1 Vpr protein are encoded by two separate genes in HIV-2/SIV(SM). *EMBO J* 15, 6155–6165.
14. Brichacek, B., Stevenson, M. (1997) Quantitative competitive RNA PCR for quantitation of virion-associated HIV-1 RNA. *Methods* 12, 294–299.
15. Butler, S. L., Hansen, M. S., Bushman, F. D. (2001) A quantitative assay for HIV DNA integration in vivo. *Nat Med* 7, 631–634.
16. Llano, M., Vanegas, M., Fregoso, O., et al. (2004) LEDGF/p75 determines cellular trafficking of diverse lentiviral but not murine oncoretroviral integrase proteins and is a component of functional lentiviral preintegration complexes. *J Virol* 78, 9524–9537.
17. Lin, C. W., Engelman, A. (2003) The barrier-to-autointegration factor is a component of functional human immunodeficiency virus type 1 preintegration complexes. *J Virol* 77, 5030–5036.
18. Simon, J. H, Fouchier, R. A., Southerling, T. E., et al. (1997) The Vif and Gag proteins of human immunodeficiency virus type 1 colocalize in infected human T cells. *J Virol* 71, 5259–5267.
19. Desire, N., Dehee, A., Schneider, V., et al. (2001) Quantification of human immunodeficiency virus type 1 proviral load by a TaqMan real-time PCR assay. *J Clin Microbiol* 39, 1303–1310.
20. Popov, S., Rexach, M., Zybarth, G., et al. (1998) Viral protein R regulates nuclear import of the HIV-1 pre-integration complex. *EMBO J* 17, 909–917.
21. O'Doherty, U., Swiggard, W.J., Malim, M.H. (2000) Human immunodeficiency virus type 1 spinoculation enhances infection through virus binding. *J Virol* 74, 10074–10080.
22. Bushman, F. D., Miller, M. D. (1997) Tethering human immunodeficiency virus type 1 preintegration complexes to target DNA promotes integration at nearby sites. *J Virol* 71, 458–464.
23. Fassati, A., Goff, S. P. (2001) Characterization of intracellular reverse transcription complexes of human immunodeficiency virus type 1. *J Virol* 75, 3626–3635.
24. Stremlau, M., Perron, M., Lee, M., et al. (2006) Specific recognition and accelerated uncoating of retroviral capsids by the TRIM5alpha restriction factor. *Proc Natl Acad Sci USA* 103, 5514–5519.
25. Graham, J., Ford, T., Rickwood, D. (1994) The preparation of subcellular organelles from mouse liver in self-generated gradients of iodixanol. *Anal Biochem* 220, 367–373.

Chapter 10

Isolation and Analysis of HIV-1 Preintegration Complexes

Alan Engelman

Abstract

A discerning feature of the retrovirus lifecycle is the covalent integration of the viral reverse transcript into a chromosome within the infected cell. Integration is required for productive infection and therefore defines the viral integrase protein of human immunodeficiency virus type 1 (HIV-1) as a bona fide target for the development of antiviral drugs in the fight against HIV/AIDS. Integrase works in the context of the viral preintegration complex (PIC), a high molecular weight nucleoprotein complex that supports the integration of its endogenous viral DNA copy made during reverse transcription into an exogenous target DNA in the test tube. PIC analyses are central to understanding the molecular mechanisms of HIV-1 integration as well as investigating the pharmacological properties of integrase inhibitors. This chapter describes techniques for isolating HIV-1 PICs from cells as well as quantifying their level of integration activity in vitro.

Key words: HIV-1; AIDS; integrase; integration; preintegration complex; DNA recombination.

1. Introduction

Retroviral particles undergo an uncoating step soon after cell entry, which leads to the formation of the reverse transcription complex. Therein, reverse transcriptase (RT) synthesizes linear double-stranded DNA containing a copy of the viral long terminal repeat (LTR) sequence at each end, which is the substrate for the viral integrase. Each LTR is comprised of U3RU5 sequences. Integrase binds to U3 sequences in the upstream LTR terminus, and to U5 sequences at the downstream DNA end *(1)* (**Fig. 10.1**). The enzyme catalyzes two distinct chemical reactions, 3′ processing and DNA strand transfer. A dinucleotide is hydrolyzed from each human immunodeficiency virus type 1 (HIV-1) LTR during 3′ processing *(2)*, which can occur

Vinayaka R. Prasad, Ganjam V. Kalpana (eds.), *HIV Protocols: Second Edition, vol. 485*

DOI 10.1007/978-1-59745-170-3_10 Springerprotocols.com

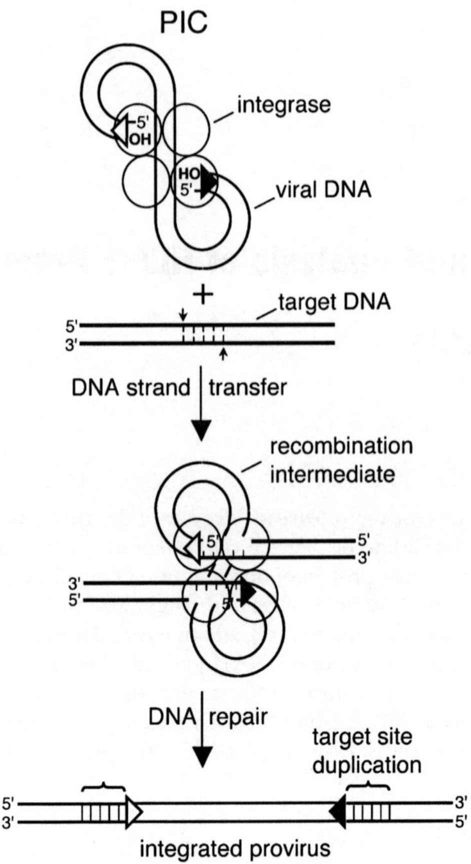

Fig. 10.1. Mechanism of HIV-1 DNA integration. The product of the 3′ processing reaction is labeled PIC. A tetramer of integrase (*open circles*) is the catalytic multimer that helps to hold the two DNA ends together in a synaptic nucleoprotein complex called the intasome *(3,17,29–34)*. The 5′ ends of viral DNA remain unattached to the integration target DNA in the strand transfer reaction product *(4–6)*. DNA repair yields the integrated provirus flanked by a 5-bp duplication of target DNA sequence *(35, 36)*. Open triangle, U3 sequences in the upstream LTR terminus essential for integrase activity; closed triangle, essential U5 sequences.

in the cell cytoplasm *(3)* and may very well define the transition from the reverse transcription complex to the preintegration complex (PIC). For DNA strand transfer, integrase employs the hydroxyl groups created during 3′ processing to make a double-stranded staggered cut in a chromosomal target DNA acceptor site (**Fig. 10.1**). This yields a recombination intermediate with the 5′-ends of the viral DNA unattached to the target DNA *(4–6)*. Cellular machinery likely repairs this structure to remove the unpaired dinucleotides from the extreme 5′-ends of the viral DNA and fill-in and seal the flanking single-strand gaps *(7)*, which duplicates the sequence of the target DNA double-stranded cut to both sides of the integrated provirus (**Fig. 10.1**). The reader should consult reference *(8)* for a more complete overview of the mechanism of retroviral DNA integration.

A landmark 1987 paper defined retroviral PICs as nucleoprotein complexes isolated from acutely infected cells that catalyze the integration of their endogenous cDNA into an added target DNA in vitro *(9)*. This initial study used recombinant *SupF*-tagged Moloney murine leukemia virus to suppress nonsense codons within bacteriophage λ gtWES, which served as the target in the in vitro integration reaction. PIC activity was genetically scored following λ DNA packaging, differential plating on two *Escherichia coli* strains, and evaluation of resulting plating efficiencies. Quantification of PIC activity soon thereafter was simplified by using Southern blotting to physically detect the integration reaction product *(4–6, 10, 11)* (**Fig. 10.2A**). More recent designs increased assay sensitivity by using real-time quantitative (RQ)-PCR to monitor the formation of viral–target DNA junctions

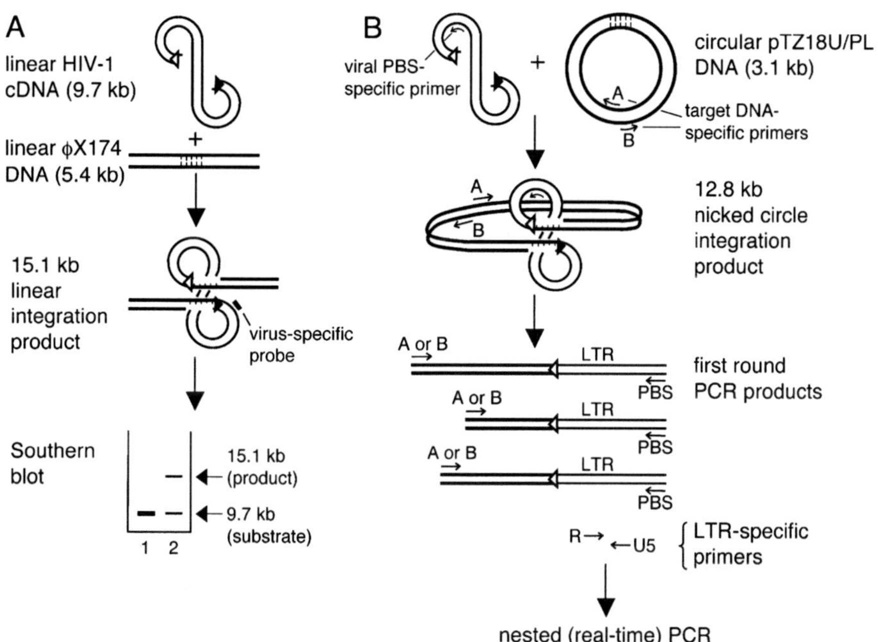

Fig. 10.2. Assay formats for quantifying HIV-1 PIC activity in vitro. (**A**) Southern blotting. The substrate is detected as a linear 9.7 kb DNA using a virus-specific probe after deproteinization and agarose gel electrophoresis (lane 1), whereas the integration product is a linear 15.1 kb species (lane 2). Integration activity is quantified as the percent of substrate DNA converted to reaction product. (**B**) RQ-PCR. This design utilizes a primer-binding site (PBS) primer that hybridizes to the viral plus-strand adjacent to the upstream LTR, and two target DNA primers, labeled **A** and **B**, that anneal to opposite strands of circular pTZ18U/PL *(14)*. First round PCR products will harbor PBS/LTR viral sequences and variable target DNA sequences whose lengths are defined by the distances between points of U3 end integration and primer A or B. First round reaction products are diluted prior to RQ-PCR, which utilizes LTR-specific primers and Taqman probe to quantify levels of first round product formation. A standard curve is generated by end-point diluting relevant integration reaction products – for example, those derived from wild-type PICs, or drug-free control reactions. Integration activities are normalized to total levels HIV-1 DNA present in the different samples.

(12–15) (**Fig. 10.2B**). Southern blotting and RQ-PCR assays are both topical, and certain parameters are taken into account when considering the use of one technique versus the other. Due to the added sensitivity of RQ-PCR, less input PIC material is required to detect integration, making the approach amenable to scale-up *(12–14)*. The main drawbacks of current RQ-PCR strategies are (i) full-length DNA molecules are not visualized and (ii) the integration of only one viral DNA end is monitored (**Fig. 10.2B**). Southern blotting, in contrast, detects full-length substrate and integration product DNAs (**Fig. 10.2A**). Monitoring the integration of one as compared to both DNA ends in vitro is representative of bona fide PIC activity, though it is worth noting that certain experimental conditions, like LTR sequence mutations, can apparently uncouple integrase's ability to integrate both DNA ends during infection *(16)*.

Preintegration complexes are isolated from acutely infected cells, and relatively high multiplicities of infection are required to yield sufficient material for detection by Southern blotting. This criterion can be met by co-culturing virus producer cells with uninfected target cells *(3, 9, 11, 17)*. Sufficiently infectious cell-free HIV-1 can be generated by transient transfection *(17)* or by pseudotyping HIV-based vectors with the vesicular stomatitis virus G (VSV-G) envelope glycoprotein *(12)*. Though the increased sensitivity of RQ-PCR detection somewhat obviates the requirement for exceedingly efficient infection systems, cell culture techniques that yield sufficient material for detection by Southern blotting are described here. These methods will help to trouble-shoot results of RQ-PCR assays in case they fail to reveal robust levels of in vitro integration activity.

2. Materials

2.1. Cell Culture, Virus Preparation, and Infection

1. Dulbecco's modified Eagle's medium (DMEM) (Invitrogen, Carlsbad, CA) supplemented to contain 10% heat-inactivated fetal bovine serum (FBS, HyClone, Ogden, UT). RPMI 1640 medium (Invitrogen) containing 10% FBS (RPMI).
2. Phosphate-buffered saline and trypsin (Mediatech, Herndon, VA).
3. MOLTIIIB *(11)*, SupT1 (AIDS Research and Reference Reagent Program [ARRRP], Germantown, MD) (*see* **Note 1**), C8166-45 (*18*), and 293T cells (American Type Culture Collection, Manassas, VA).
4. Phorbal 12-myristate 13-acetate (PMA, Sigma-Aldrich, St. Louis, MO) resuspended at 0.5 mg/mL in dimethyl sulfoxide (DMSO). Can be stored at −20 °C away from light for approximately 6 months.

5. Plasmid pNL4-3 *(19)* (ARRRP) for producing cell-free virus by transient transfection.
6. HEPES buffered saline (HBS, 2X: 280 mM NaCl, 50 mM HEPES-NaOH, pH 7.1, 1.5 mM Na_2HPO_4). Filter sterilize (*see* **Note 2**); store 10 ml aliquots at −20 °C.
7. 2.5 M $CaCl_2$.
8. TURBO DNase (Ambion, Austin, TX).

2.2. Cell Lysis and PIC Preparation

1. Buffer K−/−: 150 mM KCl, 20 mM HEPES, pH 7.6, 5 mM $MgCl_2$.
2. Digitonin (Sigma-Aldrich) dissolved at 5% (w/v) in DMSO. Store aliquots at 4 °C.
3. Buffer K+/+: buffer K−/− supplemented to contain 0.025% digitonin, 1 mM dithiothreitol, 20 μg/mL aprotinin (Sigma-Aldrich). Prepare fresh for each experiment.
4. RNase A (Sigma-Aldrich). Resuspend at 10 mg/mL in 10 mM Tris-HCl, pH 7.5, 10 mM $MgCl_2$. Boil for 10 min, cool on ice. Store aliquots at −20 °C.
5. Sucrose (60% [w/v] in buffer K−/−).

2.3. Integration Assays and DNA Recovery

1. Integration target DNAs: ϕX174 Replication Form I (Invitrogen) linearized by digestion with Xho I for Southern blotting; circular pTZ18U/PL *(20)* for RQ-PCR.
2. 10% (w/v) Sodium dodecyl sulfate (SDS).
3. 0.5 M ethylenediamine tetraacetic acid (EDTA), pH 8.0.
4. Proteinase K (20 mg/mL; Calbiochem, La Jolla, CA).
5. Phenol, pH 7.9, and chloroform (Fisher, Pittsburg, PA).
6. 3 M Na acetate.
7. Linear polyacrylamide (LPA, Sigma-Aldrich).
8. Ethanol (100% and 80%).
9. TE.1 buffer: 10 mM Tris-HCl, pH 8.0, 0.1 mM EDTA.

2.4. Southern Blotting Assay for HIV-1 PIC Activity

1. I.D.NA agarose (Cambrex, Rockland, ME).
2. Gel loading buffer (6X): 40% sucrose (w/v), 0.25% (w/v) bromophenol blue, 0.25% (w/v) xylene cyanol.
3. Gel running buffer: 40 mM Tris-acetate, pH 7.6, 1 mM EDTA (TAE). Prepare 50X stock by dissolving 242 g Tris base, 57.1 mL glacial acetic acid, and 18.6 g EDTA in 1 L of water.
4. Ethidium bromide (EtBr): 10 mg/mL in 10 mM Tris-HCl, pH 8.0, 1 mM EDTA.
5. Gel solution I: 0.24 N HCl.
6. Gel solution II: 0.6 M NaCl, 0.4 N NaOH.
7. Gel solution III: 1.5 M NaCl, 0.5 M Tris-HCl, pH 7.5.
8. Gel transfer buffer: 1.5 M NaCl, 0.15 M Na citrate (10X SSC). Prepare 20X stock by dissolving 175.3 g NaCl and 88.2 g Na citrate in 1 L of water.

9. QuikHyb hybridization solution (Stratagene, La Jolla, CA).
10. GeneScreen Plus hybridization membrane (PerkinElmer, Wellesley, MA).
11. Whatman 3 filter paper (Fisher).
12. Salmon sperm DNA, 10 mg/mL (Stratagene).
13. Labeled nucleic acid probe (*see* **Note 3**).
14. Wash solution I: 2X SSC, 0.5% SDS.
15. Wash solution II: 0.5X SSC, 0.1% SDS.
16. Wash solution III: 0.3X SSC, 0.1% SDS.
17. Phosphor screen or Kodak BioMax MR x-ray film (Eastman Kodak Co., Rochester, NY).

2.5. RQ-PCR Assay for HIV-1 PIC Activity

1. HotStarTaq DNA polymerase (QIAGEN, Valencia, CA).
2. Nucleotide mixture: 10 mM each of dATP, dCTP, dGTP, and TTP in water. Thaw aliquots from −20 °C at room temperature, place immediately on wet ice.
3. First-round PCR primers: AE2413 (5′-GTTGTTCCAGTT TGGAACAAGAGTC-3′), AE2414 (5′-ACTCAACCCTATC TCGGTCTATTC-3′), and AE2257 (5′-TTTCAGGTCCC TGTTCGGGCGCCAC-3′); primers A, B, and PBS-specific in **Fig. 10.2B**, respectively.
4. QuantiTect Probe PCR Mix (QIAGEN).
5. Second-round primers AE989 (5′-TCTGGCTAACTAGGGA ACCCA-3′) and AE990 (5′-CTGACTAAAAGGGTCTGA GG-3′), which anneal to R and U5, respectively (**Fig. 10.2B**). Taqman probe AE995, 5′-(6-Carboxyfluorescein [FAM])-TTAAGCCTCAATAAAGCTTGCCTTGAGTGC-(6-carboxytetramethylrhodamine [TAMRA])-3′ *(21)* (*see* **Note 4**).
6. Primers and Taqman probe for detecting total HIV-1 DNA (late reverse transcription [LRT] products) *(22)*: MH531, 5′-TGTGTGCCCGTCTGTTGTGT-3′; MH532, 5′-GAGTCCTGCGTCGAGAGAGC-3′; LRT-P, 5′-(FAM)-CAGTGGCGCCCGAACAGGGA-(TAMRA)-3′.

3. Methods

3.1. Cell Culture, Virus Preparation, and Infection

1. Infections are initiated by co-culturing chronically infected MOLTIIIB cells with uninfected SupT1 T-cells *(3, 11, 17)* or by using cell-free virus from transiently transfected 293T cells to infect C8166-45 T-cells *(17)* (*see* **Note 5**). Grow flasks of MOLTIIIB, SupT1, and C8166-45 cells in RPMI medium at 37 °C in a 5% CO_2-humidified incubator broad-side down for optimal gas exchange (*see* **Note 6**). Adherent 293T cells grown in DMEM medium are split by trypsinization before reaching confluency.

2. To prepare cells for co-culture, treat MOLTIIIB cells (4×10^5/mL) with PMA (10 ng/mL) for 24 h prior to infection. At the same time, prepare 20 mL of conditioned medium from unstimulated MOLTIIIB cells (10^6/mL). Infect 4×10^7 SupT1 cells by co-culturing with 4×10^6 PMA-stimulated MOLTIIIB cells in 20 mL of conditioned-medium for 5 h. Monitor the efficiency of infection by light microscopy: the majority of cells should form giant multinuclear syncytia.

3.1.1. Transient Transfection and Cell-Free Virus Infection

1. Seed 3.3×10^6 293T cells in 10 mL DMEM medium per 10 cm dish 24 h prior to transfection.
2. Transfect cells with 20 μg pNL4-3 DNA. For each transfection, carefully add 0.5 mL 2X HBS buffer to the bottom of a 15 mL conical tube. Dilute DNA to 0.45 mL using sterile water, add 50 μL 2.5 M $CaCl_2$, and vortex thoroughly. Gently layer the DNA–$CaCl_2$ mixture onto the surface of the 2X HBS buffer. Incubate at room temperature for 30 min (do not mix). Add the DNA precipitate to the cell culture medium one drop at a time. Incubate 14–17 h, discard the medium, and rinse the cell monolayer twice with serum-free DMEM medium. Add 10 mL DMEM medium containing 10 mM HEPES, pH 7.6, and incubate cells for 24 h.
3. The virus-containing cell supernatant is harvested by gravity filtration through a 0.45-μm filter. Virus yield can be determined by measuring virion-associated RT enzyme activity *(23)* or p24 capsid protein (HIV-1 p24 ELISA Kit; Perkin-Elmer). Yields should be 4–10×10^7 ^{32}P RT-cpm/mL (~ 3–10 μg p24/mL) (*see* **Note 7**).
4. To degrade the bulk of the plasmid DNA remaining after transfection, treat the virus with TURBO DNase (40 U/mL) for 1 h at 37 °C using one-tenth final volume of the 10X buffer supplied by the manufacturer.
5. Infect C8166-45 cells (3×10^7) with 15 mL cell-free virus for 7 h at 37 °C (*see* **Note 8**). Unlike the co-culture method, this method does not yield substantial syncytium formation.

3.2. Cell Lysis and PIC Preparation

1. Harvest and wash the cell pellet twice with buffer K −/−. After the second wash, use a micro pipettor to remove residual buffer.
2. Resuspend cells in 1 mL buffer K +/+ by pipetting, transfer the mixture to a 1.5-mL eppendorf centrifuge tube, and rock for 10 min at room temperature.
3. Fractionate the cell lysate by centrifugation ($1,500 \times g$ for 4 min at 4 °C).
4. Transfer the supernatant to a fresh tube and centrifuge at $16,000 \times g$ for 1 min at 4 °C.
5. Transfer the supernatant to a fresh tube, add RNase A to 20 μg/mL, and incubate at room temperature for 30 min to

degrade cellular mRNA. To store, add sucrose to 7%, flash freeze in liquid N_2, and keep at −80 °C. Integration activity may be stable for up to 6 months, though storage capacity should be determined by assaying activity at various times post-freezing.

3.3. Integration Assays and DNA Recovery

1. For Southern blotting and RQ-PCR assays, analyze 0.5 and 0.2 mL PICs, respectively. Thaw frozen PICs on wet ice. Add target DNA to 3 μg/mL, and incubate at 37 °C for 45 min. Include control samples lacking target DNA.
2. Stop reactions by adding SDS, EDTA, and proteinase K to the final concentrations of 0.5%, 8 mM, and 0.5 mg/mL, respectively. Mix thoroughly. Decontaminated tubes can now be removed from contained BSL2/BSL3 facilities. Incubate tubes overnight at 56 °C.
3. Phenol/chloroform extraction: add an equal volume of phenol, vortex vigorously, and separate fractions by centrifugation at 16,000 × *g* for 2 min. Transfer the aqueous supernatant fraction to a fresh tube, and repeat. Transfer the supernatant to a fresh tube, add an equal volume of a 1:1 mixture of phenol:chloroform, vortex, and centrifuge. Add an equal volume of chloroform to the transferred supernatant, vortex, spin, and transfer the aqueous fraction to a fresh tube.
4. To ethanol precipitate, add 1 μL LPA and Na acetate to the final concentration of 0.3 M. Mix, add 2.5 volumes of 100% ethanol, vortex, centrifuge at 16,000 × *g* for 10 min at room temperature, discard the supernatant, and rinse the pellet with 0.5 mL 80% ethanol. Lyophilize the DNA pellet to dryness. For Southern blotting and RQ-PCR, resuspend the pellet in 5 and 50 μL of TE.1 buffer, respectively.

3.4. Southern Blotting Assay for HIV-1 PIC Activity

1. Cast 60 mL of 0.6% (w/v) agarose in 1X TAE buffer into a 15-cm wide × 10-cm long tray using a thin-welled comb. Add 1 μL gel loading buffer to the DNA sample, mix, and electrophorese at 18 V for 15 h (*see* **Note 9**).
2. Stain the gel for 10 min with 0.5 μg/mL EtBr–1X TAE (*see* **Note 10**). Destain with water for 15 min. Take a picture under UV illumination to record the electrophoresis pattern.
3. Denature in Gel solution I for 15 min.
4. Neutralize for 30 min in Gel solution II.
5. Equilibrate for 30 min with Gel solution III.
6. Prepare the gel for transfer during step 5. Cut a 14 × 9 cm hole in one of the plastic masks supplied by the company, and cut one of the sponges to 15 × 10 cm (*see* **Note 11**). Cut one 15 × 10 cm piece of GeneScreen Plus hybridization membrane, and two pieces of Whatman 3 filter paper (one 15 × 10 cm and the other 17 × 12 cm). Wet the larger piece

of filter paper with 10X SSC, and place in the center of the porous membrane support pad. Briefly wet the nylon membrane with sterile water, follow by 10X SSC, and then place it atop the filter paper. Center the mask, and place the gel down-side-down such that its main body touches the membrane and approximately 0.5 cm of each edge lies atop the mask. Gently work out any bubbles between the gel and membrane with gloved fingers. Wet the second piece of Whatman 3 paper with 10X SSC, and place directly atop the gel. Thoroughly wet the sponge with 10X SSC, and place that atop the pile. Saturate the stack by pouring extra 10X SSC over the sponge. Close the lid and fasten the latches. Turn on the Pressure Control Station, use an ungloved thumb to block the hose end, and adjust the pressure to 75 mmHg. Attach the hose to the apparatus and monitor pressure buildup. It may take 1–2 min to reach 70–75 mmHg (*see* **Note 12**). If this is not attained, there is a leak in the system. Unlatch the top, carefully adjust the outer edge of the mask without effecting the gel sandwich, and re-test. Transfer for 30 min. Additional troubleshooting guides can be found in the apparatus manual (http://www.stratagene.com/products/displayproduct.aspx?pid=291). To prepare for hybridization, thaw the QuikHyb solution at room temperature, and then place at 65 °C.

7. Break down the apparatus, pierce through the gel wells with an indelible marker to mark their positions on the membrane, and rinse the membrane with 2X SSC. Stain the flattened gel with EtBr and take a picture under UV light. Efficient transfer yields a lack of detectable DNA.
8. Blot the membrane dry using Whatman 3 paper, and immediately wrap it in saran wrap. Crosslink the DNA onto the membrane using a UV Stratalinker 2400 (Stratagene) and auto cross link option.
9. Unwrap the membrane and wash thoroughly with sterile water to remove excess salt. Prehybridize the membrane in 7 mL QuikHyb hybridization solution at 65 °C for 1–18 h with rotation (*see* **Note 13**).
10. Hybridize the membrane in 7 mL QuikHyb solution containing 0.1 mg/mL salmon sperm DNA and 2×10^6 ^{32}P-cpm riboprobe/mL (*see* **Note 3**). Pre-mix the DNA and riboprobe in a 1.5-mL eppendorf centrifuge tube (*see* **Note 14**), withdraw 1 mL of warm hybridization solution from the roller bottle using a serological pipette, and carefully mix with the DNA–probe solution. Use the serological pipette to add the mixture into the roller bottle. Incubate for 1 h at 65 °C with rotation (*see* **Note 15**).
11. Remove the probe–hybridization solution mixture with a serological pipette, and discard as radioactive waste. Rinse the

bottle twice with wash solution I, and discard as radioactive waste. Incubate the membrane with 100 mL of wash solution I for 10 min at 65 °C with rotation, discard the buffer as radioactive waste then repeat. Repeat the pattern of two successive wash steps with wash solution II, then with wash solution III. Blot the membrane dry with Whatman 3 paper, and wrap it in saran wrap. Expose overnight to a phosphor screen, and develop the signal using a phosphor imager. Integration activity is defined as the percent of substrate DNA converted into reaction product (diagrammed as 50% in **Fig. 10.2A**, lane 2). Expose the blot to X-ray film at −80 °C if a phosphor imager is unavailable.

3.5. RQ-PCR Assay for HIV-1 PIC Activity

1. Make standard curve samples by fourfold diluting the reaction containing the most integration products to a final concentration of 1:4,096. For example, if the experiment is to determine the activity of an integrase mutant virus, dilute the wild-type PIC sample. If the experiment tested the activities of integrase inhibitors, dilute the no-drug control sample (*see* **Note 16**).
2. Assemble first-round PCRs: 5 μL DNA sample, 5 μL 10X HotStarTaq buffer, 1 μL of dNTP mix, 0. 5 μM final concentration of primers AE2413, AE2414, and AE2257, 0. 25 μL HotStarTaq polymerase, sterile water to 50 μL. Include a no DNA (TE.1 buffer) control. A master mix containing all components except the DNA sample can be assembled on ice. Aliquot 45 μL of the mix to separate tubes, then add 5 μL of DNA or control sample.
3. Incubate the reactions in a thermal cycler. After 5 min at 95 °C, cycle 23 repetitions of: 94 °C for 30 s, 58 °C for 30 s, and 72 °C for 4 min. Follow by a 10-min extension at 72 °C.
4. Dilute 5 μL of each first-round sample into 1 mL of sterile water, vortex to mix. Assemble second round PCRs in duplicate: 5 μL diluted DNA, 0. 3 μM each of AE989 and AE990 primers, 0. 1 μM AE995 probe, 15 μL QuantiTect Probe PCR Mix, sterile water to 30 μL. Include duplicate water (no DNA) controls. A master mix containing all the components except DNA assembled on ice can be aliquoted to empty wells before adding individual DNA or control samples.
5. Incubate the reactions in a real-time thermal cycler: 15 min at 95 °C, then 40 cycles of 94 °C for 15 s, 58 °C for 30 s, and 72 °C for 30 s.
6. Assemble reactions for determining LRT product levels in the integration reaction samples. Construct a standard curve by fivefold diluting pNL4-3 from 10^6 molecules/μL to 64 copies/μL. Assemble duplicate reactions: 2 μL of integration sample, standard curve DNA, or water-only control, 0. 3 μM each of MH531 and MH532 primers, 0. 1 μM LRT-P probe, 15 μL QuantiTect Probe PCR Mix, sterile water to 30 μL. A

no-DNA master mix assembled on ice can be aliquoted to individual wells before adding DNA or water. Cycle reactions as described in step 5 above.

7. Determine levels of integration activity by comparing unknown values to the integration standard curve. Normalize these values to HIV-1 LRT levels (*see* **Note 17**). Reactions conducted in the absence of target DNA define assay background; subtract these values from the corresponding target DNA-containing samples. It is convenient to express values as percent of wild-type integrase virus or no-drug control activities *(14)*.

4. Notes

1. The ARRRP is a valuable source of reagents for basic HIV/AIDS research. Managed by the US National Institutes of Health (NIH), qualified users must register with the agency before reagents can be obtained (http://www.aidsreagent.org/).
2. Biochemical solutions are sterilized by filtration through 0.22-μm filters.
3. RNA or DNA probes can be used to detect HIV-1 nucleic acids. Because the viral plus-strand is fragmented *(24)*, an RNA riboprobe of sense polarity that anneals to the virus minus-strand is preferred. Plasmid pSP73/XH contains the 719 bp *Xho*I-*Hin*d III fragment (nucleotides 8896–9615) from infectious clone pSVC21 *(25)* subcloned into the *Xho*I and *Hin*dIII sites of riboprobe expression vector pSP73 (Promega, Madison, WI). The radio-labeled plus-sense probe is generated from Hind III-cut pSP73/XH using [α-^{32}P]CTP (3000 Ci/mmol; PerkinElmer) and T7 RNA polymerase according to manufacturer's instructions (Riboprobe System – T7 kit; Promega). Use of radioactivity requires special training and practices that follow federal, state, as well as institutional guidelines. A labcoat, attached dosimeter, and two pairs of latex gloves are required for working with ^{32}P. Handle the isotope behind a plexiglass shield that is minimally 0.375 of an inch thick to diminish exposure. Each institution will have strict guidelines in place for radioactive waste disposal. RNA probes are relatively unstable and should be made fresh each day.
4. The original protocol described other R and U5-specific primers, and utilized SYBR Green instead of a Taqman probe to detect second round PCR products *(14)*. Subsequent unpublished observations revealed superior signal-to-noise ratios using the Taqman probe in place of SYBR Green.

5. MOLTIIIB cells produce infectious HIV-1, so it is essential to handle these under appropriate safety conditions. US Centers for Disease Control (CDC)-NIH guidelines recommend that HIV stocks are concentrated in biosafety level 3 (BSL3) facilities. The experiments described herein, encompassing cell culture and virus infection, can be performed in a contained BSL2 facility if BSL3 work practices are employed (informally referred to as BSL2+). Refer to reference *(26)* for detailed guidelines.
6. Numerous parameters, some of which are undefined, can affect PIC activity levels in in vitro integration assays. It is essential to maintain suspension cell densities at < 10^6 cells/mL; marginal overgrowth can negatively impact results. It is recommended to survey different aliquots of frozen MOLTIIIB, SupT1, and C8166-45 cells to determine those that yield ≥ 40% integration activity as defined in the Southern blotting assay. These aliquots should then be re-assayed following careful expansion and re-freezing. Fresh aliquots of cells should then be thawed for each experiment.
7. Different criteria impact the efficiency of HIV-1 production by transient transfection. It is recommended to thaw fresh aliquots of pNL4-3 plasmid DNA for each experiment. Survey different aliquots of frozen 293T cells for virus production capacity, and carefully expand the optimum sample. For example, limit the time the cells are exposed to trypsin during subculturing by preheating the enzyme solution in a 37 °C water bath. Prewarmed trypsin readies a phosphate-buffered saline-rinsed 293T cell monolayer for dislodging in < 1 min at room temperature. Briskly slap the side of the flask with your gloved hand to dislodge the cells, and quickly resuspend in DMEM medium to inhibit enzyme activity. Centrifuge the cell suspension, discard the supernatant, and thoroughly resuspend the cell pellet in fresh DMEM medium before plating into fresh flasks.
8. SupT1 cells can be substituted for C8166-45 cells, though SupT1 cells should be infected by spinoculation *(27)* as they are somewhat less susceptible to infection than are C8166-45 cells *(17)*. Resuspend the 15 mL SupT1 cell-virus mixture evenly among a six-well tissue culture plate, and centrifuge for 2 h at 480 × g in a table top centrifuge using a micro plate carrier. Transfer the plate to 37 °C for the remaining 5 h of infection. Do not spinoculate C8166-45 cells, as this reduces the efficiency by which they become infected.
9. Numerous parameters can affect the quality of Southern blotting. To control for nucleic acid transfer and hybridization, electrophorese 250 pg of *Xho*I-digested linear pNL4-3 DNA in a separate lane of the gel. This should (i) yield a hybridization signal that is clearly visible following one night of

autoradiography, and (ii) approximate the signal obtained from 0.5 mL of PICs. Include appropriately sized molecular mass markers, like phage λ DNA digested with *Hin*dIII. Labeled markers will appear on the blot. If unlabeled markers are used, align a transparent ruler adjacent to the gel during UV photography to calculate standard distances. The gel running buffer should be circulated during electrophoresis.

10. Downstream manipulations, like acid denaturation, make the gel quite flimsy. Include the gel running tray during staining and subsequent steps for ease of handling. For efficient buffer equilibration, gently dislodge the gel from the tray bed and gently agitate.
11. This procedure is for the PosiBlot 30-30 Pressure Blotter (Stratagene). See reference *(28)* for a detailed capillary transfer procedure.
12. In our experience, the second gel transfers much more smoothly when two gels are transferred in succession. We therefore conduct an initial mock transfer for 30 min at 75 mmHg using two wetted 17 × 12 cm pieces of Whatman 3 filter paper (one below the mask and the other above it) before actual gel transfer.
13. It is recommended to pre-hybridize and hybridize in as small a volume as possible. This is most easily done using roller bottles and an oven equipped with a rotisserie. It is recommended to set up prehybridization, prepare the RNA riboprobe, and then hybridize and wash. If the PosiBlot or a similar apparatus is available, Southern blotting is completed in 1 day. Prehybridization is the last step in the process that can be extended to an overnight incubation.
14. If the probe is double-stranded, denature it by boiling the salmon DNA–probe mixture for 5 min. Place the mixture directly on ice for 3 min, and then add to the QuikHyb solution.
15. The blot background will increase significantly if hybridization proceeds for more than 1 h.
16. Due to exquisite sensitivities, RQ-PCR assays are highly susceptible to amplification of minor contaminating DNAs. Work in an area that is relatively free of potential contaminating plasmid DNA sequences. Use (i) micro pipettors that are not used to manipulate plasmid stock preparations, (ii) a fresh aliquot of sterile water, and (iii) filter-plugged pipette tips. Do not mix solutions using up–down pipetting motions.
17. The assay works best when LRT values exceed 10,000 copies/μL in the integration reaction product samples, equating to approximately 50 pg of HIV-1 DNA/0.5 mL of PIC lysate. As robust infections can yield ~ 250 pg, the RQ-PCR assay is ~ 5 to 10-fold more sensitive than the Southern blotting assay. After empirically defining levels of (i) HIV-1 DNA

synthesis, (ii) PIC activity, and (iii) assay robustness, it may be possible to reduce the volume of PIC lysate analyzed by RQ-PCR from 200 μL to 50–100 μL.

Acknowledgements

I thank J.E. Daigle, K. McGee-Estrada, and N.K. Raghavendra for critically reading the manuscript. This work was supported by NIH grants AI39394, AI52014, and AI70042.

References

1. Katzman, M. and Katz, R. A. (1999) Substrate recognition by retroviral integrases. *Adv Virus Res* 52, 371–395.
2. Pauza, C. (1990) Two bases are deleted from the termini of HIV-1 linear DNA during integrative recombination. *Virology* 179, 886–889.
3. Miller, M., Farnet, C., Bushman, F. (1997) Human immunodeficiency virus type 1 preintegration complexes: studies of organization and composition. *J Virol* 71, 5382–5390.
4. Fujiwara, T., Mizuuchi, K. (1988) Retroviral DNA integration: structure of an integration intermediate. *Cell* 54, 497–504.
5. Brown, P. O., Bowerman, B., Varmus, H. E., et al. (1989) Retroviral integration: structure of the initial covalent product and its precursor, and a role for the viral IN protein. *Proc Natl Acad Sci USA* 86, 2525–2529.
6. Lee, Y. M., Coffin, J. M. (1991) Relationship of avian retrovirus DNA synthesis to integration in vitro. *Mol Cell Biol* 11, 1419–1430.
7. Yoder, K. E., Bushman, F. D. (2000) Repair of gaps in retroviral DNA integration intermediates. *J Virol* 74, 11191–11200.
8. Turlure, F., Devroe, E., Silver, P. A., et al. (2004) Human cell proteins and human immunodeficiency virus DNA integration. *Front Biosci* 9, 3187–3208.
9. Brown, P. O., Bowerman, B., Varmus, H. E., et al. (1987) Correct integration of retroviral DNA in vitro. *Cell* 49, 347–356.
10. Ellison, V., Abrams, H., Roe, T., Lifson, J., et al. (1990) Human immunodeficiency virus integration in a cell-free system. *J Virol* 64, 2711–2715.
11. Farnet, C. M., Haseltine, W. A. (1990) Integration of human immunodeficiency virus type 1 DNA in vitro. *Proc Natl Acad Sci USA* 87, 4164–4168.
12. Hansen, M. S., Smith, G. J., Kafri, T., et al. (1999) Integration complexes derived from HIV vectors for rapid assays in vitro. *Nat Biotechnol* 17, 578–582.
13. Brooun, A., Richman, D. D., Kornbluth, R. S. (2001) HIV-1 preintegration complexes preferentially integrate into longer target DNA molecules in solution as detected by a sensitive, polymerase chain reaction-based integration assay. *J Biol Chem* 276, 46946–46952.
14. Lu, R., Vandegraaff, N., Cherepanov, P., et al. (2005) Lys-34, dispensable for integrase catalysis, is required for preintegration complex function and human immunodeficiency virus type 1 replication. *J Virol* 79, 12584–12591.
15. Dismuke, D. J., Aiken, C. (2006) Evidence for a functional link between uncoating of the human immunodeficiency virus type 1 core and nuclear import of the viral preintegration complex. *J Virol* 80, 3712–3720.
16. Oh, J., Chang, K. W., Hughes, S. H. (2006) Mutations in the U5 sequences adjacent to the primer binding site do not affect tRNA cleavage by Rous sarcoma virus RNase H but do cause aberrant integrations in vivo. *J Virol* 80, 451–459.
17. Chen, H., Wei, S.-Q., Engelman, A. (1999) Multiple integrase functions are required to form the native structure of the human immunodeficiency virus type I intasome. *J Biol Chem* 274, 17358–17364.
18. Salahuddin, S. Z., Markham, P. D., Wong-Staal, F., et al. (1983) Restricted expression of human T-cell leukemia–lymphoma virus (HTLV) in transformed human umbilical cord blood lymphocytes. *Virology* 129, 51–64.
19. Adachi, A., Gendelman, H. E., Koenig, S., et al. (1986) Production of acquired immunodeficiency syndrome-associated retrovirus in human and nonhuman cells transfected with an infectious molecular clone. *J Virol* 59, 284–291.
20. Dorfman, T., Luban, J., Goff, S. P., et al. (1993) Mapping of functionally important residues of a cysteine-histidine box in the human immunodeficiency virus type 1 nucleocapsid protein. *J Virol* 67, 6159–6169.

21. Julias, J. G., Ferris, A. L., Boyer, P. L., et al. (2001) Replication of phenotypically mixed human immunodeficiency virus type 1 virions containing catalytically active and catalytically inactive reverse transcriptase. *J. Virol.* 75, 6537–6546.
22. Butler, S. L., Hansen, M. S. T., Bushman, F. D. (2001) A quantitative assay for HIV DNA integration in vivo. *Nat Med* 7, 631–634.
23. Engelman, A., Englund, G., Orenstein, J. M., et al. (1995) Multiple effects of mutations in human immunodeficiency virus type 1 integrase on viral replication. *J Virol* 69, 2729–2736.
24. Miller, M. D., Wang, B., Bushman, F. D. (1995) Human immunodeficiency virus type 1 preintegration complexes containing discontinuous plus strands are competent to integrate in vitro. *J Virol* 69, 3938–3944.
25. Harrison, G. P., Mayo, M. S., Hunter, E., et al. (1998) Pausing of reverse transcriptase on retroviral RNA templates is influenced by secondary structures both 5′ and 3′ of the catalytic site. *Nucleic Acids Res* 26, 3433–3442.
26. Byers, K. B., Engelman, A., Fontes, B. (2004) General guidelines for experimenting with HIV, in *Protocols in Immunology Supplement 59* (Coligan, J. E., Bierer, B. E., Margulies, D. H., Shevach, E. M., and Strober, W., eds.), John Wiley & Sons, Hoboken, NJ, pp. 12.1.1–12.1.9.
27. O'Doherty, U., Swiggard, W. J., Malim, M. H. (2000) Human immunodeficiency virus type 1 spinoculation enhances infection through virus binding. *J Virol* 74, 10074–10080.
28. Sambrook, J., Fritsch, E. F., Maniatis, T. (eds.) (1989) *Molecular Cloning: A Laboratory Manual*, 2nd ed., Cold Spring Harbor Laboratory Press, Plainview, NY, pp. 9.34–9.36.
29. Wei, S.-Q., Mizuuchi, K., Craigie, R. (1997) A large nucleoprotein assembly at the ends of the viral DNA mediates retroviral DNA integration. *EMBO J* 16, 7511–7520.
30. Chen, H., Engelman, A. (2001) Asymmetric processing of human immunodeficiency virus type 1 cDNA in vivo: implications for functional end coupling during the chemical steps of DNA transposition. *Mol Cell Biol* 21, 6758–6767.
31. Bao, K. K., Wang, H., Miller, J. K., Erie, D. A., Skalka, A. M., and Wong, I. (2003) Functional oligomeric state of avian sarcoma virus integrase. *J Biol Chem* 278, 1323–1327.
32. Faure, A., Calmels, C., Desjobert, C., et al. (2005) HIV-1 integrase crosslinked oligomers are active in vitro. *Nucleic Acids Res* 33, 977–986.
33. Guiot, E., Carayon, K., Delelis, O., et al. (2006) Relationship between the oligomeric status of HIV-1 integrase on DNA and enzymatic activity. *J Biol Chem* 281, 22707–22719.
34. Li, M., Mizuuchi, M., Burke, T. R. J., et al. (2006) Retroviral DNA integration: reaction pathway and critical intermediates. *EMBO J* 25, 1295–1304.
35. Vincent, K. A., York-Higgins, D., Quiroga, M., et al. (1990) Host sequences flanking the HIV provirus. *Nucleic Acids Res* 18, 6045–6047.
36. Vink, C., Groenink, M., Elgersma, Y., et al. (1990) Analysis of the junctions between human immunodeficiency virus type 1 proviral DNA and human DNA. *J Virol* 64, 5626–5627.

Chapter 11

Bisarsenical Labeling of HIV-1 for Real-Time Fluorescence Microscopy

Nathalie J. Arhel and Pierre Charneau

Abstract

Imaging studies have benefited from the development of a novel technique for non-destructive labeling of proteins within living cells, based on the use of a reagent called FlAsH-EDT2, a bisarsenical derivative of fluorescein capable of binding with high affinity and specificity to a tetracysteine motif in the protein of interest. This technique has been adapted for the stable, sensitive and specific molecular tagging of HIV-1 IN enabling the tracking of incoming viral particles inside infected living cells. Here we present the experimental steps required for the efficient labeling of HIV-1 IN, namely, molecular insertion of a tetracysteine tag, production of viruses, labeling in vitro of tagged viruses, infection of target cells and visualization of particles by fluorescence microscopy.

Key words: HIV-1, real-time imaging, infection, fluorescence microscopy

1. Introduction

The dynamic imaging of single HIV-1 particles within living cells is fundamental to a better understanding of virus–host cell interactions. Although, in theory, any HIV-1 protein may be fluorescently labeled to follow the virus within living cells, the choice of the labeled protein will determine the phase of the replication cycle that can be visualized. Unlike nonenveloped viruses, nonspecific labeling of whole HIV-1 virions by conjugation of the envelope to a fluorophore (using amine- or thiol-reactive agents) does not enable virus observation within cells since the labeled envelope is lost after fusion. HIV-1 labeling protocols have therefore relied on the insertion of a genetically encoded fluorophore such as the Green Fluorescent Protein (GFP) within

Vinayaka R. Prasad, Ganjam V. Kalpana (eds.), *HIV Protocols: Second Edition, vol. 485*

DOI 10.1007/978-1-59745-170-3_11 Springerprotocols.com

viral proteins. The tagging of Viral protein R with GFP permitted the first observation of intracellular HIV-1 complexes in living cells *(1)*, and more recently, the incorporation of Gag-GFP within virions has also been reported *(2)*. However, the large size of GFP and its derivatives can perturb the conformation, function and localization of the tagged protein, and in particular, modifications in HIV-1 *gag* or *pol*, frequently lead to disruption of viral functions and dramatic loss of infectivity *(2, 3)*. These drawbacks meant that many steps of the HIV-1 replication cycle, such as intranuclear events, remained unclear until recently.

The laboratory of Roger Tsien *(4)* demonstrated the labeling of recombinant proteins containing a very small tetracysteine tag (CCXXCC) by a bis-arsenical derivative of fluorescein called 4′,5′-bis(1,3,2-dithioarsolan-2-yl)fluorescein, or FlAsH-EDT2 (Fluorescein Arsenical Helix binder, bis-EDT adduct; Invitrogen). The fact that FlAsH-EDT2 is a membrane-permeant fluorophore that remains non-fluorescent until it binds with high affinity and specificity to the tetracysteine motif makes the tetracysteine bisarsenical labeling technique ideally suited for the fluorescent tagging of a wide variety of microorganisms. Here we provide protocols for bisarsenical labeling of HIV-1 IN protein, which recently led to the real-time imaging of a series of post-fusion events in the early phase of the HIV-1 replication cycle, allowing us to monitor both intra-cytoplasmic and intra-nuclear infectious HIV-1 complexes *(5)*. Tagging HIV-1 IN is advantageous because there are about 75 IN monomer proteins per linear DNA genome, and integrases are closely associated with HIV-1 DNA, thus allowing for a specific and improved visualization of the HIV-1 viral genome. This method may also be adapted to the labeling of other HIV-1 proteins, such as Gag *(6)* or p24 capsid.

2. Materials

2.1. Plasmids

1. The wild-type HIV-1 molecular clone (LAI) may be obtained from the NIH AIDS Reagent Program.
2. The HIV-1 LAI molecular clone deleted in *env* was made by deleting a 580 bp BglII/BglII region within the *env* gene of LAI.
3. VSV-G envelope expression plasmid, pHCMV-G *(7)*

2.2. Virus Production

1. 293T cells (ATCC).
2. Polystyrene-treated tissue culture Petri dishes, 100 mm (Falcon).
3. DMEM glutamax with Na/Pyr (Gibco Invitrogen), 100U/100μg/ml penicillin/streptomycin, 2 mM L-glutamine, 10% fetal calf serum or serum-free.

4. 2× HEPES-buffered saline pH 7.06: 140 mM NaCl, 1.5 mM $Na_2HPO4 \cdot 2H_2O$, 50 mM HEPES (Sambrook and Russel, Molecular Cloning, 3rd Edition).
5. $CaCl_2$ 1 M.
6. p24 ELISA assay (Perkin Elmer).

2.3. Labeling Mix

1. FlAsH-EDT2 (Invitrogen) 2 mM stock in 90% DMSO, 10% H_2O.
2. 2-mercaptoethanol, BME (Sigma).
3. Triscarboxyethylphosphine-HCl, TCEP (Pierce).
4. 1,2-Ethanedithiol, EDT (Sigma).
5. Hepes 1 M (Gibco) pH 7.2–7.5.
6. Polycarbonate centrifuge tubes (11 × 34 mm, Beckman).

2.4. Infection

1. P4-CCR5 indicator cells are HeLa CD4 + CXCR4 + CCR5+ cells carrying the LacZ gene under the control of the HIV-1 LTR *(8)*, (AIDS Reagent Program). These are grown in 10% DMEM supplemented with 500 μg/mL G418 (Invitrogen).
2. Glass-bottom collagen coated culture chambers (MatTek Corporation).

3. Methods

FlAsH labeling of tetracysteine tagged HIV-1 may be adapted to the labeling of both wild-type envelope HIV-1 and VSV-G (vesicular stomatitis virus glycoprotein) pseudotyped HIV-1. Although labeling of wild-type HIV-1 viruses is efficient and viruses retain infectivity, VSV-G-pseudotyped HIV-1 viruses are more stable and naturally more infectious. Furthermore, the latter are easier to label since they allow for the full recovery of viral infectivity after the critical ultracentrifugation step necessary for the removal of unbound FlAsH, as well as to achieve greater multiplicities of infection.

FlAsH-EDT2 binds with high affinity to the peptide motif CCXXCC *(4)*. We have tested both CCRECC, and the more optimal CCPGCC motif *(9)*. In our hands, both motifs are equivalent when labeling HIV-1 IN with FlAsH-EDT2, but the CCPGCC motif is optimal for ReAsH labeling. We found that the inclusion of a linker region (AGAG) next to the tetracysteine motif, whether immediately upstream, downstream, or both (AGAG-CCXXCC, CCXXCC-AGAG, or AGAG-CCXXCC-AGAG), was preferable to a tetracysteine tag alone. More recently, a thorough cell-based library screening approach led to the identification of peptide motifs most optimized for contrast of the bisarsenical-tetracysteine complex relative to background: FLNCCPGCCMEP and HRWCCPGCCKTF (*10*).

Table 11.1
PCR primers for insertion of an AGAG-CCRECC tag C-ter of IN

CCRECC tag (forward)	5′ccagtactacggttaaggc 3′
CCRECC tag (reverse)	5′gaaacatacatatgctagcagcactcccggcagcagccggcgccggcatcctcatcctgtctacttg 3′

3.1. Fusion of a CCXXCC Tag at the C-Terminus of HIV-1 Integrase

PCR is used to amplify a 547-bp fragment including the tetracysteine motif from the LAI matrix. Using the oligonucleotides shown in **Table 11.1**, the IN STOP codon and 25 downstream nucleotides belonging to the *Vif* gene (up to the NdeI site) are replaced by the tetracysteine tag sequence (**Fig. 11.1**).

As an intermediate cloning vector, the PstI/NcoI proviral fragment is cloned into a Δ *Eco*RI pBluescript II SK+. In this plasmid, the relevant *Eco*RI and *Nde*I sites become unique. The PCR amplification is digested by *Eco*RI/*Nde*I and inserted within the *Eco*RI/*Nde*I digested pBluescript II SK+proviral fragment. The PstI/NcoI fragment is then recovered from pBluescript II SK+ and cloned into the full-length proviral plasmid.

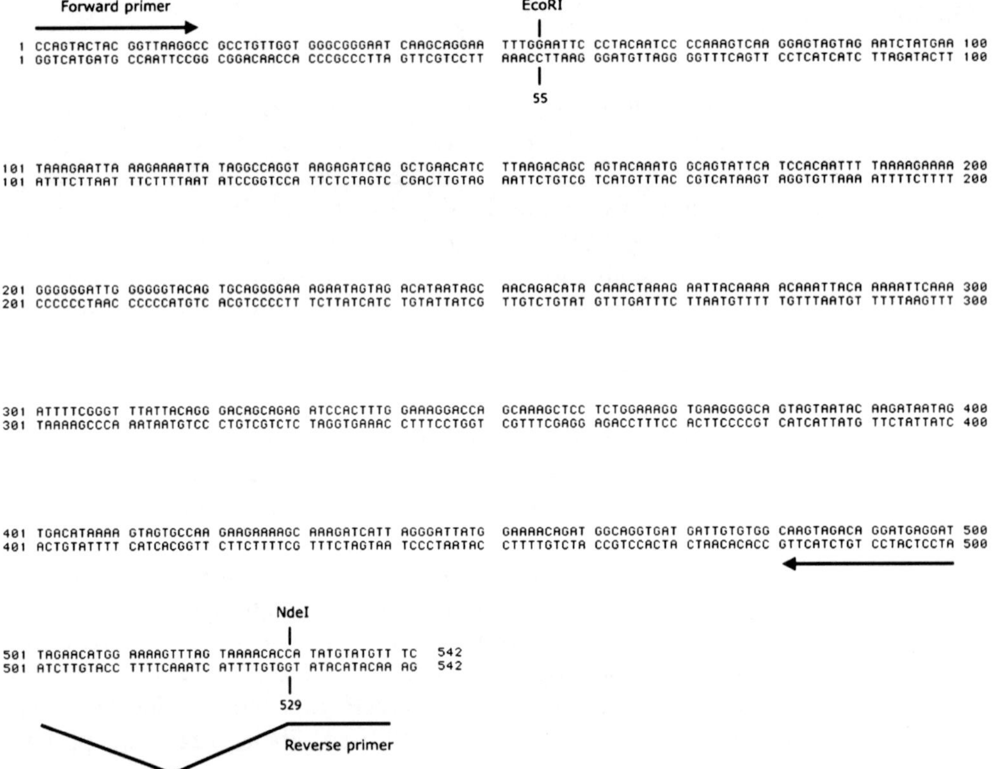

Fig. 11.1. PCR amplification fragment from the C-terminal region of LAI *IN* gene. The tetracysteine tag sequence is introduced by the downstream PCR primer.

The tetracysteine-containing proviral plasmid is amplified in bacteria and purified by MaxiPrep (Nucleobond AX PC500, Macherey-Nagel).

3.2. Virus Production

HIV-1 virus particles with wild-type envelope or VSV-G pseudotyped envelope are produced by transient transfection of 293T cells by calcium phosphate transfection method along with the proviral plasmids and the VSV-G envelope expression plasmid (pHCMV-G) in the case of Δ *env* viruses.

1. Seed 6×10^6 293T cells in Petri dishes 24 h prior to transfection.
2. Replace culture medium with ~ 8 mL fresh, prewarmed 10% DMEM medium 1 h prior to transfection to stimulate cell division (*see* **Note 1**). Cells should be no more than 80% confluent.
3. Prepare DNA mix: 10 μg of proviral plasmid (*see* **Note 2**), 5 μg VSV-G expressing plasmid (only for Δ *env* viruses), 125 μL $CaCl_2$, 350 μL deionized H_2O per Petri dish.
4. Add DNA mix drop-wise with continuous swirling onto 500 μL 2× Hepes per Petri dish (*see* **Note 3**).
5. Wait approximately 10 min for the DNA calcium phosphate precipitates to form. The solution should appear slightly cloudy and whitish.
6. Add DNA transfection mix onto cells (1 mL per Petri dish).
7. Replace culture medium with ~ 8 mL prewarmed serum-free medium 24 h post-transfection.
8. Collect supernatants 48 h post-transfection. Centrifuge supernatants at 2,500 rpm for 5 min to clarify them from cell debris. Aliquot into 1 mL aliquots and store at −80 °C. p24 viral antigen in supernatants is quantified using a p24 ELISA assay (Perkin Elmer). Viruses may also be titrated using quantitative PCR *(11)*.

3.3. FlAsH Labeling of Viruses

The experimental approach involves labeling viral particles in vitro prior to infection. This results in increased specificity of labeling and reduced background noise and toxicity linked to the use of reducing agents that are required for labeling within living cells. Tetracysteine-tagged virus particles are specifically labeled by the FlAsH ligand with a 15- to 30-fold increase in overall fluorescence above background *(5)*.

1. Prepare working solutions of each denaturing agent and FlAsH substrate in H20: 0.1 M ß-mercaptoethanol, 0.1 M TCEP, 1 mM EDT and 50 μM FlAsH-EDT2 (see **Note 4**).
2. To 1 mL viral supernatant (ideally containing 1–2 μg $p24_{CA}$) add 50 μL 1 M Hepes pH 7.5, and 10 μL of each denaturing agent, leading to final concentrations of 50 mM Hepes,

1 mM for ß-mercaptoethanol and TCEP, and 10 μM for EDT (*see* **Note 5**). Mix by inverting.

3. For control labeling reactions, replace the tetracysteine-tagged viral supernatants by non-tagged viral supernatants, or serum free medium.
4. Add 10 μl FlAsH-EDT2 solution (final concentration of 0. 5 μM). Mix by inverting.
5. Allow labeling reaction to proceed for 2 h at room temperature in the dark on a rocker.
6. Recover viruses and eliminate excess FlAsH-EDT2 by ultracentrifugation at 40,000 rpm for 30 min at 4 °C in a Beckman TL55i rotor.
7. After ultracentrifugation, remove supernatant and carefully remove residual drops from the sides of the tube using clean tissue paper.
8. Add 150 μL cold PBS onto the pellet (which would not be visible) and leave for 30 min at 4 °C to allow pellet to gently resuspend.
9. Pipet up and down to fully resuspend the pellet (*see* **Note 6**). For wild-type envelope viruses, do not pipet more than three to four times to avoid sheering the envelope.
10. Ultracentrifuged viruses may be quantified using a p24 ELISA assay, and FlAsH incorporation may be assessed by spectrofluorescence using a microplate fluorometer capable of the following settings: FlAsH excitation 500 nm, emission 535 nm. In our hands, 485–570 nm measurements were also adequate.
11. Viruses may either be used directly for infection, or may be stored for later experiments at −80 °C.

3.4. Infection and Observation Under the Fluorescence Microscope

The use of FlAsH-EDT2 to label tetracysteine-tagged HIV-1 IN has been validated for the detection of infectious HIV-1 complexes within living cells *(5)*. The fusion of a tetracysteine tag at the C-terminus of IN does not lead to loss of infectivity either in the early or in the late steps of HIV-1 replication, and labeling with the FlAsH ligand does not interfere with the normal functions of the virus.

1. Seed 0.3×10^6 HeLa P4 cells or other adherent cell type in a MatTek chamber 24 h prior to infection. Culture volume in MatTek chambers is ~ 3 mL.
2. For infection, remove the culture medium from the MatTek chamber with a 5 ml pipet, and with a 200-μL pipet completely remove the medium from the center of the chamber (containing the glass slide). Rapidly in order not to allow the cells to air-dry, add 50 μL of PBS containing FlAsH-labeled viruses (containing 100–200 ng $p24_{CA}$).

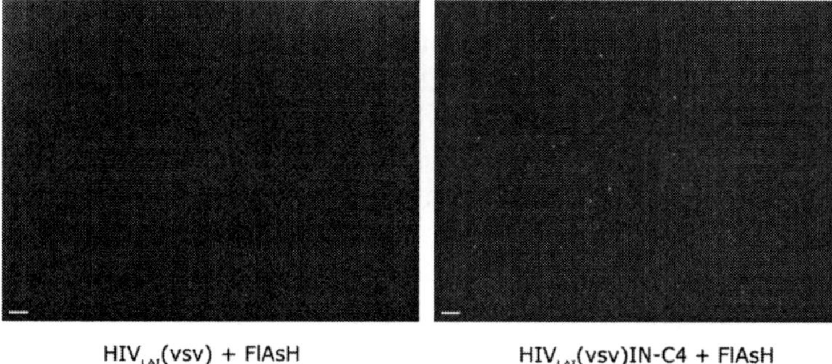

Fig. 11.2. Visualization of individual FlAsH-labeled HIV-1 particles by Total Internal Reflection Fluorescence microscopy (TIRF). HIV_{LAI}(vsv) and HIV_{LAI} (vsv)IN-C4 were incubated with FlAsH, ultracentrifuged, and spread onto a coverslip. Individual labeled particles were observed with a $100\times$ objective at 500–535 nm. Scale bars $= 5\,\mu m$.

3. Taking care not to allow the 50 μL drop to spill over into the chamber, allow infection to proceed for 1–2 h at 37 °C.
4. After infection, rinse the cells once in PBS, and add fresh culture medium.

FlAsH-labeled viruses can be observed within living cells using a 63× or 100× oil objective and high resolution

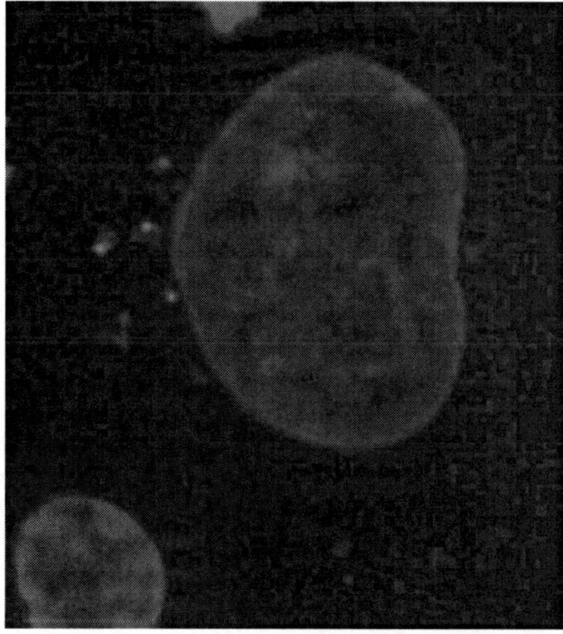

Fig. 11.3. Visualization of individual FlAsH-labeled HIV-1 particles by Apotome structured illumination fluorescence microscopy. P4-CCR5 cells were infected with FlAsH-labeled HIV_{LAI}(vsv) virus. Nuclei are visualized by DAPI staining, and viruses appear in green.

fluorescence microscopes. FlAsH-labeled virions appear as brightly fluorescent sub-resolution particles (**Note 7, Fig. 11.2**). FlAsH signals are subject to bleaching therefore good quality labeling is important in order to reduce acquisition times and laser intensities. We have obtained excellent resolution images using the following microscopes:

- Apotome (Carl Zeiss) structured illumination fluorescent imaging system mounted on an Axioplan 2 Imaging microscope equipped with a Plan-Apochromat 63×, NA 1.4 oil objective, a MRm CCD camera piloted *Axiovision* software (**Fig. 11.3**).
- Spinning disk confocal microscope (Perkin Elmer) on an Axiovert 200 microscope equipped with a 63×, NA 1.4 oil, Plan-Apochromat objective lens, a Hamamatsu ORCA ER CCD camera. Excitation laser: 488 nm, emission at 535 nm *(5)*.
- Spinning disk confocal microscope (Andor) with a DV885 EMCCD Andor camera. Excitation laser: 488 nm, emission at 535 nm.

4. Notes

1. It is important to prewarm the medium since temperature-shock will impair transfection and virus yield.
2. The quality of plasmids is critical for virus yield. These should have impeccable optical density ratios (≥ 1.8) and should ideally be loaded onto agarose gel to verify plasmid purity, high proportion of supercoiled DNA, and absence of recombination.
3. The pH of the HEPES buffer is critical for the formation of DNA calcium phosphate precipitates and for the yield of virus production. A pH value between 7.05 and 7.12 leads to good precipitate formation.
4. EDT and BME are unpleasantly odorous and are toxic by inhalation. All stock solutions should be prepared in a fume hood. When labeling the virus in a laminar flow hood, we recommend lowering the glass protection as low as possible.
5. The inclusion of disulfide bond reducing agents does not perturb viral infectivity. Their inclusion in the labeling reaction is essential since labeling will not occur if cysteines are in oxidized form. Moreover, inclusion of EDT minimizes the toxicity and nonspecific labeling that would result from free FlAsH-EDT2 binding to endogenous cysteine pairs. TCEP is aliquoted and stored at −20 °C, and a fresh aliquot is used for every labeling reaction. Old TCEP can lower the efficiency of the labeling reaction.

6. Recovery of viruses after ultracentrifugation is of approximately 40%.
7. The ultracentrifugation step needed to eliminate excess FlAsH label leads to some aggregation of particles and can induce an apparent heterogeneity in size of the particles.

References

1. McDonald, D., Vodicka, M. A., Lucero, G., et al. (2002) Visualization of the intracellular behavior of HIV in living cells. *J Cell Biol* 159, 441–452.
2. Müller, B., Daecke, J., Fackler, O. T., et al. (2004) Construction and characterization of a fluorescently labeled infectious human immunodeficiency virus type 1 derivative. *J Virol* 78, 10803–10813.
3. Engelman, A., Englund, G., Orenstein, J. M., et al. (1995) Multiple effects of mutations in human immunodeficiency virus type 1 integrase on viral replication. *J Virol* 69, 2729–2736.
4. Griffin, B. A., Adams, S. R., Tsien, R. Y. (1998) Specific covalent labeling of recombinant protein molecules inside live cells. *Science* 281, 269–272.
5. Arhel, N. J., Genovesio, G., Kim, K. A., et al. (2006) Quantitative 4D tracking of cytoplasmic and nuclear HIV-1 complexes. *Nat Meth* 3, 817–824.
6. Rudner, L., Nydegger, S., Coren, L. V., et al. (2005) Dynamic fluorescent imaging of human immunodeficiency virus type 1 gag in live cells by biarsenical labeling. *J Virol* 79, 4055–4065.
7. Yee, J. K., Miyanohara, A., LaPorte, P., et al. (1994) A general method for the generation of high-titer, pantropic retroviral vectors: highly efficient infection of primary hepatocytes. *Proc Natl Acad Sci U S A* 91, 9564–9568.
8. Charneau, P., Mirambeau, G., Roux, P., et al. (1994) HIV-1 reverse transcription. A termination step at the center of the genome. *J Mol Biol* 241, 651–662.
9. Adams, S. R., Campbell, R. E., Gross, L. A., et al. (2002) New biarsenical ligands and tetracysteine motifs for protein labeling in vitro and in vivo: Synthesis and biological applications. *J Amer Chem Soc* 124, 6063–6076.
10. Martin, B. R., Giepmans, B. N., Adams, S. R., et al. (2005) Mammalian cell-based optimization of the biarsenical-binding tetracysteine motif for improved fluorescence and affinity. *Nat Biotechnol* 23, 1308–1314.
11. Arhel, N. J., Munier, S., Souque, P., et al. (2006) Nuclear import defect of DNA Flap mutant HIV-1 viruses is not dependent on the viral strain or target cell type. *J Virol* 80, 10262–10269.

Subsection B

Late Events

Chapter 12

Methods for the Study of HIV-1 Assembly

Abdul A. Waheed, Akira Ono, and Eric O. Freed

Abstract

Virus assembly constitutes a key phase of the HIV-1 replication cycle. The assembly process is initiated by the synthesis of the Gag precursor protein, $Pr55^{Gag}$, in the cytosol of the infected cell. After its synthesis, $Pr55^{Gag}$ is rapidly transported in most cell types to the plasma membrane (PM) where it associates with the inner leaflet of the lipid bilayer. Gag-Gag interactions lead to the assembly of an electron-dense patch of Gag proteins at the membrane. The viral envelope (Env) glycoproteins associate with Gag during the assembly process. The highly multimerized Gag complex begins to bud outwardly from the PM and eventually pinches off from the cell surface. Concomitant with release, the viral protease cleaves $Pr55^{Gag}$ to the mature Gag proteins matrix, capsid, nucleocapsid and p6, leading to core condensation. The mature infectious virus particle is now able to initiate a new round of infection in a fresh target cell. Techniques have been developed in many laboratories to study each of the distinct phases of the HIV-1 assembly and release pathway. A number of these techniques are described in detail in this chapter.

Key words: Virus assembly; $Pr55^{Gag}$; membrane binding; lipid raft association; multimerization; epitope exposure, release rescue assay, punctate staining, virus budding, virus release.

1. Introduction

Retroviral Gag proteins are synthesized as polyprotein precursors in the cytoplasm of the infected cell and direct the assembly of nascent virus particles. While the Env glycoproteins and the *pol*-encoded enzymes are required for the generation of infectious virions, expression of Gag proteins alone is generally sufficient for the assembly and release of noninfectious, virus-like particles (VLPs). The HIV-1 Gag precursor, $Pr55^{Gag}$, is composed of matrix (MA), capsid (CA), nucleocapsid (NC), and p6 domains, as well as spacer peptides SP1 and SP2 (**Fig. 12.1**). Discrete regions have been identified within $Pr55^{Gag}$ that orchestrate the

Vinayaka R. Prasad, Ganjam V. Kalpana (eds.), *HIV Protocols: Second Edition, vol. 485*

DOI 10.1007/978-1-59745-170-3_12 Springerprotocols.com

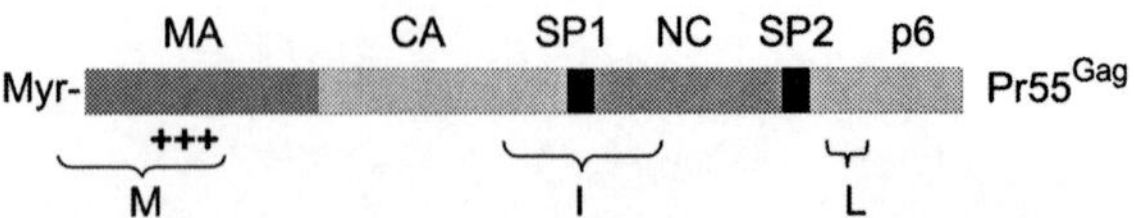

Fig. 12.1. Domain organization of $Pr55^{Gag}$. The membrane-binding domain (M) is composed of the N-terminal myristate (Myr) and the highly basic region of MA (+++). The interaction domain (I) spans the C-terminal portion of CA, SP1, and the N-terminal domain of NC. The late domain (L) consists primarily of the PTAP motif near the N-terminus of p6. Spacer peptides are shown as SP1 and SP2.

major steps in virus assembly and release: the MA domain mediates targeting of $Pr55^{Gag}$ to the plasma membrane (PM) and directs the incorporation of the Env glycoproteins into virions; the C-terminal portion of CA, SP1, and the N-terminal domain of NC promote Gag multimerization, in part via interactions between NC and the viral genomic RNA; and the p6 domain functions late in the assembly pathway to stimulate virus release from the PM. The mature Gag proteins are generated during virus release upon cleavage of $Pr55^{Gag}$ by the viral protease (PR). PR-mediated Gag processing leads to virus maturation, which is, in effect, a reassembly of CA, NC, and genomic RNA to form the mature conical core *(1)*.

In most cell types, HIV-1 assembly takes place primarily at the PM *(2, 3)*, although in some cells, including the physiologically relevant macrophage, assembly may occur in intracellular compartments known as multivesicular bodies (MVBs) *(3, 4)*. PM assembly appears to take place predominantly in cholesterol- and glycosphingolipid-enriched microdomains known as lipid rafts *(5)*. Lipid rafts can be isolated as detergent-resistant membrane (DRM) by TX-100 treatment at low temperature, followed by equilibrium flotation centrifugation. Recent findings from our lab indicate that phosphatidylinositol-*(4, 5)*-bisphosphate [PI(*4*, *5*)P_2], a member of the phosphoinositide family of lipids, directs $Pr55^{Gag}$ to the PM *(6)*, probably via a direct interaction between MA and PI(*4*, *5*)P_2 *(7–9)*. The HIV-1 assembly process can be subdivided into a series of discrete steps, which include: (1) Gag targeting to the PM; (2) membrane/lipid raft binding; (3) Env incorporation, (4) Gag multimerization; and (5) virus particle budding and release (**Fig. 12.2**). In this chapter we describe the methods to study these steps of the HIV-1 assembly and release pathway.

A number of full-length, infectious HIV-1 molecular clones are available for the study of HIV-1 assembly. Gag expression constructs can also be used to generate noninfectious VLPs. Our work has been based predominantly on the full-length HIV-1 molecular clone pNL4-3 *(10)* as well as a large number of pNL4-3 derivatives. These include Env-defective and PR-defective clones. Assembly-deficient mutants often serve as important controls in

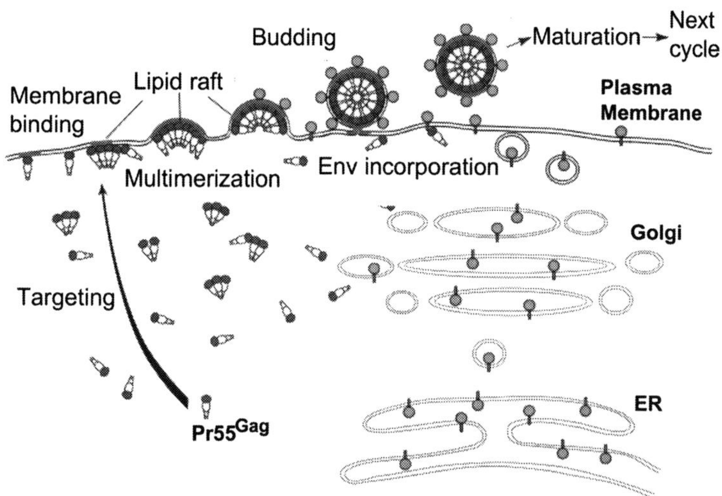

Fig. 12.2. Schematic representation of steps involved in virus assembly and release. After its synthesis, $Pr55^{Gag}$ is directed to the plasma membrane by its membrane targeting signal in the MA domain. At the plasma membrane, Gag associates with cholesterol- and sphingolipid-rich domains (lipid rafts). Gag–Gag interactions lead to the assembly of viral particles; the viral Env glycoproteins are incorporated during the assembly process. The highly multimerized Gag complex begins to bud outwardly from the plasma membrane and ultimately pinches off from the cell surface. The particle undergoes maturation upon cleavage of Gag and GagPol precursors by the viral protease.

virus assembly assays. Particularly useful are the myristylation-defective MA mutant, which is unable to bind membrane; assembly-deficient CA and NC mutants; and budding-deficient p6 mutants. Epitope-tagged Gag derivatives are also very useful, particularly for the study of Gag–Gag interactions. For this purpose, we have generated HA and FLAG-tagged derivatives.

2. Materials

2.1. Plasmids

1. Full-length HIV-1 molecular clone pNL4-3 *(10)*.
2. Env-defective clone pNL4-3/KFS contains a frameshift mutation at nucleotide 6343, which eliminates Env expression *(11)*.
3. PR-defective clone pNL4-3/PR$^-$ contains a PR active-site mutation (Asp to Asn) at PR amino acid 25 *(12)*.
4. Myristylation-deficient mutant pNL4-3/1GA contains a Gly-to-Ala change at MA amino acid 1 *(13)*. This mutant is defective in Gag binding to membrane.
5. CA dimerization mutant pNL4-3/WM184,185AA contains a two amino acid substitution at the CA dimer interface, thereby inducing a defect in Gag multimerization *(14–16)*.

6. NC basic amino acid substitution mutant pNLHX15A, in which all 15 basic residues of HIV-1 NC were mutated to alanine, is defective in higher-order Gag multimerization *(15, 17)*.
7. p6-mutant derivative pNL4-3/PTAP^{-} contains mutations in all four residues of the PTAP motif in the p6 domain of Gag *(12)*. The PTAP motif is involved in virus release and mutations in this motif result in tethering of particles at the PM *(12)*.
8. HA and FLAG-tagged derivatives pNL4-3-55HA and pNL4-3-55FLAG, in which the HA or FLAG epitope tags are fused to the C-terminus of Gag *(18, 19)*.
9. NL4-3-based GagPol expression vector pCMVNLGagPol-RRE *(20)*.
10. The vesicular stomatitis virus G glycoprotein (VSV-G) expression vector pHCMV-G *(21)*.

2.2. Cell Culture and Transfection Reagents

1. Dulbecco's modified Eagle's medium (DMEM) or RPMI (Cambrex, Walkersville, MD) supplemented with antibiotics and 5% or 10% fetal bovine serum (FBS, HyClone, Logan, UT).
2. Solution of trypsin (0.25%) and ethylenediamine tetraacetic acid (EDTA) (1 mM) from Gibco.
3. Calcium-phosphate transfection reagents.
   a. 1X HEPES buffer containing 138 mM NaCl, 5 mM KCl, 1 mM $Na_2HPO_4.7H_2O$, 5.5 mM dextrose, and 5 mM HEPES, pH adjusted to 7.05 (filtered and stored at 4 °C).
   b. 15% glycerol in 1X HEPES (stored at 4 °C).
   c. 1.25 M $CaCl_2$ in double-distilled water (DDW) (filtered and stored at 4 °C).
4. Lipofectamine™ 2000 transfection reagents.
   a. Lipofectamine™ 2000 (Invitrogen).
   b. OptiMEM-I reduced serum medium from Invitrogen to dilute Lipofectamine™ 2000 and nucleic acids before complexing.
   c. Antibiotic-free DMEM-5 or DMEM-10.
5. ExGen500 transfection reagents.
   a. ExGen 500 (Fermentas) is a sterile solution of linear polyethylenimine (PEI) molecules (22 kDa) in DDW at a concentration of 5.47 mM in terms of nitrogen residues.
   b. 0.15 M NaCl (endotoxin-free, sterile) solution for dilution of DNA.

2.3. Metabolic Labeling and Immunoprecipitation

1. RPMI-1640 medium lacking Met and Cys [Chemicon International (Temecula, CA)], with or without 2% FBS.
2. [^{35}S]Met/Cys [Perkin-Elmer (Wellesley, MA)].

3. Cell lysis buffer: 300 mM NaCl, 50 mM Tris-HCl (pH 7.5), 0.5% Triton X-100 (TX-100), 10 mM iodoacetamide, and 10 tablets of "Complete protease inhibitor" cocktail tablets (Boehringer Mannheim).
4. 2X radioimmunoprecipitation assay (RIPA) buffer: 280 mM NaCl, 16 mM Na_2HPO_4, 4 mM NaH_2PO_4, 2% NP-40, 1% sodium deoxycholate, 0.1% sodium dodecyl sulfate (SDS), 20 mM iodoacetamide, and protease inhibitors.
5. Protein A agarose beads (Invitrogen).
6. HIV-1 IgG obtained from the AIDS Research and Reference Reagent Program of the National Institutes of Health, purified from the serum of HIV-1-infected patients.
7. Triton wash buffer: 300 mM NaCl, 50 mM Tris-HCl (pH 7.5), 0.5% TX-100.
8. SDS/DOC wash buffer: 300 mM NaCl, 50 mM Tris-HCl (pH 7.5), 0.1% SDS, 0.1% deoxycholic acid.

2.4. SDS-PAGE, Autoradiography and Western Blotting

1. 29% acrylamide/1% acrylaide gel (Cambrex, Walkersville, MD) reagent for SDS-PAGE, followed by fluorography.
2. 30% acrylamide/0.8% bisacrylamide solution for SDS-PAGE, followed by Western blotting.
3. Lower gel buffer (4X): 1.5 M Tris-HCl, pH 8.7, 0.4% SDS.
4. Upper gel buffer (4X): 0.5 M Tris-HCl, pH 6.8, 0.4 % SDS.
5. 10% ammonium persulfate (APS) in DDW
6. N, N, N, N′-tetramethyl-ethylenediamine (TEMED) [Bio-Rad (Hercules, CA)].
7. Running buffer (10X): 250 mM Tris, 1.92 M glycine, 1% (w/v) SDS.
8. Prestained molecular weight markers (Invitrogen).
9. Fixative: 40% methanol and 10% acetic acid in DDW.
10. 1 M sodium salicylate in DDW.
11. Transfer buffer: 25 mM Tris, 0.19 M glycine, and 20% methanol.
12. Tris-bufferred saline containing 0.05% Tween 20 (TBS-T).
13. Blocking buffer: 5% (w/v) nonfat dry milk in TBS-T.
14. The Western lightning® chemiluminescence reagent *Plus* [Perkin-Elmer (Wellesley, MA)].

2.5. Membrane Flotation, DRM Isolation, Virion Purification

1. Teflon cell lifters (Fisher).
2. TE + Complete (hypotonic buffer): 10 mM Tris-HCl (pH 7.5), 4 mM EDTA, Complete protease inhibitor cocktail.
3. 4X TNE: 100 mM Tris-HCl (pH 7.5), 600 mM NaCl, 16 mM EDTA.
4. TNE + 0.5% [v/v] TX-100.
5. Sucrose solutions of the following concentrations (w/v) prepared in TNE: 85.5%, 65%, 60%, 50%, 40%, 30%, 20%, and 10%.

2.6. Immunofluorescence Microscopy

1. Lab-Tek eight-well glass chamber slides [Nalge Nunc (Naperville, IL)], and microscope cover slips (22 × 40 × 0. 15 mm) [Fisher (Pittsburgh, PA)].
2. Fresh 4% paraformaldehyde (w/v) solution in PBS.
3. Permeabilization solution: Methanol stored at −20 °C or 0.1% TX-100 in PBS.
4. Quench solution: 0.1 M glycine in PBS.
5. Antibody dilution buffer: 3% (w/v) BSA in PBS.
6. Secondary antibody: Alexa-488/594 [Molecular Probes (Eugene, OR)].
7. Nuclear stain: 300 nM DAPI (4.6-diamidino-2-phenylindole) in DDW.
8. Mounting medium: Antifade [Molecular Probes (Eugene, OR)].

2.7. Electron Microscopy

1. 2.5% glutaraldehyde in 0.1 M sodium cacodylate buffer, prepared fresh.
2. 0.1 M, 0.2 M, 0.4 M sodium cacodylate buffer, pH 7.2–7.4 (adjusted with 0.1 M HCl).
3. 2% osmium tetroxide in 0.2 M sodium cacodylate.
4. 0.1 N sodium acetate buffer, pH 4.5 (adjusted with acetic acid).
5. 35%, 50%, 70%, 90%, and 100% ethanol.
6. Propylene oxide (1,2 epoxy propane).
7. Epoxy resin (Epon-812 based): Mix 53.3 g of Poly/Bed® 812 resin, 20.7 g of dodecenyl succinic anhydride (E.M. grade), 26 g of nadic methyl anhydride (E.M. grade), and 1.4 mL of 2,4,6-Tri(Dimethylaminoethyl)phenol (DMP-30) in a plastic disposable beaker using a wooden spatula.
8. Uranyl acetate: 0.5% uranyl acetate in DDW (can be used for 3 months if stored in a brown glass bottle).
9. Reynold's lead citrate: Add 1.33 g lead nitrate, 1.76 g sodium citrate to 30 mL of DDW. Shake for 1 min and allow to stand for 30 min shaking occasionally. Add 8 mL of 1 M NaOH, mix and dilute to 50 mL with DDW, adjust pH to 12 (can be used for up to 6 months).

3. Methods

3.1. Preparation of Virus Stocks

Virus stocks are prepared to infect target cells and evaluate various aspects of the HIV-1 replication cycle. Virus stocks prepared by pseudotyping with VSV-G can be used to infect a wide range of target cells, including those not readily amenable to transfection or electroporation. Infectious virus stocks can be prepared as described below.

1. Plate HeLa or 293T cells (10^6 cells/60-mm dish)
2. Next day, transfect with HIV-1 molecular clone (e.g., pNL4-3 or its derivative) alone or with Gag-Pol expression vector pCMVNLGagPolRRE and VSV-G expression plasmid (pHCMV-G), using either of the transfection methods described below (*see* **Note 1**).

3.1.1. Calcium-Phosphate Transfection

Add 20 μg of DNA (total) to 0.45 mL of 1X HEPES and mix. Add 50 μL of $CaCl_2$, shake well and leave for 30 min at room temperature. Add the mixture to the cells and incubate overnight at 37 °C in a CO_2 incubator. The next morning, remove supernatant, add 4 mL of fresh DMEM-5.

3.1.2. Lipofectamine™ 2000 Transfection

Dilute 8 μg of DNA into 0.5 mL of OptiMEM-I reduced serum medium. Dilute 20 μL of Lipofectamine™ 2000 in 0.5 mL of OptiMEM-I medium. Incubate for 5 min at room temperature. Combine diluted DNA with diluted Lipofectamine™ 2000. Mix gently and incubate for 20 min at room temperature. Remove growth medium from cells and replace with 4 mL of antibiotic-free medium. Add 1 mL of the DNA-Lipofectamine complexes to 60-mm dish containing cells and mix gently. Incubate the cells at 37 °C in a CO_2 incubator for 4–8 h; medium may be replaced with fresh medium.

3.1.3. ExGen500 Transfection Reagent

Dilute 5 μg of DNA in 500 μL of 0.15 M NaCl and vortex gently. Add the 20 μL ExGen 500 and vortex the solution immediately for 10 s. Incubate for 10 min at room temperature. Add 500 μL of the ExGen 500/DNA mixture to 60-mm dish.

1. Incubate cells at 37 °C in a CO_2 incubator for 24–48 h.
2. Collect supernatant, pass through a 0.45-μm filter, aliquot into cryotubes, freeze on dry ice and store at −80 °C. Reserve a small aliquot for reverse transcriptase or p24 assay.

3.2. Infection and Transfection

Target cells are infected with virus stocks prepared as described in **Section 3.1** with appropriate input inoculum depending on the cell types used for study (discussed below). Viral proteins are often expressed by transient transfection with molecular clones as detailed in **Section 3.1**. However, infection is less cytotoxic than transfection and represents a more physiological method for transduction of the viral genome.

3.2.1. Infection of HeLa cells

1. Plate HeLa cells (0.4×10^6 cells/well of six-well plate).
2. The next morning, remove supernatant and add pseudotyped virus (amount used will depend on the purpose of the assay). Add DMEM-5 to a final volume of 0.8 mL and incubate for 6–10 h at 37 °C.

3. Remove supernatant and add 2 mL of fresh DMEM-5. Incubate for 2 days before biochemical analysis.

3.2.2. Infection of T cells

1. Place 2×10^6 cells per sample in a 15-mL tube. Spin at $260 \times g$ for 5 min; remove the supernatant.
2. Resuspend cells in a small amount of RPMI-10 (typically 0.3–0.5 mL) containing an appropriate amount of virus and incubate cells for 6–10 h in the CO_2 incubator.
3. Add 2 mL RPMI-10, spin cells, remove supernatant and resuspend cells in fresh RPMI-10 and transfer to T-25 flash and incubate for 2 days before biochemical analysis (*see* **Note 2**).

3.2.3. Transfection

1. Plate HeLa cells (0.4×10^6 cells/well in six-well plate).
2. The next day, transfect with approximately 3 μg DNA/well (ExGen500) or 6 μg/well (Lipofectamine) or 10 μg/well (calcium-phosphate) (detailed in **Section 3.1**).

3.3. Virus Release Assay

Comparing the virus release efficiency of WT vs. mutant molecular clones has provided detailed information concerning the domains in $Pr55^{Gag}$ responsible for mediating virus assembly and release. These assays can also be used to measure the effect of assembly/release inhibitors, siRNA depletion of cellular proteins, or other disruption of host cell machinery on virus particle production.

3.3.1. Cell Labeling

1. One day post-transfection or two days postinfection, starve the cells by rinsing with RPMI Cys^-/Met^- and adding 1 mL of RPMI Cys^-/Met^-. Incubate for 30 min at 37 °C.
2. Rinse cells again and add 1 mL of RPMI Cys^-/Met^- containing 2% FBS.
3. Add 250 μCi [^{35}S]Met/Cys; incubate at 37 °C, 2 h-overnight (**N**ote 3).

3.3.2. Pelleting Released Virions

1. Collect tissue culture supernatant or cell suspension; spin down at 2,000 rpm for 3 min in Sarstedt screw-cap tubes to pellet cells or remove cell debris.
2. Remove supernatant without disturbing the cell pellet and pass through 0.45-μm filter.
3. Ultracentrifuge the filtered supernatant at 35,000 rpm ($100,000 \times g$) for 45 min at 4 °C in a Sorvall ultracentrifuge using a TH-660 rotor or use comparable ultracentrifuge and rotor (e.g., Beckman SW50.1).
4. Remove supernatant by decanting, then wipe the inside wall of tube with rolled paper towel to remove as much supernatant as possible (do not touch the bottom of the tube).
5. Add 0.2 mL of PBS (to further purify the viral particles on sucrose gradient) or 0.1 mL of TX-100 cell lysis buffer. Leave

the tube at room temperature for approximately 10–15 min, and pipette several times to resuspend virus pellet.

6. Meanwhile, add 0.5 mL of lysis buffer to each well, leave for 10–15 min, and transfer cell and viral lysates to separate Sarstedt screw-cap tubes (*see* **Note 3**).

3.4. Purification of Viral Particles

Virus preparations obtained by pelleting tissue culture supernatants at $100,000 \times g$ may include contaminants such as soluble viral proteins, cell membrane-derived lipid microvesicles, and cell debris that may interfere with subsequent analysis. To increase the purity of the particle preparations, the pelleted virions can be subjected to equilibrium gradient centrifugation. A detailed protocol for purification of viral particles on sucrose gradients is described below. Other types of gradients (e.g., OptiPrep) can also be used.

1. Layer virion-containing supernatants prepared in step 5 of **Section 3.3** onto a 20% sucrose cushion; centrifuge at 35,000 rpm for 2 h at 4 °C in ultracentrifuge. Resuspend the viral pellet in 0.2 mL of PBS.
2. Layer 70, 65, 50, 40, 30, 20% [w/v] sucrose in TNE from bottom of tube (0.64 mL each) with pipetman and load 0.2 mL virus suspension on top (10%) sucrose layer.
3. Ultracentrifuge at 35,000 rpm, 16 h, at 4 °C (with slow acceleration/deceleration).
4. Remove fractions (~ 0.4 mL each) from the top of the gradient with a pipetman (try to place the tip of the pipetman just beneath the surface of the solution during fractionation). While taking the last fraction (Fr. 10), pipette several times to resuspend the (invisible) pellet.
5. Mix a portion of each fraction with sample buffer and analyze by Western blotting; store remaining material at −20 °C. An example of virion purification on a sucrose density gradient is shown in **Fig. 12.3**.
6. Pool the fractions that are enriched in viral proteins (p24CA or $Pr55^{Gag}$), dilute with TNE and spin down at 35,000 rpm for 1 h. Suspend the pellet in appropriate buffer depending on further use (*see* **Note 4**).

3.5. Immunoprecipitation and Fluorography

To evaluate the efficiency of virus production, the Gag in the viral particles produced from the metabolically labeled cells is quantified by immunoprecipitation followed by fluorography. Since accurate quantification of released particles depends on the amount of viral proteins made in the producer cells, it is essential to quantify the amount of cell associated viral protein expression. Therefore, immunoprecipitation of viral proteins from both viral

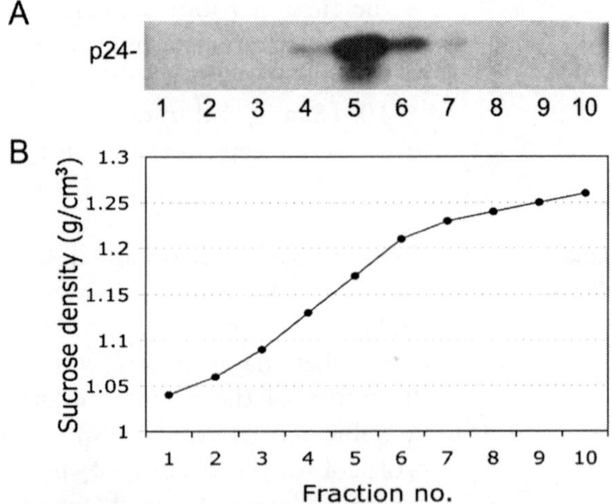

Fig. 12.3. Purification and density determination of viral particles. A, Western blot analysis of virions with HIV-Ig. Virions were concentrated by ultracentrifugation and were resuspended in 200 μL of PBS and layered onto a 20–70% (w/v) linear sucrose density gradient and subjected to ultracentrifugation in a TH-660 rotor (Sorvall) at $100,000 \times g$ for 16 h. Ten fractions were collected and analyzed by Western blotting. B, The density of sucrose is plotted against each fraction.

and cell lysates is carried out using HIV-Ig or other appropriate anti-HIV antiserum.

3.5.1. IP of Cell-Associated Material

1. Spin cell lysate for 1 min at 4 °C and transfer 50 μL of supernatant to Sarstedt tube containing 50 μL of protein A agarose beads (Gibco-BRL) and 50 μL 1% BSA in 1 mL TX-100 wash buffer.
2. Rotate for 1 h or more on an orbital rotor at 4 °C and spin for 1 min at full-speed in an Eppendorf microfuge (supernatant is the precleared lysate).
3. At the same time, prepare antibody-bead complex by combining 1 mL PBS, 0. 8 μL HIV-Ig or appropriate amount of other anti-HIV antiserum, 50 μL protein A, and 30 μL 1% BSA (per sample) in 15 or 50 mL tube (depending on the number of samples).
4. Rotate for 1 h or more on an orbital rotor at 4 °C. Spin for 5 min at 2,500 rpm and remove supernatant. Add 0.5 mL PBS/sample; aliquot 0.5 mL of this suspension to each Sarstedt tube, spin for 1 min at full speed in microfuge and remove supernatant.
5. Combine antibody-bead complex with precleared lysate and rotate for 1 h or more on an orbital rotor at 4 °C.

3.5.2. IP of Virion-Associated Material

1. Prepare antibody–bead complex as above and add half of the virion sample to antibody–bead complex. Raise volume to 1 mL with TX-100 lysis buffer.
2. Rotate for 1 h or more on an orbital rotor at 4 °C.

3.5.3. Washing of Beads

1. After incubations, wash three times with 1 mL of TX-100 wash buffer and once with 1 mL of SDS/DOC wash buffer.
2. Resuspend samples in 60 μL 1X sample buffer and boil the samples for 5 min. Mix the supernatant with 5 μL 0.1% bromophenol blue (BPB) and load samples onto acrylamide/acrylaide gel.

3.5.4. SDS-PAGE and Fluorography

1. Prepare lower gel by mixing 14 mL 30% acrylamide/0.8% bisacrylaide solution with 11 mL DDW, 9.9 mL 4X lower Tris buffer, 0.5 mL 2% APS and 40 μl TEMED. Pour the gel, leaving space for staking gel, and overlay with water-saturated isobutanol. Polymerization takes place in about 30 min.
2. Pour off the isobutanol and rinse the top of the gel with water.
3. Prepare stacking gel by mixing 2.7 mL 30% acrylamide/0.8% bisacrylamide solution with 10 mL DDW, 5 mL 4X lower Tris buffer, 0.2 mL 2% APS and 20 μL TEMED. Pour the gel and insert the comb.
4. Once the stacking gel has polymerized (~ 30 min), remove the comb and wash the wells with 1X running buffer.
5. Add 1X running buffer to the upper and lower chambers of the gel unit and load samples in wells. Also load prestained molecular weight markers in one or more wells.
6. Electrophorese until the dye front reaches the bottom of the gel.
7. Remove gel from plate and fix in 40% methanol, 10% acetic acid in water for 2 h. Rinse for 15 min in water and incubate in 250 mL of "Enhancer" (1 M salicylic acid + 5 mL glycerol) for 2 h. Place gel on paper towels to absorb excess liquid and dry in 100 °C oven for ~ 1.5 h if using AcrylAide gel, or dry in gel dryer if using conventional acrylamide gel.
8. Expose dried gel to X-ray film or phosphorimager screen for quantification. An example of a virus release assay performed with wild-type and mutant pNL4-3 molecular clone is shown in **Fig. 12.4** (*see* **Note 5**).

3.6. Western Blotting

In cases where metabolic labeling and immunoprecipitation are not required the detection of viral proteins can be carried out by Western blotting. For example, detecting p24 in equilibrium sucrose gradients to check the density of viral particles does not require radiolabeling and quantification. Also, cases in which antibodies are not available for efficient immunoprecipitation of a specific viral protein, Western blotting detection and quantification by digital imaging system can be used. In rescue assays in which epitope-tagged Gags are used, cell and virus lysates can be subjected to immunoblotting with anti-tag antibodies followed by quantification.

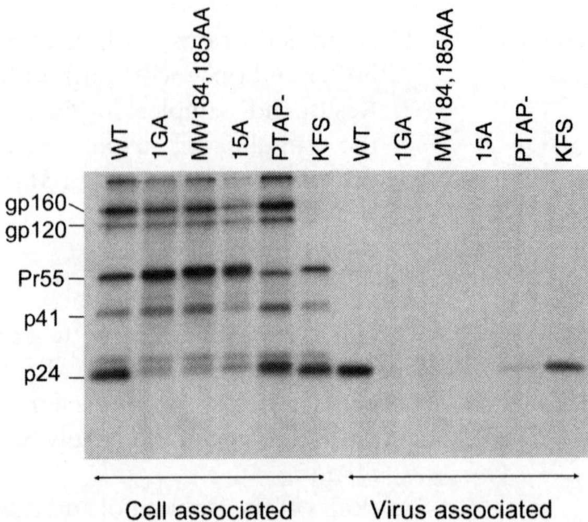

Fig. 12.4. Mutations in Gag lead to defects in virus release. HeLa cells transfected with WT pNL4-3 or the indicated Gag mutants that are defective in virus release or Env expression were metabolically labeled with [^{35}S]Met/Cys, and virions were pelleted by ultracentrifugation. Labeled viral proteins in cell and virion lysates were immunoprecipitated with HIV-Ig and analyzed by SDS-PAGE. Mutant clones are as follows: 1GA, nonmyristylated Gag mutant; WM184,185AA, CA C-terminal domain dimer interface mutant; 15A, clone in which all basic residues in NC were mutated to Ala; PTAP^{-}, mutant bearing substitutions in all four PTAP late domain residues; KFS, Env-minus molecular clone.

SDS-PAGE for Western blotting is performed as described in **Section 3.4**. Proteins separated by SDS-PAGE are transferred electrophoretically to PVDF membrane by using a semi-dry transfer apparatus.

3.6.1. Electrotransfer

1. Remove gel from glass plates and soak in transfer buffer.
2. Prepare a piece of PVDF membrane of appropriate size and soak in methanol for more than 30 s. Cut eight to 10 pieces of 3 mm Whatmann paper of the same size and soak in transfer buffer. Cut gel to the same size as PVDF membrane.
3. Place four to five pieces of 3 mm Whatmann paper, then PVDF membrane, then gel, then four to five pieces of 3 mm Whatmann paper in electroblot transfer apparatus. Avoid bubbles between PVDF membrane and gel.
4. Pour enough transfer buffer to prevent drying of the membrane and electrotransfer for 1–2 h at constant current, 1 mA/cm^2 membrane.

3.6.2. Immunoblotting

1. Incubate PVDF membrane with 50 mL of blocking buffer at room temperature for 1 h to reduce nonspecific binding.
2. Discard the blocking buffer and incubate the PVDF membrane in primary antibody appropriately diluted (2,000–10,000) with TBS-T or blocking buffer for 1 h.

3. Remove the primary antibody and wash three times for 5 min each with 50 mL of TBS-T.
4. Freshly prepare the secondary antibody (HRP-conjugated) for each experiment appropriately diluted in TBS-T or blocking buffer and add to the membrane and incubate for 1 h at room temperature.
5. Discard the secondary antibody and wash the membrane four times for 5 min each with 50 mL of TBS-T.
6. During washing, warm the Western lightning® chemiluminescence reagent *Plus* separately to room temperature, mix together and then immediately add to the membrane and incubate for 1 min.
7. Remove the membrane from the chemiluminescence reagent, wrapped with Saran wrap, and expose to X-ray film in dark for a few min. An example of Western blotting detection of $Pr55^{Gag}$ (WT or 1GA) in equilibrium flotation gradients is shown in **Fig. 12.5**.

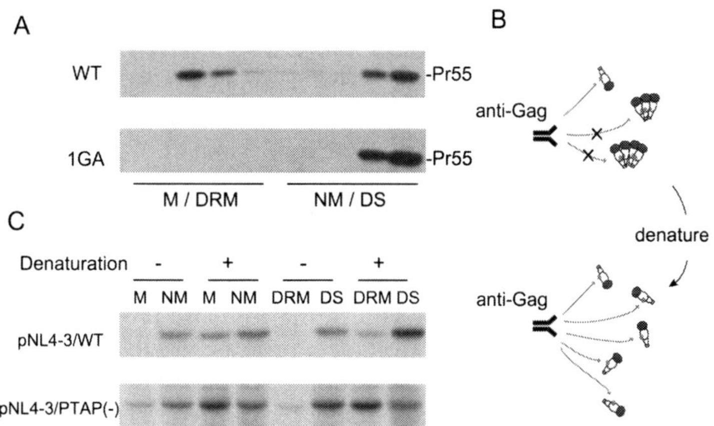

Fig. 12.5. Gag membrane binding, DRM association, and multimerization assay. A. Membrane binding of Gag. HeLa cells were transfected with pNL4-3/PR^{-} or pNL4-3/1GA/PR^{-}, PNS were prepared and subjected to membrane flotation centrifugation. The gradient fractions were subjected to Western blotting for detecting $Pr55^{Gag}$. M/DRM denotes membrane or DRM fractions; NM/DS denotes non-membrane bound or detergent soluble fractions. B. Schematic representation of epitope masking. When Gag is immunoprecipitated with anti-Gag antibodies not all the Gag is immunoprecipitated due to epitope masking resulting from higher-order multimerization. Upon denaturation with sample buffer, the multimerized Gag is converted to monomeric Gag, and the masked epitopes are exposed, increasing the efficiency of Gag immunoprecipitation. C. HeLa cells expressing either WT pNL4-3 or p6-mutant (pNL4-3/PTAP^{-}) $Pr55^{Gag}$ were pulse-labeled for 5 min and chased for 15 min. PNSs were incubated in the absence or presence of 0.25% TX-100 (final concentration) on ice for 30 min before membrane flotation. Fractions were treated with RIPA buffer and membrane (M) and nonmembrane (NM) fractions were pooled. Labeled $Pr55^{Gag}$ in each pooled fraction was recovered by immunoprecipitation either without (–) or with (+) prior denaturation.

4. Assays for Membrane Binding and DRM Association

As mentioned in the Introduction, in most cell types HIV-1 assembly takes place primarily at the PM. Specific cholesterol- and glycosphingolipid-enriched PM microdomains known as lipid rafts play an important role in HIV-1 assembly. Lipid raft-associated proteins can be isolated in DRM fractions based on the resistance of lipid rafts to solubilization in certain detergents (e.g., TX-100) at low temperature *(22)*. The analysis of membrane binding and DRM association of viral proteins thus provides important information about the late stages of the virus assembly/release process. A detailed protocol for measuring the membrane binding and DRM association of HIV-1 Gag is provided below.

1. Plate HeLa cells (10^6 cells/60-mm dish) and infect or transfect as described in **Section 3.2**.
2. One day posttransfection or two days postinfection, rinse cells with RPMI Cys^-/Met^- and add 1 ml of RPMI Cys^-/Met^-; incubate for 30 min at 37 °C.
3. Rinse the cells again and add 1 ml of RPMI Cys^-/Met^- containing 2% FBS.
4. Add 500 μCi [^{35}S]Cys/Met to the medium and incubate at 37 °C for 5 min in a CO_2 incubator.
5. Rinse the cells with DMEM-5, add the same medium, and incubate for 15 min at 37 °C in a CO_2 incubator.
6. Rinse the cells with cold PBS once and scrape with cell lifters in 1 mL cold PBS.
7. Transfer the cells to Sarstedt tubes and spin at 2,000 rpm for 2 min in microfuge.
8. Remove the PBS and resuspend the cells in 0.2 mL of TE + Complete protease inhibitor cocktail. Disrupt the cells by sonication or dounce-homogenization. Examine cells microscopically to verify complete disruption.
9. Spin the homogenate at 2,000 rpm for 3 min in a microfuge to remove the unbroken cells. The supernatant is the postnuclear supernatant (PNS).
10. Aliquot the PNS into two Sarstedt tubes (90 μL each) and add 2.7 μL of 5 M NaCl (final concentration 150 mM).
11. To the PNS, add 90 μL of either TNE or TNE + 0.5% TX-100 (prechilled), and incubate for 20–40 min on ice.
12. While samples are being solubilized, place 0.8 mL of TNE + 85.5% sucrose in each centrifuge tube. After incubation, add 160 μL of detergent-treated sample directly into the ultracentrifuge tube containing 85.5% sucrose and vortex carefully to mix the PNS (TX-100 untreated or treated) completely with sucrose (now the concentration of sucrose is 73%) (*see* **Note 6**).
13. Above the bottom 73% sucrose layer, overlay 2.4 mL of 65% sucrose in TNE without disturbing the interface, then overlay

with 0.8 mL of 10% sucrose in TNE. Carefully place the tubes in the ultracentrifuge buckets and spin at 35,000 rpm for 16 h at 4 °C

14. After ultracentrifugation, remove the tubes carefully, collect 10 fractions of 0.41 mL each from the top of each tube, and mix with 0.4 mL of 2X RIPA buffer.
15. Subject gradient fractions to immunoprecipitation and fluorography as described in **Section 3.5** or detect by Western blotting as detailed in **Section 3.6**. An example of Gag distribution in membrane and DRM fractions is shown in **Fig. 12.5** (*see* **Note** 7).

5. Gag Multimerization Assay

Membrane-bound $Pr55^{Gag}$ undergoes multimerization through Gag-Gag interactions that results in the assembly of viral particles. Gag mutants that are defective for Gag–Gag interactions do not multimerize and are thus defective in virus assembly and release. To study Gag multimerization, $Pr55^{Gag}$ derivatives bearing different C-terminal epitope tags have been constructed *(18, 19)*. FLAG-tagged nonmyristylated (1 GA) Gag is unable to bind membrane and does not assemble into VLPs when expressed alone. However, upon coexpression with HA-tagged WT Gag it is incorporated into virions via Gag–Gag interactions (**Fig. 12.6**). The extent to which the nonmyristylated mutant Gag protein is recovered in VLP fractions provides a measure of Gag–Gag interactions. Rescue of the mutant (FLAG-tagged) Gag protein into VLPs by the WT (HA-tagged) Gag protein can be monitored by quantitative immunoblotting using anti-HA and anti-FLAG antibodies. To monitor homotypic Gag multimerization in cells, our lab recently developed an assay based on multimerization-induced epitope masking *(15)*. When immunoprecipitation is carried out using anti-Gag antibodies, only a fraction of Gag is immunoprecipitated due to poor immunoprecipiation of highly assembled Gag complexes. However, when the samples are denatured prior to immunoprecipitation, Gag recovery is significantly increased (**Fig. 12.5B,C**). The difference in Gag recovery before and after denaturation thus provides a measure of higher-order Gag multimerization.

5.0.3. "Rescue-Based" Gag Multimerization Assay *(16, 18, 23)*

1. Cotransfected HeLa cells with various combinations of clones expressing epitope-tagged versions of WT $Pr55^{Gag}$ and Gag mutants (for example, HA-tagged WT Gag and a FLAG-tagged version of pNL4-3/1GA).
2. Perform virus release assay as described in **Section 3.3**.
3. Subject cell and virion lysates to SDS-PAGE and immunoblot with anti-HA and anti-FLAG antibodies; quantify levels of VLP-associated Gag. An example of rescue of 1GA mutant Gag into VLPs upon coexpression with WT Gag is shown in **Fig. 12.6**.

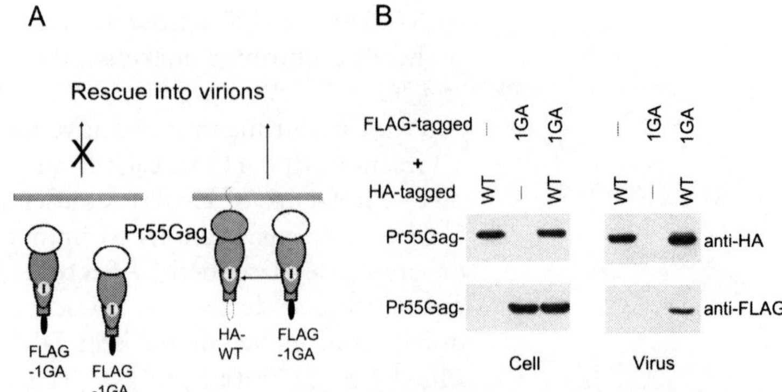

Fig. 12.6. In vivo rescue assay. (**A**) Schematic representation of in vivo rescue assay. The nonmyristylated Gag (FLAG-tagged) that is deficient in membrane binding does not produce virus particles. However, when coexpressed with HA-tagged WT Gag, the non-myristylated FLAG-tagged Gag is incorporated into virions through Gag–Gag interactions. (**B**) Coexpression of WT and mutant Gag proteins. HeLa cells were transfected with HA-tagged WT or FLAG-tagged Gag mutant alone or in combination at a 1:1 DNA. Virions were pelleted by ultracentrifugation, cell and virion lysates were analyzed by Western blotting with anti-HA or anti-FLAG antibodies. Reprinted from Joshi et al., 2006, *(16)* with permission. Copyright 2006, The American Society for Microbiology.

5.0.4. Gag Multimerization Assay Based on "Epitope Masking" (15)

1. Pool 200 μL each from upper (Frs. 1–5) and lower (Frs. 6–10) fractions from a sucrose gradient (*see* **Section 4.**) to 1.5 mL Sarstedt tubes; divide pooled fractions into two (0. 5 ml × 2).
2. Add 28. 5 μL 2X sample buffer to one of the tubes and boil for 5 min; transfer the contents to 15-mL tubes.
3. Add 3.2 mL TX-100 wash buffer to the 15 mL tubes and add 50 μL 1% BSA and 50 μL protein A agarose beads; incubate for 1 h at 4 °C.
4. Spin for 5 min at 2,500 rpm and collect supernatant for immunoprecipitation (this is the precleared lysate).
5. At the same time, combine 1 mL PBS, 0. 8 μL HIV-Ig, 50 μL protein A agarose beads, and 30 μL 1% BSA per sample in 15 or 50 mL tube (depending on the number of samples) and incubate for 1 h at 4 °C on an orbital rotor.
6. Spin for 5 min at 1, 120 × g, remove supernatant, add 0.5 mL PBS per sample, and aliquot 0.5 mL of the suspension to Sarstedt tubes and spin for 1 min at full speed in a microfuge.
7. Remove supernatant and combine the HIV-Ig-bound protein A agarose beads with precleared lysates in 15 mL tube and incubate overnight at 4 °C.
8. Spin for 5 min at 2,500 rpm, remove supernatant and wash three times with 1 mL TX-100 wash buffer, then wash once with 1 mL SDS/DOC wash buffer.
9. Resuspend samples in 60 μL of 1X sample buffer and boil the samples for 5 min. Mix the supernatant with 5 μL 0.1% BPB and subject to SDS-PAGE and fluorography as described in

Section 3.5. An example of the use of this assay to monitor Gag multimerization in membrane and DRM fractions is shown in **Fig. 12.5**.

5.1. Immunofluoresence Microscopy

Assembling virus particles have been observed either at the PM or in MVBs, depending on the cell type in which Gag is expressed. Further, we and others have described mutations that redirect Gag to intracellular sites *(13, 18, 24)*. The membrane binding assays described in **Section 4** do not distinguish Gag associated with PM from Gag bound to intracellular membrane. Thus, to identify the subcellular site(s) of virus assembly it is often highly informative to visualize Gag localization by immunofluorescence microscopy with a deconvolution or confocal microscope. The following steps are carried out for this purpose.

1. Plate HeLa cells in chamber slide. Cells should be 60–80% confluent on day 3 after plating.
2. One day after plating, transfect with appropriate amounts of DNA using one of the transfection reagents described in **Section 3.1**.
3. One day posttransfection, rapidly rinse cells once with PBS.
4. Add 4% paraformaldehyde for 10 min at room temperature to fix the cells, remove the upper plastic housing, leaving the gasket on the slide, and further incubate for 20 min. (slides can now be removed from HIV containment if using infectious material).
5. Discard the paraformaldehyde and wash the cells three to four times with PBS.
6. Permeabilize the cells by incubation in methanol (stored at −20 °C until use) for 4 min at −20 °C or with 0.1% TX-100 in PBS for 5 min at room temperature, with intermittent shaking; rinse three to four times with PBS.
7. Incubate the cells with 0.1 M glycine for 10 min, then with 3% BSA in PBS for 30 min at room temperature to block nonspecific binding.
8. Remove the blocking solution and incubate with primary antibody in 3% BSA at an appropriate dilution (1:100–1:250) for 1 h at room temperature or at 4 °C overnight.
9. Remove the primary antibody and wash the sample three times for 5 min each with PBS. Place the sample under aluminum foil and dim the lights for subsequent steps.
10. Prepare the fluorescently conjugated secondary antibody at 1:250 in 3% BSA and add to the sample for 1 h at room temperature.
11. Discard the secondary antibody and add DAPI for 10 min at room temperature to stain the DNA and identify the nuclei.
12. Wash the samples four times for 5 min each with PBS, then aspirate the excess PBS from one corner (avoid complete drying).

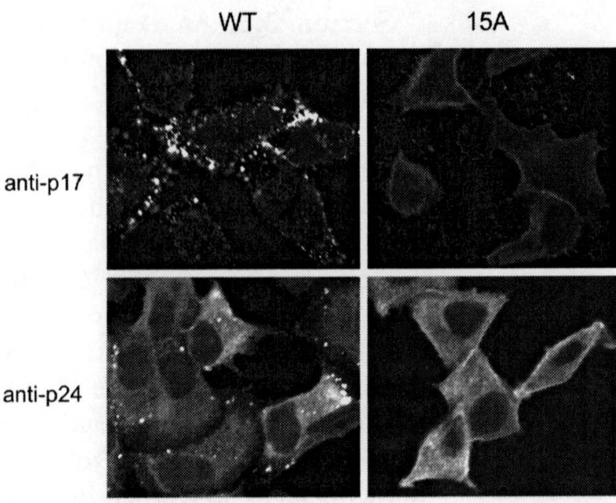

Fig. 12.7. Immunofluorescence staining showing the assembly of viral particles. HeLa cells expressing either WT or 15A-mutant $Pr55^{Gag}$ in the presence of active PR were immunostained either with monoclonal anti-p17 or -p24 antibody prelabeled with Zenon One Alexa Fluor reagent. The punctate staining pattern represents sites of virus assembly. Reprinted with permission from Ono et al., 2005, *(15)*. Copyright 2005, The American Society for Microbiology.

13. Mount the samples with mounting medium and a cover slip, seal with nail varnish and store at 4 °C for up to a month.
14. View the slides under a fluorescent/confocal microscope using appropriate excitation wavelengths. Excitation at 350 nm induces DAPI fluorescence (blue emission); excitation at 488 nm induces the FITC fluorescence (green emission); excitation at 543 nm induces Texas red fluorescence (red emission). Typical data are shown in **Fig. 12.7**.

5.2. Electron Microscopy

The fluorescence microscopy method described in **Section 5.1** provides information as to the site of Gag localization, but does not distinguish assembled from nonassembled Gag. To directly visualize the state of particle assembly, electron microscopy can be used. EM analysis also provides information crucial to discerning the morphology of assembled particles and the state of virion maturation.

1. Plate HeLa cells (0.4×10^6 cells/well of six-well plate).
2. One day post-plating, transfect with approximately 3 μg DNA/well (ExGen500), 6 μg/well (Lipofectamine), or 10 μg/well (calcium-phosphate) by using the transfection reagents detailed in **Section 3.1**. To study virus assembly and release in T-cell lines or in primary cells, infect the cells with VSV-G-pseudotyped virus prepared as described above (**Section 3.1**).
3. One day transfection Wash the cells once with PBS, and fix with 2.5% glutaraldehyde in 0.1 M sodium cacodylate buffer at room temperature, for a minimum of 2 h.

4. Remove the glutaraldehyde and wash the cells twice with 0.1 M cacodylate buffer for 10 min each time.
5. Postfix the cells in 1% osmium tetroxide in 0.2 M buffer for 1 h, and rinse twice in 0.2 M cacodylate buffer for 10 min.
6. Wash the cells once with 0.1 N acetate buffer for 10 min.
7. Incubate the cells with 0.5% uranyl acetate for 1 h.
8. Wash cells twice with 0.1 N acetate buffer for 10 min.
9. Dehydrate the cells in 35%, 50%, and 70% ethanol twice for 10 min each (samples may be stored overnight in 70% ethanol if absolutely necessary), then treat twice in 90% ethanol for 10 min, and three times with 100% ethanol for 10 min.

   Steps 10–13 are performed only with pelleted samples. For plated samples, go directly to step 14.
10. After dehydration, treat the cell or virus pellets twice with propylene oxide (1,2 epoxy propane) for 10 min.
11. Treat the samples with propylene oxide/epoxy resin mixture (50/50) for 1 h.
12. Add epoxy resin to the cell or virus pellets and leave the samples uncovered in the hood overnight to allow any remaining propylene oxide to evaporate.
13. Embed the samples in labeled capsules with freshly prepared resin and polymerize at 55 °C for 48 h.
14. Remove 100% ethanol from cells, add resin and leave the samples uncovered in the hood overnight to completely evaporate the remaining ethanol.
15. Add freshly prepared resin to the cells and polymerize at 55 °C for 48 h.

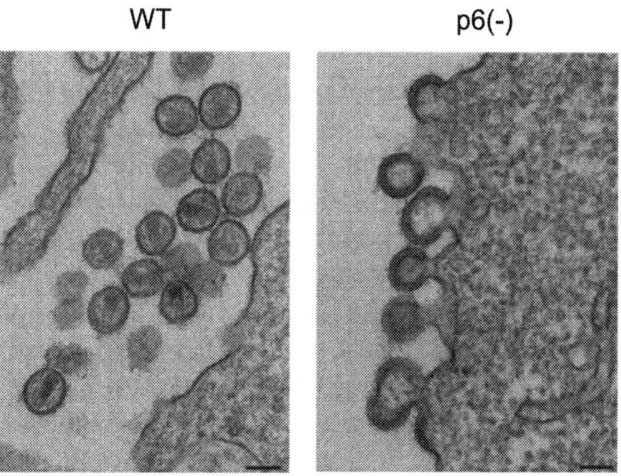

Fig. 12.8. Morphology of WT and p6-deleted HIV-1. HeLa cells transfected with WT pNL4-3 efficiently release mature virions (*left panel*), whereas cells expressing a p6-mutant molecular clone display numerous particles tethered to the plasma membrane (*right panel*). Images are visualized by transmission EM. Bar represents 100 nm (micrographs kindly provided by Kunio Nagashima. Reprinted from Freed and Martin *(25)*) Fields Virology 5th edition, copyright Lippincott Williams & Wilkins, with permission).

16. Cut sections of 100-nm thickness using a diamond knife with a ultramicrotome and the sections are picked up on carbon-coated 300 mesh copper grids.
17. Stain the sections with uranyl acetate and lead citrate, wash in DDW, and examine under an electron microscope. Typical electron micrographs of WT HIV-1 and p6-mutant derivative are shown in **Fig. 12.8** (*see* **Note 8**).

6. Notes

1. pCMVNLGagPolRRE is included in the transfection if the full-length molecular clone expresses an assembly-deficient Gag mutant or defective Pol enzymes. The VSV-G expression vector is included to increase the infectious titer of the virus stock and to allow infection of a greater range of cell types than is achievable by using the HIV-1 Env.
2. Amount of virus used in the infection will depend on the purpose of the assay and on the target cell line. For highly susceptible T-cell lines (e.g., MT-4), a smaller amount of input virus is needed relative to that used for less-susceptible lines (e.g., H9).
3. In experiments in which Env glycoproteins need to be quantified, carry out labeling with [^{35}S]Cys for more efficient Env labeling.
4. The sucrose equilibrium gradient described above is also used to measure the density of the viral particles. Densities of the fractions are determined by weighing 100 μl of each fraction on an analytical balance or by using a refractometer. The density of the viral particles corresponds to the density of the sucrose in which it is concentrated ($\sim$ 1.15–1.17 g/mL).
5. Virion-associated p24(CA) is normalized to the volume of cell lysate taken for quantification. Virus release efficiency is calculated as the amount of released p24(CA) divided by the total Gag (virion p24-CA + cell Pr55^{Gag} + cell p41^{Gag} + cell p24-CA). Since many mutations or treatments will affect the levels of cell-associated Gag, calculation of virus release efficiency based only on quantification of released p24-CA will often lead to incorrect conclusions.
6. Ultraclear (polypropylene, from Beckman) tubes are best for this application as they allow good visibility during collection of the fractions.
7. The optimal amount of TX-100 depends on the number of cells. 2X RIPA buffer preserves the multimerized Gag and inactivates any infectious particles in PNS. In cases where the

fractions are subjected to Western blotting alone, samples can be treated directly with 2X sample buffer.

8. All steps must be performed in a fume cupboard. Osmium tetroxide, propylene oxide and propylene oxide/resin waste should be collected in bottles for safe disposal.

References

1. Freed, E. O. (1998) HIV-1 gag proteins: diverse functions in the virus life cycle. *Virology* 251, 1–15.
2. Jouvenet, N., Neil, S. J., Bess, C., et al. (2006) Plasma Membrane Is the Site of Productive HIV-1 Particle Assembly. *PLoS Biol* 4, e435.
3. Ono, A., Freed, E. O. (2004) Cell-type-dependent targeting of human immunodeficiency virus type 1 assembly to the plasma membrane and the multivesicular body. *J Virol* 78, 1552–1563.
4. Pelchen-Matthews, A., Raposo, G., Marsh, M. (2004) Endosomes, exosomes and Trojan viruses. *Trends Microbiol* 12, 310–316.
5. Ono, A., Freed, E. O. (2005) Role of lipid rafts in virus replication. *Adv Virus Res* 64, 311–358.
6. Ono, A., Ablan, S. D., Lockett, S. J., et al. (2004) Phosphatidylinositol (4,5) bisphosphate regulates HIV-1 Gag targeting to the plasma membrane. *Proc Natl Acad Sci USA* 101, 14889–14894.
7. Shkriabai, N., Datta, S. A., Zhao, Z., et al. (2006) Interactions of HIV-1 Gag with assembly cofactors. *Biochemistry* 45, 4077–4083.
8. Saad, J. S., Miller, J., Tai, J., et al. (2006) Structural basis for targeting HIV-1 Gag proteins to the plasma membrane for virus assembly. *Proc Natl Acad Sci USA* 103, 11364–11369.
9. Freed, E. O. (2006) HIV-1 Gag: flipped out for PI(4,5)P(2). *Proc Natl Acad Sci USA* 103, 11101–11102.
10. Adachi, A., Gendelman, H. E., Koenig, S., et al. (1986) Production of acquired immunodeficiency syndrome-associated retrovirus in human and nonhuman cells transfected with an infectious molecular clone. *J Virol* 59, 284–291.
11. Freed, E. O., Martin, M. A. (1995) Virion incorporation of envelope glycoproteins with long but not short cytoplasmic tails is blocked by specific, single amino acid substitutions in the human immunodeficiency virus type 1 matrix. *J Virol* 69, 1984–1989.
12. Huang, M., Orenstein, J. M., Martin, M. A., et al. (1995) p6Gag is required for particle production from full-length human immunodeficiency virus type 1 molecular clones expressing protease. *J Virol* 69, 6810–6818.
13. Freed, E. O., Orenstein, J. M., Buckler-White, A. J., et al. (1994) Single amino acid changes in the human immunodeficiency virus type 1 matrix protein block virus particle production. *J Virol* 68, 5311–5320.
14. von Schwedler, U. K., Stray, K. M., Garrus, J. E., et al. (2003) Functional surfaces of the human immunodeficiency virus type 1 capsid protein. *J Virol* 77, 5439–5450.
15. Ono, A., Waheed, A. A., Joshi, A., et al. (2005) Association of human immunodeficiency virus type 1 gag with membrane does not require highly basic sequences in the nucleocapsid: use of a novel Gag multimerization assay. *J Virol* 79, 14131–14140.
16. Joshi, A., Nagashima, K., Freed, E. O. (2006) Mutation of dileucine-like motifs in the human immunodeficiency virus type 1 capsid disrupts virus assembly, gag-gag interactions, gag-membrane binding, and virion maturation. *J Virol* 80, 7939–7951.
17. Cimarelli, A., Sandin, S., Hoglund, S., et al. (2000) Basic residues in human immunodeficiency virus type 1 nucleocapsid promote virion assembly via interaction with RNA. *J Virol* 74, 3046–3057.
18. Ono, A., Orenstein, J. M., Freed, E. O. (2000) Role of the Gag matrix domain in targeting human immunodeficiency virus type 1 assembly. *J Virol* 74, 2855–2866.
19. Ono, A., Freed, E. O. (1999) Binding of human immunodeficiency virus type 1 Gag to membrane: role of the matrix amino terminus. *J Virol* 73, 4136–4144.
20. Ono, A., Freed, E. O. (2001) Plasma membrane rafts play a critical role in HIV-1 assembly and release. *Proc Natl Acad Sci USA* 98, 13925–13930.
21. Yee, J. K., Friedmann, T., Burns, J. C. (1994) Generation of high-titer pseudotyped retroviral vectors with very broad host range. *Methods Cell Biol* 43, 99–112.
22. Brown, D. A. (2006) Lipid rafts, detergent-resistant membranes, and raft targeting signals. *Physiology (Bethesda)* 21, 430–439.
23. Cimarelli, A., Luban, J. (2000) Human immunodeficiency virus type 1 virion density is

not determined by nucleocapsid basic residues. *J Virol* 74, 6734–6740.
24. Facke, M., Janetzko, A., Shoeman, R. L., et al. (1993) A large deletion in the matrix domain of the human immunodeficiency virus gag gene redirects virus particle assembly from the plasma membrane to the endoplasmic reticulum. *J Virol* 67, 4972–4980.
25. Freed, E. O., Martin, M. A. (2007) HIVs and their replication. in Fields Virology, Fourth Edition, Knipe and Howley, eds. Lippincott Williams & Wilkins.

Chapter 13

Assembly of Immature HIV-1 Capsids Using a Cell-Free System

Jaisri R. Lingappa and Beth K. Thielen

Abstract

For many years it has been known that viral capsid proteins are capable of self-assembly, but increasing evidence over the past decade indicates that in cells HIV-1 capsid assembly occurs via a complex but transient series of steps requiring multiple viral–host interactions. To better understand the biochemistry of HIV assembly, our group established a cell-free system that faithfully reconstitutes HIV-1 Gag synthesis and post-translational events of capsid assembly using cellular extracts, albeit more slowly and less efficiently. This system allowed initial identification of interactions that occur very transiently in cells but can be tracked in the cell-free system. Analysis of the cell-free system revealed that Gag progresses sequentially through a step-wise, energy-dependent series of assembly intermediates containing cellular proteins. One of these cellular proteins, the ATPase ABCE1, has been shown to play a critical role in the assembly process. The existence of this energy-dependent assembly pathway was subsequently confirmed in cellular systems, further validating the cell-free HIV-1 capsid assembly system as an excellent tool for identifying mechanisms underlying HIV-1 capsid formation. Here we describe how to assemble immature HIV-1 capsids in a cell-free system and separate assembly intermediates by velocity sedimentation.

Key words: capsid, assembly, cell-free system, wheat germ extract, in vitro translation, myristoylation, Gag, ABCE1, velocity sedimentation

1. Introduction

Immature HIV-1 capsids are protein shells that encapsidate the viral genome. These structures subsequently become enveloped by the host cell plasma membrane resulting in virus budding and release. Each immature HIV-1 capsid contains $\sim$5000 copies of the HIV-1 Gag polyprotein *(1)* that assemble into an irregular spherical capsid with a diameter of $\sim$100 nm. The events of assembly follow translation of the 55 kDa Gag polypeptide in

Vinayaka R. Prasad, Ganjam V. Kalpana (eds.), *HIV Protocols: Second Edition, vol. 485*

DOI 10.1007/978-1-59745-170-3_13 Springerprotocols.com

the cytoplasm of the host cell. Gag undergoes myristoylation, in which a 14-carbon fatty acid that is required for membrane targeting *(2–4)* is added to the N-terminus of Gag. During capsid formation, Gag polypeptides target to the cytoplasmic aspect of the host plasma membrane, where they form the immature capsid. In addition, the viral genome becomes encapsidated during the assembly process. Formation of the immature capsid is followed by capsid maturation, budding, and release of enveloped capsids from the cell (reviewed in *(5)*).

A decade ago, it was assumed that assembly occurs without the involvement of cellular factors, since purified recombinant Gag polypeptides, at high concentrations, were found to self assemble into spherical particles, albeit ones that are much smaller than immature capsids produced in cells *(6)*. Studies using the cell-free HIV-1 capsid assembly system have altered this view by revealing that in the context of cellular proteins, Gag assembly occurs through a stepwise, ordered, assembly pathway, utilizing energy as well as at least one cellular factor during post-translational events. Thus, the cell-free capsid assembly system has proven to be a powerful system for studying mechanisms of viral–host interactions during assembly, and has also contributed to an important paradigm shift in current thinking about HIV-1 capsid assembly.

The immature HIV-1 capsids produced in the cell-free system have been shown to closely resemble immature HIV-1 capsids produced in cells, by biochemical as well as electron microscopic criteria. In addition, key features of immature HIV-1 capsid formation that were initially described in cellular systems, such as the requirement for myristoylation and membranes, are faithfully reproduced in the cell-free HIV-1 capsid assembly system *(7)*. During assembly in the cell-free system and in cells, Gag progresses through a series of assembly intermediates that can be distinguished by their sedimentation value (10S, 80S/150S, 500S; *see* **Fig. 13.1**), culminating in the formation of ~ 750S completed immature HIV-1 capsids *(7)*. The finding of an energy requirement during post-translational events of capsid assembly *(7)* led to identification of a cellular ATPase (ABCE1) that appears to be critical for proper formation of immature HIV-1 capsids in the cell-free system and in cells *(8)*. ABCE1 is associated with the 80S/150S and 500S assembly intermediates but is not associated with the 10S form of Gag or with the completed immature capsid product *(8)*, which has a sedimentation value of 750S *(7)*. To date, the exact role of ABCE1 remains unclear, but it appears to chaperone events in capsid assembly, allowing efficient virus formation even when Gag is present at low concentrations in the hostile environment of the host cytoplasm.

All the features of the HIV-1 capsid assembly pathway were initially identified using the cell-free HIV-1 translation and

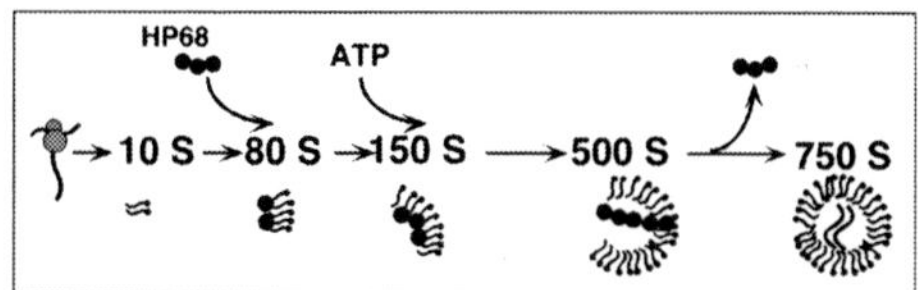

Fig. 13.1. **The HIV-1 assembly pathway**. Studies of assembly, initially in a cell-free system and subsequently in cells, has revealed that newly synthesized HIV-1 Gag progresses through a series of post-translational assembly intermediates of increasing size (10S, 80S, 150S, and 500S intermediates) before producing the completely assembled 750S immature capsid. This pathway requires ATP at a discrete step as indicated and also requires the cellular ATPase ABCE1, which associates with selected assembly intermediates.

assembly reactions described in this chapter. However, each of the key findings from the cell-free system has subsequently been verified in infected and transfected primate cells producing the HIV-1 provirus *(8--10)*. Moreover, studies suggest that all primate lentiviral Gag polypeptides assemble into immature capsids using a mechanism that closely resembles the mechanism underlying HIV-1 capsid assembly *(9, 11)*.

The methods below describe production of immature HIV-1 capsids using the HIV-1 cell-free translation and assembly reaction. The products of this reaction can then be analyzed by velocity sedimentation *(7, 8, 12)* as described below, which separates protein complexes based on their size. The cell-free reaction products can also be analyzed by immunoprecipitation *(7, 8, 13)* and transmission electron microscopy *(8)*. Notable variations of this method involve programming the cell-free translation and assembly reactions with HIV-1 Gag mutants *(7, 12)* or other primate lentiviral Gags *(9, 13)*. In addition, similar cell-free assembly reactions have been established for studying hepatitis B virus *(14)*, hepatitis C virus *(15, 16)*, and Venezuelan equine encephalitis virus (unpublished observations B. K. Thielen, K. C. Klein, and J. R. Lingappa).

2. Materials

2.1. In Vitro Transcription

1. 5X Compensating Buffer for Transcription (CB5X): 200 mM TrisAcetate, pH 7.5 (pH with Acetic Acid; *see* **Note 1**), 30 mM MgAcetate, 10 mM spermidine.
2. Nucleotide triphosphate (4NTP) mix: 5 mM ATP, 5 mM CTP, 5 mM UTP, 1 mM GTP. Adust pH to ~ 7.4 using 2 M Tris base. The NTPs are incorporated into the mRNA transcript (*see* **Note 2**).
3. 10X (5 mM) Diguanosine triphosphate (GE Healthcare Bio-Sciences Corp, Piscataway, NJ). This modified nucleotide

serves as the first nucleotide of the *in vitro* transcript and mimics the capped guanosine nucleotide that is produced by the 5′ capping reaction *in vivo*. To make 5 mM diguanosine triphosphate, dissolve 25 A250 units (1.2 mg) in 300 μL of 20 mM Tris Acetate pH 8.0.

4. 10X Dithiothreitol (DTT; Roche, Basel, Switzerland), 0.1 M.
5. 50X tRNA: Type XI bovine liver tRNA (Sigma, St. Louis, MO), 10 mg/mL.
6. 25X RNase inhibitor (Promega, Madison, WI), 40 units/μL (*see* **Note 3**).
7. 25X SP6 Polymerase, 10 units/μL (*see* **Note 4**).
8. Linearized DNA template. In the case of HIV Gag plasmid, this should be used at final concentration of 0.8–1.0 mg/mL. The DNA template for the *in vitro* transcription reaction is the SP6 HIV-1 Gag plasmid, which contains the SP6 promoter, followed by the 5′ untranslated region for xenopus oocytes and the coding region for HIV-1 Gag (from the SF 2 strain of HIV-1) (*see* **Note 5**).

2.2. Wheat Germ Translation

1. Wheat germ extract (WGE), made by the method of Erickson et al. *(17)* but without cyclohexane flotation (*see* **Note 6**). WGE made by this method should be subjected to centrifugation in a TLA 100 rotor (Beckman-Coulter, Fullerton, CA) at 50,000 rpm (96,000 × *g*) for 15 min, and the supernatent removed and flash frozen in liquid nitrogen (*see* **Note 7**). Typically, WGE should constitute 20% of the final reaction volume (i.e., wheat germ stock is 5X), although this could vary with specific batch of WGE.
2. 10X Compensating Buffer for Translation (10X CB) contains salts needed for the final reaction (adjusted for salts introduced by WGE used at a 20% volume in the final reaction): 40 mM Hepes, pH 7.6 (from a stock of 1.0 M Hepes, pH 7.6), 1.2 M potassium acetate (from a 4-M Potassium acetate stock), 2 mM EDTA.
3. Energy and Amino Acid Mix (5X Emix): 5 mM ATP, 5 mM GTP, 60 mM creatine phosphate, and 0.2 mM each of all amino acids except methionine, and 1 mCi of ^{35}Smet/cys Translabel (MP Biomedicals, Solon, OH) (*see* **Note 8**).
4. Myristoyl Coenzyme A lithium salt (Sigma), made as a 125 μM stock in 20 mM Tris Acetate, pH 7.4 (*see* **Note 9**).
5. Creatine kinase (Roche), made as a 4 mg/mL stock in 20 mM Tris Acetate, pH 7.4 with 50% glycerol and stored at −20 °C. Creatine kinase allows for rapid regeneration of ATP by transferring a phosphate to ADP from creatine phosphate.
6. RNase inhibitor (Promega), 40 units/μL.
7. Type XI bovine liver tRNA (Sigma), 10 mg/mL.

8. In vitro transcription reaction, 20% of final in vitro translation reaction volume.

All stocks for transcription and translation except enzymes are best stored as aliquots at −80 °C. Thaw at room temperature, keep on ice for short times, and flash freeze in liquid nitrogen for storage as soon as possible. Most reagents can tolerate many flash freezes, but it is best to aliquot WGE so that each aliquot is thawed fewer than ∼ 20 times. The enzymes (SP6 polymerase, RNase inhibitor, and creatine kinase) should be stored at −20 °C in 50% glycerol.

2.3. Velocity Sedimentation Gradients

1. 10X NP40 Buffer: 500 mM KAc, 0.1 M Tris Acetate 3.0% NP-40 (Roche), 1 M NaCl, 40 mM MgAc.
2. Sucrose solutions: 10%, 15%, 20%, 40%, 50%, 66%, and 80% wt/vol solutions in 1XNP40 buffer (can be stored at −20 °C).
3. Optima TL centrifuge and MLS 50 rotor (Beckman-Coulter)
4. 5 mL polyallomer gradient tubes (Beckman-Coulter)

3. Methods

The cell-free HIV-1 capsid assembly system involves two linked reactions: in vitro transcription and cell-free translation. A plasmid encoding the SP6 polymerase and the *Xenopus* globin 5′ untranslated region *(18)* upstream from the HIV-1 Gag coding region constitutes the template used to program in vitro synthesis of the HIV Gag mRNA transcript *(7)*. Components that are added to the in vitro transcription reaction include the ribonucleotide triphosphates (rNTPs) that form the mRNA transcript, as well as the modified rGTP diguanosine nucleotide triphosphate that forms the 5′ mRNA cap. Additionally, the reaction includes an RNase inhibitor, the SP6 polymerase enzyme, and bovine tRNA, which also acts to inhibit RNases. The buffers in this in vitro transcription reaction are adjusted to be compatible with the cell-free translation/assembly reaction, thereby eliminating the need to purify the mRNA transcript. Thus, the in vitro transcription reaction can be added directly to the cell-free translation/assembly reaction at a final volume of 20%. WGE is used in the cell-free translation/assembly reaction as a source of cellular factors, while amino acids and energy substrates (ATP and GTP) are all added exogenously (**Fig. 13.2**), since they are removed during a desalting step that occurs during preparation of WGE. Cellular factors present in the WGE include initiation and elongation factors required for translation *(17)*, membrane vesicles required for Gag targeting *(7, 11)*, as well as cellular proteins required for assembly such as ABCE1 *(8–10)*. In addition, the cellular extract contains

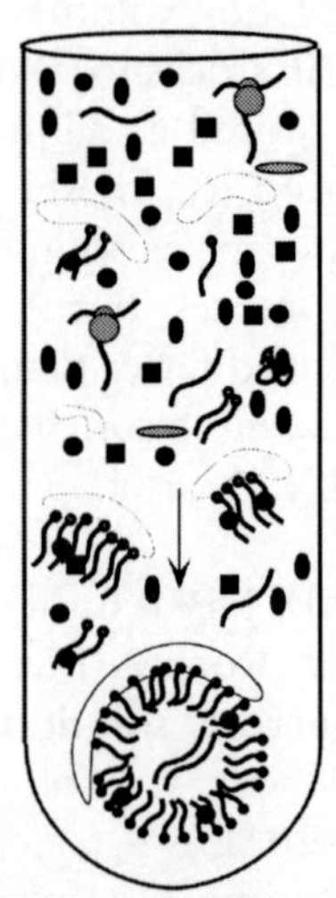

Fig. 13.2. **Components of the cell-free HIV-1 capsid assembly system and events reconstituted by this system.** The system is programmed using an in vitro mRNA transcript that codes for HIV-1 Gag. Wheat germ extract (WGE) provides cytosolic factors that are critical for translation and assembly, as well as membrane vesicles important for Gag targeting during assembly. Unlabeled amino acids, a radiolabeled amino acid (mainly methionine), and energy substrates (ATP and GTP) are added exogenously since they are lost from wheat germ extract during buffer exchange. Creatine phosphate and creatine kinase are added to the system to allow regeneration of ATP and GTP. The enzyme N myristoyl transferase, which myristoylates Gag, is present in wheat germ extract, but myristoyl coA must be added since it is lost during preparation of wheat germ extract. The cell-free system reconstitutes translation of Gag, co-translational myristoylation of Gag, membrane targeting of Gag, and multimerization of Gag in a manner that closely resembles these events in infected cells.

membrane vesicles derived from various organelles during preparation of the WGE (*see* **Note 10**).

While this cell-free system faithfully reconstitutes the events of Gag multimerization that occur in infected cells, it should be noted that there are a number of limitations to this system. First, the final product of the cell-free translation/assembly reaction is radiolabeled HIV-1 Gag (also called p55), while in infected cells, both Gag and GagPol are produced. Notably, the GagPol polyprotein encodes the HIV-1 protease which cleaves Gag and Pol during maturation, so in the absence of GagPol, Gag remains unprocessed and spherical immature HIV-1 capsids (rather than cone-shaped mature capsids) are produced. Additionally, in infected cells, many other viral proteins are produced, while in the cell-free system Gag is the only viral protein present. Moreover, only 20–40% of newly-synthesized Gag polypeptides assemble into completed immature HIV-1 capsids in the cell-free system. The remainder of newly synthesized radiolabeled Gag polypeptides are arrested in the form of various assembly intermediates. Finally, although immature capsid formation is reproduced with great fidelity in the cell-free translation/assembly reaction, neither budding nor envelopment is reconstituted in these cell-free reactions. Nevertheless, since production of immature capsids is a necessary prerequisite for proper budding, envelopment, release, and maturation of virus particles, this reaction can be used to study mechanisms involved in the critical first stage of virus formation. Moreover, by altering the plasmid used to generate the in vitro transcript, the reactions described here can also be used to study the effect of mutations in HIV-1 Gag on assembly *(7, 9, 11, 12)*, or to study assembly of other primate lentivirus immature capsids *(9, 11)*.

3.1. HIV-1 Gag in Vitro Transcription

For each translation, prepare the following 6.0 μL in vitro transcription Master Mix from components listed in 2.1 above:

5X CB	2 μl
4NTP mix	1.0 μL
CAP	1.0 μL
1 M DTT	1.0 μL
SP6 polymerase	4 units (0.4 μL)
RNase inhibitor	10 units (0.4 μL)
bovine liver tRNA	2 μg (0.2 μL)

To the 6.0 μL Master Mix, add 4.0 μL linearized HIV-1 Gag plasmid (2 or 2.5 μg/μL).

Incubate the 10-μL reaction for 1.25 h at 40 °C. At 50 min into the incubation, add an additional 2 units of SP6 polymerase to replenish the enzyme in the reaction. Stop the reaction by placing it in an ice bath.

3.2. Cell-Free Translation/Assembly Reaction

For a standard velocity sedimentation reaction, set up the following 25 μL cell-free translation/assembly reaction (also called the final reaction):

CB 10X	2.5 μL
5X EMIX containing 35S met/cys	5.0 μL
WGE	5.0 μL
Myristoyl CoA	2.0 μL
RNase inhibitor	0.25 μL
Creatine Kinase	0.25 μL
tRNA	0.25 μL
transcript (from 3.1)	5.0 μL

We have previously shown that capsid assembly in the cell-free system is highly concentration dependent *(11)*. Thus, anything that reduces the amount of Gag protein synthesized will result in a dramatic reduction in capsid assembly (*see* **Note 11**).

3.3. Velocity Sedimentation Gradients

To make 10/15/20/40/50/66/80% sucrose step gradients, first make each sucrose solution by dissolving the appropriate amount of sucrose into 1X NP40 buffer. Form step gradients in 5-mL centrifuge tubes by using a pipetman to layer 675 μL of each sucrose/NP40 solution, starting with the 80% sucrose and overlayering with progressively lighter solutions (*see* **Note 12**).

Dilute the 25-μL cell-free reaction into a final volume of 100 μL using 1XNP 40 buffer. Layer this sample gently onto the 10% (top) sucrose layer of the gradient. Centrifuge in an MLS 50 rotor (Beckman-Coulter) for 45 min at 45,000 rpm (163,000 × *g*). Fractionate into serial 200 μL fractions at the end of the spin (*see* **Note 13**).

3.4. SDS-PAGE and Autoradiography

Subaliquot 20 μL from each fraction into loading buffer containing SDS and DTT, and load serially into lanes of a 10% or 12% SDS-PAGE gel. When gel has finished running, fix the proteins into the gel by agitating in a solution of 30% methanol/10% acetic acid, and dry it onto filter paper using a gel dryer. The dried gel can be exposed to Biomax MR film (Kodak, Rochester, NY) to generate autoradiographs showing the amount of 35-S labeled HIV-1 Gag in different fractions of the velocity sedimentation gradients. Alternatively, a phosphorimager can be used to visualize radiolabeled Gag in gradient fractions. When using the gradients described above, the 10S Gag-containing complex (or assembly intermediate) is present in fractions 1 and 2; the 80S assembly intermediate is typically present in fractions 4–6; the 150S assembly intermediate is typically present in fractions 7–9; the 500S assembly intermediate is typically present in fraction 16–18; and the 750S completed capsid is typically present in fractions 21–24 (*see* **Note 14**). Approximately 20–40% of newly synthesized Gag is typically found in fractions 21–24. The pellet typically contains aggregates of virus particles or denatured Gag.

4. Notes

1. Chloride ions inhibit cell-free translation in the wheat germ system, thus acetic acid, rather than HCl, should be used to pH solutions.
2. 10X 4NTP mix is made from separate 0.1 M stocks of each rNTP (sodium or lithium salts dissolved in water). These stocks can be kept frozen at −80 and thawed to make the 4NTP mix. Note that the amount of GTP is lower because of the presence of capped GTP in the reaction (see below). SP6 RNA polymerase incorporates capped GTP at the 5′ end of the newly synthesized transcripts, thereby mimicking the capping reaction that occurs post-transcriptionally in vivo.
3. RNase inhibitor is sufficient for inhibiting RNase A activity in these reactions, so use of DEPC treated water and other reagents for inactivating RNAse A are not required. However,

use of glassware that has been baked for 6 h at 180 °C to inactivate RNase A is recommended while making solutions.

4. While other polymerases can be used for *in vitro* transcription, the system described here is set up as a linked transcription/translation system in which the transcription is performed using SP6 polymerase.
5. For best results, plasmid should be linearized at any restriction site that is downstream from the Gag coding region and creates 5′ overhangs, and linearized plasmid should be repurified using phenol/chloroform extraction and dissolved in water at concentration of 2.0–2.5 mg/mL (as determined by OD_{260}), with an $OD_{260/280}$ ratio ≥ 1.5 indicating sufficient DNA purity. In the absence of linearization, somewhat less transcript may be produced. The 5′ UTR is important for optimizing translation.
6. Production of WGE has been described in detail previously *(17)*. Our group always uses WGE that we produce from fresh wheat germ ourselves (as opposed to commercial WGE) because that allows us to control the quality of the WGE we use.
7. WGE contains an as yet unidentified inhibitory factor that is removed by the centrifugation described. In addition, some lots of WGE support translation and capsid assembly better than others, so multiple batches are usually made and tested for their quality. WGE should be kept thawed on ice for as short a time as possible. Aliquots can be repeatedly flash frozen.
8. The nonradioactive components of the Emix can be set up in advance as a mix of unlabeled components. The unlabeled component mix consists of:
   10 μL of 01.M ATP (5 mM final concentration); 10 μL of 01.M GTP (5 mM final concentration); 30 μL 0.4 M creatine phosphate (60 mM final concentration); and 40 μL of a 19 amino acid mix in which each amino acid except methionine is at 0.4 mM.
   Bring this to a volume of 90 μL and store at −80. Just before the first use of this Emix, thaw the unlabeled component mix, add 1 mCi (~ 100 μL) of ^{35}Smet/cys Translabel, adjust the pH to ~ 7.5 with 2 M Tris base, bring final volume to 200 μL with water, mix, aliquot, flash freeze, and store at −80 °C.
9. Myristoyl CoA is the substrate for myristoylation and must be added exogenously because it is lost upon preparation of WGE during desalting. The enzyme responsible for myristoylation, N-myristoyl transferase, is present in WGE. Myristoylation is critical for proper targeting of Gag in the cell-free system *(7)* and in cells *(2, 4)*. The percent of newly synthesized Gag that undergoes myristoylation is not known.

10. Historically, cell-free translations have been performed using WGE or rabbit reticulocyte lysate (RRL). However our group always uses WGE for cell-free translation/assembly because to date we have not found that assembly occurs efficiently in RRL. It has also been reported that RRL can be used for cell-free assembly of HIV-1 capsids *(19)*, but the authors of this study did not comment on the percent of newly synthesized Gag that forms fully assembled capsids when RRL is used (i.e., reaction efficiency).
11. The efficiency of immature capsid assembly is dependent on the concentration of Gag produced in the cell-free assembly reaction. At maximal efficiency, up to 35% of HIV-1 Gag assembles into immature capsids in the cell-free translation/assembly reaction although more typically, ~ 25% of radiolabeled Gag assembles *(7)*. We demonstrated this by programming cell-free translation/assembly reactions with diluted in vitro mRNA transcript and demonstrating that this resulted in a proportional reduction in both the amount of radiolabeled Gag synthesized as well as the amount of capsid formation *(11)*. On a practical level, this means that suboptimal in vitro transcription or poor quality WGE will result in low levels of capsid production.
12. Note that the gradients that are used to analyze these reactions are not linear. This is because the assembly pathway contains three relatively small complexes that are similar to each other in size (the 10S, 80S, and 150S assembly intermediates) and two large complexes (the 500S assembly intermediate and the 750S completed capsid) that are also similar in size, with a relatively large gap in between the 150S and 500S complexes. To optimally separate all of these complexes we use a step gradient that maximizes resolution in both the 10–150S and the 500–750S regions. S values of assembly intermediates are calculated S values (see reference 7 for details on how to calculate S values).
13. Our group uses either fractionation by hand from top to bottom, or fractionation using a Haake–Buchler automatic gradient fractionator (Labconco Corporation, Kansas City, MI) from top to bottom. However, there are many other approaches to gradient fractionation and the method described here can be adapted to any other approach including bottom to top fractionation.
14. To mark the position of immature capsids in this gradient, we collect 2.0 mL of medium containing virus from cells transfected to express Gag, concentrate the medium 10-fold using an Amicon concentrator, bring to 0.6% NP40 to remove virus envelopes, and layer these authentic HIV-1 capsids onto the sucrose gradients described in **Section 3.3** above in parallel with cell-free reactions.

Acknowledgments

This work was supported by NIH R01AI48389 to JRL; BKT was supported by the Medical Scientist Training Program NIGMS 5 T32 GM07266, an STD/AIDS Research Training Fellowship NIH T32 AI007140, and a Poncin Award.

References

1. Briggs J. A., Simon M. N., Gross I, et al. (2004) The stoichiometry of Gag protein in HIV-1. *Nat Struct Mol Biol* 11, 672–675.
2. Bryant M., Ratner L. (1990) Myristoylation-dependent replication and assembly of human immunodeficiency virus 1. *Proc Natl Acad Sci USA* 87, 523–527.
3. Gheysen D., Jacobs E., de Foresta F., et al. (1989) Assembly and release of HIV-1 precursor Pr55gag virus-like particles from recombinant baculovirus – infected insect cells. *Cell* 59, 103–112.
4. Gottlinger H. G., Sodroski J. G., Haseltine W. A. (1989) Role of capsid precursor processing and myristoylation in morphogenesis and infectivity of human immunodeficiency virus type 1. *Proc Natl Acad Sci USA* 86, 5781–5785.
5. Klein K., Reed J. C., Lingappa J. R. (2007) Intracellular destinies: degradation, targeting, assembly, and endocytosis of HIV Gag. *AIDS Rev* 9, 150–161.
6. Campbell S., Rein A. (1999) *In vitro* assembly properties of human immunodeficiency virus type 1 Gag protein lacking the p6 domain. *J Virol* 73, 2270–2279.
7. Lingappa J. R., Hill R. L., Wong M. L., et al. 1997) A multistep, ATP-dependent pathway for assembly of human immunodeficiency virus capsids in a cell – free system. *J Cell Biol* 136, 567–581.
8. Zimmerman C., Klein K. C., Kiser P. K., et al. (2002) Identification of a host protein essential for assembly of immature HIV-1 capsids. *Nature* 415, 88–92.
9. Dooher J. E., Lingappa J. R. (2004) Conservation of a step-wise, energy-sensitive pathway involving HP68 for assembly of primate lentiviral capsids in cells. *J Virol* 78, 1645–56.
10. Dooher J. E., Schneider B. L., Reed J. C., et al. (2007) Host ABCE1 is at plasma membrane HIV assembly sites and its dissociation from Gag is linked to subsequent events of virus production. *Traffic* 8, 195–211.
11. Lingappa J. R., Newman M. A., Klein K. C., et al. (2005) Comparing capsid assembly of primate lentiviruses and hepatitis B virus using cell-free systems. *Virology* 333, 114–123.
12. Singh A. R., Hill R. L., Lingappa J. R. (2001) Effect of mutations in Gag on assembly of immature human immunodeficiency virus type 1 capsids in a cell-free system. *Virology* 279, 257–270.
13. Dooher J. E., Lingappa J. R. (2004) Cell-free capsid assembly of primate lentiviruses from three different lineages. *J Med Primatol* 33, 272–280.
14. Lingappa J. R., Martin R. L., Wong M. L., et al. (1994) A eukaryotic cytosolic chaperonin is associated with a high molecular weight intermediate in the assembly of hepatitis B virus capsid, a multimeric particle. *J Cell Biol* 125, 99–111.
15. Klein K. C., Dellos S. R., Lingappa J. R. (2005) Identification of residues in the hepatitis C virus core protein that are critical for capsid assembly in a cell-free system. *J Virol* 79, 6814–6826.
16. Klein K. C., Polyak S. J., Lingappa J. R. (2004) Unique features of Hepatitis C Virus capsid formation revealed by de novo cell-free assembly. *J Virol* 78, 9257–9269.
17. Erickson A. H., Blobel G. (1983) Cell-free translation of messenger RNA in a wheat germ system. *Methods Enzymol* 96, 38–50.
18. Melton D. A., Krieg P. A., Rebagliati M. R., et al. (1984) Efficient *in vitro* synthesis of biologically active RNA and RNA hybridization probes from plasmids containing a bacteriophage SP6 promoter. *Nucleic Acids Res* 12, 7035–7056.
19. Spearman P., Ratner L. (1996) Human immunodeficiency virus type 1 capsid formation in reticulocyte lysates. *J Virol* 70, 8187–8194.

Chapter 14

Preparation of Recombinant HIV-1 Gag Protein and Assembly of Virus-Like Particles In Vitro

Siddhartha A.K. Datta and Alan Rein

Abstract

The mechanism of assembly of retroviruses is not fully understood. Purification of retroviral Gag protein and studying its solution state and assembly properties might provide insights into retroviral assembly mechanisms. Here we describe a rapid method for the purification of Gag and its subsequent assembly into virus-like particles in a defined system in vitro. The purification scheme does not use affinity tags, but purifies the native protein by virtue of its high affinity for phosphocellulose, a property presumably related to the affinity of Gag proteins for nucleic acids.

Key words: Gag, HIV-1, virus assembly, recombinant protein purification, virus-like particle, electron microscopy.

1. Introduction

Expression of a retroviral Gag protein in mammalian cells leads to the production and release of virus-like particles (VLPs) from the cells, even in the absence of other viral gene products or packageable genomic RNA. Isolation of the protein in pure form allows one to characterize it biochemically, and to assess possible contributions from the intracellular environment to the assembly process. As first shown by Campbell and Vogt *(1)*, retroviral Gag proteins expressed in bacteria can be readily purified in native form; these proteins can then assemble into VLPs under defined conditions. This chapter will describe the purification of an HIV-1-derived Gag protein and its use in an in vitro assembly reaction.

The protein we have used in these experiments is Δ16–99 Δp6 Gag from Gross et al. *(2)*. These investigators

Vinayaka R. Prasad, Ganjam V. Kalpana (eds.), *HIV Protocols: Second Edition, vol. 485*

DOI 10.1007/978-1-59745-170-3_14 Springerprotocols.com

termed the protein ΔMA-CA-NC-SP2; it contains sequences from both the NL4-3 and BH10 clones of HIV-1. We find that the deletion of residues 16–99 enhances the efficiency with which purified Gag protein assembles in vitro. The p6 domain at the extreme C-terminus of authentic HIV-1 Gag protein has also been deleted from the protein to avoid cleavage by bacterial proteases; to our knowledge, the absence of p6 has no significant effect on the assembly properties of the protein *(2)*. Finally, the protein we describe here lacks the N-terminal myristate modification found on HIV-1 Gag produced in eukaryotic cells. We have found that myristoylated Gag protein has very low solubility, and thus we do not know how the myristyl group might affect the assembly properties of recombinant Gag protein. For convenience, we will refer to the protein described here simply as "Gag".

We present below a rapid procedure for the purification of HIV-1 Gag protein. This was first described by Campbell and Vogt *(3)*. This protocol will typically yield protein of 80–90% purity; this is sufficient for in vitro assembly experiments and electron microscopy of the VLPs, as well as for many biochemical analyses and for use of the protein in immunoblotting, etc. For applications requiring higher purity, the preparation can be "polished" by other techniques such as size exclusion chromatography. With minor modifications, the same procedure can be used for the preparation of Gag proteins from other retroviruses: the modifications would probably be in the ammonium sulfate concentration used at the initial purification step, and in the NaCl concentrations used in the phosphocellulose chromatography steps. It should be noted that we do not use $(His)_6$ or other affinity tags, to avoid the possibility that these "extra" residues affect the properties of the protein.

We recommend that small aliquots be saved from multiple steps along the purification protocol; analysis of these aliquots by SDS-PAGE (and/or immunoblotting) will provide invaluable information for trouble-shooting if the ultimate yield is low; if the protein has suffered degradation; or if it is still contaminated by significant amounts of bacterial proteins.

2. Materials

2.1. Protein Expression

1. Plasmid (pET series) (Novagen) (we have used pET3xc).
2. BL21(DE3)pLysS *Escherichia coli*
3. Luria–Bertani broth (LB), supplemented in all cases with appropriate antibiotics.
4. Isopropyl-beta-D-thiogalactopyranoside (IPTG).

2.2. Protein Purification

1. A 500-W sonicator with probe (Sonics VCX-500, or similar product).
2. Ammonium sulfate (saturated solution at room temperature and crystals).
3. Fibrous cellulose phosphate (Sigma).
4. Buffer A: 20 mM Tris–HCl (pH 7.4), 1 mM phenylmethylsulfonyl fluoride (PMSF), 10 mM β–mercaptoethanol (or 1 mM Tris[2-carboxyethyl] phosphine [TCEP]),1 μM $ZnCl_2$ (*see* **Note 1**).
5. Buffers B/C/D/E are buffer A with 0.1/0.2/0.5/1.0 M NaCl respectively.
6. Buffer F: 20 mM Tris–HCl pH 8.0, 1 mM DTT.
7. Lysis buffer: Buffer A with 0.75 M NaCl and 10% w/v glycerol. Optional additions include protease-inhibitors of choice (Complete Protease inhibitor (Roche), as recommended by the manufacturer), and 0.05% Triton X-100 (w/v) (Roche).
8. Reagents for SDS-PAGE including 4X loading buffer, Coomassie brilliant blue stain, and MW markers.
9. Reagents for Bradford protein assay with IgG standard.

2.3. In Vitro Assembly

1. Dialysis tubing or Slide-A-Lyzer (Pierce) or equivalent microdialysis apparatus.
2. Yeast tRNA (Ambion), 1 mg/ml stock in 20 mM Tris–HCl, pH 7.4.

2.4. Electron Microscopy

1. Parafilm.
2. Carbon-coated formvar grids (Electron Microscopy Sciences # FCF400-Cu) (glow-discharged just before use).
3. Whatman No.1 paper, cut in thin slivers.
4. 2% aqueous uranyl acetate (Polysciences Inc.) solution filtered through 0. 22-μm filter (*see* **Note 2**).

3. Methods

3.1. Preparation of Phosphocellulose (*4*)

1. Weigh out 2 g of PC resin (fibrous cellulose phosphate) in a 50-mL Falcon tube.
2. Add 0.5 M NaOH to the 50 mL mark and resuspend PC by swirling. Do not use a stir bar or rod as this will generate resin fines. Leave the PC in NaOH for exactly 4 min.
3. Pellet the resin by spinning for 30 s at 1,000 × *g* in a swinging bucket rotor. Pour off supernatant. It should have a strong ammonia odor.
4. Suspend the resin in 50 mL of water. Pellet the resin and measure the pH of the water supernatant and then decant it off. Repeat this step till the pH of the water supernatant is <10 (—five to six times), then proceed to step 5. These steps

should all be performed rapidly to avoid chemical degradation of the PC at high pH.

5. Add 0.5 M HCl to the 50 mL mark and resuspend the resin by swirling. Leave the PC in HCl for exactly 4 min.
6. Pellet the resin and pour off the supernatant.
7. Suspend the resin in 50 mL of water. Pellet the resin and measure the pH of the water supernatant and then decant it off. Repeat this step till the pH of the water supernatant is >4 (—five to six times), then proceed to step 8. These steps should all be performed rapidly.
8. Suspend the resin in 50 mL of 0.2 M Tris–HCl, pH 7.4 buffer and equilibrate 30 min. Pellet the resin and repeat step 8 twice.
9. Suspend the resin in 50 mL of 20 mM Tris–HCl and store at 4 °C (*see* **Note 3**).

3.2. Selection of a High-Producer Bacterial Colony (See Note 4)

1. Transform a culture of BL21(DE3)pLysS bacteria with the pET expression plasmid. Plate on agarose plates with appropriate antibiotics and incubate at 37 °C overnight.
2. Pick individual colonies and streak each of them on sectors of two more plates, one without and one supplemented with IPTG (0.4 mM).
3. Select a colony that grows well on the IPTG-free plate but inefficiently on the IPTG-containing plate. Proceed to **Section 3.3**.

3.3. Expression of Gag in Bacteria

1. Grow 200 mL of overnight culture from the selected colony in LB at 37 °C.
2. Transfer 50 mL of the overnight culture to each of four 2-L baffled Erlenmeyer flasks, each containing 1 L LB. Incubate at 37 °C with shaking at 250 rpm. Monitor the growth of the culture by measuring its O.D. at 600 nm.
3. When the O.D. of the culture has reached 0.8, add IPTG to 0.1 g/L.
4. Continue the incubation with shaking for 4–5 h (*see* **Note 5**).
5. Harvest the bacteria by centrifugation at 3,500 × g for 7 min at 4 °C in preweighed bottles. It may be convenient to centrifuge multiple aliquots of the culture in the same bottle, accumulating all of the bacteria in a single pellet.
6. Drain the bottle by inversion and freeze the bacterial pellet at −80 °C.

3.4. Purification of Gag Protein (See Note 6)

1. Thaw the bacterial pellet without letting it warm up above ice temperature (*see* **Note 7**).
2. Weigh the bottle again to determine the added weight due to the bacterial pellet. A good yield is around 3 g of bacteria per liter of induced culture.
3. Resuspend the pellet in 10 mL of lysis buffer per gram of pellet. Disperse the pelleted material by repeated pipetting. The lysate should have sheen (*see* **Note 8**).

4. Sonicate the lysate on ice. Continue the sonication until the lysate is freely pipettable with a micropipette tip (*see* **Note 9**).
5. Remove insoluble debris from the sonicated lysate by centrifuging at 12,000 × *g* for 15 min. Decant the supernatant into a clean beaker and measure its volume. Place the beaker in an ice bath (*see* **Note 10**). Resuspend the pellet in 10 mL buffer A and save for subsequent analysis.
6. Add one-half volume (relative to the supernatant) of a saturated ammonium sulfate solution to the beaker with constant stirring (*see* **Note 11**).
7. Turn off the stirrer and leave the mixture on ice for an additional 30 min to allow the protein precipitate to accumulate.
8. Centrifuge at 12,000 × *g* for 15 min. Carefully remove the supernatant by decanting (*see* **Note 12**) and save it until the fractions have been analyzed by SDS-PAGE.
9. Resuspend the pellet in 15–20 mL buffer D by gentle pipetting. Avoid frothing during the pipetting. If a large fraction of the pellet fails to redissolve, the volume of buffer D can be increased. However, it is likely that a minor portion of the pellet will not be solublized at this step.
10. Remove any undissolved material from the solution by centrifuging at 12,000 × *g* for 15 min. The supernatant from this centrifugation should be clear; if not, recentrifuge for an additional 20–30 min.
11. Remove the supernatant to a fresh 50 mL Falcon tube by pipetting. It is important to avoid transferring any material from the pellet at this step.
12. Pour 10 mL settled PC resin (*see* **Note 13**) from a 50 ml Falcon tube into the solution. Thoroughly mix the PC slurry with the solution by pouring the mixture back and forth between the two 50 mL Falcon tubes.
13. Divide the slurry equally between the two tubes and add buffer A to reduce the NaCl concentration from the 0.5 M present in buffer D to ~ 0.1 M. Mix by inverting the tubes.
14. Leave the tubes on ice for 15–30 min to allow the Gag protein to bind to the PC. Gently swirl the contents every few min to re-disperse the PC.
15. Centrifuge at 700 × *g* for 2 min in a swinging bucket rotor. Pour off and discard the supernatant, retaining an aliquot for subsequent analysis.
16. Fill the tube with buffer B. Mix the contents by inverting the tube, pellet the PC as in step 15, and discard the supernatant, retaining an aliquot for analysis. Repeat.
17. Fill the tube with buffer C and wash the PC twice with this buffer as in step 16.
18. Wash with 20 mL of buffer D. Some protein may start eluting at this stage. This is evidenced by increased tendency of the buffer to form bubbles when the tube is inverted. Remove

the supernatant with a Pipetman and save in a Corning glass tube (*see* **Note 14**).

19. Elute the Gag protein from the PC by adding 10 mL of buffer E. Centrifuge the PC as in step 15, and save the supernatant in a second tube as in step 18.
20. Repeat step 19 with a fresh addition of 10 mL buffer E to the PC pellet. Save the supernatant from this step in a third tube.
21. Repeat step 19 with 7 mL buffer E. Save the supernatant in a fourth tube.
22. Repeat step 19 with 5 mL buffer E. Save the supernatant in a fifth tube.
23. To each of the tubes collected in steps 18–22, add ammonium sulfate crystals to a final concentration of ~ 60% saturation (0.375 g/mL) and dissolve by gentle mixing. This will precipitate the protein in the tubes; it will be obvious from the visible precipitates which tubes contain significant amounts of protein. Incubate the tubes on ice for 1–2 h (*see* **Note 15**).
24. Isolate the precipitated Gag protein by centrifuging those tubes containing visible precipitate at 12,000 × *g* for 15 min.
25. Redissolve the precipitates in a small volume of buffer D with 10% w/v glycerol (*see* **Note 16**), by gentle pipetting. Centrifuge at 20,000 × *g* for 15 min to remove any material that failed to redissolve.
26. Dialyze the protein against additional buffer with 10% w/v glycerol.
27. Estimate the concentration and yield of the protein (*see* **Note 17**). This estimate can be obtained by inspection of the SDS-PAGE gel; by the optical density of the protein at 280 nm, using a molar absorbance coefficient based on the amino acid composition of the protein; or from a Bradford assay. In the latter case, IgG rather than BSA should be used as the standard in the assay. If assembly experiments will be performed by direct dilution rather than dialysis (see below), it is desirable that Gag be at a concentration of at least 5 mg/mL.
28. Freeze the protein prep in aliquots at −80 °C. Analyze an aliquot of this final preparation, and the aliquots saved over the course of the purification, by SDS-PAGE and Coomassie brilliant blue staining (*see* **Note 18**).

3.5. In Vitro Assembly of Virus-Like Particles from Recombinant Gag Protein (See Note 19)

1. Thaw an aliquot of purified Gag protein by gently shaking the tube in a room-temperature water-bath.
2. Remove any precipitated material by centrifugation at 20,000 × *g* at 4 °C for 15 min. Estimate the protein concentration in the supernatant by measuring its absorbance. There should not be more than 5% difference in the protein concentration estimates before and after thawing.

3.5.1. Use of Dilution to Achieve Assembly Conditions

1. Dispense 50 μg Gag into one or more microcentrifuge tubes. Add 2 μg yeast tRNA to each tube.
2. Add enough buffer F to bring the NaCl concentration to 0.1 M. The final protein concentration should be around 1 mg/mL. The buffer should be added to the Gag–RNA mixture gradually, with gentle mixing during the stepwise addition.
3. Incubate at 4 °C for 2–4 h.

3.5.2. Use of Dialysis to Achieve Assembly Conditions

1. If the NaCl concentration is to be lowered by dialysis, mix 50 μg Gag protein with 2 μg tRNA as in point 1 of **Section 3.5.1**. If necessary, dilute the mixture with buffer F containing 0.5 M NaCl to a final Gag concentration of 1 mg/mL.
2. Place the mixture in a microdialysis apparatus and dialyze against buffer F containing 0.1 M NaCl at 4 °C for 2–4 h. The volume of the mixture will increase as the low-salt buffer accumulates inside the dialysis chamber.
3. Collect the material from the dialysis chamber. Be sure to remove any material stuck to the walls, as assembled VLPs are no longer in solution.
4. Analyze the mixture for the presence of VLPs (*see* **Note 20**).

3.6. Assessment of Assembly

3.6.1. Assessment of Assembly by Centrifugation

1. Centrifuge the assembly mixture at 20,000 × *g* in an Eppendorf centrifuge at 4 °C for 30 min. Remove and save the supernatant in a microfuge tube.
2. Add 17 μL of 4X loading buffer to 50 μL of supernatant.
3. Dissolve the pellet in 67 μL of 1X SDS-PAGE loading buffer.
4. Load 25 μL of the supernatant and pellet fractions onto an SDS-PAGE gel. After electrophoresis, visualize the Gag in the fractions by Coomassie brilliant blue staining.

3.6.2. Assessment of Assembly by Electron Microscopy

1. Place 10 μL of the assembly reaction onto clean Parafilm. Place a carbon-coated Formvar film grid (that has been recently glow-discharged to increase the surface charge density) on the droplet, carbon side down. Allow material from the assembly reaction to adsorb to the grid 2–5 min at room temperature. Wick off excess sample from the grid with thin slivers of Whatman no.1 filter paper.
2. Air-dry the grid in a grid box for 5 min at room temperature.
3. Place 10 μL of 2% uranyl acetate on the same side of the grid (this can be accomplished by placing the droplet of uranyl acetate on Parafilm and placing the grid on the droplet, as in 3.6.2 step 1). Allow this staining reaction to proceed for 1–5 min (*see* **Note 21**) and again remove excess moisture with Whatman paper wicks. An example of negatively stained VLPs is in **Fig. 14.1**.

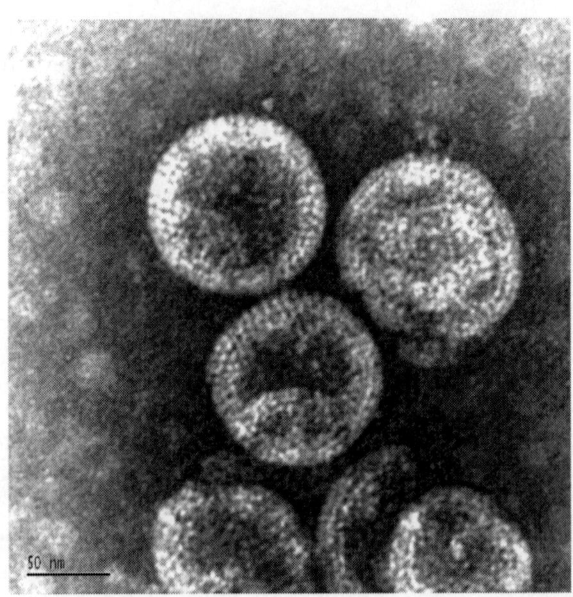

Fig. 14.1. Electron micrograph of VLPs assembled from Δ16–99Δp6 Gag by dilution method, as described. Samples were negatively stained with 2% uranyl acetate. Note the heterogeneity in morphology, including some incompletely assembled VLPs.

4. Examine the material on the grid in a transmission electron microscope (*see* **Note 22**).

4. Notes

1. Unless stated, all solutions should be prepared in water with a resistivity of at least 18.2 MΩ and total organic content of less than five parts per billion. Where used, PMSF and β-mercaptoethanol (or TCEP) should be added to buffers just prior to use.
2. Store the filtered uranyl acetate in the dark as exposure to light causes precipitate to accumulate.
3. Suspensions of cellulose phosphate may be stored at 2–8 °C in the presence of a bacteriostat (0.02% w/v sodium azide) for up to 1 year. However, if Gag fails to bind to the PC in 0.1 M NaCl (step 3.4.15), it may be necessary to make a fresh PC preparation, or even obtain a fresh batch of dry PC.
4. For reasons we do not understand, sometimes individual colonies of bacteria transformed with identical plasmid molecules vary markedly in their expression of protein following induction. Colonies with high expression can be recognized by their inefficient growth on agar plates supplemented with the inducer.

5. Shorter inductions are also acceptable. The protein is generally visible by SDS-PAGE of the total bacterial lysate after only ~ 2.5 h.
6. All steps from here on are at 4 °C unless otherwise specified. Buffers used should be prechilled to 4 °C before the purification.
7. The safest way to thaw the pellet is on ice, but one can also thaw it in a room-temperature water bath with frequent agitation. In this case, it should be placed on ice as soon as it has thawed. The surface of the thawed pellet should look glassy. This is an indication that the bacteria have been successfully lysed. Another indication of lysis is if the pellet begins to run or slide as the supernatant is decanted. If neither of these signs is observed, it is likely that many of the bacteria did not lyse. In this case, the bottle should be re-frozen and re-thawed.
8. If the lysate looks "chalky", the lysis was probably incomplete. If this is a persistent problem, the lysis buffer could be supplemented with detergents such as Triton X-100 at 0.05% w/v or with lysozyme at 0.5–2.0 mg/mL. While adding detergents, it is important to consider whether these might interfere with subsequent experiments; in this case, further steps may be necessary to eliminate traces of detergent.
9. In sonicating the bacterial lysate, it is of critical importance to keep the lysate chilled at all times. The vibrating probe of the sonicator produces heat within the lysate. Thus the lysate should be surrounded by ice, but further steps should also be taken to avoid localized heating within the lysate. The sonicator should be on pulse-mode and the lysate should be allowed to re-cool between the sonication pulses. Here, use of a sonicator probe with a temperature sensor helps monitor actual temperature in the lysate. If necessary, the lysate should be sonicated in smaller 50 ml aliquots rather than all at once, so that the entire lysate has a high surface: volume ratio and thus good contact with the ice-bath. The wattage of pulses used, number of pulses delivered, volume of each batch of lysate sonicated, and time allowed to cool for maximum recovery of protein must be determined empirically. We use 12–15 pulses of 0.5 s, each interspersed with 0.7 s gaps, using a 13-mm probe with maximum power in a 500-W ultrasonicator. This regimen is repeated with 1-min intervals in between to cool, until the sample is freely pipettable.
10. It is important that the supernatant be poured off carefully, so that very little pelleted material is brought along with it. The pellet will not be firm at this point.
11. The lysate should be stirred constantly as the ammonium sulfate is added, and the ammonium sulfate should be added gradually, in a dropwise fashion, at no more than 0.5–1.0 mL/min. These are precautions to ensure that *local*

concentrations of ammonium sulfate never rise above the target concentration, as other proteins from the bacteria will precipitate at higher ammonium sulfate concentrations. We indicate here that a volume of saturated ammonium sulfate equal to half the volume of the lysate should be added, bringing the final ammonium sulfate concentration to 33% saturation. It should be noted that this concentration can be varied if necessary. For example, 40% saturation is required for precipitation of Moloney murine leukemia virus Gag protein. If the yield of HIV-1 Gag protein is particularly high, it may be advantageous to reduce the final ammonium sulfate concentration somewhat, e.g., to 28–30% saturation.

12. After decanting, remove as much of the remaining supernatant as possible by pipetting and drain the pellet by inverting the tube. This is to remove as much of the ammonium sulfate as possible and will help resolublize the pellet.
13. Keep the PC equilibrated in buffer A and temperature-equilibrated on ice, prior to adding it to the protein.
14. All steps after step 18 should use Corning glass tubes: the protein is now quite dilute (and relatively purer, as contaminating bacterial proteins have been largely removed) and will tend to stick to plastic, resulting in significant losses.
15. If the protein yield is low, the tubes can be incubated on ice overnight to improve the recovery.
16. The pellets at this stage should look white. This is usually a sign of good purity with relatively less *E. coli* contaminating proteins. Add sequential small amounts of buffer D with 10% glycerol (e.g., 100 μL) to the pellets and try dissolving the pellets in the smallest volumes possible.
17. The yield of protein and the relative purity, for different Gag constructs varies. Yields of Gag are usually in the range of 2–5 mg/L of induced bacterial culture with purity of 75–90%, as determined by SDS-PAGE and Coomassie staining.
18. Evaluation of the purity and yield from the purification will help optimize the purification scheme for different Gag constructs. *See* **Fig. 14.2** for an example of such an analysis. It shows the progressive purification of Gag at different stages of the scheme. Note that if a substantial fraction of the Gag protein being expressed is not solubilized in the initial lysis (i.e., is in the pellet saved at step 3.4.5), this may be due to inefficient lysis of the bacteria (*see* **Note 8**), but it may also mean that the protein itself has low solubility. In the latter case, temperatures of 30 °C or below should be tried for growth and induction of the bacteria. Addition of detergent or lysozyme can also help increase the recovery of proteins from bacteria; *see* **Note 8**. If it is found that a substantial fraction of the Gag remained in the supernatant after ammonium sulfate precipitation (step

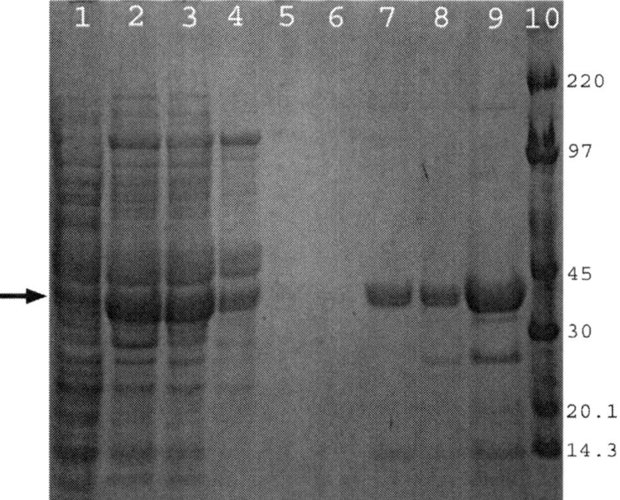

Fig. 14.2. SDS-PAGE analysis of different stages of the purification procedure. The arrow on the left shows the position of Gag on the gel. Lanes 1: supernatant from step 3.4.5 after sonication and centrifugation (8 μL) ; 2: step 3.4.9 before centrifugation (5 μL); 3: step 3.4.10 after centrifugation (5 μL); 4: supernatant from step 3.4.15 – unbound material from PC (25 μL); 5: step 3.4.16 –0. 1 M NaCl wash of PC (25 μL); 6: step 3.4.17 –0. 2 M NaCl wash of PC (25 μL); 7: step 3.4.18 –0. 5 M NaCl wash of PC (20 μL); 8 (4 μL) and 9 (20 μL) of pooled fractions from step 3.4.19 to 22 –1 M NaCl washes; 10: molecular weight markers. M.W. indicated next to bands in kDa. Note that some Gag failed to bind the PC at step 3.4.15, but this loss is offset by the fact that a very large fraction of the contaminating proteins has been eliminated here (lane 4).

3.4.8), then the ammonium sulfate concentration used for the initial precipitation (step 3.4.6) may need to be adjusted.

19. Assembly of the Gag protein into VLPs is induced by addition of nucleic acid, and simultaneously reducing the NaCl concentration from 0.5 to 0.1 M. At this reduced ionic strength, the protein has a much higher affinity for the nucleic acid than it does at 0.5 M NaCl. The dilution to reduce the NaCl concentration in the presence of nucleic acid can either be performed by direct dilution, or more gradually by dialysis. We describe both methods below: direct dilution as procedure 3.5.1 and dialysis as procedure 3.5.2. If the starting protein concentration is low, we suggest using the dialysis procedure.
20. In turn, assembly of the Gag protein into VLPs can be assessed either by a pelleting assay or by electron microscopy. We describe the pelleting as procedure 3.6.1 and the electron microscopy as procedure 3.6.2 here.
21. The length of time that stain is applied to the grid can be varied. Grids can be stained for different times and examined to decide on the optimal staining protocol.

22. If there is too much stain on the grid, so that everything is very dark, then excess stain can be removed by applying a droplet of water to the grid for 15 s. It should then be removed with a Whatman paper wick and the grid can be re-examined.

Acknowledgments

This work was supported by the Intramural Research Program of the NIH, National Cancer Institute, Center for Cancer Research.

References

1. Campbell S., Vogt, V. M. (1995) Self-assembly in vitro of purified CA-NC proteins from Rous sarcoma virus and human immunodeficiency virus type 1. *J Virol* 69, 6487–6497.
2. Gross, I., Hohenberg, H., Wilk, T., et al. (2000) A conformational switch controlling HIV-1 morphogenesis. *EMBO J* 19, 103–113.
3. Campbell, S., Vogt, V. M. (1997) In vitro assembly of virus-like particles with Rous sarcoma virus Gag deletion mutants: identification of the p10 domain as a morphological determinant in the formation of spherical particles. *J Virol* 71, 4425–4435.
4. Peterson, E. A., Sober, H. (1962) Column chromatography of proteins: substituted celluloses. *Methods Enzymol* 5, 3–27.

Chapter 15

Methods for the Analysis of HIV-1 Nucleocapsid Protein Interactions with Oligonucleotides

Andrew G. Stephen and Robert J. Fisher

Abstract

HIV-1 Nucleocapsid protein (NC) is a small basic protein that contains two retroviral zinc fingers. It is a highly effective nucleic acid chaperone that plays a critical role in viral replication acting as a cofactor in reverse transcription as well as other aspects of the viral lifecycle. We have used a variety of biophysical techniques to characterize the high affinity binding of NC to a short deoxyoligonucleotide ($d(TG)_4$). Here we outline in detail the use of fluorescence anisotropy and surface plasmon resonance spectroscopy to study the binding of NC to $d(TG)_4$.

Key words: Nucleic acid binding, fluorescence anisotropy, surface plasmon resonance, binding kinetics, equilibrium binding constant.

1. Introduction

The Nucleocapsid protein (NC) protein of HIV-1 is a very small basic protein that contains two retroviral zinc fingers ($C\text{-}X_2\text{-}C\text{-}X_4\text{-}H\text{-}X_4\text{-}C$). NC is initially synthesized as a part of the Gag polyprotein. As a domain of Gag, NC functions in the selecting and encapsidation of the viral genome *(1)*. Acting as a nucleic acid chaperone it unwinds the cellular $tRNA^{Lys,3}$ and anneals it to the viral genome where it serves as a primer for reverse transcription *(2, 3)*. After release of the viral particle from the infected cell the Gag polyprotein is cleaved by the viral protease, resulting in the release of NC from the polyprotein *(4)*. After maturation NC stabilizes the dimeric linkage between the genomic RNA molecules. During the next cycle of infection NC acts as a nucleic acid chaperone during reverse transcription *(5, 6)*. The

Vinayaka R. Prasad, Ganjam V. Kalpana (eds.), *HIV Protocols: Second Edition, vol. 485*

DOI 10.1007/978-1-59745-170-3_15 Springerprotocols.com

essential role of NC in the viral replication process means it is important to understand the details of these interactions. To gain an insight into NC's nucleic acid binding properties we chose to study the binding to short oligonucleotides using surface plasmon resonance (SPR) spectroscopy *(7)*. NC binds with high affinity to the repeating sequence $d(TG)_n$. Recently we continued our investigations of NC binding to $d(TG)_4$ using a variety of biophysical techniques and demonstrated that more than one NC can bind to one $d(TG)_4$ molecule and more than one $d(TG)_4$ can bind to one NC molecule *(8)*. In this chapter we describe in detail methods to study the binding of NC to deoxyoligonucleotides using fluorescence anisotropy and SPR spectroscopy.

Fluorescence anisotropy is based on the observation that when a fluorescent molecule is excited with plane-polarized light, it emits polarized fluorescent light into a fixed plane if the molecules are stationary between excitation and emission *(9)*. However molecules are not stationary but tumble and rotate and so the fluorescence anisotropy of the molecule is proportional to the extent of molecular rotations and hence the molecule's molecular volume. This relationship is described by the Perrin Equation shown in (15.1),

$$r = \frac{r_0}{1 + \left(\frac{\tau}{\theta}\right)} \tag{15.1}$$

r is the observed anisotropy, r_0 the fundamental anisotropy (determined in the absence of rotational diffusion), τ is the fluorescence lifetime and θ is the rotational correlation time, given by $\theta = \eta V/RT$ (where η is the viscosity, T is the temperature in kelvins, R is the gas constant and V is the volume of the rotating molecule).

Small molecules rotate fast but when they interact with large molecules their rotation is slowed, resulting in an increase in the anisotropy of the small molecule. This phenomenon can be used to study molecular interactions. The sample is excited with vertically polarized light and the intensity of the emitted light is measured with polarizers that are perpendicular or parallel to the excitation light. The fluorescence anisotropy is a dimensionless quantity that is independent of the total intensity of the sample. The term fluorescence polarization is often used and polarization and anisotropy values can be interchanged. However anisotropy is generally preferred because theoretical considerations are much simpler when expressed in terms of this parameter *(9)*.

In protein–nucleic acid interactions the oligonucleotide is usually labeled with a fluorescent molecule (as it is generally the smaller molecule) and any change in anisotropy is recorded after addition of the protein. Anisotropy values are calculated by the instrument software and relatively easy to measure and record. However care needs to be taken when measuring and interpreting data as changes in fluorescence intensity, temperature and viscosity will all affect the anisotropy value.

Surface plasmon resonance is an optical resonance effect where the back of a glass slide coated with gold affects the angle at which a minimum of reflected light exits. The SPR signal is very sensitive to any change in the dielectric constant of the medium adjacent to the gold surface of the sensor. The change in dielectric constant may be due to adsorption of molecular material in this area (or changes in refractive index of the buffer, see later) and so can be used to study molecular interactions close to the sensor surface. Biacore Inc has provided commercial SRP biosensors since 1991, although additional SPR biosensors are becoming available from other sources (such as IBIS or BioRad). In the Biacore instrument the gold surface of the sensor chip has noncrosslinked carboxymethylated dextran covalently attached at a density of 2%. This layer acts as the attachment point for the ligand molecule. An analyte (binding partner) is introduced to the flow cell via a fluidic system and binding to the ligand is followed in real time. Any changes in the dielectric constant (nonspecific binding effects, buffer refractive index mis-matches) will be observed as an increase in signal so appropriate controls need to be included in the experimental design. The response signal is recorded as "Response Units" which is defined as $1/10,000°$ change in the angle of reflected light. The sensor chip has four different flow cells that can be addressed independently, so several of these can be used as reference surfaces. After the injection any bound analyte is removed from the immobilized ligand by surface regeneration. In this flow-based system the analyte moves from the bulk flow to the ligand surface by diffusion and convection. The process of diffusion is called material transport and it is important in kinetic studies to ensure the effects of this transport effect are minimized. The use of a low ligand density coupled with high flow rates (both are determined empirically) can minimize the transport effects. Once a concentration range of analyte has been flowed over the ligand, the data are then processed to remove refractive index and nonspecific binding events *(10)*. The processed sensorgrams can then be analyzed to provide information about the stoichiometry of the binding, the Equilibrium binding constant K_D or the microscopic rate constants k_a and k_d *(11)*.

2. Materials

2.1. Fluorescence Anisotropy

1. Fluorescein labeled oligonucleotide. We have used 3′ fluorescein labeled deoxyribonucleotides that are synthesized in house using 3′-(6-Fluorescein) CPG (Glen Research). How-

ever a large number of commercial vendors can supply such oligos (*see* **Note 1**).
2. HIV-1 Nucleocapid protein provided by Dr. Robert Gorelick, SAIC-Frederick, Inc.
3. Assay buffer: 20 mM Hepes pH7.5, 150 mM NaCl, 1 μM zinc chloride, 100 μM Tris (2-carboymethyl)phosphine hydrochloride (TCEP, Pierce), 5 mM β-mercaptoethanol (*see* **Note 2**).
4. 96-well polypropylene U bottom plates (Costar).
5. 386-well black polypropylene or low bind polystyrene plates (Costar).
6. Plate seals (Biacore).
7. Tecan Ultra plate reader equipped with excitation and emission polarizers and filters to detect fluorescein (excitation wavelength 485 nm, Emission wavelength 535 nm) (*see* **Note 3**).
8. 96-well reagent block (Costar) (*see* **Note 4**).
9. Multichannel pipette.

2.2. Surface Plasmon Resonance Spectroscopy

1. A suitable biosensor such as the Biacore instruments is required: for our studies a Biacore 3000 or S51 was used.
2. Neutravidin: 40 μg/mL in 10 mM sodium acetate pH 4.5 (Pierce).
3. Oligonucleotide (labeled at the 3′ end with biotin) approximately 0. 2 μM in10 mM Tris-HCl pH 7.5, 300 mM NaCl, 1 mM EDTA.
4. Amine coupling kit from Biacore: 1 M ethanolamine hydrochloride-NaOH pH8.5, 11.5 mg/mL N-hydroxysuccinate (NHS), 75 mg/mL 1-Ethyl-3-(3-dimethylaminopropyl) carbodiimide hydrochloride (EDC).
5. Immobilization buffer (HBS): 10 mM Hepes-HCl pH7.5, 150 mM NaCl.
6. CM5 research grade chips from Biacore.
7. Regeneration buffer: 0.1% SDS, 3.4 mM EDTA pH 8.0.
8. NC protein provided by Dr. Robert Gorelick, SAIC-Frederick.
9. Microspin G25 columns from Amersham Biosciences.
10. Running buffer: 10 mM Hepes-HCl, 150 mM NaCl, 1 μM zinc chloride, 100 μM Tris (2-carboymethyl)phosphine hydrochloride (TCEP, Pierce), 5 mM β-mercaptoethanol, 0.005% Tween-20.

3. Methods

3.1. Resuspension of NC

1. Add 50 μL of 20 mM Hepes pH7.5, 150 mM NaCl, 100 μM TCEP to 100 μg of lyophilized NC protein and mix by

pipetting up and down (this assumes 2.1 equivalents of Zn is present in the lyophilized NC preparation) (*see* **Note 2**).
2. Leave on ice for 5 min to allow formation of zinc fingers.
3. Add additional 456 μL of 20 mM Hepes pH7.5, 150 mM NaCl, 100 μM TCEP for a final concentration of 30 μM.
4. Make aliquots and store at −80°C.

3.2. Fluorescence Anisotropy Plate assay

1. Make a10 nM solution containing fluorescein labeled oligo in 20 mM Hepes, pH7.5, 150 mM NaCl, 1 μM $ZnCl_2$, 100 μM TCEP, 5 mM mercaptoethanol.
2. Aliquot 50 μL of the 10 nM oligo across 11 wells of a row.
3. In 12th well add 96.7 μL of the labeled oligo and then add 3.3 μL of 30 μM NC stock into this well. This gives a final NC concentration of 1 μM.
4. Mix regents by pipetting up and down and transfer 50 μL from the NC-oligo mixture to the following well. Continue to do this across the entire row. You should now end up with 0.488 nM NC and 10 nM oligo in 100 μL in well 1.
5. Seal the plate, cover with aluminum foil and allow the mixture to equilibrate for 10 min.
6. Remove 40 μL from each well and place it into a 384 well plate.
7. Add 40 μL of 10 nM oligo solution (without added NC) to the 384-well plate.
8. Read the plate in the Tecan Ultra plate reader (*see* **Note 5**).

3.3. Data Analysis of Fluorescence Anisotropy

1. The anisotropy and the fluorescence intensity data should be inspected (*see* **Note 6**). If the data are satisfactory it is imported into a suitable graphical package, such as Graph pad or Sigma plot.
2. Representative data are shown in **Fig. 15.1**. For our initial analysis of the data we usually use a simple binding model that assumes one NC binds to one oligo shown by Eq. (15.2). (*see* **Note 7**).

$$r = \frac{(R\,r_b - r_f)\left(\frac{(K_D+C_T+L_T)-\sqrt{(K_D+C_T+L_T)^2-4L_T C_T}}{2\,L_T}\right) + r_f}{(R-1)\left(\frac{(K_D+C_T+L_T)-\sqrt{(K_D+C_T+L_T)^2-4L_T C_T}}{2\,L_T}\right) + 1} \tag{15.2}$$

R is the bound intensity/free intensity, r_b is the anisotropy of the oligo when it is bound by NC, r_f is the anisotropy of the oligo when it is free, K_D is the equilibrium disassociation constant, L_T is the concentration of the oligo and C_T is the NC concentration.

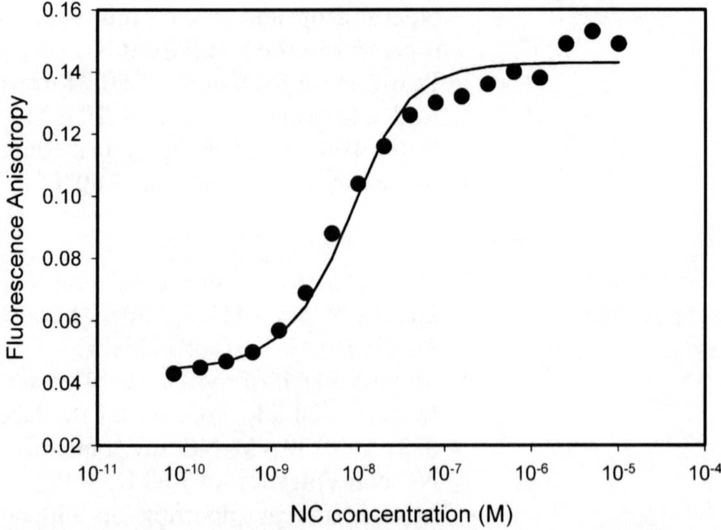

Fig. 15.1. Fluorescence anisotropy of NC binding to $d(TG)_4$. A 10-μM solution of NC was serially titrated into a 10-nM solution of fluoresceinated $d(TG)_4$ and the anisotropy of the oligo was measured using a Tecan Ultra Plate reader. The data were fit using Eq. (15.2) ($R = 1.7$). It is clear that this binding model assuming a 1:1 stoichiometry does not fit the data, although it does give a reasonable estimation of the Equilibrium binding constant (K_D is 7 nM). A superior fit of the data is achieved when terms that describe two molecules of NC binding one oligo and two oligos bind one NC are included in the model *(8)*.

3.4. Preparation of Oligonucleotide Immobilized Chip for SPR Sepctroscopy

1. Take the biotinylated oligo stock solution and remove any free biotin using the MicroSpin G-25 columns from Amersham Biosciences. Specifically, place the column in a 1.5-mL eppendorf tube and centrifuge for 735 × *g* for 1 min. Discard the old eppendorf tube and replace it with a clean one. Carefully load the oligo on to the column and centrifuge for 2 min at 735 × *g*. Determine the oligo concentration and then dilute it to 0. 2 μM in10 mM Tris-HCl pH 7.5, 300 mM NaCl, 1 mM EDTA.
2. Insert the CM5 sensor chip into the Biacore 3000. All Biacore experiments were performed at 25°C.
3. Prime the instrument with filtered and degassed MilliQ water.
4. Precondition the chip with two sequential injections of 10 mM HCl, 10 mM NaOH, 0.1% SDS and finally water each with a contact time of 10s.
5. Replace the water with the Immobilization buffer (HBS) and prime the instrument.
6. Use the Biacore immobilization wizard to amine couple 500 Response Unit's (RU's) of neutravidin on to Flow cells 1 and 2 of the sensor chip.

7. The wizard automatically runs the immobilization procedure. The binding rate of neutravidin is estimated by a preconcentration step. The surface is activated with a 1:1 mixture of NHS and EDC for 7 min (flow rate of 5 μL/min). A series of neutravidin injections are made until 500 RU's are immobilized. The reaction is quenched by 7 min injection of ethanolamine.
8. Once completed the wizard reports the amount of neutravidin captured.
9. Any noncovalently attached neutravidin is removed by injecting Regeneration buffer over flow cell 2 using a flow rate of 20 μL/min and a contact time of 30s (use the QuickInject function with the Extra clean option).
10. Allow the baseline to stabilize by flowing Immobilization buffer for 5 min at 20 μL/min (*see* **Note 8**).
11. Inject the biotinylated oligo (0.2 μM) at 20 μL/min with a contact time of 30 s. Monitor the amount of oligo captured on the neutravidin surface. Make further injections until approximately 5 RU's of oligo have been captured.
12. Any nonspecifically bound oligo is removed using Regeneration buffer (use a QuickInject with Extraclean and a contact time of 30s).
13. Allow the baseline to stabilize by flowing the Immobilization buffer for 5 min at 20 μL/min and record the final amount of oligo captured.

3.5. Measure Binding of NC to Immobilized Oligo by SPR

1. Serial dilutions of the NC stock are made in Running buffer. 120-μL solutions of 200-1.7 nM of NC are prepared.
2. Using the Kinject option NC solutions are injected over flow cell 1 and 2 at a flow rate of 64 μL/min with a 1-min contact time and a 1-min disassociation time. The injection order was from lowest to highest concentration and all injections are done in duplicate.
3. After each NC injection any bound NC is removed by a 30s Quickinject (with Extraclean) of Regeneration buffer.
4. Each NC injection is followed by a Running Buffer injection using the same injection parameters for the NC injections. This allows for adequate correction of machine effects (such as the opening and closing of valves) and slight mismatches between the refractive index of the NC sample and the Running buffer.
5. After completion of the run the senor chip is removed, replaced with a Maintenance chip and the instrument is primed with filtered and degassed MilliQ water. The instrument is then cleaned to remove any NC protein remaining in the fluidics system (*see* **Note 9**).

3.6. Processing and Analysis of SPR Data

1. After the run is completed all the sensorgrams are imported into a data analysis program. We usually use Scrubber which is available from Dr. David Myszka (University of Utah) (*see* **Note 10**), alternatively the software provided by Biacore (Biaevaluation) can be used.
2. All sensorgrams are *x* and *y* transformed and any data not to be used is cropped. At this point it also advisable to review the sensorgrams and remove any that look abnormal.
3. Any bulk shifts or nonspecific binding are removed by subtracting sensorgrams from the reference flow cells (flow cell 1 in this experiment, *see* **Note 11**).
4. Buffer injections should be evaluated for reproducibility of signal and those which deviate should be removed. The remaining buffer sensorgrams are then subtracted from the NC sensorgrams. This process is known as double referencing.
5. At this point the data are ready for analysis. There is much information contained within the NC sensorgrams, the stoichiometry of binding, the Equilibrium binding constant (K_D) and the microscopic rate constants (k_a and k_d). NC binding to low surface densities of $d(TG)_4$ saturates at a stoichiometry of 1 molecule of NC per 1 molecule of $d(TG)_4$ (**Fig. 15.2A**) and so we fit these data using a 1:1 binding model as shown in Eq. (15.3).

$$\begin{pmatrix} \frac{d}{dt}F(t) \\ \frac{d}{dt}B(t) \end{pmatrix} = \begin{pmatrix} k_t\,(C - F(t)) - (k_v\,((R_{max} - B(t))\,F(t) - K_D B(t))) \\ k_v\,((R_{max} - B(t))\,F(t) - K_D B(t)) \end{pmatrix} \tag{15.3}$$

$F(t)$ is the NC concentration in the reaction volume, $B(t)$ is the NC-oligo complex, $C =$ is the NC concentration at flow-cell injection, k_t is the NC transport term, k_v is the Reaction rate $((k_v(R_{max}) >> k_t$ for the NC system), K_D is the equilibrium dissociation constant (*see* Note **12**).

6. When the individual NC sensorgrams reach a steady sate (as they do in **Fig. 15.2A**) it is possible to calculate the total amount of NC bound to the immobilized oligo and plot it as a function of NC concentration (**Fig. 15.2B**)(*see* **Note 13**). This binding curve can be then fit using Eq. (15.4) to determine the Equilibrium binding constant (K_D).

$$\text{NC} - \text{oligo Concentration} = \frac{R_{max}\ (\text{NC Concentration})}{K_D + (\text{NC Concentration})} \tag{15.4}$$

K_D is the equilibrium dissociation constant, R_{max} is the maximum binding response.

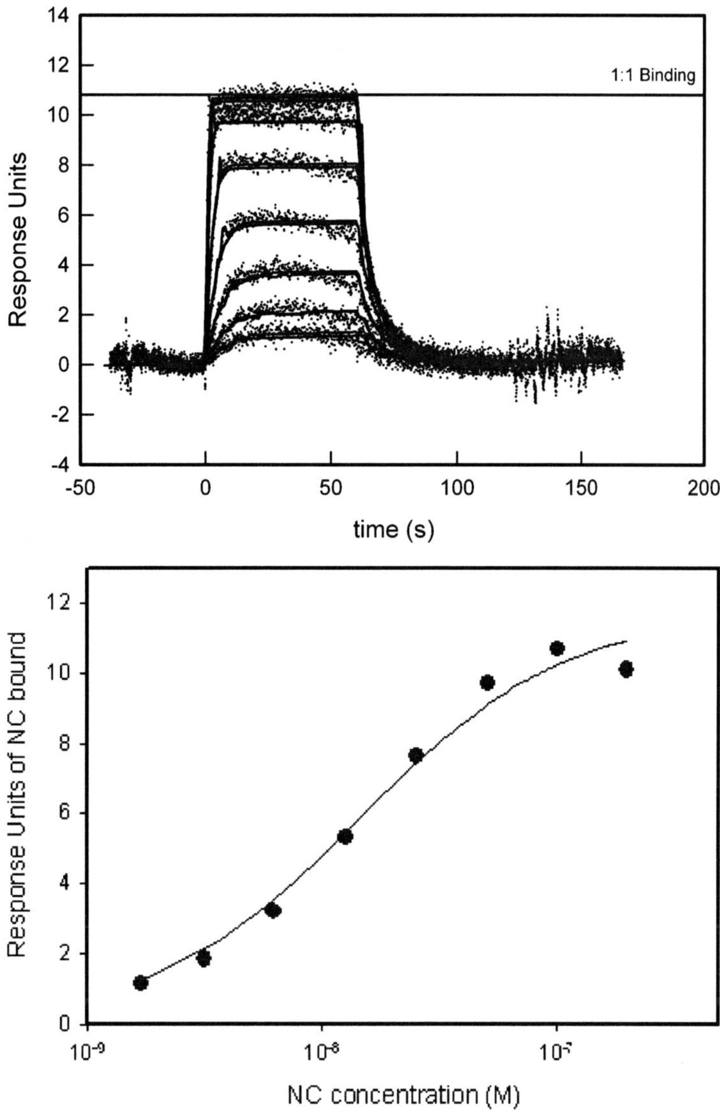

Fig. 15.2. Binding of NC to immobilized $\mathrm{d(TG)_4}$ using SPR. (**A**) Kinetic binding of NC to a 5.4 RU surface of $\mathrm{d(TG)_4}$ on a Biacore S51. Solutions of 1.7, 3.13, 6.25, 12.5, 25, 50, 100, and 200 nM of NC were flowed over the surface as $100\,\mu\mathrm{L/min}$. The horizontal line represents the expected binding response for a 1:1 complex of NC and $\mathrm{d(TG)_4}$. The black lines are the fit of the data assuming a simple 1:1 binding model using Eq. 15.3. The K_D is 2.9 nM (reproduced from Ref. 8 with permission from Oxford University Press). (**B**) Steady state binding of NC to a 5.4 RU surface of $\mathrm{d(TG)_4}$. The total amount of NC bound to the 5.4 RU surface of $\mathrm{d(TG)_4}$ is calculated and plot as a function of NC concentration. This data fit well with Eq. (15.4). The calculated K_d is 14 nM and is in reasonable agreement with that calculated from the kinetic analysis.

4. Notes

1. Labeled DNA should be purified by hplc to ensure no unincorporated fluorescein remains.
2. The activity of NC is depends on reduced cysteines that coordinate the zinc to form the zinc fingers. NC is susceptible to inactivation by air oxidation and so appropriate reducing agents must be present in the buffers to maintain activity. TCEP is included in the buffer as this reducing agent is particularly resistant to air oxidation unlike other reducing agents such as dithiotheritol or β-mercaptoethanol.
3. We routinely use our Aminco Bowman Series 2 Fluorescence Spectrophotometer equipped with excitation and emission polarizers with a 1-mL quartz cuvette to optimize our fluorescence anisotropy assays before we reformat them into plate-based assays. The advantage of using a 96-well plate-based format for this assay is the decreased sample requirement and increased through-put of experimental parameters that can be investigated.
4. Using this plate assay format multiple reaction conditions (such as varying salt conditions, oligo concentration) can be evaluated on one plate. In this case the different reaction buffers are placed in a separate well of a 96-well reagent block and 50-μL aliquots are then removed and transferred to the 96-well plate using a multichannel pipette.
5. The settings for measuring fluorescence anisotropy on the Tecan Ultra require an integration time of 40 μs and 10 flashes per read (although useable data are collected with five flashes per read). The *G* factor is measured every 3 months or so using a 10-nM solution of fluorescein. This corrects for machine specific effects in collecting data from vertically and horizontally polarized light. The "gain" is set to "optimal" this allows for any changes in fluorescence intensity that may occur when NC binds to the fluorescein labeled oligo (*see* **Note 6**).
6. We have observed changes in the fluorescence intensity with the binding of NC to the $d(TG)_n$ oligos as this can affect the anisotropy value. Any change in the fluorescence intensity can effect the fluorescence lifetime (*see* **Eq. 15.1**). This can be corrected for during the data fitting stage *(9)* by including the term *R* in Eq. 15.1 (*R* is the intensity of bound oligo divided by the intensity of the free oligo). This increase in fluorescence intensity can be attributed to the quenching of fluorescein by the adjacent guanine base from the oligo *(12)*. The binding of

NC to the oligo reduces this quenching effect and so results in an increase in the fluorescence intensity of the fluorescein. During the course of our work it became clear that fluorescein is not an ideal dye to use in fluorescence anisotropy studies. Not only is the fluorescence intensity of fluorescein greatly affected by the pH it is also quenched relatively easily. In addition fluorescein is negatively charged at neutral pH which acts to repel the label away from the phosphate backbone of the oligo and into the solvent. This increases the independent rotation of fluorescein with respect to the oligo itself and is known as the "propeller effect". Thus fluorescein now reports quite poorly on the "global" tumbling of the oligo and generally reports lower anisotropy values than are predicted based on the size of the oligo to which it is attached *(13)*. Molecular Probes offers a variety of Alexa dyes that offer many advantages over fluorescein.

7. We have determined the binding of NC to oligo is not described adequately by assuming a simple 1:1 binding model. NC is able to participate is ternary complex formation which leads to higher order complexes. The details of these models are beyond the scope of this work and for a further explanation we encourage the reader to refer to *(8)*.
8. We have observed that Regeneration buffers containing 0.1% SDS often cause the baseline to decrease followed by a upward drift in the baseline (~ 5 RU's/min). A stable baseline is required in order to calculate the total amount of oligo capture on the surface.
9. A rigorous cleaning protocol is essential to enable the collection of the best quality data. In the instrument software the Working Tools section lists a series of routine maintenance commands that should be followed. We routinely run a Desorb every week and Super Clean and Sanitize every month. More frequent cleanings can be performed if the needed.
10. The Scrubber software is freely available to academic and government institutions and for a fee to commercial institutions. We have found Scrubber to be a very easy and rapid way of processing large data sets. After processing the data can be further analyzed in Scrubber or exported as a text file and analyzed in Biaevaluation or other modeling software.
11. In preliminary experiments the extent of nonspecific binding to carboxymethly dextran or neutravidin is determined. If background binding is a problem increasing the salt concentration, addition of soluble dextran (1–5 mg/mL final concentration from Fluka), or addition of detergents (tween-20) to the Running buffer can reduce it. However it is not always possible to completely eliminate nonspecific binding. If the non-specific binding gives a regular response then it is acceptable to subtract it.

12. An accurate measurement of the microscopic rate constants for NC binding an immobilized oligo is not possible using SPR. This is because the binding is diffusion limited and so NC binds to the oligo as quickly as it can be delivered there. In Eq. (15.3) we include a transport term and do not fit the microscopic rate constants, but mathematically determine the Equilibrium disassociation constant *(8)*.
13. On oligo surfaces of greater than 10 RUs NC no longer binds with a stoichiometry of 1:1 but forms additional NC-oligo complexes. These interactions are exemplified by low concentrations of NC not reaching a steady state, higher concentrations of NC not completely disassociating from the immobilized oligo and binding stoichiometries greater than 1:1. We have developed models to fit these data that account for additional NC-oligo complexes and for a more detailed discussion of this approach we refer the reader to *(8)*. We would highly recommend using immobilized oligos at low densities to enable a simple analysis of this binding reaction.

Acknowledgments

We are very grateful to continued support by Matt Fivash on modeling of NC-oligo interactions. This project has been funded in whole or in part with federal funds from the National Cancer Institute, National Institutes of Health, under contract N01-CO-12400. The content of this publication does not necessarily reflect the views or policies of the Department of Health and Human Services, nor does mention of trade names, commercial products, or organizations imply endorsement by the U.S. Government. This Research was supported [in part] by the Intramural Research Program and of the NIH, National Cancer Institute, Center for Cancer Research and the Developmental Therapeutics Program in the Division of Cancer Treatment and Diagnosis of the National Cancer Institute.

References

1. Berkowitz, R., Fisher, J., Goff, S. P. (1996) RNA packaging. Curr. Top. Microbiol. *Immunol* 214, 177–218
2. Feng, Y. X., Campbell, S., Harvin, D., et al. (1999) The human immunodeficiency virus type 1 Gag polyprotein has nucleic acid chaperone activity: possible role in dimerization of genomic RNA. *J Virol* 73, 4251–4256
3. Cen, S., Huang, Y., Khorchid, A., et al. (1999) The role of Pr55 (gag) in the annealing of tRNA3Lys to human immunodeficiency virus type 1 genomic RNA. *J Virol* 73, 4485–4488
4. Swanstrom, R., Wills, J. W. (1997) Synthesis, assembly, and processing of viral proteins, in *Retroviruses*. (Coffin JM, Hugcs SH, Varmus HE. ed.) Cold Spring Harbor Laboratory Press, Plainview, NY, pp. 263–334
5. Rein, A., Henderson, L. E., Levin, J. G. (1998) Nucleic acid chaperone activity of retroviral nucleocapsid proteins: significance for viral replication. *Trends Biochem Sci* 23, 297–301
6. Levin, J. G., Guo, J., Rouzina, I., et al. (2005) Nucleic acid chaperone activity of

HIV-1 nucleocapsid protein: critical role in reverse transcription and molecular mechanism. *Prog Nucleic Acid Res Mol Biol* 80, 217–286
7. Fisher, R. J., Rein, A., Fivash, M. J., et al. (1998) Sequence-specific binding of human immunodeficiency virus type 1 nucleocapsid protein to short oligonucleotides. *J Virol* 72, 1902–1909
8. Fisher, R. J., Fivash, M. J., Stephen, A. G., et al. (2006) Complex interactions of HIV-1 nucleocapsid protein with oligonucleotides. *Nucleic Acids Res* 34, 472–484
9. Lakowicz, J. R. (1999) Fluorescence Anisotropy, in Principles of Fluorescence Spectroscopy, 2nd ed., (Lakowicz JR, ed) Kluwer Academic/Plenum Publishers, New York, pp. 291–320
10. Myszka, D. G. (1999) Improving biosensor analysis. *J Mol Rec* 12, 279–284
11. Simpson, J. T., Fisher, R. J. (2005) Measurement of protein-protein interactions using surface plasmon resonance spectroscopy, in *Protein-Protein Interactions, 2nd ed.* (Golemis EA, Adams PD, ed) Cold Spring Harbor Laboratory Press, Plainview, NY pp. 355–376
12. Seidel, C. A. M., Schulz, A., Sauer, M. H. M. (1996) Nucleobase-specific quenching of fluorescent dyes. 1. Nucleobase one-electron redox potentials and their correlation with static and dynamic quenching efficiencies. *J Phys Chem* 100, 5541–5553
13. Unruh, J. R., Gokulrangan, G., Lushington, G. H., et al. (2005) Orientational dynamics and dye-DNA interactions in a dye-labeled DNA aptamer. *Biophys J* 88, 3455–3465

Chapter 16

Methods for Analysis of Incorporation and Annealing of tRNA^{Lys} in HIV-1

Shan Cen, Fei Guo, and Lawrence Kleiman

Abstract

In HIV-1, tRNA^{Lys3} serves as the primer for reverse transcription of minus strand strong stop cDNA. During viral assembly, the tRNA^{Lys} isoacceptors, tRNA^{Lys1,2} and tRNA^{Lys3}, are selectively packaged into the virion. The selectively packaging of tRNA^{Lys3} facilitates the annealing of tRNA^{Lys3} to the viral genome and the initiation of reverse transcription. We describe herein a set of experimental approaches for studying the mechanism by which tRNA^{Lys} is selectively incorporated into HIV-1 and investigate how primer tRNA^{Lys3} is annealed to viral genome. The methods described will also help in the analysis of cellular RNAs packaged in the virus particles.

Key words: tRNA^{Lys}, selective tRNA packaging, two-dimensional polyacrylamide gel electrophoresis, primer tRNA^{Lys3} extension.

1. Introduction

A select number of tRNAs have been identified as primer tRNAs in retroviruses. For example, tRNA^{Trp} is the primer for all members of the avian sarcoma and leucosis virus group examined to date. There are three major tRNA^{Lys} isoacceptors in mammalian cells. tRNA^{Lys1,2}, representing two tRNA^{Lys} isoacceptors differing by one base pair in the anticodon stem, is the primer tRNA for several retroviruses, including Mason–Pfizer monkey virus and human foamy virus. tRNA^{Lys3} serves as the primer for mouse mammary tumor virus, and lentiviruses such as equine infectious anemia virus, simian immunodeficiency virus, and human immunodeficiency virus type 1 and 2 (HIV-1 and HIV-2). Primer tRNAs are selectively incorporated into avian retroviruses, MMTV, and

Vinayaka R. Prasad, Ganjam V. Kalpana (eds.), *HIV Protocols: Second Edition, vol. 485*

DOI 10.1007/978-1-59745-170-3_16 Springerprotocols.com

HIV-1 during their assembly. In general, the term "selectively packaging of tRNA" refers to the increase in the percentage of the low molecular weight RNA population representing primer tRNA in going from the cytoplasm to the virus. For example, in HIV-1 produced from COS7 cells transfected with HIV-1 proviral DNA, the major $tRNA^{Lys}$ isoacceptors, including $tRNA^{Lys3}$ and $tRNA^{Lys1,2}$, are selectively packaged, and the relative concentration of $tRNA^{Lys}$ changes from changes from 5–6% in the cytoplasm to 50–60% in the virus *(1)*. Evidence has been generated that the formation of a $tRNA^{Lys}$ packaging complex, whose components include Gag, GagPol, viral genomic RNA, lysyl-tRNA synthetase (LysRS) and $tRNA^{Lys}$, is required for selectively packaging of $tRNA^{Lys}$ *(2–4)*.

As an increased concentration of $tRNA^{Lys}$ in HIV-1 directly affects the degree of annealing of primer $tRNA^{Lys3}$ to viral genome as well as viral infectivity *(5, 6)*, the selective packaging of $tRNA^{Lys}$ represents a key event in the life cycle of HIV-1. We describe herein a set of experimental approaches to study the mechanisms involved in $tRNA^{Lys}$ selective packaging and annealing to the viral RNA, including the specific procedures to isolate virus and extract viral RNA, quantify viral $tRNA^{Lys}$ by using a dot blot assay, determine the selectivity of incorporated tRNA in virions by a two-dimensional polyacrylamide gel electrophoresis (2D-PAGE), and measure $tRNA^{Lys3}$ priming efficiency by a primer extension assay.

2. Materials

2.1. Preparation of Total Viral RNA Sample

1. Phosphate-buffered saline (PBS): 137 mM NaCl, 10 mM Phosphate, 2.7 mM KCl, pH 7.4. Store at 4 °C.
2. TE: 10 mM Tris-HCl (pH 8.0), and 1 mM EDTA. Store at room temperature.
3. 15% sucrose (*w/v*): 15 g of sucrose in 100 mL (final volume) of PBS. Store at 4 °C.
4. 65% sucrose (*w/v*): 65 g of sucrose in 100 mL (final volume) of PBS. Store at 4 °C.
5. TNE (10 ×): 10 mM Tris-HCl (pH 7.4), 100 mM NaCl and 1 mM EDTA-Na. Store at room temperature.
6. Solution D (5 mL): 2.36 g of Guanidium thiocyanate, 1.25 mL of 100 mM sodium citrate (pH 4.0), 85 μL of Sarcosyl (30 %, *w/v*) and 36 μL of freshly made β-mercaptoethnol (*see* **Note 1**). Store at 4 °C and avoid light.

2.2. Two-Dimensional Polyacrylamide Gel Electrophoresis

1. 11% polyacrylamide-7 M urea gel solution (1 L): 50 mL of 10X TBE, 109.8 g of acrylamide, 5.75 g of BIS acrylamide, and 420 g of urea in ddH2O. Filter with Whatman 1 paper. Light sensitive, store at 4 °C.

2. 20% polyacrylamide-7 M urea gel solution (1 L): 50 mL of 10X TBE, 199.8 g of acrylamide, 10.5 g of BIS acrylamide, and 420 g of urea in ddH2O. Filter with Whatman 1 paper. Light sensitive, store at 4 °C.
3. 1× TBE: 50 mM Tris, 5 mM boric acid, plus 1 mM EDTA-Na_2. Store at room temperature.

2.3. Quantification of $tRNA^{Lys}$ Isoacceptors in HIV-1

1. 20X SSC: 150 mM NaCl, 15 mM sodium citrate in water. Adjust pH to 7.0 with HCl, and store at 4 °C.
2. Denaturation buffer (29 μL) : 20 μL of deionized formamide, 7 μL of formaldehyde (37%), and 2 μL of 20X SSC.
3. 18-mer DNA probe specific for $tRNA^{Lys3}$: 5′-TGGCGCCC GAACAGGGAC-3′.
4. 18-mer DNA probe specific for $tRNA^{Lys1,2}$: 5′-TGGCGCCC AACGTGGGGC-3′.
5. 17-mer DNA probe specific for the 5′ end of HIV-1 genomic RNA: upstream of the PBS, 5′-CTGACGCTCTCGCACCC-3′.

2.4. Primer Extension Assay

1. RT buffer (5X): 250 mM Tris-HCl (pH 7.5), 300 mM KCl, 15 mM $MgCl_2$, 50 mM dithiothreitol. Store in aliquots at −20 °C.
2. G solution (10X): 2 mM dCTP, 2 mM dTTP, and 0.5 mM ddATP. Store in aliquots at −20 °C.
3. Termination buffer (180 μL) : 100 μL of isopropanol, 10 μL of NaOAc, 1 μg Linear Acrylamide (Ambion), and 70 μL of ddH_2O.
4. 6% polyacrylamide-7 M urea gel solution (1 L): 420 g of urea, 30 mL of acrylamide (40%, Bio-Rad), 10 mL of 10X TBE in water. Store at 4 °C and avoid light.

3. Methods

3.1. Preparation of Total Viral RNA Sample

3.1.1. Isolation of HIV-1 Virus

1. COS7 cells (10 or 20 100-mm plates) are transfected with 10 μg proviral DNA, using the calcium phosphate method as previously described *(7)*. Supernatant is collected 63 h post-transfection.
2. The supernatant is centrifuged in a Beckman GS-6R rotor at 3000 rpm for 30 min at 4 °C, and then filtered through a 0. 2-μm-pore-size filter. This step will remove the dead cells and debris.
3. Viruses are pelleted from culture medium by centrifugation in a Beckman 45 Ti rotor at 35,000 rpm for 1 h at 4 °C.
4. The viral pellets are resuspended in 1 mL of TE (pH 8.0) and then purified by centrifugation in a Beckman SW41 rotor at

26,500 rpm for 1 h at 4 °C through 15% sucrose (4 mL) onto a 65% sucrose (2 mL) cushion.

5. The band corresponding to purified virus is removed and pelleted in 1X TNE in a Beckman 45 Ti rotor at 40,000 rpm for 1 h at 4 °C.
6. The viral pellets are resuspended into 100–500 μL of TE (pH 8.0). A portion of virus supernatant is used for the analysis of viral protein, RT activity, and viral infectivity. The remaining can be used for viral RNA extraction.

3.1.2. Extraction of Total Viral RNA

1. Virus samples (from step 6) are mixed sequentially with 500 μL of fresh solution D, 50 μL of 2 M sodium acetate (pH 4.0), 500 μL of phenol:ddH_2O and 170 μL of chloroform (*see* **Note 2**).
2. The mixture is centrifuged in eppendorf tubes at 10,000 rpm for 30 min at 4 °C. Aqueous phase is removed and transferred to a fresh eppendorf tube.
3. After adding an equal volume of isopropanol, total viral RNA is precipitated at −80 °C overnight.
4. Total viral RNA is pelleted by the centrifugation at 10,000 rpm for 30 min at 4 °C.
5. The RNA pellet is resuspended in 300 μL of solution D. Then the precipitation and centrifugation steps (steps 9 and 10) are repeated once more.
6. The RNA pellet is washed twice with 70% ethanol.
7. After vacuum drying, total viral RNA is resuspend in 20 μL of ddH_2O and stored at −80 °C.

3.2. Two-Dimensional Polyacrylamide Gel Electrophoresis

Low-molecular-weight RNA in virions is analyzed by first 3′-end-labeling total viral RNA with [^{32}P]pCp, and resolving this labeled RNA using 2D-PAGE.

3.2.1. 3′-End-Labeling Total Viral RNA with [^{32}P]pCp

The desired amount of tRNA sample (usually 3–4 μL) is mixed with 4 μL of [^{32}P]pCp, 1 μL of RNA ligase (Ambion, 10 U/μL), and 1 μL of reaction buffer (10X, Ambion), then adding ddH2O up to 10 μL. The mixture is incubated at 4 °C overnight, and then added 3. 3 μL of RNA loading buffer (Ambion).

3.2.2. Two-Dimensional Polyacrylamide Gel Electrophoresis

1. Electrophoresis of [^{32}P]pCp-labeled viral RNA was carried out in 0.5X TBE at 4 °C with the Hoeffer SE620 gel electrophoresis apparatus. The gel size was 14 by 32 cm.
2. The first dimension was run in an 11% polyacrylamide-7 M urea gel for 16 h at 800 V, until the bromophenol blue dye was beginning to elute from the bottom of the gel.
3. After autoradiography, the piece of gel containing RNA was cut out and embedded in a second gel (20% polyacrylamide-7 M urea) at an orientation of 90° to the original run, and electrophoresed for 30 h (25 W, limiting).

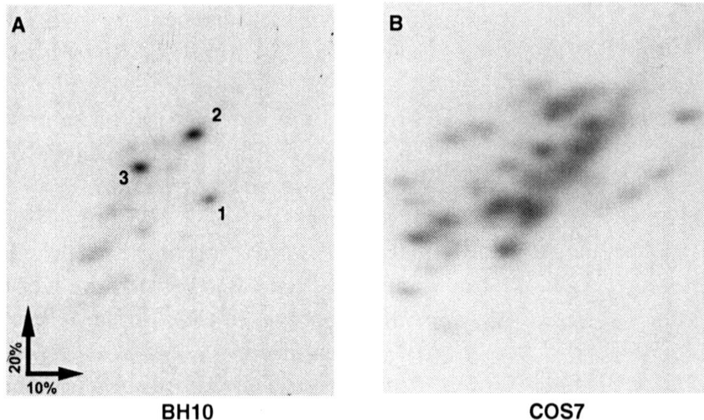

Fig. 16.1. Two-dimensional PAGE patterns of cellular and viral tRNAs. Electrophoretic conditions are as described in the text. Viral RNA (**A**) and cellular RNA (**B**) were extracted from the virions produced from COS7 cells transfected with HIV-1 proviral DNA BH10 and COS7 cells, respectively. Spots 1 and 2 are collectively referred to as $tRNA^{Lys1,2}$, while spot 3 is referred to as $tRNA^{Lys3}$. These tRNA species have been identified as previously *(8)*.

4. After autoradiography, the electrophoretic gel patterns will show only the low-molecular-weight viral RNA, since the high-molecular-weight viral genomic RNA cannot enter into the polyacrylamide gels.

Figure 16.1 shows 2D-PAGE patterns of low-molecular-weight RNA found in the virions produced from COS7 cells transfected with the proviral DNA BH10 (panel A) and in COS7 cells (panel B). The labeled spots have previously been identified; spot 3 is $tRNA^{Lys3}$, the primer tRNA for RT, and spots 1 and 2 represent $tRNA^{Lys1,2}$ *(8)*.

3.3. Quantification of $tRNA^{Lys}$ Isoacceptors in HIV-1

The relative amount of $tRNA^{Lys}$ isoacceptor (i.e., both $tRNA^{Lys3}$ and $tRNA^{Lys1,2}$) per copy of HIV-1 genomic RNA is determined by dot blot hybridization. Briefly, each sample of total viral RNA is blotted onto nitrocellulose membrane and is probed with $5'$-^{32}P-end-labeled DNA probes specific for different $tRNA^{Lys}$ isoacceptors. The relative amounts of $tRNA^{Lys}$ isoacceptor per sample were analyzed by using phosphorimaging. The blots were then stripped according to the manufacturer's instructions, and were reprobed with a $5'$ ^{32}P-end-labeled DNA probe specific for HIV-1 genomic RNA. Phosphorimaging was used to quantitate the relative amount of HIV-1 genomic RNA per sample and the relative amount of $tRNA^{Lys}$ isoacceptor or $tRNA^{Lys}$ isoacceptor copy of HIV 1 genomic RNA was determined. The amount of total viral RNA used in these determinations contained 3×10^8 to 10×10^8 copies of genomic RNA, an amount producing signals within the linear range of measurement for hybridization of both $tRNA^{Lys}$ isoacceptors and genomic RNA (*see* **Note 3**).

3.3.1. Dot Blot of Viral RNA

1. Prewet a piece of nitrocellulose membrane (Hybond, Amersham Pharmacia) in ddH_2O for 15 min and then soak it in 10X SSC for another 15 min.
2. Place one sheet of 3M Whatman paper previously wet with 10X SSC, on the top of the vacuum unit of the microfiltration apparatus (Bio-Rad). Place the wet nitrocellulose membrane on the bottom of the sample wells cut into the upper section of the manifold.
3. Smooth away any air bubbles trapped between the upper section of the manifold and the nitrocellulose membrane, and then assemble the manifold together.
4. Fill all the slots with 10X SSC and apply gentle suction until all of the fluid has passed through the nitrocellulose membrane. Turn off the vacuum and refill the slots with 10X SSC.
5. Mix the RNA dissolved in 10 μL of ddH_2O with 29 μL of denaturation buffer. Incubate the mixture for 15 min at 68 °C, and then cool the samples on ice.
6. Add two volumes of 20X SSC (78 μL) to each of the samples.
7. Apply gentle suction to the manifold until the 10X SSC in the slots has passed through the nitrocellulose filter, then turn off the suction.
8. Load the samples into the slots and apply gentle suction. After all of the samples have passed through the filter, continue suction for 5 min to dry the nitrocellulose membrane.
9. Remove the membrane from the manifold, and allow it to dry completely at 37 °C.
10. UV cross-link the RNA to the nitrocellulose membrane under UV light for 3 min. The nitrocellulose membrane is ready for hybridization.

3.3.2. DNA Probe Radiolabeling Using T4 Polynucleotide Kinase

1. Mix the following: 1 μL of probe (50 ng/μL) specific for HIV-1 genomic RNA or $tRNA^{Lys}$ described previously, 1 μL of T4 polynucleotide kinase, 5 μL of [γ-^{32}P] ATP, 4 μL of reaction buffer (5X forward buffer: Invitrogen), then add ddH2O up to 20 μL. Incubate at 37 °C for 30 min.
2. Purify the probe using the MicroSpin G-25 columns (Amersham Pharmacia) according to the manufacturer's instructions.

3.3.3. Hybridization

1. Warm up the hybridization buffer (Ambion) at 37 °C for a few minutes.
2. Place the membranes in a blue capped Falcon tube along with the hybridization buffer. This tube should be placed on the rotating platform of a hybridization oven (VWR), and then prehybridize the membranes for 1–2 h.

3. Mix the probe with prehybridized membrane in the oven and incubate between 16 and 24 h at 42 °C.
4. Wash the membrane for 20 min at room temperature in 1X SSC, 0.1% SDS followed by three washes of 20 mn each at 42 °C in 0.2X SSC, 0.1 % SDS. For $tRNA^{Lys}$ isoacceptors determination, instead of three washes at 42 °C one secondary wash at room temperature in 0.2X SSC, 0.1 % SDS is sufficient.
5. Expose the nitrocellulose membrane to film or phosphorimager screen.

3.4. $tRNA^{Lys3}$ Primer Extension Assay

To measure the amount of $tRNA^{Lys3}$ annealed to genomic RNA, $tRNA^{Lys3}$-primed initiation of reverse transcription is measured using equal amounts of total viral RNA (determined by dot blot hybridization) as the source of primer tRNA/viral RNA template in an in vitro HIV-1 reverse transcription reaction which will extend the primer by 6 deoxyribonucleotides (six base primer extension assay).

3.4.1. Preparation of a Control Priming Complex – Purified $tRNA^{Lys3}$ Annealed to Synthetic Viral Genomic RNA

1. Synthetic HIV-1 genomic RNA (497 bases) is made from Acc I-linearized plasmid pHIV-PBS *(9)* with the MEGAscript transcription system (Ambion). The synthetic genomic RNA comprises the complete U5 region, the PBS, and a part of the Gag-coding region ($HIV\text{-}1_{IIIB}$ DNA sequence positions 473–958).
2. The purification of $tRNA^{Lys3}$ from human placenta was performed as previously described *(8)*, using standard chromatography procedures (sequentially, DEAE-Sephadex A-50, reverse-phase chromatography (RPC-5), and Porex C4).
3. To anneal placental $tRNA^{Lys3}$ to synthetic HIV-1 genomic RNA, 0.5 pmol of synthetic genomic RNA was incubated with 0.5 pmol of $tRNA^{Lys3}$ in RT buffer at 85 °C for 2.5 min, at 50 °C for 8 min, and then at 37 °C for 10 min.
4. The annealed $tRNA^{Lys3}$ primer/viral RNA complex was stored at −80 °C for future use.

3.4.2. Six-Base Primer Extension Assay

1. Total viral RNA (containing approximately 0.5×10^8 molecules of viral genomic RNA) was brought to a volume of 10 μL with ice-cold water.
2. The RNA sample was then mixed with 10 μL of a premixed solution containing: 4 μL Reverse transcriptase buffer (5X), 1 μL HIV-1 Reverse transcriptase (50 ng/μL), 0. 5 μL [α-^{32}P]dGTP (Dupont; 3,000 Ci/mmol, 10 mCi/ml), 0. 5 μL RNase inhibitor, 2 μL G Solution and 2 μL H_2O (*see* **Note 4**). Incubate at 37 °C for 15 min.
3. The sequence of the first six deoxynucleoside triphosphates incorporated is CTGCTA. The extension of primer $tRNA^{Lys3}$ is terminated by adding ddATP, resulting in a six-

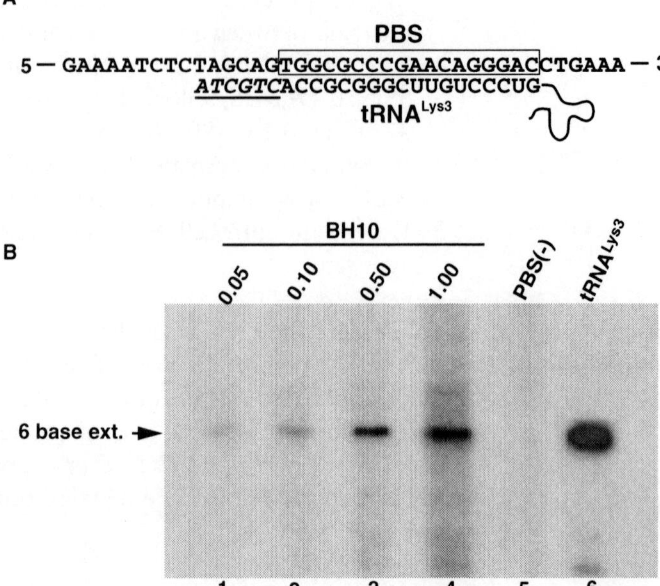

Fig. 16.2. Initiation of *in vitro* reverse transcription total viral RNA as the source of primer tRNA^{Lys3} annealed to the viral genomic RNA template. (**A**) Diagram showing the first six dNTPs incorporated into tRNA^{Lys3} during the initiation of reverse transcription. (**B**) 1D PAGE of tRNA^{Lys3} extension products.

base extended tRNA^{Lys3} product, as graphically shown in **Fig. 16.2A**.

4. Stop the reaction by adding 180 μL termination buffer, then mix well and place the samples at −80 °C for 1 h, to precipitate the reverse transcription product.
5. Spin at 15,000 rpm at 4 °C for 15 min, and then remove the supernatant. Wash the pellet once with 70% ethanol, and dry at room temperature for 15 min.
6. Add 10 μL of loading buffer (1:1 dilution with H_2O) and vortex. Boil in water bath for 10 min and put on ice.
7. Prepare a 6% PAGE (7 M urea, 1X TBE) sequence gel and prerun the gel at 50–60 W for 30 min or longer. Wash the wells with 1X TBE and load the sample.
8. Electrophorese the sample at 50–60 W (volts around 1,000) until the second dye reaches the middle of the gel, and then dry the gel at 80 °C for 30 min.

Figure 16.2B shows an example of the six-base extension products of reverse transcription as resolved by 1D PAGE. Equal amounts of viral genomic RNA (0.5×10^8 copies) isolated from wild-type and mutant HIV-1 are used in each reaction to examine tRNA^{Lys3} annealed to viral genome. A standard curve is also generated (lanes 1–4), in which 0.05, 0.1, 0.5 and 1 times this amount of genomic RNA were used. Lane 5 (negative control) shows the absence of priming when RNA isolated from a virion lacking the PBS was used, while lane 6 (positive control) represents a reaction in which purified $tRNA^{Lys3}$ annealed in vitro to synthetic genomic RNA served as the source of primer-template.

4. Notes

1. Solution D has to be freshly made, with the remaining solution stored at 4 °C and only used for resuspending viral RNA pellets (step 11). After that, it should be discarded.
2. Although commercial kits are available for the extraction of viral RNA, such as Triazol (Invitrogen), the extraction procedure described herein produces higher purity of viral genomic RNA and higher yields of low-molecular-weight RNA.
3. To determine the absolute amount of $tRNA^{Lys3}$ (or $tRNA^{Lys1,2}$) and viral genomic RNA in a sample, known amounts of purified or in vitro transcribed $tRNA^{Lys}$ (0.2–10 ng) or synthetic HIV-1 genomic RNA ($0.15–10 \times 10^8$ molecule) are used to establish a standard RNA concentration curve. This curve is used to establish the concentration of viral RNA in the HIV-1 RNA sample tested, and also to show that your hybridization signals are in the linear range.
4. In HIV-1, $tRNA^{Lys3}$ annealed to viral genome exists in two major forms: unextended or in a form extended by the first two DNA bases incorporated, C and T. The two-base extended form of the $tRNA^{Lys3}$ is considered an indicator of endogenous initiation of reverse transcription in HIV-1 (10). A modified primer extension assay is used to determine endogenous initiation of reverse transcription in HIV-1. To resolve unextended and 2-base-extended $tRNA^{Lys3}$, the reaction mixture contains 5 μCi each of [α-^{32}P]dGTP or [α-^{32}P]dCTP (Dupont; 3,000 Ci/mmol, 10 mCi/mL), and will produce a $tRNA^{Lys3}$ extended one or 3 bases, respectively, through the addition of one more base, G, to the initially 2-base-extended $tRNA^{Lys3}$, or C, to the unextended $tRNA^{Lys3}$.

References

1. Mak, J., Jiang, M., Wainberg, M. A., et al. (1994) Role of $Pr160^{gag\text{-}pol}$ in mediating the selective incorporation of $tRNA^{Lys}$ into human immunodeficiency virus type 1 particles. *J Virol* 68, 2065–2072.
2. Cen, S., Khorchid, A., Javanbakht, H., et al. (2001) Incorporation of lysyl-tRNA synthetase into HIV-1. *J Virol* 75, 5043–5048.
3. Cen, S., Javanbakht, H., Niu, M., et al. (2004) Ability of wild-type and mutant lysyl-tRNA synthetase to facilitate tRNA(Lys) incorporation into human immunodeficiency virus type 1. *J Virol* 78, 1595–1601.

4. Khorchid, A., Javanbakht, H., Parniak, M. A., et al. (2000) Sequences within Pr160$^{gag-pol}$ affecting the selective packaging of tRNA^{Lys} into HIV-1. *J Mol Biol* 299, 17–26.
5. Gabor, J., Cen, S., Javanbakht, H., et al. (2002) Effect of altering the tRNA(Lys)(3) concentration in human immunodeficiency virus type 1 upon its annealing to viral RNA, GagPol incorporation, and viral infectivity. *J Virol* 76, 9096–9102.
6. Guo, F., Cen, S., Niu, M., et al. (2003) Specific inhibition of the synthesis of human lysyl-tRNA synthetase results in decreases in tRNA^{Lys} incorporation, tRNA^{Lys} annealing to viral RNA, and viral infectivity in HIV-1. *J Virol* 77, 9817–9822.
7. Jiang, M., Mak, J., Wainberg, M. A., et al. (1992) A variable tRNA content in HIV-1 3B. *Biochem Biophys Res Commun* 185, 1005–1015.
8. Jiang, M., Mak, J., Ladha, A., et al. (1993) Identification of tRNAs incorporated into wild-type and mutant human immunodeficiency virus type 1. *J Virol* 67, 3246–3253.
9. Cen, S., Huang, Y., Khorchid, A., et al. (1999) The role of Pr55^{gag} in the annealing of tRNA^{Lys3} to human immunodeficiency virus type 1 genomic RNA. *J Virol* 73, 4485–4488.
10. Huang, Y., Wang, J., Shalom, A., et al. (1997) Primer tRNA^{Lys3} on the viral genome exists in unextended and two base- extended forms within mature Human Immunodeficiency Virus Type 1. *J Virol* 71, 726–28.

Section III

Specialized Approaches to Study HIV-1 Biology and Pathogenesis

Chapter 17

Somatic Cell Genetic Analyses to Identify HIV-1 Host Restriction Factors

Susana T. Valente and Stephen P. Goff

Abstract

Cellular proteins are critically involved in all steps of the human immunodeficiency virus type 1 (HIV-1) life cycle. Disruption of host functions essential for virus replication or discovery of new proteins that block viral replication may provide novel antiviral approaches. In recent years, genetic selections for and against genes carried by retroviral vectors have become increasingly powerful, allowing for the isolation of cells with altered susceptibility to virus infection. Screening complementary DNA libraries for clones able to induce resistance to infection by recombinant HIV-1 genomes, has proved to be an excellent tool to identify new interfering factors. The restriction factors TRIM5α *(1)*, the Zinc Finger Antiviral Protein (ZAP) *(2)* as well as the dominant negative factor N-86-HnRNPU *(3)* have all been discovered by means of such genetic screens. Here we report the strategy and techniques to prepare a library and isolate HIV antiviral genes, using the identification of N-86-HnRNPU as an example.

Key words: HIV-1, cDNA libraries, Genetic Screen, restriction factors.

1. Introduction

The strategy developed to select and recover antiviral genes from a library involves: expression of a complementary DNA (cDNA) library in cells highly susceptible to viral infection; the efficient transduction of these cells with a virus carrying a counterselectable genetic marker; the elimination of the infected cells using a drug targeting the genetic marker; and the isolation and recovery of the resistant clones. In this chapter we describe each of these steps. The methodology was actually used for the identification of N-86-HnRNPU. In our example described here, a cDNA expression library was constructed using the MLV-based vector pBabe-HAZ *(2)*. The cDNAs were expressed as fusion

Vinayaka R. Prasad, Ganjam V. Kalpana (eds.), *HIV Protocols: Second Edition, vol. 485*

DOI 10.1007/978-1-59745-170-3_17 Springerprotocols.com

proteins with a hemagglutinin (HA) epitope tag at the 5′ end and the zeocin resistance marker (Zeo) at the 3′ end. The retroviral vector contains a LoxP site in the 3′ LTR, which after reverse transcription becomes duplicated onto both LTRs such that the provirus can be excised from the genome by the Cre recombinase. The cDNA library was introduced into the human rhabdomyosarcoma cell line, TE671, highly susceptible to human immunodeficiency virus type 1 (HIV-1) infection. Pools of the transduced cells were selected for Zeocin resistance, ensuring expression of at least one member of the cDNA library per cell. To recover any HIV-resistant clones, pools of the transduced TE671 cells were repeatedly challenged with VSV-G pseudotyped HIV-1 retroviral particles expressing the Herpes Simplex virus (HSV) thymidine kinase (*TK*) gene, to infect as many susceptible cells as possible. The virus-susceptible cells, having become HSV-TK+, were then eliminated by growth in media containing Ganciclovir (GCV), and the rare HSV-TK-negative clones were recovered as potential HIV-resistant clones. We will also describe the methods to retest and confirm the ability of the recovered cDNA to confer virus resistance.

2. Materials

2.1. Tissue Culture

1. Eukaryotic cells grown at 37 °C in a humidified atmosphere containing 5% CO_2.
2. Dulbecco's modified eagle's medium (DMEM; Gibco).
3. Fetal bovine serum (FBS; Atlanta).
4. 200 mM L-glutamine (L-Gln) solution in saline (Gibco).
5. Penicillin: solution at 10,000 U/mL (Gibco).
6. Streptomycin: 10, 000 μg/mL Streptomycin sulfate in 0.85% saline (Gibco-BRL).
7. TE671 (ATCC HTB-139) and 293T fibroblast cells were obtained from the American Type Culture Collection.

2.2. Transfections

1. Fugene 6 (Roche Molecular Biochemical's)
2. FBS free OptiMEM (Gibco)

2.3. Retroviral Vectors and Plasmids

1. pBabe-HAZ: MLV based retroviral vector *(2)* modified to SalI-NotI *(3)*
2. pMD.G: VSV-G envelope protein *(4)*.
3. p8.91: HIV Gag-Pol expression vector *(4)*.
4. pHit60: MLV Gag-Pol expression vector *(5)*.
5. pCre: Cre recombinase expression vector *(2)*.

2.4. cDNA Library Construction

1. A 16 and 37 °C water bath.
2. RNA extraction kits, Rneasy Midi kit (Qiagen).

3. mRNA purification kit, Oligotex kit (Qiagen).
4. RNase-free β-mercaptoethanol.
5. RNase-free water.
6. RNase-free ethanol 100% and 70% (v/v).
7. RNase-free NH_4OAc 7.5 M (filtered though a sterile 0. 2-μm nitrocellulose filter).
8. 1–10 μL [α-^{32}P]dCTP (1 μCi/μL).
9. Phenol:chloroform:isoamyl alcohol (25:24:1).
10. Electrocompetent cells, Stbl2 (Invitrogen).
11. Restriction Enzymes: Sal I and Not I.

2.5. Production and Titrations of VLPs

1. Fugene-6 (Roche).
2. TaqMan PCR Mix (Invitrogen).
3. Zeocin (Invitrogen).

2.6. Genetic Screen

1. Ganciclovir (Sigma).
2. Zeocin (Invitrogen).

2.7. Recovery of cDNA Insert

1. DNeasy Tissue Kit (Qiagen).
2. High Fidelity kit (Roche).

2.8. Strategies to Retest cDNA Ability to Confer Resistance to HIV Infection

1. G418 (Sigma).
2. Puromycin (Sigma).

3. Methods

3.1. Preparation of cDNA Library in pBabe-HAZ

Complementary DNA libraries can be directional (oriented for sense-strand expression) or random. All members of a directional library are usually constructed to drive expression of the cloned gene, by a controllable promoter contributed by the vector. Unlike random libraries, in which only 50% of the clones are likely to contain cDNA inserts oriented properly for expression, all members of a directional library are potentially able to express the insert and therefore, a given cDNA can be recovered by screening half as many clones. We will be using as an example a replication-defective retroviral vector based on the Moloney Murine Leukemia Virus genome, designated pBabe-HAZ, that expresses the cDNA inserts under the control of a constitutive promoter (**Fig. 17.1**). This vector allows expression of genes, or portions of genes, as fusion proteins with the zeocin resistance marker (Zeo), and has several advantages: (1) the ability to be packaged using a variety of envelope proteins, allowing delivery to a wide range of cell types; (2) the ability to allow easy recovery of introduced genes from transduced cells; (3) the presence

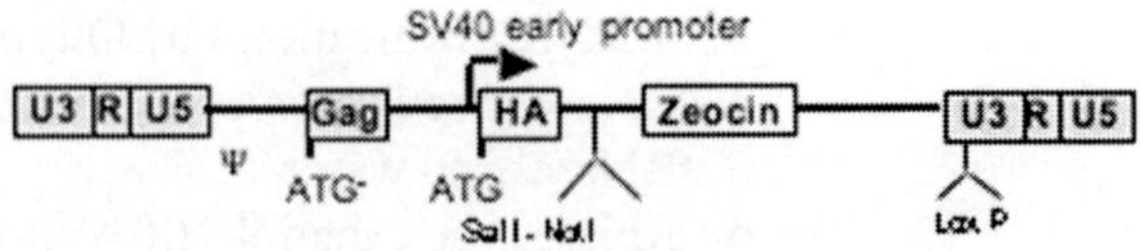

Fig. 17.1. Schematic representation of the pBabe-HAZ construct. Y, MuMLV viral RNA packaging signal; SalI-NotI, linker sequence containing SalI and NotI sites; LoxP, LoxP sequence for recognition by Cre recombinase.

of LoxP sites such that the cDNA can be excised for assessment of its functionality; and (4) expression of a HA tag on the fusion protein to facilitate biochemical studies.

3.1.1. mRNA Preparation

The first step in cDNA construction is the isolation of the cellular mRNA. RNA is extremely sensitive to RNases and should be prepared with care. Very small amounts of RNases are sufficient to degrade RNA. The most common sources of RNases are from hands and dust particles. Therefore extra caution is needed for the preparation and storage of RNA. The tubes, tips, and water should be RNase-free and gloves should be worn all times.

1. Count 50×10^6 cells of the desired cell type you wish to make the cDNA library from and wash once in PBS 1X. The cells should be sub-confluent and growing in log phase.
2. Extract RNA according to the manufacturer's protocol (RNeasy Midi Kit; Qiagen):
   1. Use 4 mL of RLT buffer (lysis buffer) with 10 μL β-mercaptoethanol/mL.
   2. Homogenize cells by passing the cells though a 19 G needle at least 20 times.
   3. Elute twice in 250 μL of RNase-free water.
   4. Use 5 μL of total RNA in 995 μL of RNase-free H_2O to determine concentration by optical density (OD) at 260 nm.
3. Purify mRNA from total RNA using Oligotex kit (Qiagen).
   1. Adjust volume of total RNA to 500 μL.
   2. Use 55 μL of Oligotex suspension beads.
   3. Elute twice with 30 μL of elution buffer (OEB). Pool fractions.
   4. Run 3 μL on a gel to check the quality of the RNAs (*see* **Note 1**). Keep mRNA at −80 °C.
4. Precipitate mRNA and resuspend as follows:
   1. Mix: 50 μL volume of mRNA.
      5 μL 7.5 M NH_4OAc (1/10 volume).
      110 μL ethanol 100% (2 volumes).
   2. Incubate for 30 min at 4 °C to allow precipitation.
   3. Centrifuge at maximum (max) speed for 30 min.
   4. Resuspend in 200 μL of Ethanol 70%.
   5. Centrifuge at max speed for 30 min to pellet mRNA.

6. Remove supernatant and allow pellet to air-dry.
7. Recover in 6 μL RNase-free water.
8. Use 1 μL in 99 μL of RNase-free water to determine OD at 260 nm.

3.1.2. cDNA Synthesis and cDNA Cloning

For cDNA synthesis and cloning use Invitrogen's kit "SUPERSCRIPT™ Plasmid System GATEWAY™ Technology for cDNA Synthesis and Cloning". We adapted the kit to the following protocol:

3.1.2.1. First Strand Synthesis

Start with 5 μg of mRNA (unless you would like to favor the synthesis of short cDNA fragments, and in that case use 15 μg), in a 5- μL final volume.

1. Add 2 μL of Not I primer-adapter.
2. Heat to 70 °C for 10 min, quick chill on ice, and add:
   4 μL 5X First Strand Buffer
   2 μL 0.1 M DTT
   1 μL 10 mM dNTP Mix.
3. Leave 2 min at room temperature to equilibrate temperature, and then add:
   5 μL of Superscript II RT (200 units/μg of mRNA for 1–5 μg of mRNA).
4. Mix gently and incubate for 1 h at 37 °C.
5. Place on ice to terminate reaction.

3.1.2.2. Second Strand Synthesis

1. On ice add the following reagents in order:
   92 μL DEPC water (0.1% diethyl pyrocarbonate in water).
   30 μL 5X second strand buffer.
   3 μL 10 mM dNTP Mix.
   1 μL *Escherichia coli* DNA ligase.
   4 μL DNA polymerase (10 units/μL)
   1 μL RNase H
   1 μL [α-^{32}P]dCTP (1 μCi/μl)
2. Gently mix and incubate 2 h at 16 °C.
3. Add 2 μL of T4 DNA polymerase and continue incubating for 5 min.
4. Place reaction on ice and add 10 μL 0.5 M EDTA.
5. Add 150 μL phenol:chloroform:isoamyl:alcohol (25:24:1), vortex, and centrifuge at maximum speed at room temperature for 5 min.
6. Remove 140 μL of upper aqueous solution and transfer to 1.5 mL microcentrifuge tube.
7. Add 70 μL of 7.5 M NH_4OAc and 0.5 mL absolute ethanol.
8. Vortex and centrifuge at room temperature for 5 min at max speed.
9. Remove supernatant carefully.
10. Overlay the pellet with 0.5 mL 70 °C ethanol.

11. Centrifuge for 2 min at 14,000 × *g*.
12. Briefly dry the cDNA pellet to remove residual ethanol (not too long, otherwise it is hard to resuspend).

3.1.2.3. Sal I Adaptor Addition

1. Add the following reagents on ice:
   25 μL DEPC water
   10 μL 5X T4 DNA ligase buffer
   10 μL Sal I adaptors
   5 μL T4 DNA ligase
   50 μL final volume.
2. Mix gently and allow to incubate overnight at 16 °C.
3. Add 50 μL phenol:chlorophorm:isoamyl alcohol (25:24:1), vortex and centrifuge for 5 min at max speed.
   - Carefully remove 45 μL from top layer to a new microcentrifuge tube.
   - Add 25 μL of 7.5 M HN_4OAc and 150 μL alcohol 100%.
   - Vortex and centrifuge for 20 min at maximum speed.
   - Remove ethanol and let air-dry.
   - Resuspend in double the volume mentioned in the protocol:
     82 μL DEPC H_2O
     10 μL RE_{ACT} 3 buffer
     8 μL Not I
     100 μL final volume
   - Incubate for 2 h at 37 °C.
   - Precipitate with 100 μL of phenol:chlorophorm:isoamyl alcohol (25:24:1), vortex and centrifuge at room temperature for 5 min at 14,000 × *g*.
   - Carefully remove 90 μL of top phase, and transfer it to a new tube.
   - Add 50 μL 7.5 M NH_4OAc and 300 μL of absolute ethanol, vortex and centrifuge for 20 min at max speed.
   - Remove supernatant, wash pellet with 500 μL of ethanol 70%.
   - Centrifuge at max speed for 5 min.
   - Remove supernatant and let the pellet air-dry.

3.1.2.4. Column Chromatography

This procedure optimizes size fractionation for larger cDNA fragments and makes cloning of larger insert more probable.

1. Resuspend pellet in 100 μL TEN buffer [10 mM Tris-HCl (pH 7.5), 0.1 mM EDTA, 25 mM NaCl; autoclaved].
2. Open the caps of columns and let the buffer go thru. Add 0.8 mL TEN and let it drain completely. Repeat this procedure four times.
3. Add 100 μL of cDNA, and recover 100 μL into a new microcentrifuge tube.

4. Add 100 μL of TEN to the column and recover 100 μL effluent in a second tube. Let the column drain completely and add a new 100 μL TEN aliquot.
5. Collect single drops in individual tubes continue adding 100 μL TEN aliquots until you collected 18 tubes with approximately 35 μL per tube (drop volume).
6. To maximize the range of the cloned inserts you can pool all tubes from the first radioactivity fraction to the fraction with the peak of radioactivity (measure radioactivity with the Geiger counter). Determine the amount of DNA with the spot technique. (**Note 2**).

3.1.2.5. Ligation of cDNA to the Vector

1. Prepare the vector (pBabe-HAZ) DNA by digestion with Sal I and Not I for 2–3 h at 37 °C followed by gel isolation. Determine the concentration of the linearized vector.
2. You can use more cDNA than the amount suggested by the manufacturer. You should keep an insert to vector DNA ratio of approximately 1:2.
3. Before ligation, precipitate the desired amounts of vector and insert together using:
   Half volume of absolute ethanol
   Two volumes of 7.5 M NH_4OAc
   1 μL glycogen (to better see the pellet)
4. Vortex and centrifuge for 30 min at room temperature at maximum speed.
5. Add 150 μL of 70% ethanol at room temperature for 2 min.
6. Centrifuge for 5 min at max speed, air-dry the pellet and resuspend in 8 μL of water.
7. For ligation, use the rapid ligation kit from Roche. Add:
   2 μL DNA dilution buffer 5X
   10 μL T4 DNA ligase buffer
   1 μL T4 ligase.
8. Incubate at 37 °C for 2 min.
9. Leave at room temperature for 5 min and then add 20 μL of phenol chloroform
10. Centrifuge at maximum speed for 3 min, and remove top layer to a new tube, then precipitate by adding:
    10 μL 7.5M NH_4OAc
    60 μL of absolute ethanol.
11. Vortex and spin at maximum speed for 20 min.
12. Remove supernatant and wash pellet with 200 μL of 70% ethanol.
13. Centrifuge at maximum speed for 2 min, air-dry the pellet and resuspend in 5 μL of H_2O.

3.1.2.6. Introduction of Ligated cDNA into *E. coli* by Electroporation

For the electroporation, use competent cells such as Stbl2 (Invitrogen). These cells are suitable for stable cloning of direct repeat and retroviral sequences.

1. Prepare three tubes each with 30 μL of Stbl2 suspension and 1.5 μL of ligation product.
2. Add each mixture to three independent electroporation cuvettes and electroporate at 1.8 kV, 25 μF and 200 Ω.
3. Immediately add 1 mL of Super Optimal Catabolite (SOC) medium to the electroporated cells, transfer the mixture to 14-mL round bottom tubes and incubate 1 h at 37 °C with vigorous aeration.
4. Plate 150 μL of cells per 15 cm LB plate containing 100 μg/mL ampicillin. Use approximately 20 plates per library. Prepare a test plate by diluting 1 μL of cells in 1 mL of SOC and plating 200 μL. (**Note 3**).
5. Incubate overnight, count the colonies on the test plate, and calculate the number of colonies that would result from plating the entire 1 mL of cells. This is the number of colonies per μl of electroporated cells resuspended in 1 mL of SOC (call it *Υ*). To determine the number of bacterial colonies in your library: multiply *Υ* by 150 (which is the volume plated per plate) by the number of plates used to plate the entire library. This will define the complexity of the library.
6. Pick 10 colonies from the plate as samples for examining inserts, and grow them in 4 mL LB with 100 μg/mL ampicillin overnight.
7. To recover pools of the colonies, add 10 mL of LB to the plates with the transformed cells. Shake very slowly for 15 min to allow cells to detach from agar and scrape the cells.
8. Preserve the transformed cells by preparing 10 vials glycerol stocks: add 750 μL of transformed cells to 250 μL of 50% glycerol mix and store at −80 °C.
9. Recover plasmid DNA from remaining cells using a Maxiprep kit (Qiagen). Determine concentration at 260 nm.
10. Recover plasmid DNA from the 10 minicultures grown overnight with Miniprep kit (Qiagen), digest the DNA with the enzymes Sal I and Not I, run the digested product in a 1% agarose gel to determine the range of the inserted fragments.

3.2. Production of Virus-Like Particles

To produce a library of transducing viruses, the cDNA library cloned into a retroviral vector is transfected along with a Gag-Pol expression vector (Gag and Pol proteins corresponding to the same virus as the backbone of the retroviral vector used for the cDNA library construction) and the pantropic VSV-G *env* gene from the Vesicular Stomatitis Virus. The mixture of the three DNAs is transfected into 293T cells in a 10-cm culture dish at approximately 70% confluence with Fugene (Roche). Other transfection methods may be utilized as well. In this protocol, transfect the cDNA library cloned in pBabe-HAZ as follows:

1. Mix in a total of 15 μL of TE the plasmids:
   2 μg of cDNA library
   1 μg of pHit60
   1 μg of pMD.G.
2. Prepare 200 μL of OptiMEM without serum or antibiotics and add 18 μL of fugene. Do not allow Fugene to touch the sides of the tube; mix by flicking.
3. Add the DNA mixture.
4. Incubate for 20 min at room temperature.
5. Change the media of the cells with 8 mL of complete media (DMEM, 10%FCS, Penicillin/Streptomycin).
6. Add the DNA mix to the cells drop wise, and swirl the plate. Return to incubator.
7. Change media next day, but it is not essential. (**Note 4**).
8. Collect supernatant at 48, 72 and 96 h post-transfection. At each collection replace the media with fresh 8 mL complete media.
9. Filter with a 0.45-μm filter, aliquot and store at −80 °C. Avoid freeze–thaw cycles, as infection efficiency decreases. Add Hepes pH 7.5–50 mM final concentration.

Any type of Virus-like particles (VLP), not just cDNA carrying VLPs, can be produced in the same manner; for example for the preparation of VSV-HIV-TK, or VSV-HIV-Puro, the 2 μg of cDNA library would be replaced respectively by an HIV retroviral vector expressing thymidine kinase (pHIV-TK) or the puromycin resistance gene (pHIV-Puro) and the structural vector should encode HIV Gag and Pol (p8.91) *(3)*.

3.2.1. Titration of VLPs

Depending on the type of VLP produced, different titration methods can be used. If the retroviral vector encodes a resistance gene, you can take advantage of that to titer your viral supernatant. If no resistance marker is expressed, the titer of your virus can be determined by real-time PCR, measurement of the viral DNA made after infection.

When determining the titer of a VLP cDNA expression library, titration by drug selection is preferred because it will give you the number of functional cDNAs in your expression library. When titrating by real time PCR you should keep in mind that by measuring the number of integrated proviruses, you are not determining the precise number of cDNAs that will actually be expressed as protein in the cell.

3.2.1.1. Titration by Drug Selection

1. Day 1: Plate HIV sensitive cells (for example, TE671 cells), at 1×10^5 cell per six-well plate.
2. Day 2: Next day, infect the cells with serial 10-fold dilutions of the virus to be titered according to the table below:

Well	1	2	3	4	5	6
Virus (μL)	100	10	1	0.1	0.01	0
Cell #	10^5	10^5	10^5	10^5	10^5	10^5
Dilution:	200 μL virus	200 μL dil.1	200 μL dil.2	200 μL dil.3	200 μL dil. 4 media	
	1.8 mL media*	1.8 mL media	1.8 mL media	1.8 mL media	1.8 mL media	2 mL media

Note: *The complete media used may contain 1/1000 dilution of Polybrene (stock 5 mg/mL). Polybrene increases the efficiency of the VSV-G envelope entry into the target cells. In some cell types, however, polybrene might actually decrease infection efficiency. Test your cells first.

3. Use 1 mL for each infection (several dilutions are performed, because you do not know at which dilution you will be able to count the number of colonies).
4. Day 4: forty-eight hours postinfection, add the selective drug to the media. Some average concentrations used for human cells, for the most popular drugs are:
   Neomycin – 400 μg/mL
   Puromycin – 1–5 μg/mL
   Zeocin – 50–100 μg/mL
   Hygromycin – 50 μg to 1 mg/mL.
   However, sensitivities vary with cell line, and therefore you should perform a killing curve on the type of cells you are using.
5. Day 5: Change the media and add fresh media with drug whenever cells start to die. Then keep changing every 3 days. Noninfected cells usually die after about 2 weeks postinfection. **Note 5**
   When visible colonies can be observed, stain the plates with blue Giemsa, and count the colonies as follows:
   Empty media and add 100% methanol at –20 °C for 10 min room temperature.
   Add Blue dye (1/20 giemsa: H20; 10% methanol) for 1 h at room temperature.
   Wash plates in large volumes of water several times and allow to dry.
   Count the colonies.
6. The virus titer is equal to the number of colonies on the plate times the dilution factor.

3.2.1.2. Titration of VLP by Quantitative PCR

1. Day 1: Seed sensitive cells (ex: TE671), at 1×10^5 cells per well of six-well plate.
2. Day 2: Next day infect the cells with serial dilutions of the virus to be titered as described above.

3. Day 3: Remove media, wash cells with 1X PBS, and add fresh complete media.
4. Day 4: Wash once in 1X PBS,
   Trypsinize cells, wash again in 1X PBS.
   Prepare total DNA using DNeasy Kit (Qiagen), following manufacturer's instructions.
   Recover DNA in 50 μL elution buffer.
   Determine DNA concentration by OD at 260 nm.

Rationale for the standard curves for real time PCR:

For example, for the HIV-Puro expressing vector (pSCPW is 9542 nt):

50 μg/mL of pSCPW is about 4.79×10^{12} molecules/mL.

Prepare serial dilutions of your viral vector, as the following example:

Start with your stock vector at 100 ng/μL, for pSCPW 100 ng/μl is 9.58×10^{9} molecules/mL.

10.4 μL of stock in1000 μL H_2O – 10^8 mol/μL	Do not take to PCR area
100 μL of 1 in 1000 μL H_2O – 10^7 mol/μL	
100 μL of 2 in 1000 μL H_2O – 10^6 mol/μL	Use 5 μl per tube for PCR reaction
100 μL of 3 in 1000 μL H_2O – 10^5 mol/μL	
100 μL of 4 in 1000 μL H_2O – 10^4 mol/μL	
100 μL of 5 in 1000 μL H_2O – 10^3 mol/μL	
100 μL of 6 in 1000 μL H_2O – 10^2 mol/μL	

Primer pairs and probes that can be used to quantify an MLV- or HIV-based VLP are summarized below (they encompass the packaging sequence):

Primers	**sequence (5′ → 3′)**
5′HIV	TGGGCAAGCAGGGAGCTA
3′HIV	TCCTGTCTGAAGGGATGGTTGT
5′MLV	CATGGACACCCAGACCAGG
3′MLV	CGGATGGAGGAAGAGGAGG
Probes	
HIV	[DFAM]ACCGATTCGCAGTTAATCCTGGC CTGTT[DTAM]
MLV	[DFAM]GCCCCTACATCGTGACCTGGGAA GCC[DTAM]

Mix per reaction (using TaqMan):

1.5 μL of 1/10 dilution of stock primer 5′ (50 μM)
1.5 μL of 1/10 dilution of stock primer 3′ (50 μM)
0.75 μL of 1/10 dilution of stock probe (5′FAM and 3′TAMRA) (50 μM)
12.5 μL PCR Mix (TaqMan Universal PCR Master Mix, Applied Biosystems)

2.75 μL of water

19 μL total

Calculate amounts for number of samples: *x* samples
10 standards
2 blanks
+2 (for error)

y point reactions

Add 19 μL per well of PCR plate for standards and blanks
Add 5 μL of each standard dilution starting on 10^5 molecules
Add Carrier to the remaining mixture (x + 2)*5 μL carrier (1/100 dilution of salmon sperm DNA)
Add 24 μL of this final mixture in PCR plate for the x number of samples
Add 2 μL of DNA template

Run PCR program:
95 °C – 10 min – 1 cycle

95 °C – 15 s
60 °C – 1 min } 40 cycles

4 °C – 10 min

Viral copies should be normalized per total DNA input, and you can define the titer as viral copy number per 100 ng genomic DNA. Otherwise, assuming that one cell contains in average a total of 6 pg, determine the number of provirus per cell, and from there, knowing the number of plated cells, and volume used, calculate number of particles per milliliter of viral supernatant.

3.3. Production of a Population of Cells Transduced with Expressed cDNAs Cloned Into pBabe-HAZ

A single member of the cDNA library should be expressed per cell, to avoid screening clones with several cDNAs, because it would be harder to identify the clone responsible for the resistance. To make sure that one cDNA is expressed per cell, choose a multiplicity of infection (MOI) of approximate 0.25, or one virus per four cells.

In order to determine the starting cell number to infect with the VLP-cDNA library, a choice has to be made of how many clones to screen at a time. Keep in consideration that if you titered your library by real-time PCR, not all of the proviruses measured will actually express any fusion protein. Therefore, estimate the titer of the library to be at least 1/3 less, due to the possible cloning of the cDNA in any of the three reading frames with only one forming a functional zeocin fusion protein. However, if resistance to zeocin was used to measure the titer you have the accurate titer.

When considering the type of eukaryotic cells to use to express the cDNA library, and perform the screen, keep in mind that these cells must have the least possible level of background resistance, not only toward the HIV infection, but also, to the drug used to kill the cells that became infected. A killing curve should therefore be performed on the cells you chose to use, to assess their infectability. **Note 6**.

The authors Gao et al. *(2)* and Stremlau et al. *(1)*, used Rat 2 and HeLa cells to identify ZAP and TRIM5α, respectively. In this example, we used the rhabdomyosarcoma cell line, TE671 that is highly sensitive to HIV infection. A population of cells expressing the cDNA library was established by selection with zeocin as follows:

1. Set up the number of cells equivalent to four times the number of cDNA clones to be screened. It is preferable to plate the cells in 15 cm culture dishes at 30–40% confluence. If cells are plated at high density, you will probably need to split the cells very soon after. Lower densities will allow for the selection to start without having to split the cells right away.
2. Next day, in the presence or absence of 0. 5 μg/mL of polybrene (according to the cells used), dilute your virus to the appropriate concentration in the minimum volume possible (12 mL per plate should be enough) and incubate the virus for at least 4 h.
3. Add complete media to a total of 15–20 mL per plate.
4. Forty hours after infection, add zeocin at 50 μg/mL in complete media.
5. Keep the cells under selection for approximately 2 weeks, changing the media regularly to remove the dead cells.

   The selection should be carried in the smallest amount of time, to avoid loss of cDNA representation in the population due to overgrowth of rapidly growing clones over slower ones.
6. Make frozen stocks of this population in 15% DMSO in FCS. Store at −160 °C.

3.4. Screening of the Library and Isolation of Clones Conferring Resistance to HIV Infection

The rationale for selecting cDNA fragments capable of conferring resistance to HIV infection is to submit the cDNA expressing cells to several rounds of infection with a genetically marked HIV particle, treat the cells with a drug that will kill the infected cells, and recover the resistant clones that would be possibly resistant due to the presence of the cDNA.

In this example, we used the genetically modified virus VSV-HIV-TK, a retrovirus particle encoding the Herpes Simplex Thymidine Kinase (HSV-1-TK) under the control of a constitutive promoter, and pseudotyped with the pantropic VSV-G envelope protein. The virus can be prepared and titered as described in **Section 3.2**.

To eliminate the infected cells we chose to treat cells with GCV. GCV is a synthetic nucleoside analog of 2′-deoxyguanosine that can be enzymatically phosphorylated into an active triphosphate analog by HSV-TK. The phosphorylated product is incorporated into the DNA causing death of the infected cell. GCV may sometimes cause a bystander effect, that is, toxicity in the cells that are in the vicinity of an HIV-TK infected cell. Therefore a killing curve should be performed in the cell type of your choice, to determine the maximum amount of death by the drug without affecting noninfected cells. The rare TK cells surviving selection in GCV are then recovered as possible virus-resistant cells.

A key point of this screen methodology is the amount of challenging VSV-HIV-TK to be used to infect the pools of cDNA expressing cells. The majority of the cellular restriction factors discovered to date, tend to be saturable, meaning that if the cell is infected with too many viruses, eventually all the restriction factor will be bound by the initial virus, and the following incoming virus will not face any restriction. Therefore, if too much challenging virus is used you may end with no resistant clones, and if you use too little virus you may end with too many clones, most of them false positives.

We have found that several rounds of infection at a moderate MOI, is sufficient to challenge most of the cells, without overcoming any eventual block.

For TE671 cells expressing the cDNA library in pBabeHAZ we use the following method:

1. Day 1: Set up the pool of cDNA-expressing TE671 cells at approximately 25% confluent in 15-cm culture dishes.
2. Day 2: Infect the cells with VSV-HIV-TK at a MOI of about 0.1 in the minimum volume possible (12 mL should be enough per plate). Incubate for 4 h.
3. Aspirate the media and repeat the infection two more times.
4. Day 3: Next day repeat the infections three to four times.
5. Day 4: Add GCV at 50 μM in complete media. The killing by GCV is not immediate; it might take a few days to observe cytotoxicity. When cell death is observed, the culture supernatant should be replaced with new GCV-containing media. In the beginning of the screen you might need to change the selection media every day, and then, when not that many cells are left alive, it is better to leave the media for longer as growth factors will enrich the media and facilitate the growing of individual clones. If you have many cells in the plate, you should split them, and keep the plates at 30–40% confluence. GCV treatment is more efficient when cells are actively replicating. Upon a week or two of GVC treatment there are three possible outcomes:
   a. There are still a lot of cells in the plate after the GCV treatment, most probably background resistance, and therefore another round of infections with VSV-HIV-TK at low MOI is required. Repeat the infection and GCV selection.

b. There are no cells left in the plates. Two things could have happened: (1) there are no restricting cDNA fragments in the batch of pooled cDNA-expressing cells tested, or (2) too much of the VSV-HIV-TK challenge virus was used and the block was overcome. In both the cases, the screen has to be restarted again with another batch of pooled TE671 cDNA library expressing cells.

c. There are a few cells in the plates that are struggling to survive. In this case, add conditioned media to the plates. Use media recovered from plates growing the same type of cells at a good confluence; this media is enriched with growth factors that will help boost growth of single clones. Filter it with 0.45-μm filter and add to the plates in the presence of GCV. When the clones achieve a visible size, recover the clones and gradually amplify by growing them in culture dishes of increasing size. Be very careful at this stage not to lose any potential resistant clones.

There is a fine line between recovering too many clones that are just false positives, and overcoming the cellular block to the virus and ending with no resistant clones at all. However, if a strong consistent cellular block to the virus is desired a reasonable amount of virus has to be used. This will be the longest part of the process.

There are other strategies to recover cDNA clones resistant to HIV that may be less time consuming. For example, the screen used in the identification of TRIM5α, involved sorting the cells by Fluorescent Assortment Cell Sorter (FACS). The cDNA library expressing cells were challenged with an HIV vector expressing green fluorescent protein (GFP), VSV-HIV-GFP. Sufficient virus was used to infect at least 99% of the cells. About 0.5% of the cells were selected for absence of fluorescence by FACS. These cells were allowed to grow and then subjected to a second round of VSV-HIV-GFP infection at a higher MOI. GFP-negative colonies were identified by fluorescent microscopy, cloned and expanded.

Even though this strategy might be less time consuming, it is subject to the same considerations mentioned above about the amount of virus used to challenge the library. Moreover, the limit of GFP detection of the FACS machine may lead to different outcomes. If the cut off is too low it may result in the exclusion of some not so strong blocks, but potentially interesting (for example blocks in the mRNA export), if the cut off is too high it might lead to increased background. Also, keep in mind that some cell lines are autofluorescent and therefore not appropriate for this methodology.

3.5. Recovery of the cDNA Insert in Resistant Clones

Once the resistant clones have been amplified to about a confluent well of a six-well plate, use part of these cells to prepare total DNA and recover the cDNA insert by PCR. The rest of the cells should be amplified and frozen vials should be prepared with 15% DMSO in FCS and kept at −160 °C.

From each resistant clone recovered:

1. Prepare total genomic DNA using the DNeasy Tissue kit (Qiagen), elute in 100 μL of buffer AE, and repeat the elution using the first eluate to increase final concentration.
2. Prepare PCR mix (using High Fidelity kit, Roche):

1 μg	Total DNA as template
5 μL	5′primer – 5′ GCTTATCCATATGATGTTCCAGATT 3′
5 μL	3′Primer – 5′ GCACCGGAACGGCACTGGTCAACTT
5 μL	$MgCl_2$
5 μL	Buffer 3
4 μL	dNTP
0. 5 μL	Taq polymerase (High Fidelity, 1.75 U)
50 μL	Final volume

Run PCR program with the following parameters:

94 °C – 10 min – 1 cycle

94 °C – 15 s
50 °C – 30 s
72 °C – 60 s
] 10 cycles

94 °C – 15 s
55 °C – 30 s
72 °C – 60 + 5 s each cycle
] 20 cycles

4 °C – 10 min

Once you recover the PCR product, sequence the DNA fragment to determine the identity of your cDNA insert.

3.6. Strategies to Retest the Ability of the cDNAs to Confer Resistance to HIV Infection

Once the putative resistance clone is recovered, the functionality of the overexpressed cDNA has to be determined. The resistance to viral infection could be due to the cDNA fusion protein expression, or to somatic mutations incurred during the selection process.

Two approaches can be taken to assess the role of the cDNA in the observed resistance; (a) excise the cDNA fragment from the resistant clone and test if the sensitivity to HIV infection is recovered, and (b) reclone the cDNA fragment onto pBabe-HAZ, express the fusion protein in fresh cells and test if cDNA-expressing clones become resistant to infection. These assays will ensure that the cDNA is sufficient and necessary for the observed restriction.

3.6.1. Excision of the cDNA Insert

The vector pBabe-HAZ contains a lox P site in the 3′LTR. Upon reverse transcription of the vector, another Lox P site is duplicated into the 5′ LTR. These two sites can be used to excise the provirus using the Cre recombinase.

3.6.1.1. Transfection and Selection of Cre Expressing Clones

1. Transfect 1 μg of the pCre expressing vector along with 25 ng of a selectable marker DNA encoding drug resistance of your choice, other than zeocin^R, into the resistant clones. We used a Neomycin resistance marker.
2. 48 h later, add 400 μg/mL of G418 in complete DMEM.
3. Select for G418 resistant clones. Set up 10 cm plates with dilutions of the resistant population at 1/100, 1/200 and 1/500. Keep the cells under selection, and recover individual clones. Gradually amplify them.
4. Once the G418 resistant clones have been expanded determine if the cDNA fragment has been excised. Extract total DNA from each clone and determine the presence or absence of the cDNA insert by PCR as described in **Section 3.5**.

3.6.1.2. Test Cre-Lox Clones for HIV Infectivity

To test if the newly established clones have regained their sensitivity to infection, they should be challenged with a HIV-VPL. It is important to use another reporter gene for this assay, to ensure that the observed resistance developed toward virus and not merely against the reporter.

1. Chose four or five clones in which cDNA excision was successful and seed 1 × 10^5 cells per 10-cm dish. Prepare three dishes per clone. In parallel, prepare three plates with 1 × 10^5 cells of the wild-type cells expressing the empty vector, and three plates with the initial resistant clones expressing the cDNA insert. These will be, respectively, the positive and negative controls.
2. Next day, infect the cells with another reporter HIV VLP at three different MOIs, for example, VSV-HIV-Puro. The number of puromycin resistant colonies at three different MOIs will help score the sensitivity to infection. When choosing the three MOIs to use, keep in mind that 1 × 10^5 cells were plated and that a maximum of approximately 500 colonies can be counted per 10-cm plate (this figure is dependent on the cell type). Therefore the highest MOI to be used should be around 5 × 10^{-3}.
3. 48 h later add 2–5 μg/mL of puromycin and follow same protocol as for virus titration, **Section 3.2.1.1**.

If the cDNA insert is responsible for the observed HIV-1 resistance, the number of colonies observed for the excised cDNA clones should approach the number of resistant colonies found on the empty vector control plates. In contrast, the number of colonies determined for the cDNA expressing resistant clones should be much lower, at least 10 times lower, to be worth pursuing its study.

3.6.2. Expression of the cDNA in Fresh Cells

1. Amplify the cDNA insert from the genomic DNA as described in **Section 3.5**. Clone this PCR product into the vector pBabe-HAZ by using the restriction sites Sal I/Not I. Package the cDNA into a VLP as described in **Section 3.2**, or else transfect the vectors directly into fresh cells TE671.
2. Select cDNA expressing clones by selecting cells with zeocin and prepare individual clones as described above. Prepare empty vector clones in parallel as controls.
3. Once cDNA expressing clones have been amplified, extract total genomic DNA and determine the presence of cDNA insert by PCR and as in **Section 3.5**.
4. Test the sensitivity to infection in the same way as described in 3.6.1.b. This time, one would expect that the cDNA-expressing clones should yield the same low number of colonies as the original cDNA-expressing clone used as negative control.

3.7. Test Replication of wt HIV

It is important to determine if the resistant clones obtained are effective in blocking wild-type HIV replication. There is always the possibility that the cDNA targets a part of the HIV-VLP that is not actually present in the wild-type particles. We will determine virus replication by measuring reverse transcriptase RT activity.

To test wt HIV replication, the target cells need to express the CD4 and CXCR4 receptors. If the cDNA expressing cells do not express any of these receptors, as in the case of TE671 cells ($CXCR4^+$ only) then the gene for the missing receptor has to be introduced, or the resistance inducing cDNA can be expressed in cells that already express the two receptors. One can use adherent or nonadherent cell lines to express the cDNA. Some examples are: HeLa CD4/CXCR4, CEM or Jurkat. Make cellular clones expressing the cDNA fragment and expressing both receptors in any of these ways.

3.7.1. Wt HIV Infections

1. HIV-1 particles can be prepared by transient transfection of 293T cells with the proviral HIV-1 pNL4-3 DNA *(6)*. Virus stocks for infections are produced by amplification of the virus by acutely infecting CEM SS cells with HIV-1 pNL4-3 and concentration by ultracentrifugation of the cell supernatant.
2. To initiate infection:

   Day 1: Seed 1×10^5 of the target cells per well in a six-well plate. Prepare a control plate with the corresponding cell type expressing empty vector lacking the cDNA.

   Day 2: Infect the cells. Use serial 10-fold dilutions in complete DMEM of the concentrated virus in six-well plates, with 1 mL per well of virus dilution. If a nonadherent cell line is used, the cells are seeded and infected on the same day. Count the cells, microcentrifuge, and recover the cell pellets directly in the viral dilutions. Use complete RPMI media for nonadherent cells.

Day 3: Aspirate the media and replace with 1.5 mL of complete DMEM. For nonadherent cell lines, microcentrifuge the cells, discard supernatant and resuspend the cells in 1.5 mL of complete RPMI.

3. From next day on, recover 30 μL of virus supernatant. Depending on the quality and concentration of the virus used, the cytopathic effects can appear as soon as 24 h after infection or as much as 2 weeks later if very low MOIs were used. Because serial dilutions are performed, it is advisable to start with a highly concentrated stock of virus. Even if there is a risk of overcoming the block with too much virus in the most concentrated wells, differences in viral production at lower MOIs can still be observed.

3.7.2. RT Assay

1. For the exogenous RT assay, load 10 μL of each of the recovered undiluted cell culture supernatants into each well of a round bottom 96-well plate.
2. Add 50 μL of following RT cocktail per point:
   60 mM Tris-HCl (pH 8.0)
   72 mM NaCl, 12 mM $MgCl_2$
   2 mM dithiothreitol, 0.06% NP40
   6 μg/mL oligo(dT)
   12 μg/mL poly(rA)
   0.012 mM dTTP
   0. 05 μM [α-P^{32}] dTTP (800Ci/mmol)].
3. Incubate 1 h at room temperature.
4. Spot 5 μL of reaction mixtures onto DEAE paper cut to replicate a 96-well plate.
5. Wash two times with 2 × SSC (300 mM NaCl, 30 mM sodium citrate (pH 7)) for 20 min.
6. Rinse the DEAE paper with 95% ethanol and dry under a heat lamp or with an air dryer.
7. Cover paper with plastic wrap and expose to a phosphorimager cassette overnight.
8. Use PhosphorImager to scan and quantitate the radioactivity incorporated.

   Plot RT activity vs. days postinfection. If a true resistant clone was discovered, a drastic reduction in viral production or a significant delay of viral replication should be observed.

4. Notes

1. Wash the electrophoresis tank with bleach, and rinse with DEPC water. Use DEPC water to make the gel, and RNase free agarose.

2. Spot Technique: Start with a 20 ng/μL dilution of stock DNA ladder. Prepare the following dilutions: 10, 8, 6, 4, 2, 1 ng/μL.

   Prepare a 10-cm culture dish covered with 8 mL of 1% Agarose with 1 μg/mL ethidium bromide. Dot 1 μL of each ladder dilution, as well as 1 μL of your cDNA sample. Measure intensities under the UV and estimate concentration of your samples by comparison with ladder concentrations.
3. Alternatively, incubate test plate overnight at 37 °C and store the remaining transformed cells at 4 °C overnight. Count the colonies on the plate, calculate the number of colonies that would result from plating the entire 1 mL of cells, and plate them at a density suitable for screening. Some loss of viability may occur upon storage.

   If the number of colonies obtained is not sufficient for your needs, you can electroporate an additional 1. 5 μL of ligation mixture and repeat the procedure.
4. If you need to prepare VLPs with no viral DNA contamination, you should change the media, and moreover you should treat the collected virus with DNase prior to infection. For this add 75 U of DNase per ml of virus in the presence of 10 mM $MgCl_2$ for 1 h at 37 °C.
5. Some drugs take longer to kill the cells. Puromycin is one of the fastest killing drugs. The day after adding puromycin to cells, usually more than 80% of the cells die and 100% the next day. Neomycin and zeocin take a little longer, sometimes up to 7 days. You can increase the concentration of the drug if the killing is too slow.
6. The most appropriate cells to perform the screen have to be extremely sensitive to infection, and readily killed by the drug you use to eliminate infected cells. Therefore, test first which cell type has the highest sensitivity to infection. Seed identical amounts of cells of the lines you want to test and infect with three different MOIs of an HIV-VLP (for example VSV-HIV-GFP). Assess the percentage of infected cells (by FACS in this example). Choose the cell line for which the lowest amount of virus achieved the maximum amount of infected cells.

Next, determine the amount of drug needed to eliminate infected cells. Prepare 10 plates with 1×10^5 cells per 10-cm plate. Infect five plates with the challenging virus (in this example VSV-HIV-TK), in order to cover approximately 100% of the cells. Leave the other five plates uninfected, to serve as the uninfected control for each concentration of drug to be tested. Forty hours later, prepare five different drug concentrations to be tested for elimination of infected cells (in this example GCV). Add each concentration of drug to a pair of plates, one infected and one uninfected. Determine which concentration of drug allowed the fastest killing of the infected cells, without being toxic to uninfected cells. Chose this concentration to perform the genetic screen.

References

1. Stremlau, M., Owens, C. M., Perron, M. J., et al. (2004) *Nature* 427, 848–853.
2. Gao, G., Guo, X., Goff, S. P. (2002) *Science* 297, 1703–1706.
3. Valente, S., Goff, S. P. (2006) *Mol Cell* 23, 597–605.
4. Naldini, L., Blomer, U., Gallay, P., et al. (1996) *Science* 272, 263–267.
5. Markowitz, D., Goff, S., Bank, A. (1988) *Virology* 167, 400–406.
6. Adachi, A., Gendelman, H. E., Koenig, S., et al. (1986) *J Virol* 59, 284–291.

Chapter 18

Rapid, Controlled and Intensive Lentiviral Vector-Based RNAi

Manuel Llano, Natassia Gaznick, and Eric M. Poeschla

Abstract

RNA interference (RNAi) is a powerful technology for studying the functional significance of genes. The technique is more accessible than gene knockout methods, and is directly applicable to diverse human cells. However, inadequate reductions in target mRNAs can reduce the utility of RNAi and insufficiently rigorous controls can lead to spurious conclusions. Optimally combining pol III promoters to drive short hairpin RNA expression with the gene transfer capabilities of lentiviral vectors has led to ways to perform especially effective and convincing RNAi, which we review here. We detail practical methods, including one-step vector construction. Deep, stable knockdowns to trace mRNA levels are readily achieved in T cell lines, which can then be subjected to comprehensive HIV challenge studies. Rescue of preknockdown phenotype by RNAi-resistant gene re-expression is a critical validating step. The methods can also be applied to primary T cells and macrophages. The time from thinking of a target to initial data read-out can be a few weeks.

Key words: RNAi, lentiviral vector, short hairpin RNA (shRNA), T cell, HIV.

1. Introduction

This protocol is designed to allow a scientist familiar with basic molecular biology and cell culture to rapidly and effectively apply two powerful technologies: RNA interference (RNAi) and lentiviral vectors. The biologies that underlie each of these technologies rest on complex foundations, and the development of RNAi is particularly in flux at present, such that comprehensive review is beyond the scope of this protocol. Rather, after brief review of the necessary principles, our emphasis is on practical application such that the biomedical researcher can attain quickly the goals of stringent knockdown and appropriate control validation

Vinayaka R. Prasad, Ganjam V. Kalpana (eds.), *HIV Protocols: Second Edition, vol. 485*

DOI 10.1007/978-1-59745-170-3_18 Springerprotocols.com

as achieved with the intensified lentiviral vector RNAi (ilvRNAi) of Llano et al. *(1)*. We also describe an update that allows one-step construction of ilvRNAi transfer vectors.

RNA interference, a post-transcriptional gene silencing mechanism evolutionarily conserved from *C. elegans* to *H. sapiens*, is triggered by double-stranded RNAs (dsRNA), which are processed by a ribonuclease III-type protein into short double-stranded 21–23 nucleotide fragments called small interfering RNAs (siRNAs) *(2–7)*. siRNAs become incorporated into a multi-protein complex containing RNA helicase and nuclease activities, the RNA-induced silencing complex (RISC). The helicase activity converts the RNA duplex into a single-stranded RNA that hybridizes to its target RNA and directs the RNA endonucleolytic activity. The catalytic subunit of RISC, Argonaute2, has recently been identified *(8,9)*. This process results in a reduction of specific mRNA levels and hence decreased specific gene expression.

The process can be initiated by intracellular dsRNA synthesis rather than exogenous siRNA addition if appropriate promoters are used to express sense and anti-sense sequences. The two strands can be expressed separately but a more convenient and efficient method is to link them in tandem with a short intervening loop of variable length. The resulting transcript anneals by Watson–Crick base-paring into a short hairpin RNA (shRNA) which is then processed by the RNAi pathway nuclease Dicer to form a fully functional siRNA. This process mimics the cellular microRNA (miRNA) pathway *(10–12)*.

While siRNAs or plasmid based expression of shRNAs can be quite effective, in those not uncommon situations where small amounts of the target protein retain significant biological activity, more powerful RNAi systems must be pursued or a gene-level knockout obtained. In addition, siRNA oligonuleootides can be toxic, have poor uptake in many cells, and present problems with nonspecific inhibition.

An option that can address a number of these problems is to introduce shRNA expression cassettes into viral vectors. Lentiviral vectors are retroviral vectors that can transduce nondividing cells *(13–15)*. We and others have found that shRNA-expressing lentiviral vectors are highly effective for either transient or stable RNAi. In a recent series of experiments to target the lentiviral integrase-binding protein LEDGF/p75, use of optimized lentiviral vectors in human T cell lines allowed both a graded intensification of RNAi, and much deeper knockdown than is obtainable with either siRNAs or selectable marker-bearing plasmids *(1)*. Moreover, inclusion of fluorescent proteins in the same lentiviral vector allowed enrichment of the most intensively knocked down cells by fluorescence-activated cell sorting (FACS). These RNAi-enhancing maneuvers proved crucial to understanding the role of LEDGF/p75 in HIV replication, because a small residuum of this protein (a fraction of wild-type levels) persisted in a chromatin-bound form after conventional siRNA transfection or plasmid-based shRNA expression (whether transient or stable). Since the protein is efficiently chromatin-trapped, residual protein was effectively concentrated with one substrate of the integration reaction (chromatin). Only T cells that had no detectable LEDGF/p75 in the detergent-resistant chromatin-bound fraction were blocked for HIV replication. The extra knockdown efficacy was highly revealing in this case and may well be in other biological applications. These experiments were also revealing from the perspective of RNAi methodology *(1)*. They helped delineate the conditions needed to push well-controlled RNAi to this level, by testing it in what was in effect a highly stringent bioassay with a convincing endpoint (blocking HIV infection).

This chapter describes construction, production, and use of lentiviral vectors to perform highly effective stable knockdowns in T cells, analyses of the extent of knockdown, use of control shRNAs, and the important control of re-expressing the target protein by retroviral transduction of shRNA resistant cDNAs. Co-expression of other proteins in the same vector (in our case, fluorescent protein-fused dominant-interfering proteins) is also reviewed.

2. Materials

Active siRNAs are typically composed of two hybridized 21 mer RNAs with 19 complementary nucleotides and 3′ dinucleotide overhangs, so the design process begins with identification of candidate 21 nt sequences in a given RNA target. Longer shRNA sequences have also been used, however, and may improve DICER processing and RNAi efficacy (e.g., see *(16)*, although sequences longer than 29 nt risk inducing a PKR response *(17)*. The intervening loop is not strictly constrained and we have simply continued to use the 9 nt loop TTCAAGAGA described by Brummelkamp et al. *(10)*.

Various algorithms for specific shRNA design are available (e.g., *(18)* or http://sfold.wadsworth.org/. The Ambion web platform is a convenient and accessible tool for choosing candidate shRNAs http://www.ambion.com. **Fig. 18.3A** shows the human U6 promoter arrangement we used in ref. *(1)* and **Fig. 18.3B** a subsequently re-engineered version that allows one-step insertion of an shRNA adaptor into the lentiviral vector.

3. Methods

3.1. Overview

The basic lentiviral vectors are now adapted for one-step cloning. An overview of the methodology is depicted in **Fig. 18.1**. There are seven main steps:

1. Identify the target sequence and synthesize and anneal two oligonucleotides to form an shRNA-encoding adaptor.
2. Insert the adaptor into the lentiviral vector using standard recombinant DNA methods. One step cloning is now possible with current versions. Other proteins or GFP fusions can also be expressed in the vector (*see* **Note 1**).
3. Produce particles by transient co-transfection with a lentiviral packaging plasmid and a viral envelope pseudotyping plasmid.
4. Transduce target cells with the lentiviral vector particles. A control shRNA-encoding lentiviral vector should be used to derive a control cell line.
5. Assay directly for phenotype or if a more stringent knockdown is desired, first enrich by sorting for higher mCherry or eGFP expression in both active and control shRNA lines.

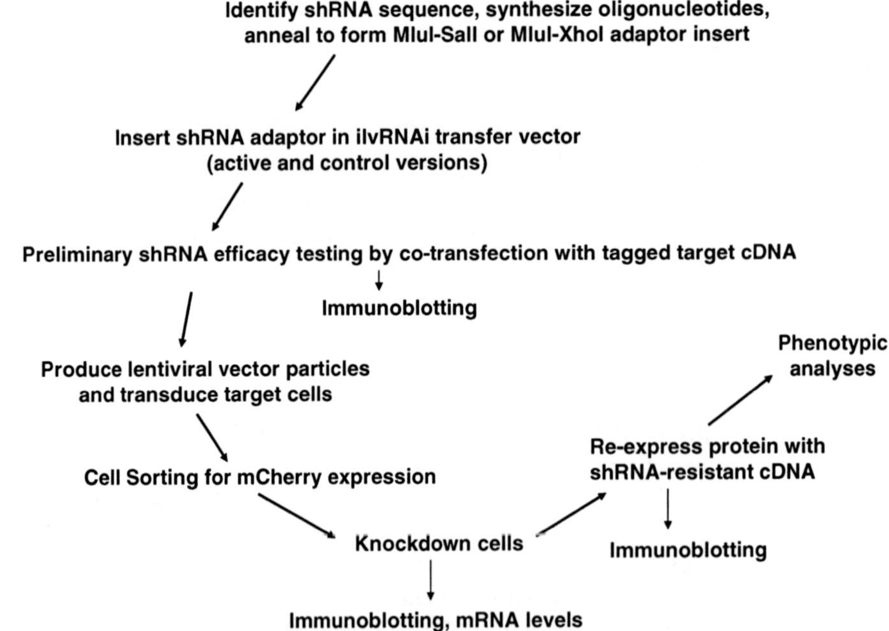

Fig. 18.1. ilvRNAi process. T cell lines are transduced with shRNA expressing lentiviral vectors and maximally transduced cells are selected by cell sorting for mCherry expression (other *X*FPs are readily used, e.g., eGFP as in *(1)*). Knockdown cells depleted of the endogenous mRNA are then engineered to re-express the gene using an shRNA-resistant allele. For the latter step we have typically used gamma-retroviral (i.e., murine type C) vectors that co-encode a selectable marker such as *neo*.

6. Re-express the gene in both lines with an shRNA-resistant cDNA allele (i.e., one with synonymous mutations at the shRNA target site) to ensure that the effects are shown to be specific by rescue of pre-knockdown function. Re-expression in the cell line expressing control shRNA controls for over-expression of the protein.
7. Assay phenotype in all four lines.

The techniques are readily adapted by laboratories conversant in basic molecular biology and cell culture. LEDGF/p75 knockdown (as in *(1)*) is used as an example and we emphasize methods we have found highly effective in human T cells because of our interest in HIV biology but the lentiviral vectors with broad tropism can be generated by pseudotyping and are easily adapted to other RNA targets, cell lines and purposes. Detailed discussion of the burgeoning RNAi armamentarium – the subject of a literature explosion in the past 2 years – is beyond our purview here. In addition, other iterations of lentiviral vector-based RNAi that include regulated shRNA expression have recently been published *(19–22)*.

3.2. Basic shRNA Expression Parameters

Pol III promoters are useful for shRNA expression because the defined transcriptional start and precise termination signals (a polythymidine tract) generate the requisite RNA termini. Certain exogenously introduced pol III promoters are highly active in human T cell lines and in our hands the human U6 promoter was superior to the H1 promoter (*see* **Note 2**). Reverse orientation with respect to the lentiviral vector transfer vector transcript and location upstream of internal pol II marker gene segments also proved ideal (**Fig. 18.2**).

These directions assume use of the newer transfer vector illustrated in **Fig. 18.3B**, which allows one-step cloning of a

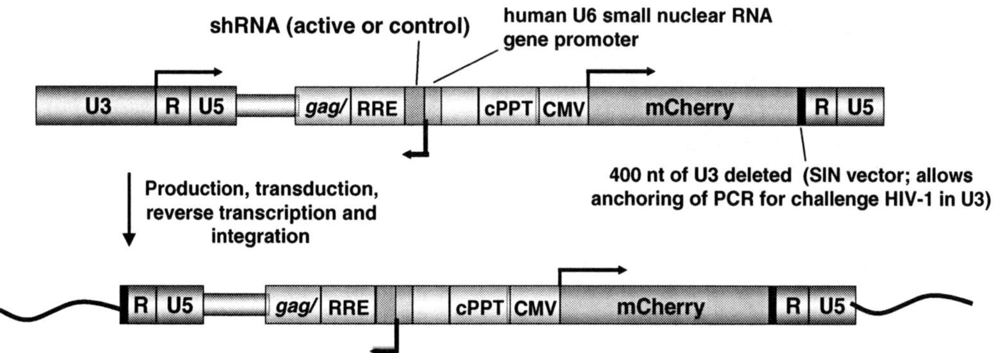

Fig. 18.2. ilvRNAi vector. 400 nt were deleted from U3 to permit U3-anchored PCR assays for challenge virus DNA intermediates (e.g., Alu-PCR and 2-LTR circle assays). The key components are readily inserted in FIV- or EAIV-based lentiviral vectors as well, but HIV-derived lentiviral vectors have superior efficacy compared to these nonprimate lentiviral vectors in human T cell lines.

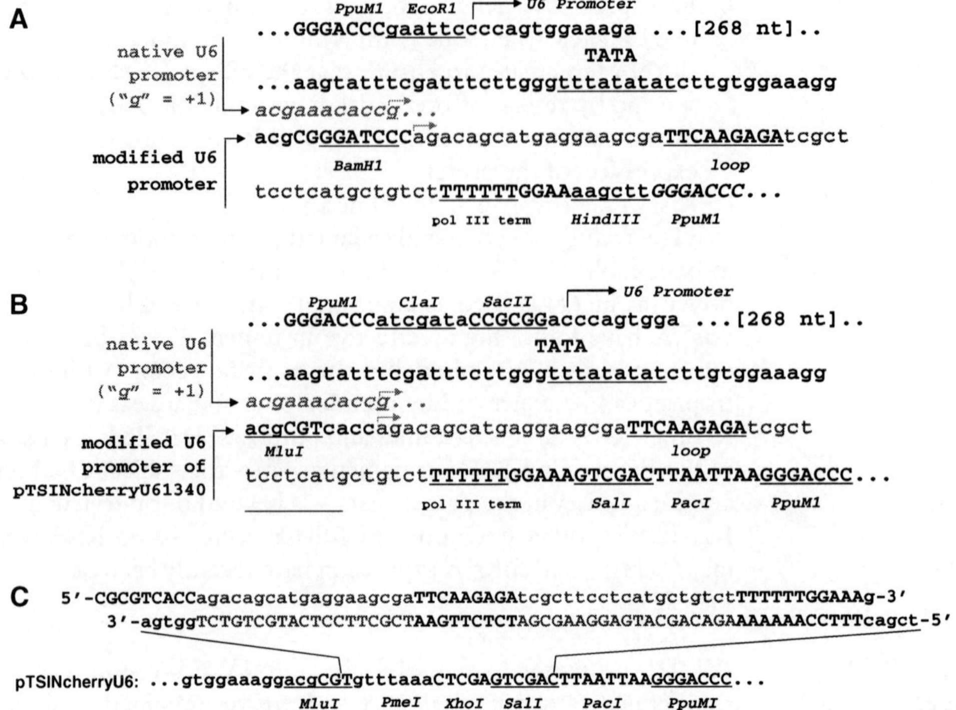

Fig. 18.3. Promoter-shRNA arrangements in ilvRNAi vectors employing the human pol III promoter. **(A)** In the lentiviral vectors used in Llano et al. *(1)*, 9 nt (CGGATCCC) replace the native 7 nt (aaacacc) preceding the transcriptional start site in the human U6 promoter. Transcription normally begins 23 nt downstream of the TATA box, at a "G" preceded by a pyrimidine. Here, transcription begins 25 nt downstream of the TATA box. The Bam H1 and Hind III sites used for shRNA adaptor insertion originated in the U6 promoter cassette of the Ambion pSilencer series. The "1340" p75-specific shRNA is shown in red, with an intervening 9 nt loop as an example (the numbering refers to nt 1340 of the p75 reading frame). Because the *Bam*H1 and *Hin*dIII sites for oligonucotide insertion were not unique in the lentiviral vector, this version requires insertion of the annealed shRNA oligonucleotides into an initial U6 promoter-containing template plasmid (such as in the pSilencer series) followed by PCR amplification of the PpuMI flanked region and insertion of the amplicon into the PpuMI site of the lentiviral vector. The version in (**B**) simplifies the process. **(B)** Re-engineered pTSINcherryU6 transfer vector. The vector allows one-step insertion of annealed shRNAs directly into the lentiviral vector. A unique MluI site is used for left hand adaptor insertion and either SalI or XhoI (not shown, but located upstream of SalI in the acceptor plasmid) is used for right-hand adaptor insertion. In addition, this version preserves native 23 nt spacing from the TATA box to the start of U6 promoter transcription and, by using MluI, only three promoter nucleotides (CGT) differ from the normal human U6 promoter. pTSINcherryU6 construction was carried out by amplifying the human U6 promoter with PCR primers ATATGGGtCCCatcgataCCGCGGaCCCAGTGGAAAGACGCGCAGG and AAAAGGGaCCCTTAATTAAGTCGACTCGAGtttaaacACGCGTCCTTTCCACAAGATATATAAACCC. The resulting PpuM1-tailed amplicon was inserted into the PpuM1 site of TSINcherry, placing it between the RRE and the cPPT-CTS; this resulted in directional insertion with the U6 promoter in reverse orientation to the lentiviral vector. The resulting transfer vector plasmid is pTSINcherryU6 (**Fig. 18.3C**). MluI at the left end, and either XhoI or SalI at the right end, can be used for shRNA oligonucleotide adaptor insertion and PmeI can be used to select ligations against parental pTSINcherryU6. **(C)** A p75 specific shRNA adaptor formed by annealing the two oligonucleotides is shown at top; this was then inserted into the MluI – SalI backbone of pTSINcherryU6 (relevant sequence shown below) to generate the pTSINcherryU61340 vector shown in panel B. To synthesize oligonucleotides that target a different gene, the nt illustrated in red here should be substituted with the appropriate genes-specific complementary sequences. As noted above, rather than the 19 nt insert shown here, some workers prefer longer (27 nt) inserts to produce a better DICER substrate *(16, 32, 33)*.

gene-specific transfer vector. Previously *(1)*, a cumbersome insertion of shRNAs into *Bam*H1 and *Hin*dIII sites of a donor plasmid, followed by PCR amplification of the U6-shRNA cassette and insertion into the lentiviral transfer vector was needed. Here, a single cloning step is followed by lentiviral particle production.

3.3. Construction of shRNA Expression Plasmid

1. Dilute the sense and the antisense shRNA templates in TE buffer to 1 μg/μL by determining the absorbance at 260 nM and multiplying the A260 by the dilution factor and the extinction coefficient (consider this to be −33 μg/μL; *see* **Note 3**!).
2. Mix 2 μL of each oligonucleotide with 46 μL of 1X DNA annealing solution (10 mM Tris, pH 7.5–8.0, 50 mM NaCl, 1 mM EDTA; the annealing buffer supplied by Ambion in pSilencer kits is also effective).
   Incubate at 90 °C for 3 min followed by cooling to 37 °C in a thermocycler, with a plateau at 37 °C for 1 h.
3. Dilute 5 μL of the annealed oligonucleotides in 45 μL of nuclease-free water and ligate 4 μL of diluted annealed oligonucleotides with 1 μL of Mlu+Sal-linearized pTSIN cherryU6 (*see* **Note 4**) using 5 U of DNA T4 ligase in a 10 μL reaction overnight at 16 °C. The ligation can be selected against wild type first (*see* **Note 5**) or transformed immediately into *Escherichia coli* with 5 μL of the ligation and plate on ampicillin LBA plates.
4. Screen 10 colonies by DNA restriction digest (*see* **Note 6**).
5. Sequence one or more of the positive constructs with an appropriate primer. (5′-TCATAATGATAGTAGGAGGC-3′, which is 147 n upstream of the U6 termination signal, works well).

3.4. Validation of Knockdown Activity

1. Co-transfect shRNA expression plasmid and epitope-tagged target cDNA in 293T cells and evaluate target protein levels by immunoblotting and/or mRNA levels by RT-PCR. This preliminary screening step is easier than making lentiviral vector particles and allows one to prioritize the most effective shRNAs. Generally about one in five shRNAs is effective but this is highly variable. Various algorithms are available to predict efficacy of candidate shRNAs, but we have good success simply by choosing 21 nt sequences that (i) begin with an AA dinucleotide (the subsequent 19 nt are inserted in the vector); (ii) have no more than 50% GC content; and (iii) do not have homonucleotide runs > 3.
2. Plate 293T cells at up to 60% confluence in DMEM supplemented with 10% FCS and penicillin, and streptomycin (DMEM culture medium). Harvest cells by trypsinization the

next day and plate them at 0.45×10^6 cells in 2 mL per well in a six-well plate.

3. The following day, transfect the cells via the calcium-phosphate co-precipitation method, using all transfection reagents at room temperature. For transfection of each well, dilute a total of 2 μg of the shRNA expression plasmid in 1 mM Tris, pH 7.4, 0.1 mM EDTA up to a final volume of 121.5 μL. To this mix add 13.5 μL of 2.5 M $CaCl_2$, vortex briefly. Add 135 μL of 2X HBS (50 mM Hepes pH 7.1, 280 mM NaCl, 1.5 mM Na_2HPO_4; *see* **Note 7** for preparation) and vortex immediately for 1 s. Incubate the transfection mix for 5 min at room temperature and then distribute evenly in the well by droplet pipetting.
4. Sixteen to 18 h after transfection, remove the cell culture medium gently and carefully add 2 mL of fresh DMEM culture medium per well so as not to disrupt the loosely adherent 293T cell monolayer.
5. Culture the transfected cells for an additional 48 h and then determine by immunoblotting, the expression levels of the targeted gene (*see* **Note 8**).

3.5. Production and Titration of ilvRNAi Vectors

The basic procedure is to co-transfect a packaging plasmid that expresses Gag/Pol, an ilvRNAi transfer vector derived from pTSINcherryU6, and an envelope construct. The latter is generally vesicular stomatitis virus glycoprotein G (VSV-G) because of the high titers, broad tropism and stability it affords *(23)*. VSV-G-pseudotyped three or four (the latter express Rev from a separate construct) plasmid lentiviral vector systems should be used at the appropriate biosafety level determined by the institutional safety review board of your institution. Generally these are used at the BL2+ level. Several good HIV packaging plasmids are available *(22, 24)*.

3.5.1. Calcium-Phosphate Co-Transfection in 293T Cells

1. Harvest 293T cells (60% confluence) and plate 3×10^6 cells in 12 mL in a 75 cm^2 flask. Plate eight flasks per vector. Scaled up procedures using high area cell factories are described in *(25)* but generally are unnecessary for deriving knocked down cell lines.
2. Transfect the cells the next day by the calcium-phosphate co-precipitation method, using all transfection reagents at room temperature. For transfection of each 75 cm^2 flask dilute 10 μg of the control shRNA-expressing (*see* **Note 9**) or the specific shRNA-expressing transfer plasmid, 10 μg of an HIV vector packaging plasmid (*see* **Note 10**) and 3.3 μg of a VSV-G expression plasmid (e.g., pMD.G) in 1 mM Tris up to a final volume of 720 μL. To this mix add 80 μL of 2.5 M $CaCl_2$, vortex for 1 s and then add 800 μL of 2X HBS pH 7.1 and vortex

immediately for 1 s. (Note that this process can be batched for convenience if desired, with precipitate for all eight flasks prepared in one larger tube.) Incubate the transfection mix for 5 min at room temperature and then add it to the medium drop-wise; swirl very gently to mix without dislodging the 293T monolayer.

3. Remove the cell culture supernatant 16–18 h after transfection and add 12 mL of fresh culture medium per flask.
4. Culture transfected cells for an additional 48–60 h and then harvest the tissue culture supernatant (vector particle supernatant).
5. Clarify viral supernatant by centrifugation at 800 × *g* for 10 min and then filter through a 0.45-μm membrane. Layer 3 volumes of clarified supernatant on to one volume of 20% sucrose in 50 mM Tris pH 7.4, 100 mM NaCl. Ultracentrifuge at 100,000–125,000 × *g* for 2 h in a swinging rotor at 4 °C. Re-suspend each vector pellets in 1 mL of RPMI 1640 10% FCS. Store concentrated vector at −80 °C (*see* **Note 11**).

3.5.2. Titration of Lentiviral Vectors

Transduce 10^5 target cells with 3, 10, 30 and 90 μL of the concentrated vector supernatant. Sixteen hours after transduction, wash cells to remove the input vector. Analyze mCherry expression by flow cytometry 72 h later.

3.5.3. Generation of Stable Cell Lines

Target cell lines can be generated at different multiplicity of infections (MOIs), cryopreserved and the phenotype analyzed. The efficiency of the vectors, their integrated state, and, if desired, the possibility of immediate enrichment by FACS, allow stable cell lines with graded intensities of RNAi to be derived. However, stable cell lines may not be feasible or desired in some cases, and immediate analyses can be performed (*see* **Note 12**).

1. Plate 10^5 target cells and transduce them at a range of MOIs (e.g., 10, 50, 150, 300) with the specific shRNA-expressing vectors. Do the same in parallel with the control shRNA vector. Sixteen hours after transduction wash to remove the input vector, and expand the culture until 10^7 cells are available.
2. Using FACS for mCherry, select both active and control shRNA cells for the 10% brightest cells. Expand cells to the appropriated number for phenotype analysis (*see* **Note 13**).
3. Verify knockdown levels by immunoblotting and/or RT-PCR and also analyze the relevant phenotype of the cells for your particular experimental needs, pairing control and active shRNA cells in each measurement (*see* **Note 14**). A complete set of data requires that re-expression of the target protein rescue parental function (step IV below).

3.6. Back-Complementation of Knockdown Cell Lines

RNA interference is powerful but vulnerable to mis-interpretation if not properly controlled. To demonstrate the specificity of the observed phenotype and exclude off-target effects, knockdown cells should be engineered to re-express the targeted protein *(26)*. The importance of this control cannot be over-emphasized. We describe here the use of Murine leukemia virus (MLV)-derived vectors as an approach for efficient back-complementation (*see* **Note 15**).

1. A cDNA containing synonymous mutations in the shRNA target sequence (we mutated 7 within the 19 nt) is inserted into the MLV transfer vector diagramed in **Fig. 18.4**.
2. MLV vector production. We then follow steps III.1–III.5 with the following variations. We use the MLV packaging cell line Phoenix A (obtained from G. Nolan, Stanford). We transfect 3×10^6 cells with 10 μg of the MLV transfer vector and 3 μg of pMD.G. Supernatant from a total of eight 75 cm^2 flasks (96 mL) is concentrated on a 20% sucrose cushion, and resuspended in one ml of 10% RPMI 1640.
3. Generation of the complemented cell lines.
   i. Baseline G418 kill curve. Plate 10^5 knocked down cells/mL in a 24-well plate and add 1 mL of culture medium containing different amounts of G418; use as control cells cultured in the absence of G418. Select the minimal dose that kills all the cells in 1 week.
   ii. Transduce 10^5 knocked down cells with 30, 90 and 270 μL of the concentrated MLV vector (*see* **Note 16**). Sixteen hours after transduction remove the input vector by washing and culture the cells for 48 h before adding G418 at the determined level. Expand cells in selection medium to the number needed for phenotypic analysis and cryopreservation.
   iii. Evaluate re-expression by immunoblotting. Select cells expressing levels of targeted protein that are similar to the unmodified original cell line.

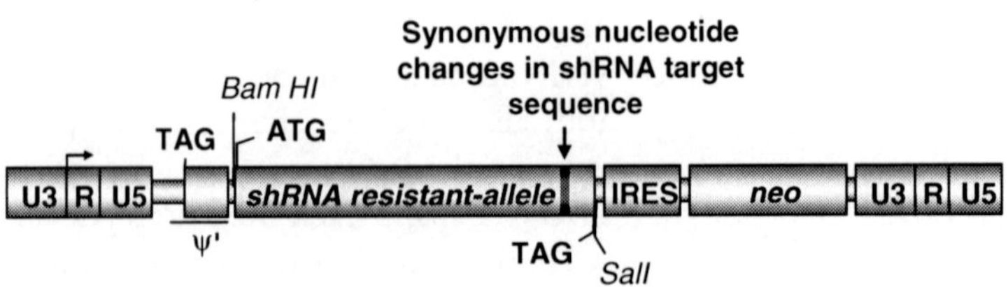

Fig. 18.4. MLV transfer vector for re-expressing a knocked down gene.

4. Notes

1. We used single vector expression of both an shRNA and a dominant-interfering GFP fusion to the LEDGF/p75 integrase-binding domain (IBD), by using GFP-IBD in place of the mCherry marker. The combination of RNAi plus DI protein results in over three logs inhibition of HIV-1 replication (**Fig. 18.3E** of *(1)* and unpublished data).
2. The H1 promoter may have more activity depending on the cell target; in addition, various other options including pol II expression have been useful in other's hands *(27,28)*. We have not observed U6 transcript toxicity in CD4+ human T cell lines, but this has been reported to occur in primary human lymphocytes *(29)*.
3. HPLC-purified oligonucleotides are preferred.
4. Alternatively, the XhoI site in the pTSINcherryU6 polylinker, shown in **Fig. 18.3**, can be used, provided the adaptor is similarly modified to have an XhoI rather than a SalI overhang.
5. Option: Before transformation, the ligation can be heated to 65 °C for ten minutes and then digested with PmeI (see **Fig. 18.3C**) to select against ligated plasmids lacking an insert. XhoI can be used for this purpose as well if the SalI site is used for the shRNA insertion.
6. Screening of the colonies is facilitated if the cloned shRNA sequence has an informative restriction site but this is unusual. For visualizing the insert directly (the 68 bp band resulting from Mlu I + Sal I) use about one-third of the miniprepped DNA for digestion. An easier method is to screen with SacII + SalI which will drop out a 397-bp band.
7. 2XHBS stock: dibasic Na_2HPO_4 : 52.5 g Na_2HPO_4, 5000 mL H_2O. 2xHBS: 80 g NaCl, 65 g HEPES (sodium salt), 100 mL Na_2HPO_4. pH optimization of 2XHBS is an important variable. A frequently used method is to prepare several batches and adjust the pH with 1 N NaOH to several set points within a narrow range, e.g., pH 6.95, 7 and 7.05. Bring to final volume of 5000 mL. Optimal pH is generally 6.95 in our hands but final preferred stocks will depend on empirical determination of transfection efficiency obtained. Store frozen (− 20 °C). The pH will shift with time, as may percent transfection obtained. We generally prepare it new every 3–6 months. Optimal transfection reagents will yield 90–95% transfection of 293T cells.
8. For validation of the activity of the shRNA, the expression of the endogenous target gene can be evaluated in cells transiently expressing the shRNA expression plasmid. This approach is especially useful if the target gene is robustly

expressed. Alternatively equal amounts of plasmids expressing the shRNA and the target gene can be transiently co-expressed in cells. This method is particularly advantageous in the case that the target gene is poorly expressed. Also co-expression of the shRNA and its epitope-tagged or GFP-tagged target will minimize the diluting effects of low transfection efficiency on the validation of the shRNA.

9. The Ambion control "scrambled" shRNA (actaccgttgttatag-gtg) is functionally neutral even when transduced at high MOI.
10. We have generally used pCMVΔ R8.9 (encodes Gag/Pol, Tat and Rev) *(24)*; the combination of pLP1 and pLP2 (Invitrogen) or pCHGP plus pCMV-Rev from Rossi and colleagues to supply Rev and Gag/Pol expression separately *(22)*. For FIV or EIAV vectors, the PpuM1 cassette in **Fig. 18.3B** can be used within an FIV or EIAV transfer vector; however for human T cell lines, HIV vectors are more efficient *(25, 30)*. Detailed description of lentiviral vector production can also be found in *(25)*.
11. The vector pellet is usually not visible. To resuspend it, distribute 1 mL of ice-cold RPMI 10% FCS among all the tubes and incubate the tubes on ice for 15 min, flick the tubes occasionally. Then gently pipette up and down with an automatic pipette approximately 10 times, while distributing the medium around the bottom of the tube; avoid foaming. Then collect the entire vector-containing medium and distribute it into 100-μL aliquots.
12. Stable cell lines greatly facilitate rigorous downstream analyses and reproducibility. However, some knockdowns result in loss of cell viability or other toxicity. In that case analyses can be done within a shorter time frame similar to that of an siRNA transfection (but the nonspecific chemical toxicity of siRNAs is avoided). The same methods can be used, including sorting for high-expressers.
13. Transduced cells will express both the shRNA and the mCherry fluorescent protein at proportionate levels. Thus, cells with a higher level knockdown can be obtained by sorting the 10% brightest cells. Sorting should be performed on control and specific shRNA expressing cells using the same gating criteria, preferably in parallel on the same day, using the same apparatus.
14. Control and specific knockdown cultures are expected to have similar growth rates, which is easily assayed by cell counting or other assays. For immunoblotting analysis of chromatin bound proteins it is advisable to directly interrogate this compartment *(1)*; similar considerations may apply to other subcellular compartments.

15. Re-expression can also be achieved by stable plasmid transfection *(31)*.
16. The most complete approach will be to also conduct parallel expression with the MLV vector in control cells to assess confounding effects of over-expression or of exogenous expression.

Acknowledgments

We thank I. Kemler for assisting N. G. with pTSINcherryU6 and pTSINcherryU61340 construction and D. T. Saenz and M. Meehan for assistance with multiple aspects of lentiviral vector experiments.

References

1. Llano, M., Saenz, D. T., Meehan, A., et al. (2006) An Essential Role for LEDGF/p75 in HIV Integration. *Science* 314, 461–464.
2. Fire, A., Xu, S., Montgomery, M.K., et al. (1998) Potent and specific genetic interference by double-stranded RNA in Caenorhabditis elegans. *Nature* 391, 806–811.
3. Hamilton, A. J., Baulcombe, D. C. (1999) A species of small antisense RNA in post-transcriptional gene silencing in plants. *Science* 286, 950–952.
4. Zamore, P. D., Tuschl, T., Sharp, P. A., et al. (2000) RNAi: double-stranded RNA directs the ATP-dependent cleavage of mRNA at 21 to 23 nucleotide intervals. *Cell* 101, 25–33.
5. Hammond, S. M., Bernstein, E., Beach, D., et al. (2000) An RNA-directed nuclease mediates post-transcriptional gene silencing in Drosophila cells. *Nature* 404, 293–296.
6. Hammond, S. M., Caudy, A. A., Hannon, G. J. (2001) Post-transcriptional gene silencing by double-stranded RNA. *Nat Rev Genet* 2, 110–119.
7. Bernstein, E., Caudy, A. A., Hammond, S. M., et al. (2001) Role for a bidentate ribonuclease in the initiation step of RNA interference. *Nature* 409, 363–366.
8. Rivas, F. V., Tolia, N. H., Song, J. J., et al. (2005) Purified Argonaute2 and an siRNA form recombinant human RISC. *Nat Struct Mol Biol* 12, 340–349.
9. Liu, J., Carmell, M. A., Rivas, F. V., et al. (2004) Argonaute2 is the catalytic engine of mammalian RNAi. *Science* 305, 1437–1441.
10. Brummelkamp, T. R., Bernards, R., Agami, R. (2002) A system for stable expression of short interfering RNAs in mammalian cells. *Science* 296, 550–553.
11. Paddison, P. J., Caudy, A. A., Bernstein, E., et al. (2002) Short hairpin RNAs (shRNAs) induce sequence-specific silencing in mammalian cells. *Genes Dev* 16, 948–958.
12. Zeng, Y., Wagner, E. J., Cullen, B.R. (2002) Both natural and designed micro RNAs can inhibit the expression of cognate mRNAs when expressed in human cells. *Mol Cell* 9, 1327–1333.
13. Naldini, L., Bloemer, U., Gallay, P., et al. (1996) In vivo gene delivery and stable transduction of nondividing cells by a lentiviral vector. *Science* 272, 263–267.
14. Poeschla, E., Wong-Staal, F., Looney, D. (1998) Efficient transduction of nondividing cells by feline immunodeficiency virus lentiviral vectors. *Nature Medicine* 4, 354–357.
15. Olsen, J. C. (1998) Gene transfer vectors derived from equine infectious anemia virus. *Gene Therapy* 5, 1481–1487.
16. Siolas, D., Lerner, C., Burchard, J., et al. (2005) Synthetic shRNAs as potent RNAi triggers. *Nat Biotechnol* 23, 227–231.
17. Sledz, C. A., Holko, M., de Veer, M. J., et al. (2003) Activation of the interferon system by short-interfering RNAs. *Nat Cell Biol* 5, 834–839.
18. Reynolds, A., Leake, D., Boese, Q., et al. (2004) Rational siRNA design for RNA interference. *Nat Biotechnol* 22, 326–330.
19. Morris, K. V., Rossi, J. J. (2006) Lentiviral-mediated delivery of siRNAs for antiviral therapy. *Gene Ther* 13, 553–558.
20. Szulc, J., Wiznerowicz, M., Sauvain, M. O., et al. (2006) A versatile tool for conditional

gene expression and knockdown. *Nat Methods* 3, 109–116.
21. Wiznerowicz, M., Szulc, J., Trono, D. (2006) Tuning silence: conditional systems for RNA interference. *Nat Methods* 3, 682–688.
22. Aagaard, L., Amarzguioui, M., Sun, G., et al. (2007) A facile lentiviral vector system for expression of doxycycline-inducible shRNAs: Knockdown of the Pre-miRNA processing enzyme drosha. *Mol Ther* 15, 938–945.
23. Burns, J. C., Friedmann, T., Driever, W., et al. (1993) Vesicular stomatitis virus G glycoprotein pseudotyped retroviral vectors: concentration to very high titer and efficient gene transfer into mammalian and nonmammalian cells. *Proc Natl Acad Sci USA* 90, 8033–8037.
24. Zufferey, R., Nagy, D., Mandel, R. J., et al. (1997) Multiply attenuated lentiviral vector achieves efficient gene delivery in vivo. *Nat Biotechnol* 15, 871–875.
25. Saenz, D. T., Barraza, R., Loewen, N., et al. (2006) In Rossi, J. and Friedman, T. (eds.), *Gene Transfer: A Cold Spring Harbor Laboratory Manual.* Cold Spring Harbor Laboratory Press, Cold Spring Harbor, NY, pp. 57–74.
26. (2003) Whither RNAi? *Nat Cell Biol* 5, 489–490.
27. McManus, M. T., Petersen, C. P., Haines, B. B., et al. (2002) Gene silencing using micro-RNA designed hairpins. *Rna* 8, 842–850.
28. Stegmeier, F., Hu, G., Rickles, R. J., et al. (2005) A lentiviral microRNA-based system for single-copy polymerase II-regulated RNA interference in mammalian cells. *Proc Natl Acad Sci USA* 102, 13212–13217.
29. An, D. S., Qin, F. X., Auyeung, V. C., et al. (2006) Optimization and functional effects of stable short hairpin RNA expression in primary human lymphocytes via lentiviral vectors. *Mol Ther* 14, 494–504.
30. Saenz, D. T., Teo, W., Olsen, J. C., et al. (2005) Restriction of Feline Immunodeficiency Virus by Ref1, LV1 and Primate TRIM5a Proteins. *J Virol* 79, 15175–15188.
31. Llano, M., Vanegas, M., Fregoso, O., et al. (2004) LEDGF/p75 determines cellular trafficking of diverse lentiviral but not murine oncoretroviral integrase proteins and is a component of functional lentiviral pre-integration complexes. *J Virol*, 78, 9524–9537.
32. Rose, S. D., Kim, D. H., Amarzguioui, M., et al. (2005) Functional polarity is introduced by Dicer processing of short substrate RNAs. *Nucleic Acids Res* 33, 4140–4156.
33. Amarzguioui, M., Lundberg, P., Cantin, E., et al. (2006) Rational design and in vitro and in vivo delivery of Dicer substrate siRNA. *Nat Protoc* 1, 508–517.

Chapter 19

Reverse Two-Hybrid Screening to Analyze Protein–Protein Interaction of HIV-1 Viral and Cellular Proteins

Supratik Das and Ganjam V. Kalpana

Abstract

HIV-1 replication involves a complex network of multiple protein–protein interactions. HIV-1 viral proteins exhibit both homomeric interactions among themselves and heteromeric interactions with other viral or cellular proteins. Identification and characterization of these protein–protein interactions have provided a wealth of information about the biology of the virus. Precise information about the residues involved in interaction is valuable in understanding the functional significance of these interactions, and can be determined relatively easily for proteins whose three-dimensional structure is known. However, the lack of three-dimensional structural information for several host proteins makes it harder to carry out detailed biochemical and functional studies. Reverse-two-hybrid system, a variation of the yeast-two-hybrid system can be used to genetically isolate mutants of a protein that are defective for specific protein–protein interactions. The strategy is to create a library of random mutations in one of the interacting partners and from among this library, screen for those that are defective for interaction using yeast two-hybrid system. In this review, we will describe a method to efficiently generate a library of random mutations and to further screen this library using the simple color scheme of using *LacZ* as a reporter gene. Once the mutants are isolated, they are tested in other biochemical systems and can be subjected to further functional and virological studies.

Key words: Protein–Protein interactions, reverse two-hybrid system, *lacZ*, interaction-defective-mutants.

1. Introduction

HIV-1 proteins exhibit both homomeric and heteromeric interactions that are essential for structural integrity and function of the virus during its replication cycle. HIV-1 proteins such as Env, Gag, PR, RT, and IN have been demonstrated to be homomeric proteins that minimally exhibit dimeric interactions. In addition to homomeric interactions, these proteins exhibit

Vinayaka R. Prasad, Ganjam V. Kalpana (eds.), *HIV Protocols: Second Edition, vol. 485*

DOI 10.1007/978-1-59745-170-3_19 Springerprotocols.com

Table 19.1
Some of the Host proteins interacting with HIV-1 Viral proteins

HIV-1 Viral proteins	Host proteins
ENV	CD4, CxCR4, CCR5, DC-SIGN
IN	INI1, LEDGF, Polycomb EED, XRCC5 (Ku autoantigen), Karyopherin, ATR, ATM
MA	Calmodulin, Karyopherin, Histidyl-t-RNA synthetase like (HO3), VAN/NAF1
CA	Cyclophilin A, Trim5α
NC	Actin, HP68
p6	Nedd4, TSG101, ALIX
Nef	AIP, ASK1, PACS-1, PAK, lac, PI3K
Rev	Crm1, p32
Tat	Cyclin T1, TBP
Vpr	Uracil DNA-glycosylase, Karyopherin, Wee1
Vif	APOBEC3G

heteromeric interactions either with other HIV-1 proteins [for example, RT interacts with IN *(1–3)*; or with host proteins for example, IN interacts with INI1/hSNF5 and LEDGF, *(4, 5)*]. Some of the interactions of HIV-1 viral proteins with host proteins are listed in **Table 19.1**.

Many of these interactions have been shown to be functionally significant *(6)*. The first set of host proteins that were identified as binding partners for HIV-1 proteins using yeast-two hybrid system include Cyclophilin A and INI1/hSNF5 *(5, 7)*. HIV-1 Env protein interaction with host cell receptor CD4 and co-receptors CXCR4 or CCR5 are an essential step during viral entry *(8)*. After post-entry events, viral proteins and nucleic acids encounter host restriction factors that need to be overcome in order to progress through viral replication *(9, 10)*. The interplay between the virus and host restriction factors is likely to be mediated by protein–protein interactions. During HIV-1 reverse transcription, host cell proteins that are likely to be associated within a large reverse transcription complex may play a role in the process *(11)*. Once the reverse transcription is completed, the newly formed preintegration complex (PIC) needs to be translocated into the nucleus, where cellular import machinery plays an important role. The importin/karyopherin pathway has been implicated in this process *(12)*. These import factors bind to viral or host proteins in the PIC. Some of the host proteins that are incorporated into the RTC and PIC include INI1/hSNF5 and LEDGF, respectively, the two cellular proteins that bind to IN *(13–15)*. Once the viral PIC enters the nucleus, the viral DNA integrates into the host chromosome facilitated by the chromatin tethering of LEDGF [reviewed in **Ref.** *(16)*]. Subsequent to the formation of proviral DNA, LTR-mediated transcription is initiated by Tat and the associated cellular proteins forming a complex P-TEF-b, that includes Cyclin T and cdk9 *(17, 18)* and cellular transcription factors such as SP1, INI1 and other proteins *(19–22)*. Several proteins are required to mediate nuclear export of viral RNA, and the viral Rev protein is essential for transport of unspliced RNA from the nucleus into the cytoplasm *(23)*. Subsequent to nuclear export of viral genomic and mRNAs, and translation of the mRNAs, the viral RNA and proteins assemble at the plasma membrane and subsequently bud off from the plasma membrane. Several host factors have been demonstrated to play major roles in assembly and budding of viral proteins including the proteins HP68 and TSG101 *(24, 25)*. The PTAP motif of the Gag p6 was demonstrated to bind to TSG101 coupled with Vps28 and Vps4 that are components of the endosomal sorting machinery. Recent data suggests that the multifunctional protein ALIX also binds to p6 *(26)*. Both TSG101 and ALIX in turn recruit the ESCORT complex required for the pinching-off of viral buds *(26)*. These varied interactions indicate the significance of specific protein-protein interactions during HIV-1 replication.

The details of protein–protein interactions and the residues involved in these interactions are essential to understand the mechanistic and functional implications of these interactions. The precise information about the residues involved in interaction can be determined relatively easily for proteins whose three-dimensional structure is known. However, the lack of three-dimensional structural information for several host proteins makes it harder to carry out detailed biochemical and functional studies. Isolation and characterization of mutants of these proteins defective for specific protein–protein interactions will be valuable to determine the functional significance of such interactions. The 'yeast two-hybrid system' represents an efficient in vivo genetic system for identification and analysis of potential protein–protein interaction partners *(27)*. Additionally, modifications of the system can be used to detect other macromolecular interactions such as DNA-protein interaction (one-hybrid), RNA-protein interaction (RNA-based three-hybrid) and small molecule–protein interaction (ligand based three-hybrid), trimeric interactions and post-translational modifications *(28–30, 31–33)*. Furthermore, the yeast two-hybrid system allows the efficient isolation of interaction defective mutants by 'reverse two-hybrid system' *(34, 35)*.

In this review, we have outlined the detailed steps by which interaction-defective mutants of a protein is isolated by using reverse two-hybrid screen. This scheme has been used for studying HIV-1 IN and INI1 interactions. However, it can be adapted to other protein-protein interaction partners. The functional role of IN and the tumor suppressor Integrase Interactor 1 (INI1) in HIV-1 replication and cancer formation has been studied by using both the forward and reverse yeast two-hybrid systems *(5,36–40)*. The reverse two-hybrid system was successfully used to isolate mutants of: (i) HIV-1 IN that were defective for oligomerization [V260E, **Ref.** *(41)*]; (ii) IN that were defective for binding INI1 [H12Y, in **Ref.** *(40)*]; and (iii) S6 (the transdominant inhibitory fragment derived from INI1, amino acid residues 183–294), that was defective for binding to IN [S6, D225G, in **Ref.** *(40)*]. Virus carrying the V260E mutation was found to be blocked in HIV-1 replication at the step of integration demonstrating that integrase multimerization was important for its activity *in vivo (41)*. A mutant D225G in the INI1 fragment S6 that was unable to interact with HIV-1 integrase was unable to block late events of HIV-1 replication unlike the wild-type S6 transdominant fragment *(40)*.

The reverse two-hybrid system provides a unique opportunity for efficient and rapid screening of a large number of mutants deficient in protein–protein interaction. In the yeast 'two-hybrid system' a pair of interacting proteins (X and Y) bring a transcription activation domain (AC) in close proximity of a DNA binding domain (DB) to regulate the expression of an adjacent reporter gene *(42)*. Expression of the reporter gene is monitored by simple analytical assays. In the 'reverse two-hybrid system' the two interacting proteins are already known and a library of mutants of one of the interacting proteins X is transformed into yeast cells along with wild-type Y and the inability of the mutant X protein to activate the transcription of the reporter gene in contrast to its wild-type counterpart is scored. When using *lacZ*, as read out for protein-protein interactions, blue color indicates interaction and white color indicates a lack of it. A two-hybrid system involving a selectable marker such as HIS3 would not allow the isolation of interaction-negative mutants, as the mutants defective for interaction will not lead to prototrophy and hence these colonies will be eliminated from the screen. However, a counter selectable marker such as URA3 has been employed to isolate mutants defective for protein–protein interaction *(43)*. The principle here is that the URA3 marker when expressed causes lethality under specific growth conditions. The presence of mutants that are defective for interaction will not allow the expression of URA3 and hence those colonies expressing these mutants will survive. Other assays for protein-protein interactions such as GST-pull down assay or co-immunoprecipitations will only allow testing of selected individual mutations and not a screen of an entire library of mutations. We will describe below the simplest screen for the isolation of interaction defective mutants using *lacZ* as marker.

1.1. Strategy for Isolating Interaction-Negative Mutants

The strategy described here involves the use of a two-hybrid yeast strain, which has a *lacZ* reporter gene downstream of a *LexA* operator site. The plasmids encode the LexADB-X or GAL4AC-X PCR-based mutants, which takes advantage of the error prone nature of the Taq polymerase. By altering the concentration of template and dNTPs, the rate of mutagenesis, number of mutants obtained within an insert and kind of mutations can be monitored (Genemorph II kit; Strategene). These libraries of mutagenized plasmids will serve as starting material to isolate interaction-defective mutants. The mutagenized pool of LexADB-X* (or GAL4AC-X*) plasmids will be co-transformed with unmutagenized plasmid encoding respective GAL4AC-Y (or LexADB-Y) fusions into yeast with *LexA-lacZ* reporter gene and a large number of yeast transformants are screened for the presence of white colonies. Since the wild type LexADB-X protein interacts with GAL4AC-Y to give a blue colony, the presence of white colonies indicates the presence of LexADB-X* mutants that are no longer capable of protein–protein interaction. The mutagenized plasmids are rescued from the white colony and sequenced to determine the nature of the mutation. Once the plasmid bearing the mutant gene is isolated, it is tested to see if it is defective for specific or multiple protein–protein interactions by simply transforming it into yeast reporter strain with the respective GAL4AC-fusion protein. For example, IN mutants defective for interactions with INI1 are transformed with GAL4AC-IN to determine if they are also negative for IN-IN interactions.

One obstacle in this method is the presence of false positive white colonies, which could result from mutations in regions other than the insert (X or Y region) on the plasmid or mutation in the yeast strain. To eliminate the mutations in regions other than the insert, only the coding region is subjected to PCR mutagenesis and cloned into the unmutagenized vector backbone avoiding mutagenesis outside the insert. However, a low frequency of non-recombinants in the ligation reaction would also result in white colonies. Nevertheless, presence and absence of inserts can be easily distinguished using restriction analysis of the plasmid. The lack of stability of mutant protein could be another issue. Screening for stable protein expression early on in the strategy by mini-protein preparation of each white colony and by Western could eliminate the analysis of false positive clones that result either from lack of inserts or from unstable mutant proteins. Interaction negative mutant proteins with multiple amino acid changes could also be obtained. In such case, a subsequent analysis of segregating each mutation and testing for the interaction negative phenotype is required. However, by using the mutagenesis protocol outlines below, we have been able to obtain high frequency of negative clones with one or two mutations. Negative mutants with single amino acid change and not those with multiple mutations are of significance in this regard. If the mutant with single-amino acid change is defective for one or more interactions but is able to interact with at least one of the fusion proteins, it indicates that the protein is stable, folded properly and interaction-negative phenotype is a true phenotype.

In the method outlined below, creation of the random mutagenic library and the screening of interaction negative mutants have been outlined using the following separate procedures. (i) Preparation of electrocompetent *Escherichia coli* for efficient transformation of mutagenized libraries; (ii) construction of library of random mutations; (iii) transformation of yeast strain with library of random mutations; (iv) Screening of transformants using colony filter lift β-galactosidase assay; (v) Isolation of plasmid DNA from yeast; (vi) transformation of HB101 cells and isolation of and restriction analysis of plasmid DNA; (vii) making glycerol stocks of yeast and *E. coli* harboring the mutagenized plasmid; (viii) yeast protein extraction; (ix) analysis of proteins expressed in yeast by SDS-PAGE and Western blot analysis. Once the plasmid is isolated, the mutation is identified by sequence analysis and the inability of the mutant to interact with its wild-type partner is then confirmed by biochemical techniques.

2. Materials

2.1. Preparation of Electrocompetent Cells

1. Luria Broth (LB) media and LB agar plates (10 g Bacto-tryptone, 5 g Bacto-yeast extract, 10 g NaCl, (pH 7.0); for plates add 20 g Agar).
2. Autoclaved water kept at 4 °C or in cold room (1 L).
3. 10% glycerol solution prepared by mixing 50 mL of glycerol into 450 mL of de-ionised water. Autoclave. Store at 4 °C or in cold room.
4. *E. coli* strains: DH10B (F- *end*A1 *rec*A1 *gal*U *gal*K *deo*R *nup*G *rps*L Str^r Δ*lac*X74 Φ80*lac*ZΔM15 *ara*Δ139 Δ(*ara,leu*)7697 *mcr*A Δ(*mrr-hsd*RMS-*mcr*BC) λ-); and HB101 (F- hsdS20(r_B- m_B-) *rec*A13 *leu*B6 *ara*-14 *pro*A2 *lac*Y1 *gal*K2 *xyl*-5 *mtl*-1 *rps*L20(Str^r) *sup*E44 λ-).

2.2. Construction of Library of Random Mutants of Interacting Partner for Reverse Two-Hybrid Screening

1. Template DNA: pSP72-INI1 and cloning yeast two hybrid vector pGADNot.
2. Forward and reverse PCR primers for amplifying the entire INI1 gene. Use the following primers: INI1-forward primer 5′GC*GGATCC*GTATGGCGCTGAGCAAGACC3′ and

INI1-reverse primer 5′ACGC*GTCGAC*TTACCAGGCCGG GCCCG3′.

3. GeneMorph II Random Mutagenesis Kit (Stratagene Cat. # 200550).
4. MircoAmp Reaction Tubes (Applied Biosystems) and MicroAmp caps (Applied Biosystems).
5. Qiagen Kits. QIAprep Spin Miniprep Kit (Cat. # 27106), QIAEX II Gel Extraction Kit (Cat. # 20021), QIAGEN Plasmid Maxi Kit (Cat. # 12163).
6. Restriction enzymes *Bam*H1, *Sal*1, Calf Intestinal Alkaline Phosphatase, T4 DNA ligase are from New England Biolabs.
7. Agarose GPG/LE from American Bioanalytical (Cat. # AB00972-00500).
8. 25X TAE: Tris base 968 g, Glacial Acetic Acid 228 mL, 400 mL 0.5 M EDTA (pH 8) in 8 L of deionized water.
9. 1 Kb DNA ladder (Invitrogen Cat. # 15615-016) and λ*Hin*dIII Fragments (Invitrogen Cat. # 15612-013).
10. DH10B electrocompetent cells.
11. Bio-Rad Gene Pulser 2 mm Cuvette (Cat. # 1652086).
12. LB medium and LB-ampicillin (35 μg/mL) agar plates.
13. 80% glycerol: 80 mL of 100 % glycerol mixed with 20 mL of deionized water and autoclaved.
14. Nunc disposable inoculating needles (Cat. # 254437).
15. Prepare 3 M Na-acetate (pH 5.2) and 70% ethanol by mixing 350 mL of 200-proof ethanol with 150 mL of autoclaved deionized water. Buy 200-proof ethanol from Pharmco-AAPER (Cat. # 111ACS200).
16. Miracloth (Calbiochem Cat. # 475855).

2.3. Transformation of Yeast With Random Mutation Library By Lithium Acetate Method

1. YAPD liquid medium. 20 g Bacto-peptone, 10 g of Yeast extract per liter of medium, stock solution of filter sterilized 40% glucose in deionized water, stock solution of filter sterilized L-Tryptophan (SIGMA) at 1 g/100 mL and filter sterilized Adenine hemisulphate (SIGMA) at 5 mg/mL.
2. Drop-out medium. 2 g Drop out mix (*see* **Note 1**), 1.45 g Yeast nitrogen base, 5 g Ammonium sulphate, 20 g Bacto-agar, filter sterilized 500 mL of 40% w/v glucose (do not autoclave).
3. Prepare 100 mM LiAc in TE from autoclaved stock solution of 1 M Lithium Acetate (pH 7.5) and 10X TE [100 mM Tris-HCl (pH 8.0) and 10 mM EDTA (pH 8.0)], by mixing 1/10th each of 1 M LiAc and 10X TE with 8/10th of autoclaved deionized water.
4. Prepare 40% PEG in 100 mM LiAc-1XTE freshly from stock solutions of freshly prepared 50% PEG 4000 or 3500 (*see* **Note 2**) in autoclaved deionized water, 1 M LiAc (pH 7.5) and 10X TE. To make this solution, mix 1/10th volume of 1 M LiAc pH 7.5, 1/10 volume of 10X TE and 8/10th volume of 50% PEG and filter sterilize.

5. Sonicated and phenol/choroform extracted Salmon Sperm DNA [also available from vendors such as Sigma, Invitrogen or Stratagene (*see* **Note 3**)].
6. Make 5 mg/mL stock solution of Adenine hemisulfate in water and filter sterilize. Prepare YAPD liquid medium by adding Adenine hemisulfate to 30 μg/mL final concentration in 1X YEPD (*see* **Note 4**).
7. Prepare selective medium (SC drop out) plates using synthetic complete medium omitting specific amino acids. In the case of transformation with pSH2 and pGADNot plasmids, prepare SC-Leu-His plates.

2.4. Colony Filter-Lift β-galactosidase Assay

1. Prepare Z-buffer by dissolving 16.1 g of $Na_2HPO_4.7H_2O$, 5.5 g of $NaH_2PO_4.H_2O$, 0.75 g of KCl and 0.24 g of $MgSO_4.7H_2O$ in water and make up volume to 1 L. Filter sterilize and store at room temperature. Do not autoclave (*see* **Note 5**).
2. Prepare X-Gal solution by dissolving 5-bromo-4-chloro-3-indolyl-β-D-galactoside at 20 mg/mL in DMF (Dimethyl formamide). Store at −20 °C (*see* **Note 6**).
3. β-mercapto ethanol (β-ME) (SIGMA). Stored at room temperature with stabilizer in amber colored bottles.
4. Nitrocellulose circles (Schleicher & Schuell, Protran pure nitrocellulose membrane, BA85, Cat # 10402579).
5. 3 MM Whatman filter paper circles to fit the inner circumference of Petri dish.
6. Plastic Petri plates.

2.5. Isolation of Plasmid DNA from Yeast Cells

1. SC-leucine drop-out medium (for isolating pGDNot plasmid).
2. Triton solution: 2% Triton X-100, 1% SDS, 100 mM NaCl, 10 mM Tris pH 8.0, 1 mM EDTA.
3. Tris-saturated phenol, prepare by mixing 100 mL of 1M Tris-HCl (pH 8.0) with phenol (> 99% purity) and shake vigorously. Store at 4 °C in amber colored bottle.
4. Phenol:chloroform:iso-amyl alchohol (25:24:1) solution −50 mL. Prepare by adding 25 mL of Tris-saturated phenol, 24 mL of chloroform and 1 mL of iso-amyl alchohol and shake vigorously to mix and store at 4 °C.
5. Chloroform:iso-amylalchohol (24:1) solution −50 mL. Prepare by adding 2 mL of iso-amyl alchohol to 48 mL of chloroform and mix vigorously. Prepare fresh before use.
6. Glass beads of size 425–600 μm (30–40 U.S. sieve) from SIGMA (Cat # G8772) and acid wash them as described below.
7. 3M Na-acetate solution (pH 5.2).
8. 70% ethanol solution. Prepare by adding 350 mL of 200-proof absolute, anhydrous ethyl alcohol to 150 mL of

autoclaved deionized water and thorough mixing. Store at room temperature.

9. Prepare 80% glycerol stock solution by mixing 80 mL of 100% glycerol and 20 mL of deionized water. Autoclave and store at room temperature.
10. Prepare 1 L of stock solution of TE [10 mM Tris-HCl (pH8.0), 1 mM EDTA] by adding 10 mL of 1 M Tris-HCl (pH 8.0), 2 mL of 0.5 M EDTA (pH 8.0) to 988 mL of deionized water. Autoclave and store at room temperature.

2.6. Minimal Medium (BA1) Plates for Selection of Leucine Prototrophy in HB101 Strain of E. coli

1. This is a modification of M9 minimal medium. Make a 5X stock solution of salts containing 64 g Na_2HPO_4. 7 H_2O, 15 g KH_2PO_4, 2.5 g of NaCl, 5 g NH_4Cl. Dissolve salts in deionized water to a final volume of 1 L. Divide the salt solution into 200-mL aliquots and autoclave.
2. Prepare 100 mL stock solutions (1,000X) of each of the following: Proline 23 g/100 mL, Threonine 4 g/100 mL, Thiamine 4 g/100 mL and Histidine 4 g/100 mL and autoclave them separately.
3. Prepare 100 mL each of 1 M $MgSO_4$ and 1 M $CaCl_2$ and autoclave them separately.
4. Prepare 100 mL of 20% glucose and filter sterilize.
5. Prepare 'Solution A' by adding 15 g of agar to 800 mL of deionized water in a 2-L flask and autoclave it.
6. Prepare 'Solution B' by adding 1 mL each of the stock solution of Proline, Threonine, Thiamine and Histidine to 200 mL of prewarmed (keep in water bath at 45–50 °C) salt solution.
7. When the agar solution cools down to appropriate temperature (~ 45 °C), mix solution A and solution B and add 2 mL of the stock solution of $MgSO_4$, 0.1 mL of the stock solution of $CaCl_2$ and 20 mL of the stock solution of glucose. Swirl gently and pour the contents into 100 × 15 mm polystyrene bacterial growth plates (25 mL of mixture per plate).

2.7. Preparation of Glycerol Stocks of Yeast and E. coli

1. Prepare LB liquid media and LB-Ampicillin (35 μg/mL) plates (for *E. coli* stocks) and SC-Leu media and SC-Leu plates (for yeast stocks). See above for composition.
2. *E. coli* and yeast cells.
3. 80% glycerol in deionised water. Autoclave and store at room temperature.

2.8. Yeast Protein Extract Preparation

1. Breaking buffer: 50 mM Tris-HCl (pH 7.5), 10% glycerol, 10 mM $MgCl_2$, 1 mM EDTA, 100 mM NaCl, 1 mM DTT, 1 mM PMSF and 1X protease inhibitors. To prepare 10 mL of breaking buffer, add 500 μL of 1 M Tris-HCl (pH 7.5), 2 mL of 50% glycerol, 100 μL of 1 M $MgCl_2$, 50 μL of 0.2 M EDTA (pH 8.0), 200 μL of 5 M NaCl, 10 μL of 1 M

DTT, 1X protease inhibitor mix (Roche Diagnostics Complete Mini EDTA-free tablets) and 100 μL of 0.1 M PMSF in 7.04 mL of deionized water.

2. Acid-washed glass beads are prepared from glass beads (Thomas Scientific, Cat # 5663R50). Immerse one bag of beads in Nitric Acid (~ 400 mL) in a beaker. Stir and let sit a few hours to overnight in a hood. Pour off the Nitric acid and wash multiple times with deionized water, for hours until the pH is neutral. Pour the beads into Buckner Funnel hooked to vacuum on top of a flask, rinse one more time with water. Dry and bake for 4 h.

2.9. SDS-Polyacrylamide Gel Electrophoresis and Western Blotting

1. Stabilized 30% Acrylamide solution (Cat. # EC-810) and 2% N, N′-Methylene bisacrylamide solution (Cat. # EC-820) from National Diagnostics. N,N,N′,N′-Tetramethylethylenediamine (TEMED) from SIGMA (Cat. # T9281).
2. 10% Ammonium persulphate (APS) solution. Add 1 g APS to final volume 10 mL deionized water. Vortex to mix. Store in 200 ml aliquots at −20 °C.
3. 10% SDS solution. Add 10 g of SDS powder to 100 mL of deionized water. Keep at 50 °C till the SDS dissolves. Store at room temperature.
4. 1.5 M Tris-HCl (pH 8.3) and 0.5 M Tris-HCl (pH 6.8).
5. 5X loading buffer (15 mL): 7.5 mL of 100% glycerol, 3.75 mL of β-mercaptoethanol, 3.75 mL of 1 M Tris-HCl (pH 6.8), 1.5 g of SDS powder and 0.015 g of bromophenol blue.
6. Running buffer. Prepare 1X working solution from 10X stock (150 g Tris-base, 720 g glycine in 5 L deionsed water), add 10 mL/L 10% SDS and make up volume with deionized water.
7. Western blot transfer buffer. Prepare 1X working solution from 10X stock (150 g Tris-base, 720 g glycine in 5 L deionized water).
8. Benchmark prestained protein ladder from Invitrogen (Cat. # 10748-010).
9. Bio-Rad Trans-Blot Transfer Medium (pure nitrocellulose membrane, 0.45 μm, Cat. # 162-0115) and Whatman 3 MM filter paper.
10. TBST and Blocking buffer. Prepare 1X working solution of TBS from 10X stock (400 mL 1M Tris-HCl (pH 7.5), 350.64 g NaCl in 4 L) and add 0.1% Tween-20. To prepare blocking buffer add 5% w/v fat-free dry milk to 1X TBST. To prepare buffer for antibody dilution use 0.5% w/v fat-free dry milk in 1X TBST.
11. Monoclonal antibody against GAL4 activation domain (Clontech Cat. # 630402).

12. Secondary Goat anti-mouse HRP-conjugated antibody from PIERCE (Cat. # 1858413).
13. Supersignal West Pico chemiluminescent Substrate from PIERCE (Cat. # 34080) and Kodak BioMax Light Film (Chemiluminescence, Cat. # 2008–11).

3. Methods

3.1. Preparation of Electrocompetent Cells

Electrocompetant cells yield highest transformation efficiency to obtain a large number of mutant clones in the library.

1. All steps must be carried out under sterile conditions. Streak BH10B or HB101 cells from a glycerol stock onto an LB plate. Grow overnight at 37 °C.
2. Pick one large single colony from plate and grow in 100 ml of LB media overnight at 37 °C.
3. Inoculate the starter culture into 1 L of LB media such that A_{600} is ~ 0. 1. Grow for 3 h at 37 °C till A_{600} reaches ~ 0. 5.
4. Place conical flask containing culture in a container filled with ice-water. Let it sit for 15 min. Then swirl the container in the ice-water bath to cool down the culture. Pour culture into 500-mL bottles and centrifuge using Sorvall rotor (GS-3) at 6000 rpm for 20 min.
5. Decant supernatant and suspend pellet in 250 mL of cold water. Centrifuge to pellet.
6. Decant supernatant and suspend pellet in 125 mL of cold water. Centrifuge to pellet.
7. Decant supernatant and suspend pellet in 10 mL (per 500 mL culture) of 10% glycerol. Centrifuge to pellet at 3500 rpm for 15 min in a table-top centrifuge.
8. Resuspend pellet in 1 mL of 10% glycerol. Make small aliquots and quick freeze in dry ice-ethanol bath. Store aliquots at −80 °C till further use.

3.2. Construction of Library for Reverse Two-Hybrid Screening

Genemorph II random mutagenesis kit from Stratagene is based on a novel error prone PCR enzyme blend containing a mixture of the Mutazyme I polymerase and the *Taq* polymerase to produce a less biased mutational spectrum with equivalent mutation rates at A's and T's vs. G's and C's. Furthermore, by altering the template concentration and number of cycles used in the PCR reaction, approximate number of expected mutations per kb of template DNA can be monitored. In general, to get 0–4.5 mutations per kb of DNA, 500–1000 ng of template DNA and 30 PCR reaction cycle are used (Genemorph II, mutagenesis kit manual). Construction of the library is carried out in several steps as follows.

3.2.1. PCR-amplification and Purification of the Insert

1. Prepare 50 μL of reaction mixture using the GeneMorph II Random Mutagenesis Kit (Stratagene) by adding 41.5 μL of autoclaved deionized water, 5 μL of 10X Mutazyme II reaction buffer, 1 μL of 40 mM dNTP mix (200 mM each final concentration), 0.5 μL of primer mix (250 ng/μL of each primer), 1 μL of Mutazyme II DNA polymerase (2.5 U/μL) and 1 μL of pSP72-INI1 DNA (500–1000 ng/μL) in MircoAmp Reaction Tubes (Applied Biosystems). Mix thoroughly, cover tubes with MicroAmp caps (Applied Biosystems) and place tubes in thermal cycler.
2. Run PCR reaction using the following PCR program: one cycle of denaturation at 95 °C for 2 min; 30 cycles of denaturation at 95 °C for 30 s, annealing at 60 °C for 30 s, amplification at 72 °C for 1 min and finally amplification at 72 °C for 10 min. After PCR amplification keep sample at 4 °C (*see* **Note 7**).
3. Purify PCR amplified DNA using the Qiagen PCR Purification Kit (Cat. # 28104). Add 5X volume of Buffer PB1 to sample and mix thoroughly. Add mixture to purification column. Centrifuge for 30 s in microfuge at 10,000 rpm. Discard flow through. Add 0.75 mL of wash-buffer (Buffer PE). Centrifuge for 30 s in microfuge at 10,000 rpm. Discard flow through. Centrifuge at 13,000 rpm for 1 min in a microfuge. Take column and place it on a fresh sterile tube. Add 40–50 μL of water. Incubate at room temperature for 5 min (*see* **Note 7**). Centrifuge at 13,000 rpm for 1 min in microfuge. The flow-through is the purified PCR preparation.

3.2.2. Restriction Digestion of the Insert and the Vector

1. Digest pGADNot vector and PCR reaction products containing the mutagenized inserts with *Bam*H1 and *Sal*I restriction enzymes in 1X Buffer 3 (stock 10X) and 1X Bovine Serum Albumin (BSA) (stock 100X) for 1–2 h at 37 °C.

3.2.3. Alkaline Phosphatase Treatment of the Vector

After digestion, add Calf Intestinal Alkaline Phosphatase (stock 10 U/μL) at 0.5 U/μg of DNA to the reaction mixture containing pGADNot. Incubate for an additional 1 h at 37 °C.

3.2.4. Gel Purification of the Vector and Insert

1. Prepare 1% agarose gel by suspending 1 g of dry agarose powder into 100 mL of 1X TAE. Heat in microwave for 2 min till the solution becomes clear. Let it cool to ~ 65 °C. Add ethidium bromide (stock 10 mg/mL) to 1:1000 dilution. Swirl and pour into gel cassette. Leave it at room temperature for 30 min to 1 h. Once the gel has solidified, put it into gel tank and add 1X TAE buffer till gel is completely submerged.
2. Load digested DNA samples as well as 1 kb DNA ladder into wells and run gel electrophoresis at constant voltage of 80 V. After the DNAs are separated well (45 min to 1 h), guided by

the long wavelength UV light, cut out DNA bands according to size using a clean razor and put in individually marked and preweighed sterile Eppendorff tubes.

3. Weigh the tubes to determine weight of the gel slice, subtracting the weight of the tubes before and after the addition of the gel slice.
4. Use QIAEX II Gel Extraction Kit to purify DNA from cut agarose gel. Add 3X volume of Buffer QX1 and 10 μL of QIAEX II suspension. Vortex to mix and incubate in a 50 °C water bath for 10 min with occasional vortexing. Once the gel melts, briefly centrifuge in microfuge. Discard supernatant. Wash 1X with Buffer QX1 and 2X with Buffer PE by vortexing and centrifugation. Discard supernatant each time and finally remove all liquid. Dry beads under vacuum in a Speed-Vac for 5 min (*see* **Note 7**). Add 15 μL of autoclaved water to beads and vortex. Incubate at room temperature for 5 min. Spin four times in microfuge at 13,000 rpm for 1 min each to completely separate liquid from beads. The final supernatant contains the required DNA fragments. Quantitate isolated DNA by running a 1% agarose gel and λ HindIII marker.

3.2.5. Ligation of Insert and Vector to Create Library

Set up ligation reaction using 150 ng of digested vector plasmid, 450 ng of digested PCR fragment, 3 μL of 10X T4 Ligation buffer, 3 μL of 10 mM ATP, and 3 μL of T4 DNA ligase (2000 U/mL). Make up volume to 30 μL with autoclaved water. Incubate at 16 °C overnight.

3.2.6. Transformation into *E. coli* with the Ligated DNA to Generate Random Mutation Library

1. Transform 1 μL of ligation product by electroporation (2.5 kV constant voltage and 25 μF capacitance) into 40 μL of DH10B electrocompetent cells in a 2-mm Gene Pulser cuvette using a Bio-Rad Gene Pulser II electroporation apparatus.
2. Following transformation take out the bacterial cells from the cuvette and add to 1 mL of LB and grow for 1 h at 37 °C. Plate 0.2 mL of culture each onto 100 × 15 mm polystyrene plates containing LB-agar and ampicillin (35 μg/mL final concentration). Use ~ 50 plates to plate transformants for each library.
3. Incubate plates overnight at 37 °C.
4. Pick 10–20 individual colonies for the analysis of frequency of recombination, which is done by preparing the plasmid DNA from individual colonies, and restriction analysis with an enzyme to release the insert. Percentage of colonies with plasmids carrying an insert will provide recombination frequency, which can be used to calculate background.

5. Pool remaining colonies from 50 plates in 50 mL of LB liquid media containing 35 μg/mL ampicillin by scraping the colonies with a glass pipette with bent end (*see* **Note 8**).
6. Add 200 μL of 80% autoclaved glycerol to each of 800-μL aliquots of pools of colonies to make several 1 mL of stocks.
7. Quick freeze in ice-ethanol bath and store aliquots at −80 °C.

3.2.7. Isolation of DNA from Library of Mutants Transformed into E. coli

1. To get adequate library DNA for yeast cell transformation entire glycerol stock (0.5 or 1 ml) from one vial is inoculated into 500 mL of LB media containing 35 μg/mL of ampicillin at 37 °C with shaking. The entire glycerol stock from one vial is inoculated to minimize the bias that is introduced to maintain the complexity of the library.
2. Grow until the OD_{600} reaches 0.8–1.0.
3. Pellet cells in 500-mL bottles by centrifugation at 6,000 rpm for 20 min using a Sorvall rotor (GS-3). Discard supernatant.
4. Purify plasmid DNA using a Qiagen Maxi Prep Kit or using any plasmid DNA purification method *(44)*.

3.3. Transformation of Yeast with Library of Random Mutants

1. Prepare 1 L of YAPD medium by adding 20 g Bactopeptone, 10 g Yeast extract into 940 ml of deionized water. Autoclave and cool to about 45–55 °C.
2. To the cooled down solution, add 50 mL of 40% glucose, 2 mL of L-Tryptophan stock solution and 6 mL of Adenine hemisulphate stock solution. Mix and filter sterilize before using (*see* **Note 9**).
3. Make plates using the dry drop-out mix as follows. Take one 2-L flask and one 1-L flask. In the 2-L flask add 20 g Bacto-Agar to 450 mL of deionized water (solution A). In the 1-L flask add 2 g Drop-out mix, 1.45 g of Yeast nitrogen base and 5 g ammonium sulfate to 500 mL of deionized water (solution B). Autoclave both the flasks separately. Cool down the autoclaved solutions to lukewarm temperature (~ 45 °C). Transfer solution B into solution A. Add 2 mL of L-Tryptophan stock solution. Add 50 mL of filter sterilized glucose solution to this mixture. Mix and pour plates (*see* **Note 10**).
4. Streak CTY5d-10 yeast strain onto a YAPD plate and incubate the plate at 30 °C for 3–4 days till colonies of decent size appears.
5. Inoculate a massive colony of yeast from the plate into 100 mL of YAPD medium the day prior to transformation. Incubate at 30 °C with shaking overnight. This is the starter culture.
6. Inoculate this culture early in the morning into 1 L of fresh YAPD media such that A_{600} is at 0.2.
7. Grow yeast cells to a A_{600} of 0.5 (it usually takes ~ 3 h).

8. Pellet cells at 2,000 rpm in a table-top centrifuge for 5 min at room temperature.
9. Wash pelleted cells with 1/10th volume (to the original culture volume) of 100 mM LiAc in TE.
10. Pellet cells at 2,000 rpm for 5 min at room temperature and resuspend pelleted cells in 1/200th volume of 100 mM LiAC in TE.
11. Shake for one hour at 30 °C (*see* **Note 11**). Set up a heating block at 42 °C.
12. Next, set up the following reagents in a sterile Eppendorf tube: 200 μL yeast cells, 1–10 μg of each plasmid DNA and 200 μg of Salmon sperm carrier DNA (20 μL of 10 mg/mL). Mix it by pipetting up and down (*see* **Note 12**).
13. Add 1.2 mL of 40% PEG in 100 mM LiAc/TE.
14. Incubate at 30 °C with agitation for 30 min.
15. Add 10% final DMSO, mix by tapping (will increase the efficiency to 5 fold).
16. Heat shock at 42 °C for 15 min.
17. Spin the cultures after heat shock for 10 min in the table-top centrifuge at 2,000 rpm.
18. Remove the supernatant by aspiration and resuspend the pellet in 1 mL of YEPD.
19. Incubate at 30 °C for 1 h with shaking. This step increases the efficiency by 5–10 fold.
20. Spin down the cells as before and remove supernatant by aspiration. Wash the cells once with 1 mL of TE.
21. Resuspend the pellet in 100–200 μL of TE and plate on SC-Leu and His selective plates.
22. Incubate the cells at 30 °C until distinct colonies appear, which will take about 3–4 days.

3.4. Colony filter lift β-galactosidase assay

1. Place premarked nitrocellulose filter circles on yeast colonies. If one is not screening a library and is looking to see if all the colonies in a plate become blue or white, then one can use sectors of nitrocellulose instead of whole circles. Make sure that the filter becomes wet evenly throughout (*see* **Note 13**).
2. Remove the filters, place colony side up on a 3 MM Whatman paper. Place the entire ensemble in −80 °C for 10 min.
3. Bring the filters to the bench and leave it for about 5 min at room temperature to allow it to thaw.
4. In the mean time, prepare as many Petri dishes as the number of nitrocellulose filters that have to be tested (If you are using nitrocellulose filter quarters, then prepare one Petri dish for four sectors). Place two circles of Whatman 3MM paper in the bottom half of a Petri dish. Add 5 mL of Z buffer containing X-gal solution and β-mercaptoethanol (add 30 μL of

20 mg/mL X-gal solution and 2.7 μL of β-mercaptoethanol per milliliter of Z buffer to be used) (*see* **Note 14**).

5. Place the filter circles colony side up and close the lid. Ensure that the nitrocellulose filter circle is making complete contact with the Whatman paper and is evenly wet (*see* **Note 15**).
6. Incubate at 30 °C till blue color appears. The color should appear in 30 min to overnight.
7. While most of the colonies appear blue, look for distinct white colonies among the blue colonies. Pick white colonies and restreak them onto fresh SC-His-Leu plates and repeat the X-gal assay to confirm the lack of *LacZ* expression in the white colonies.

3.5. Isolation of Plasmid DNA from Yeast Cells

1. Pick a massive white yeast colony from SC-Leu-His plates and inoculate it into 2 mL of SC-Leu (when using pGADNot-X* plasmid) medium in a 15-mL round bottom snap cap tube.
2. Incubate at 30 °C with vigorous shaking until the culture reaches an absorbance at 600nm (A_{600}) of 0.5–0.8) (*see* **Note 16**).
3. Decant 1.5 mL of the culture into an Eppendorf tube (*see* **Note 17**). Spin in the microfuge at 6,000 rpm for 2 min. Discard supernatant.
4. Resuspend the cell pellet in 200 μL of triton solution and 300 μL of phenol:chloroform:iso-amyl alcohol (25:24:1) solution.
5. Add about 0.3 g of acid washed glass beads into each suspension (*see* **Note 18**).
6. Vortex the tube for 2 min. Spin in the cold room in a microfuge at 13,000 rpm for 10 min.
7. Collect the top most aqueous layer into a separate Eppendorf tube.
8. Add equal volume of Tris-saturated phenol, invert vigorously to make emulsion and spin in a microfuge in the cold room at 13,000 rpm for 5 min (phenol extraction).
9. Collect the top aqueous layer into another Eppendorf tube, add equal volume of chloroform:iso-amyl alcohol (24:1 v/v) solution, invert vigorously and spin in the microfuge in the cold room at 13,000 rpm for 5 min.
10. Collect the top aqueous layer into another Eppendorf tube. To ethanol precipitate the DNA with carrier, add 1/10th volume of 3M Na-acetate, 1 μL of 10 mg/mL glycogen, and two volumes of 200-proof ethanol. Mix and keep at −80 °C for 20 min.
11. After 20 min, spin in a microfuge in the cold room at 13,000 rpm for 15 min. Carefully remove the supernatant by decantation.
12. Wash the pellet twice with 70% ethanol. Each time add 1 mL of 70% ethanol solution, mix by inversion and spin in a

microfuge in the cold room at 13,000 rpm for 5 min. Decant the supernatant. Wash the pellet once again as before.

13. Remove the supernatant by decantation. Dry the pellet under vacuum in a Speed-Vac for 10 min. Resuspend the pellet in 3–5 μL of TE (*see* **Note 19**).

3.6. Transformation of HB101 Cells and Isolation and Restriction Analysis of Plasmid DNA

This is a simple and additional rapid step that can be used to obtain single isolated plasmid of desired type (pGADNot-X* or pSH2-X*) from yeast by using the ability of markers on the pSH2 (*His3*) and pGADNot (*Leu2*) plasmids to complement Histidine or Leucine auxotrophy of the HB101 strain *(36)*. The method described here is for complementing Leucine auxotrophy and for isolating pGADNot plasmid that has *Leu* marker.

1. Take 1 μL of plasmid DNA isolated from yeast and mix with 40 μL of HB101 electrocompetent cells. Electroporate at 2.5 kV voltage and 25 μF capacitance in a 2 mm electroporation cuvette.
2. Transfer cells to 1 mL LB media and grow at 37 °C for 1 h. Plate on LB-ampicillin (35 μg/mL) agar plates. Grow overnight at 37 °C. Replica plate onto BA1-Leu minimal media plates to select for leucine prototrophy. Grow overnight at 37 °C.
3. Pick single colonies and grow in LB-ampicillin (35 μg/mL) media.
4. Isolate plasmid DNA using the QIAprep Spin Miniprep Kit or any other method of plasmid isolation.
5. Restriction digest 5–10 μL of DNA sample with desired enzymes (New England Biolabs) in 1X Buffer. Separate digested fragments and 1 Kb DNA ladder (Invitrogen) in a 1% agarose gel using 1X TAE. Look for the presence of the desired insert in the pGADNot plasmid.
6. Submit the clones for sequencing to identify the mutations.

3.7. Making Glycerol Stocks of Yeast

It is necessary to preserve the original clones to ensure getting the correct mutation.

1. Grow an overnight culture of the yeast strain by picking the colony from the original plate that needs to be stocked, in SC-Leu medium (for the yeast cell is carrying the pGADNot-X* mutant plasmid).
2. To a 3 mL of the culture, add 1 mL of 80% glycerol so that the final glycerol concentration is 20%. Mix well by pippetting up and down.
3. Clearly label the freezing vials with permanent marker (two for each strain). Write the name of the yeast strain, plasmid in the yeast and the date. At the same time write in a note-book that a stock of a particular yeast strain containing a particular plasmid was made on that date. After labeling the tubes, place them in dry ice–ethanol bath (*see* **Note 20**).

4. Mix the culture and aliquot 1.8 mL of each culture into two prelabeled and prechilled freezing vials.
5. Once the cultures are frozen, place them in the strain collection boxes and store at −80 °C freezer (*see* **Note 21**).

3.8. Yeast Protein Extraction

Often the mutant that is defective for protein-protein interaction is not well expressed or has reduced stability. Therefore, it is essential to ensure the stable expression of mutant protein.

1. Pick a colony of yeast and grow overnight in drop-out liquid media. Inoculate 10 mL of YAPD liquid media at A_{600} of ~ 0.2 and grow culture until the A_{600} is ~ 0.8, to get ~ 1×10^7 yeast cells/mL.
2. Harvest cells at 2000 rpm for 5 min at 4 °C in a table-top centrifuge. Pour out most of the media.
3. Vortex cells in remaining media, and transfer cell suspension to an Eppendorf tube.
4. Pellet cells in a microfuge for 20 s via quick spin. Remove supernatant.
5. Freeze the cell pellet at −80 °C or in dry ice–ethanol bath.
6. Thaw cells. Resuspend in 250 μL of breaking buffer.
7. Add 250 μL worth of acid washed glass beads.
8. Vortex 1 min in cold room then leave it for 1 min to prevent excess heating of your sample. The Tekmar multitube vortexer works well for multiple samples.
9. Repeat step #8 five more times.
10. Check lysis of your cells under the microscope. Greater than 90% lysis is expected. If poor lysis occurs, continue vortexing until most cells are lysed.
11. Remove liquid to a fresh tube. Try to avoid glass beads in the pipette tip.
12. Add 250 μL of breaking buffer to the glass beads.
13. Vortex 1 min to wash the beads (and increase the yield).
14. Remove liquid from the beads, and combine with the previous supernatant.
15. Centrifuge at 13,000 rpm for 10 min in a microfuge in the cold room to pellet debris and unlysed cells from the soluble extract.
16. Remove clarified supernatant to a new tube.
17. Determine protein concentration using Bio-Rad Protein Assay reagent (Cat. # 500-0006) using 1–2 μL of the clarified extract. Add 1–2 μl of extract to 800 μL of water and 200 μL of reagent. Use different dilutions of BSA as positive control and extraction buffer as blank. Mix by inversion. Incubate at room temperature for 10 min. Take reading at A_{595} after calibrating for blank. Measure protein concentration in comparison to BSA control (*see* **Note 22**).
18. Once the protein is extracted from yeast, any method of SDS-PAGE and Western blotting technique can be used to

determine the expression of the mutant protein *(45)*. Use of antibody against GAD4AD (Clontech Cat. # 630402) is recommended to determine the expression of AD-fusion protein.

4. Notes

1. Drop-out mix: This can be prepared in two different ways. First method involves creating 100 mL stocks of individual amino acids and adding indicated amounts of the stocks of individual amino acids to the media. Amount in mg per 100 mL of stock solution/ml stock per liter: adenine hemisulfate* (200 mg/10), uracil* (200 mg/10), L-tryptophan (1 g/2), L-histidine-HCl (1 g/2), L- arginine-HCl (1 g/2), L-methionine (1 g/2), L-tyrosine (200 mg/15), L-leucine (1 g/3), L-isoleucine (1 g/3), L-lysine-HCl (1 g/3), L-phenylalanine (1 g/5), L-glutamic acid* (1 g/10), L-aspartic acid*[a] (1 g/10), L-valine (3 g/5), L-threonine*[a] (4 g/5), L-serine (8 g/5).
   [*Store at room temperature, [a] Add after autoclaving the media.]
   A second method is creating dry drop-out mixture (powder). When creating a drop-out mix, weigh out amounts indicated for 100 mL stock (or scaled up quantities) of all the amino acids needed in a plastic bottle (for example, take 200 mg of adenine hemisulfate, 200 mg of uracil, 1 g of L-tryptophan etc.). Add —three to four autoclaved marbles and shake the amino acid mix for about 30–45 min. Leave out the specific amino acids from the drop out mix. For example, to select for leucine prototrophy, leave out leucine from the drop out mix. Store this in the room temperature. If you need to add additional amino acids to this mixture, make a 100 ml stock of amino acid as indicated above and add designated amount of stock solution to the autoclaved medium before pouring the plates.
2. Do not use PEG 8000. To dissolve PEG into water warm mixture to 50 °C till a clear solution appears.
3. We have found that Salmon sperm DNA from Sigma works best. Alternatively, you can purchase ready-to-use Salmon sperm DNA from Invitrogen (Cat. # 15632-011).
4. Yeast cultures can get easily contaminated with *E. coli* when growing overnight. To avoid contamination, prepare 1X YEPD liquid media without glucose/dextrose in 500- or 1,000-mL bottles, autoclave and store. Just before use, add both Adenine hemisulfate and glucose to the required concentration and filter sterilize the entire medium once again using

0.45-μm filter units. This filter sterilization step enormously decreases the frequency of contamination of yeast cultures.

5. For long term storage, keep Z buffer at 4 °C.
6. Store tube containing X-Gal solution wrapped in aluminum foil.
7. It is not advisable to store PCR products for long either at 4 or −20 °C without purification. It is not advisable to incubate the column with water for more than 5 min at room temperature. Do not over dry DNA-bound QIAEXII beads.
8. Bend a Pasteur pipette over a flame to make a large loop with which the colonies can be scraped.
9. This protocol is for liquid medium. To make plates, add 20 g of BactoAgar prior to autoclaving. After the solution is cooled down, add the required amount of glucose, L-tryptophan and adenine hemisulfate solutions and pour plates.
10. This protocol is for making the plates. To make the liquid medium, add everything except Bacto-Agar into the same 2-L flask.
11. It is not good to shake for more than one hour as the cells will starve and die.
12. When doing a library transformation, combine reagents in a round bottom 15 mL snap cap tube. When doing the transformation of a cloned plasmid DNA, and a high efficiency of transformation is not necessary, then use a sterile Eppendorf tube to combine reagents.
13. There is no need to press the filters hard against the colonies. Just make sure that the filters are in contact with the surface of the plate medium. It is best to do this as soon as the plates are out of the incubator. However, if it is impossible to do it right away, one can store the plates with colonies at room temperature for up to a day.
14. If the Z buffer was kept in the cold room it is necessary to bring it to room temperature before adding other reagents. Always drain excess buffer off the plates as otherwise the colonies could be lost through flooding.
15. Make sure that there are no bubbles between the nitrocellulose membrane and the filter paper to ensure that all colonies are equally in contact with the buffer. This can be achieved by gently folding the nitrocellulose membrane at the center before placing it on to the wet filter paper and once the center of the membrane wets gently release the sides till the periphery of the membrane touches the filter paper.
16. Overnight to 2 days of incubation with shaking is sufficient.
17. Save 0.5 mL of the culture and make glycerol stocks.
18. In order to be efficient, prepare several Eppendorf tubes containing 0.3 g of glass beads ahead of time. At this point, you could simply decant the beads into the Triton solution.
19. Use 1 μL of the resuspended DNA pellet for electroporation.

20. It is very important to snap-freeze the glycerol stocks.
21. It is not a good practice to go to the glycerol stocks every time you need to grow an overnight culture. You can streak the yeast cultures from the glycerol stocks on YAPD plates or SC drop-out plates. This plate can be kept for a maximum of one month in the cold room, wrapped. Take a single colony from this streak to grow an overnight culture for your experiment. After a month, if you need the strain again, go back to the glycerol stocks. When you need to use the glycerol stocks you need to use extreme precaution. Do not ever thaw the glycerol stocks. Before streaking, place the vial you need on dry ice. Using a thin, long, autoclaved wooden stick scrape out some frozen stock in front of the flame and streak on your plate. Close the cap of the vial and return it to the freezer.
22. Expect a total yield of 500 μg–1 mg of protein per 10^8 cells, depending on strain, extent of lysis and total number of cells. *(38)*

References

1. Hehl, E. A., Joshi, P., Kalpana, G. V., et al. (2004) Interaction between human immunodeficiency virus type 1 reverse transcriptase and integrase proteins. *J Virol* 78, 5056–67.
2. Wu, X., Liu, H., Xiao, H., et al. (1999) Human immunodeficiency virus type 1 integrase protein promotes reverse transcription through specific interactions with the nucleoprotein reverse transcription complex. *J Virol* 73, 2126–35.
3. Zhu, K., Dobard, C., Chow, S. A. (2004) Requirement for integrase during reverse transcription of human immunodeficiency virus type 1 and the effect of cysteine mutations of integrase on its interactions with reverse transcriptase. *J Virol* 78, 5045–55.
4. Cherepanov, P., Maertens, G., Proost, P., et al. (2003) HIV-1 integrase forms stable tetramers and associates with LEDGF/p75 protein in human cells. *J Biol Chem* 278, 372–81.
5. Kalpana, G. V., Marmon, S., Wang, W., et al. (1994) Binding and stimulation of HIV-1 integrase by a human homolog of yeast transcription factor SNF5. *Science* 266, 2002–6.
6. Goff, S. P. (2001) Intracellular trafficking of retroviral genomes during the early phase of infection: viral exploitation of cellular pathways. *J Gene Med* 3, 517–28.
7. Luban, J., Bossolt, K. L., Franke, E. K., et al. (1993) Human immunodeficiency virus type 1 Gag protein binds to cyclophilins A and B. *Cell* 73, 1067–78.
8. Berger, E. A., Murphy, P. M., Farber, J. M. (1999) Chemokine receptors as HIV-1 coreceptors: roles in viral entry, tropism, and disease *Annu Rev Immunol* 17, 657–700.
9. Goff, S. P. (2004) Genetic control of retrovirus susceptibility in mammalian cells. *Annu Rev Genet* 38, 61–85.
10. Sorin, M., Kalpana, G. V. (2006) Dynamics of virus-host interplay in HIV-1 replication. *Curr HIV Res* 4, 117–30.
11. Nermut, M. V., Fassati, A. (2003) Structural analyses of purified human immunodeficiency virus type 1 intracellular reverse transcription complexes. *J Virol* 77, 8196–206.
12. Bukrinsky, M. (2004) A hard way to the nucleus. *Mol Med* 10, 1–5.
13. Iordanskiy, S., Berro, R., Altieri, M., et al. (2006) Intracytoplasmic maturation of the human immunodeficiency virus type 1 reverse transcription complexes determines their capacity to integrate into chromatin. *Retrovirology* 3, 4.
14. Turelli, P., Doucas, V., Craig, E., et al. (2001) Cytoplasmic recruitment of INI1 and PML on incoming HIV preintegration complexes: interference with early steps of viral replication. *Mol Cell* 7, 1245–54.
15. Vandegraaff, N., Devroe, E., Turlure, F., et al. (2006) Biochemical and genetic analyses of integrase-interacting proteins lens epithelium-derived growth factor (LEDGF)/p75 and hepatoma-derived growth factor related protein 2 (HRP2) in preintegration complex

function and HIV-1 replication. *Virology* 346, 415–26.
16. Ciuffi, A., Bushman, F. D. (2006) Retroviral DNA integration: HIV and the role of LEDGF/p75. *Trends Genet* 22, 388–95.
17. Garber, M. E., Wei, P., Jones, K. A. (1998) HIV-1 Tat interacts with cyclin T1 to direct the P-TEFb CTD kinase complex to TAR RNA. *Cold Spring Harb Symp Quant Biol* 63, 371–80.
18. Wei, P., Garber, M. E., Fang, S. M., et al. (1998) A novel CDK9-associated C-type cyclin interacts directly with HIV-1 Tat and mediates its high-affinity, loop-specific binding to TAR RNA. *Cell* 92, 451–62.
19. Ariumi, Y., Serhan, F., Turelli, P., et al. (2006) The integrase interactor 1 (INI1) proteins facilitate Tat-mediated human immunodeficiency virus type 1 transcription. *Retrovirology* 3, 47.
20. Kamine, J., Chinnadurai, G. (1992) Synergistic activation of the human immunodeficiency virus type 1 promoter by the viral Tat protein and cellular transcription factor Sp1. *J Virol* 66, 3932–6.
21. Mahmoudi, T., Parra, M., Vries, R. G., et al. (2006) The SWI/SNF chromatin-remodeling complex is a cofactor for Tat transactivation of the HIV promoter. *J Biol Chem* 281, 19960–8.
22. Nekhai, S., Jeang, K. T. (2006) Transcriptional and post-transcriptional regulation of HIV-1 gene expression: role of cellular factors for Tat and Rev. *Future Microbiol* 1, 417–26.
23. Pollard, V. W., Malim, M. H. (1998) The HIV-1 Rev protein. *Annu Rev Microbiol* 52, 491–532.
24. Garrus, J. E., von Schwedler, U. K., Pornillos, O. W., et al. (2001) Tsg101 and the vacuolar protein sorting pathway are essential for HIV-1 budding. *Cell* 107, 55–65.
25. Zimmerman, C., Klein, K. C., Kiser, P. K., et al. (2002) Identification of a host protein essential for assembly of immature HIV-1 capsids. *Nature* 415, 88–92.
26. Fujii, K., Hurley, J. H., Freed, E. O. (2007) Beyond Tsg101: the role of Alix in 'ESCRTing' HIV-1. *Nat Rev Microbiol* 5, 912–6.
27. Chien, C. T., Bartel, P. L., Sternglanz, R., et al. (1991) The two-hybrid system: a method to identify and clone genes for proteins that interact with a protein of interest. *Proc Natl Acad Sci U S A* 88, 9578–82.
28. Chong, J. A., Mandel, G. (1997) in (Bartel, P. L., and Fields, S., Eds.), *The Yeast Two-Hybrid System*, pp. 289–97, Oxford University Press, New York, NY.
29. Licitra, E. J., Liu, J. O. (1996) A three-hybrid system for detecting small ligand-protein receptor interactions. *Proc Natl Acad Sci U S A* 93, 12817–21.
30. Zhang, B., Kraemer, B., Sengupta, D., et al. (1997) *in* "The Yeast Two-Hybrid System" (Bartel, P. L., and Fields, S., Eds.), pp. 298–315, Oxford University Press, New York, NY.
31. Osborne, M. A., Dalton, S., Kochan, J. P. (1995) The yeast tribrid system–genetic detection of trans-phosphorylated ITAM-SH2-interactions. *Biotechnology (N Y)* 13, 1474–8.
32. Tirode, F., Malaguti, C., Romero, F., et al. (1997) A conditionally expressed third partner stabilizes or prevents the formation of a transcriptional activator in a three-hybrid system. *J Biol Chem* 272, 22995–9.
33. Wang, H., Peters, G. A., Zeng, X., et al. (1995) Yeast two-hybrid system demonstrates that estrogen receptor dimerization is ligand-dependent in vivo. *J Biol Chem* 270, 23322–9.
34. Vidal, M., Brachmann, R. K., Fattaey, A., et al. (1996) Reverse two-hybrid and one-hybrid systems to detect dissociation of protein-protein and DNA-protein interactions. *Proc Natl Acad Sci U S A* 93, 10315–20.
35. Yamada, K., Senju, S., Shinohara, T., et al. (2001) Humoral immune response directed against LEDGF in patients with VKH. *Immunol Lett* 78, 161–8.
36. Cheng, S. W., Kalpana, G. V. Unpublished observations.
37. Cheng, S. W., Davies, K. P., Yung, E., et al. (1999) c-MYC interacts with INI1/hSNF5 and requires the SWI/SNF complex for transactivation function. *Nat. Genet.* 22, 102–5.
38. Lee, D., Sohn, H., Kalpana, G. V., et al. (1999) Interaction of E1 and hSNF5 proteins stimulates replication of human papillomavirus DNA. *Nature* 399, 487–91.
39. Morozov A, Y. E., Kalpana GV (1998) Structure-function analysis of integrase interactor 1/hSNF5L1 reveals differential properties of two repeat motifs present in the highly conserved region. *Proc Natl Acad Sci U S A* 95, 1120–5.
40. Yung, E., Sorin, M., Pal, A., et al. (2001) Inhibition of HIV-1 virion production by a transdominant mutant of integrase interactor 1. *Nat Med* 7, 920–6.
41. Kalpana, G. V., Reicin, A., Cheng, G. S., et al. (1999) Isolation and characterization of an oligomerization-negative mutant of HIV-1 integrase. *Virology* 259, 274–85.
42. Fields, S. (2005) High-throughput two-hybrid analysis. The promise and the peril. *Febs J* 272, 5391–9.

43. Bouhamdan, M., Benichou, S., Rey, F., et al. (1996) Human immunodeficiency virus type 1 Vpr protein binds to the uracil DNA glycosylase DNA repair enzyme *J Virol* 70, 697–704.
44. Engebrecht, J., Brent, R., Kaderbhai, M. A. (1991) in (Ausubel, F. M., Brent, R., Kingston, R. E., Moore, D. D., Seidman, J. G., Smith, J. A., and Struhl, K., eds.), *Current Protocols in Molecular Biology*, John Wiley and Sons.
45. Scopes, R. K., Smith, J. A. (2006) in (Ausbel, F. M., Brent, R., Kingston, R. E., Moore, D. D., Seidman, J. G., Smith, J. A., and Struhl, K., eds.), *Current Protocols in Molecular Biology*, pp. 10.0.1, John Wiley and Sons, Inc.

Chapter 20

Methods to Study Monocyte Migration Induced by HIV-Infected Cells

Vasudev R. Rao, Eliseo A. Eugenin, Joan W. Berman, and Vinayaka R. Prasad

Abstract

HIV-associated dementia (HAD) is a multi-factorial disease set in motion by the presence of HIV-infected cells in the brain. A characteristic feature of HAD is the infiltration of mononuclear phagocytes into the brain, which is aided by HIV-1 Tat protein and other chemokines secreted by both HIV-infected cells and uninfected cells in their vicinity. Both direct and indirect chemokine activity of HIV-1 Tat protein has been demonstrated employing purified recombinant Tat protein. However, a corroboration of a key role for Tat or other chemokines in monocyte migration, in the context of HIV-infection, has not yet been demonstrated. Here we describe methods, to measure the role of soluble factors, such as chemokines and Tat, released by HIV-infected cells or uninfected cells in their vicinity, in monocyte migration in vitro.

Key words: HIV-associated dementia/monocyte migration/monocyte chemotaxis/HIV-1 Tat/monocyte chemotaxis protein-1/chemokines

1. Introduction

HIV-associated dementia (HAD) affects a significant portion of HIV-infected individuals *(1)*. Although the incidence of HAD has decreased with the advent of HAART *(2, 3)*, the increased survival rates of HIV-infected individuals has led to an increase in the prevalence of cognitive impairment and a greater risk of HAD, in part as a result of chronic exposure of central nervous system (CNS) to HIV *(2, 4, 5)*. Infiltration of monocytes into the brain is a key mechanism of neuronal injury. HIV-infected monocytes, upon their migration to the brain, further enhance monocyte infiltration both by direct monocyte chemotaxis and by inducing

Vinayaka R. Prasad, Ganjam V. Kalpana (eds.), *HIV Protocols: Second Edition, vol. 485*

DOI 10.1007/978-1-59745-170-3_20 Springerprotocols.com

the synthesis of chemokines such as CCL2 (MCP-1, monocyte chemoattractant protein-1) *(6–8)*. HIV-infected cells within the CNS also promote enhanced transmigration of HIV-infected monocytes across the blood brain barrier (BBB) by up regulation of CCR2, the receptor for CCL2, that allows HIV-infected leukocytes to sense lower amounts of CCL2 in the brain. *(9)*. The presence of HIV-infected monocytes in the brain leads to the infection of resident CNS cells, such as microglia and astrocytes. Although neurons are not productively infected by HIV, neuronal damage results from HIV-infected macrophage/microglia and astrocytes *(10, 11)*. In response to these infected CNS cells, inflammatory cascades are propagated within the brain, including the production of excitotoxins, chemokines, cytokines and viral proteins from infected and/or activated cells in the CNS. This virally mediated inflammation is a major contributor to neuronal damage and subsequent BBB compromise *(12–15)*. These factors collectively lead to chronic damage to the CNS circuits resulting in alterations in behavior and cognitive impairment typically observed in HIV-infected individuals.

One of these neurotoxins is the HIV transcriptional transactivator, Tat, that is essential for viral replication *(16–23)* and is secreted or released from infected cells into the extracellular environment. Tat has been demonstrated to play a key role in the development of HAD, including by both direct and indirect mechanisms on monocyte/macrophage chemotaxis. Tat contains a dicysteine motif that has been previously shown to be responsible for its direct monocyte chemotaxis activity in in vitro assays employing recombinant purified Tat *(24–26)*. In addition, treatment with Tat protein induces astrocytes and the cells of the monocyte/macrophage lineage to secrete chemokines like CCL2 *(25–28)*.

Recently, we demonstrated that the Tat protein of subtype C HIV-1 is divergent and is characterized by a specific set of signature residues characteristic to this subtype, but distinct from Tat proteins found in nonsubtype C HIV-1 isolates such as those prevalent in the US (subtype B). Importantly, the dicysteine motif (residues 30 and 31) is highly conserved in all HIV-1 subtypes examined except subtype C, which exhibited a C31S polymorphism that is conserved in $\sim 90\%$ of the sequences examined. We hypothesized that this alteration in Tat sequence renders it defective for chemokine function, reducing or preventing CNS inflammation and chemotaxis of leukocytes into the brain *(29)*. In vitro studies using purified recombinant Tat proteins proved that the subtype C Tat protein was indeed defective for monocyte chemotaxis and that this could be reversed by an S31C mutation that restores the dicysteine motif in subtype C Tat protein. Biological evidence in human studies or animal models is yet to be obtained to demonstrate a direct role for this polymorphism in disease incidence.

Although the in vitro results with Tat protein are consistent with a lower neuropathogenesis by subtype C virus, several questions remain. For example, it is not known what are the levels of Tat protein released by HIV-infected cells in vitro or whether the levels that accumulate in the medium are sufficient to induce monocyte chemotaxis. There are no sensitive methods to accurately measure extracellular Tat protein. We have taken a qualitative approach to measure the ability of HIV-infected cells to recruit monocytes and to study whether monocyte migration is mediated by Tat and/or other soluble factors. We describe below, an in vitro procedure to measure chemotaxis in the context of HIV-infected cells. We also describe procedures to measure monocyte chemotaxis induced by the media from HIV-infected cells to identify the soluble factors in the medium that stimulate chemotaxis. Such migration assays should stimulate future studies to delineate viral or virally induced factors important for chemotaxis as well as to measure differences among HIV isolates that contain polymorphisms in viral proteins involved in chemotaxis.

2. Materials

2.1. Plasmids

Infectious molecular clones of HIV-1 to be compared can be obtained from the NIH AIDS reagent program. If molecular clones are not available, primary biological virus isolates can be obtained. Such isolates are likely to be genetically more heterogeneous. Thus, one needs to be sure to use the same stocks for all migration experiments.

2.2. Virus Production and Quantitation of Viral Stock

1. 293T cells (ATCC CRL-11268).
2. GenePORTER® 2 Transfection Reagent (Genlantis, Inc., San Diego, CA, USA).
3. DMEM Tissue culture medium with L-glutamine, glucose and Sodium Pyruvate (Mediatech, USA).
4. Pooled fetal bovine serum (Hyclone, USA).
5. Penicillin–Streptomycin [10,000 International Units (IU)/mL Penicillin, 10,000 μg/mL Streptomycin] (Mediatech, USA).
6. HIV-p24 ELISA kit (Perkin Elmer, USA).
7. GHOST® cells- X4/R5 obtained from NIH AIDS reagent program (Catalog number, 3492).

2.3. Macrophage Differentiation and HIV-Infection

1. Elutriated primary human monocytes (Innovative Research, Inc., Novi, MI, USA).

2. Monocyte colony stimulating factor (MCSF) (Sigma-Aldrich, USA).
3. Pooled human AB serum (Sigma-Aldrich)
4. 24-well polysterene-treated tissue culture dishes (Fisher Scientific, USA).
5. Nalgene® Teflon-coated 250 mL flasks (Fisher Scientific).
6. DMEM Tissue culture medium with glucose and Sodium Pyruvate (Mediatech, USA).
7. L-Glutamine-200 mM (Mediatech).
8. Penicillin–Streptomycin-(10,000 IU/mL Penicillin, 10,000 μg/mL Streptomycin, Mediatech, USA).

2.4. HIV-p24 Staining and Elisa to Determine Viral Load

1. 24-well polysterene-treated tissue culture dishes (Fisher Scientific, USA).
2. Round cover slips (Fisher Scientific).
3. Phosphate buffered saline (PBS)(Hyclone, USA).
4. Staining tray.
5. Cold 70% ethanol.
6. Block solution (1 mL 0.5 M EDTA, 100 μL fish gelatin, 0.1 g Immunoglobulin-free 1% BSA, 100 μL 1% Horse serum, 9 mL ddH_2O).
7. HIV-p24 antibody (NIH Repository, USA, catalog No. 4121).
8. Anti-mouse antibody conjugated to FITC (Sigma, USA, catalog No. F2883).
9. Prolong Gold antifade reagent with DAPI (Invitrogen, USA, Catalog No. P36931).
10. HIV-p24 ELISA Kit (Perkin-Elmer, USA, Catalog No. NEK050A001KT).

2.5. Monocyte Migration and Blocking

1. 24-well polystyrene-treated tissue culture dishes (Fisher Scientific).
2. 3-μm pore filter insert (Becton Dickinson Labware, Franklin Lakes, NJ).
3. Carboxy Fluorescein-diacetate Succininidyl Ester (CFSE) (Sigma).
4. Human monocytes (Innovative Research, Inc., Novi, MI, USA).
5. DMEM tissue culture medium with L-glutamine, glucose and Sodium Pyruvate (Mediatech, USA)
6. Neutralizing anti-Tat antibody (P. Venkatesh P., Ranga U. unpublished observations).
7. Neutralizing anti-MCP-1/CCL2 antibody (MAB 279; R & D Systems).
8. IgG isotype matched control antibody (Cappel Pharmaceuticals, Aurora, OH).
9. Pansorbin beads (Calbiochem).

3. Methods

3.1. Virus Production

Infectious HIV-1 virus particles are produced by transient transfection of 293T cells using Geneporter® transfection reagent as follows:

1. Seed 4 × 10^6 293T cells in 100-mm culture dishes 24 h prior to transfection.
2. Replace culture medium with ~ 8 mL fresh, prewarmed 10% DMEM medium 1 h prior to transfection to stimulate cell division. Cells should be no more than 60% confluent.
3. Prepare DNA/Geneporter reagent mix: 10 μg of proviral plasmid DNA in 2.5 mL of serum-free DMEM Medium and 45 μL Geneporter® transfection reagent in 2.5 mL of serum free DMEM Medium are incubated separately for 5 min at room temperature. Then mix well and incubate at room temperature for 45 min.
4. Add the 5 ml DNA, Geneporter®, DMEM mix to 100-mm culture dish with 293T cells.
5. Incubate for 4 h at 37 °C with 5% CO_2.
6. Add 5 mL 20% FBS DMEM medium to the 100-mm culture dish with 293T cells.
7. After 12 h, replace culture medium with 10 mL prewarmed 10% FBS DMEM.
8. Collect supernatants 48 h post-transfection. Centrifuge supernatants at 2,500 rpm for 5 min to clear the cell debris. Add 10 μL of 1 M HEPES Buffer each to 1-mL aliquots of the supernatant and store at −80 °C for future use (*see* **Note 1**).

3.2. Titering the Virus Stocks

Virus stocks are titered using GHOST X4-R5 indicator cell line obtained from the NIH AIDS Repository. GHOST cells are human osteosarcoma *(30)* cells that constitutively express human CD4, and have been genetically engineered to contain the GFP gene linked to the HIV-2 promoter and stably express both CXCR4 (GHOST-X4) and CCR5 (GHOST-R5) These cells are ideal indicators to measure IU of virus preparation obtained from the above step. HIV-1 p24 antigen assays of the supernatant can be done using Perkin-Elmer HIV-p24 ELISA kit (see protocol below). The titer of the viral stocks prepared by transient transfection of 293T cells is measured as follows:

1. Plate 50,000 GHOST X4-R5 cells per well in a six-well plate with the recommended medium containing 90% DMEM, 10% FBS, G418 (500 μg/mL), Hygromycin (100 μg/mL), Penicillin (100 units/mL), Streptomycin (100 μg/mL) and Incubate it at 37 °C overnight with 5% CO_2 (*see* **Note 2**).

2. Add 10, 25, 50 and 100 μL (in triplicates) of the virus supernatant to each well and let the virus adsorption proceed for 2 h. Use three wells with no virus added as uninfected controls.
3. Wash cells with PBS three times following the 2 h adsorption and replace the medium.
4. At 48 h postinfection, trypsinize cells (0.05% Trypsin, 0.5 mM EDTA for 5 min at 37 °C), wash with PBS and resuspend them in 500 μL PBS in 5-mL tubes to be used for flow cytometry.
5. Count the total number of cells in the three uninfected control wells using a hemocytometer and obtain the average number of cells present from the three wells. This number is used as representative of the total number of cells present in each well.
6. Analyze the triplicate HIV-infected cell samples by flow cytometry to obtain the average number of GFP positive cells for each of the virus input volumes (5, 10, 25, 50 and 100 μL).
7. Using the average number of IU observed, the infectious titer (number of infectious virus particles per milliliter) of the virus supernatant is calculated based on all four volumes tested as follows.

The number of infectious virus particles in the sample equals the total number of GFP-positive cells. Therefore, if the number of GFP-positive cells obtained in the 10-μL aliquot tested is z, then the infectious titer per milliliter can be calculated by multiplying z by a factor of 100. Similarly, for the input volumes of 5, 25, 50 and 100, z is multiplied by the factors of 200, 40, 20 or 10 respectively to obtain the putative infectious titer expressed as infectious virus particles per ml. If z turned out to be 11,250 and it was obtained from 10 μL of virus supernatant, then the titer would be 1.125×10^6. The mean titers (x) per triplicates of input volumes tested (y), thus obtained are plotted on a graph and the actual titer is deduced from the linear portion of the resulting graph.

3.3. Inducing Monocyte Differentiation, Procedures for HIV Infection and Titering Virus Infectivity

Monocytes can be obtained from the sources mentioned in the Materials section (from PBMCs or elutriated monocytes). Once a known number of monocytes are obtained, they are induced to differentiate into macrophages by adherence before infecting them with HIV. Prior to performing migration experiments, the infectious titer of the virus should be obtained, especially if different virus isolates are being compared. The replication potential of different HIV-1 isolates, strains or subtypes vary. While comparing the potential of different HIV-1 isolates to induce monocyte migration, it is crucial to use macrophages preparations that have equal viral loads and similar proportions of infected cells. This requires special attention. Described below are the protocols to differentiate, infect and quantify infected macrophages prior to setting up the migration experiments.

1. Human monocytes are incubated at 37 °C for 7 days in DMEM, 10% Human serum, Penicillin-Streptomycin (100 Units/mL, 100 μg/mL), 1X L-glutamine (2 mM) and monocyte colony stimulating factor (M-CSF; Sigma) at 6.6 ng/mL in Teflon coated flasks with media changes every third day. At day 7, monocytes in suspension are counted and equal numbers of cells are plated as described below to allow their differentiation into macrophages under adherent conditions. At this stage, 1×10^5 monocytes/well are plated on 24-well plates in triplicates.
2. Twenty-four hours later adherent macrophages are infected at several multiplicities of infection (MOIs) in triplicate; it is also important to try different adsorption times of 1–4 h for each MOI. The effect of the MOI and adsorption time on the viability of macrophages should be monitored and recorded.
3. Following HIV-infection, the macrophages are incubated for 7–14 days and media changes are done every third day. Supernatant from media changes are collected for measuring p24 levels using Perkin-Elmer p24 ELISA kit. Also, after 7 days the cells are stained daily for p24 using the immunocytochemistry protocol described below to monitor the numbers of HIV-infected cells.
4. The virus titers, infection and incubation times that reproducibly give similar infectivity levels in terms of the p24 amounts in the supernatant and the percentage p24-positive cells with the least harm to the viability of the macrophages are the ones selected to perform migration experiments. In comparing two different isolates of HIV-1 in monocyte migration, it is important to select the conditions that lead to comparable p24 levels in the medium and similar proportions of p24-positive cells (*see* **Note 3**).

3.4. Staining of Macrophages for Intracellular HIV-1 p24 and quantifying Extracellular p24

To compare chemotactic abilities of two or more virus subtypes, it is crucial to employ HIV-infected macrophages that contain equal numbers of HIV-infected cells using HIV-1 p24 staining and similar viral loads as determined by ELISA as markers of infectivity. Below is the protocol for p24 staining of the HIV-infected macrophages that enables the quantification of the number of HIV-infected cells. Also described is a brief protocol for HIV-1 p24 ELISA.

3.4.1. HIV-1 p24 Staining Protocol

1. 1×10^5 HIV-1-infected human macrophages, prepared as above, are plated in 24-well plates on cover slips for at least 24 h
2. Aspirate media and wash cells once with 1X PBS.
3. Add 1 mL cold 70% Ethanol to fix and permeabilize cells for at least 20 min at −20 °C.

4. Wash each well —three to five times in 1X PBS. Fixed cells can be stored at 4 °C for 1 month.
5. Prepare a humid staining tray; lay out paper towels on a plastic tray and soak in 1X PBS. Make sure to flatten all ridges/bumps in the towels. Lay out two thick strips of wax paper/parafilm on top of the towels.
6. Label wax paper to keep track of cover slips.
7. 100-μL droplet of Block Solution (prepared as mentioned in the Materials section) is placed on to the wax paper for each cover slip.
8. Using forceps, transfer the cover slips from the wells, on to the corresponding block droplet, making sure that the cells are facing down and come in contact with block solution.
9. Block for at least 1 h at room temperature or overnight.
10. Prepare 1:50 dilution of HIV-1 p24 antibody (NIH Repository) in the block solution during the 1 h incubation and add 100-μL droplets to a labeled wax strip placed on a different staining tray. It is important to include an isotype matched control antibody to confirm the specificity of the staining.
11. Carefully move the cover slips from the blocking tray to the appropriate droplet of antibody. Incubate overnight at 4 °C.
12. Move the cover slips back to the 24-well plate and wash them 5X times with PBS to eliminate the excess or unbound primary antibody.
13. Prepare the staining tray with a 1:400 dilution of the secondary mouse conjugated FITC antibody (Sigma) in a similar manner as with the primary p24 antibody. Using forceps transfer the cover slips from the wells, on to the corresponding secondary FITC droplet.
14. Incubate for 3 h at room temperature in the dark.
15. Move the cover slips back to the 24-well plate and wash them 5X times with PBS to eliminate the excess or unbound secondary fluorescent antibody.
16. Mount cover slips on glass slides using a drop of Prolong Gold Antifade Reagent with DAPI (Molecular Probes).
17. Using a fluorescent microscope, calculate the percentage of p24 positive cells versus the total number of cells in the field determined by the DAPI staining to determine infectivity of the macrophage culture.

3.4.2. HIV p-24 ELISA Protocol (Summarized from Perkin-Elmer Manual)

1. Supernatants from post-infection media changes stored at −80 °C are used to measure HIV p-24 levels using HIV-1 p24 ELISA Kit (Perkin-Elmer, USA).
2. Thaw the samples on ice in a cell culture hood. Equilibrate all kit components to room temperature. Prepare multiple dilutions of the positive control to obtain a standard curve required for calculating the final p24 values from

the supernatants. Prepare 1:500 and 1:2,500 dilutions of the supernatant in order to get values in the range of the standard curve.

3. Prior to starting the assay, calculate the number of wells needed for the different samples and add 20 μL Triton-X 100 to all wells except the negative substrate blank. Add 200 μL of the positive control and sample dilutions to the appropriate wells and incubate for 2 h at 37 °C.
4. Wash plate with diluted wash buffer for 6 times and add 100 μL detector antibody except the negative substrate blank. Seal and incubate for 1 h at 37 °C.
5. Repeat the washes following the incubation. Dilute Streptavidin-HRP concentrate in the diluent provided and add 100 μL of the diluted SA-HRP to the wells except the negative substrate blank. Seal and incubate the plate at room temperature for 30 min.
6. Repeat the washes following the incubation. Prepare OPD substrate solution using the components of the kit. Add 100 μL of the OPD substrate solution to each of the wells and seal the plate. Incubate in dark at room temperature for 30 min.
7. Add 100 μL of the Stop solution and read the plate at 490 nm using an ELISA plate reader. Plot a standard curve from and extrapolate the values for individual samples (*see* **Note 4**).

Percentage of p24 positive cells as determined by p24 staining and p24 ELISA values are used to determine equality of viral loads prior to performing migration experiments to compare differential chemotactic abilities of two or more viruses.

3.5. Monocyte Migration

HIV-infected macrophage cultures with comparable levels of infectivity and virus replication as determined by p24 staining of infected cells and HIV-p24 ELISA of the medium, respectively, are used to examine migration using different virus isolates or subtypes. In our experiments, we compared two different subtypes of HIV-1: subtype B (HIV-1_{ADA}), and subtype C (HIV-$1_{IndieC1}$).

Migration assays are performed in a Boyden chamber, by using a 3-μm filter insert to separate the two chambers to create a gradient that will allow monocytes to migrate to the chamber where the concentration of the chemokine is highest (lower chamber) (see **Fig. 20.1**). The filter size chosen reflects the fact that the average size of the cells in the upper chamber (monocytes - ~ 10 μm) is much larger than the filter pore size. Thus, the set up requires active podocytosis to allow cells to migrate to the lower chamber.

1. Approximately 1×10^5 HIV-1 infected human macrophages, prepared as above, are plated in 24-well plates. These

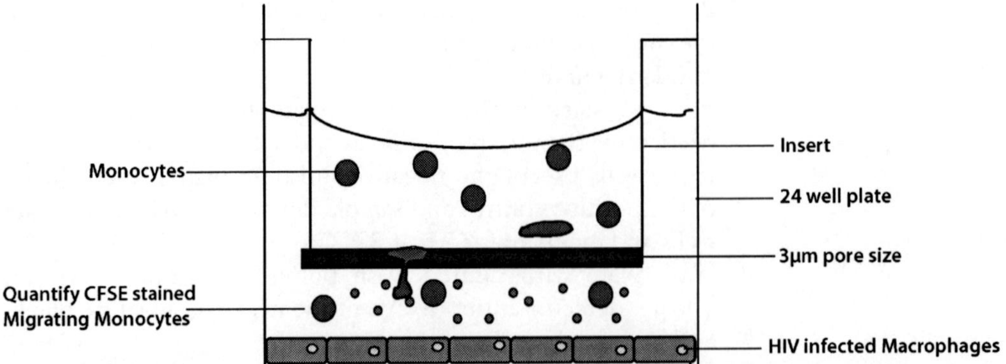

Fig. 20.1. Schematic overview of the monocyte migration assay. Uninfected Monocytes imbued with CFSE dye, actively migrate from the *upper chamber* to the *lower chamber* along a chemoattractant concentration gradient induced by the HIV-infected macrophages plated in the *bottom chamber*. The average size of the monocytes is 7–12 μm. The pore size of the filter between the two chambers is 3 μm, which ensures that the cells that are detected in the lower chamber are due to active migration.

uninfected or HIV-infected adherent cells and the factors released by them, including Tat and CCL2 are the chemotactic factors that will promote the migration of monocytes from the upper chamber to the lower chamber (see **Fig. 20.1**).

2. At this point two different experimental approaches can be used. First, one can use the macrophages directly and test their ability to cause chemotaxis. Second, one can use the conditioned medium from these macrophage cultures and use them to perform the chemotaxis. The second approach is better for characterizing specific factors and design neutralizing protocols, because you will know the concentration of the factors that will be examined (see below).
3. In the upper chambers (using inserts), 2×10^5 undifferentiated fluorescent monocytes are added (see **Fig. 20.1**) (*see* **Note 5**).
4. After 24 h incubation to allow migration of cells from the upper to the lower chamber, cells in the lower chamber are examined in a fluorescence microscope to identify and quantify the numbers of migrated cells. Also FACS to detect fluorescent cells can be used to quantify the numbers of transmigrated cells.
5. The nonadherent cells in the supernatant from the lower chamber are counted first under a fluorescent microscope using a hemocytometer and lower chamber is observed under a fluorescent microscope to make sure that there are no adherent fluorescent cells.
6. Average numbers of migrating cells are calculated and the numbers of cells migrated in response to the presence of macrophages infected by different viral strains are compared with each other (see **Fig.** 20.2 for typical data on differential

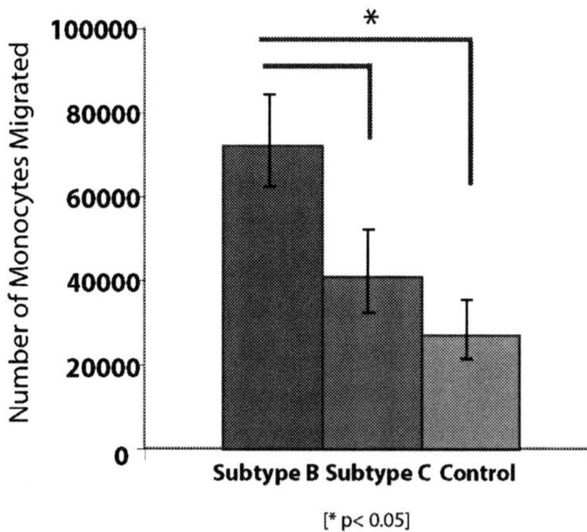

Fig. 20.2. Monocyte migration assays in the presence of HIV-1 infected monocyte-derived macrophages (MDM). Monocyte migration caused by subtype-B and subtype-C HIV-infected MDM. The numbers plotted reflect the monocytes detected in the lower chamber by counting the CFSE-positive cells. Subtype-B HIV recruits twice the number of monocytes when compared to subtype-C.

induction of migration data induced by the two representative viral strains HIV-1_{ADA} vs. HIV-$1_{IndieC1}$).

3.6. Immunodepletion of Soluble Factors that Promote Monocyte Migration

Blocking the migration of monocytes using antibodies against chemokines and/or Tat will help determine the relative contributions of the many important chemoattractants in the migration process. Neutralization experiments will help one better understand the differential migration induction abilities of different viral strains and also in devising strategies to control virally induced monocyte infiltration. Below, we detail the experimental approach to examine the role of chemokine- and Tat-induced migration, using specific antibodies.

1. Uninfected and HIV-infected cultures of human adherent macrophages are cultured in 24-well plates for 2–5 days (refer to **Section 3.3** for HIV-infection protocol). Every day, conditioned medium is obtained from these cultures and centrifuged (2,000 rpm, 3 min, 4 °C) to eliminate cell debris. Supernatants are frozen and stored for migration experiments (*see* **Note 6**).
2. To obtain the conditioned medium, low serum is required, because human serum contains high amounts of CCL2 and/or other factors that may induce the production of CCL2 or other monocyte chemoattractants by macrophages unrelated to HIV-infection. To perform the migration experiments, slowly defrost the frozen supernatant on ice. The medium is precleared by adding 20 μL of prewashed Pansorbin beads (Calbiochem) followed by 20 min incubation on ice. This step

is critical to clear the conditioned medium of any proteins that can bind nonspecifically to the beads.

3. After pre-clearing, the medium containing the beads is centrifuged at 1,000 rpm for 2 min at 4 °C and the supernatant is collected and the pellet is discarded.
4. 350 μL of conditioned medium is incubated for 1 h on ice with Pansorbin beads (Calbiochem) to which either anti-Tat antibodies E1.1 (250 ng), which neutralizes both subtype B and C Tat proteins (P. Venkatesh and U. Ranga, personal communication) or anti-CCL2 (250 ng, MAb 279; R & D Systems) neutralizing antibodies have been pre-bound (*see* **Note 7**). The tubes are centrifuged to pellet the Pansorbin-neutralizing antibody complexed with the respective antigen. The immuno-depleted supernatant, with minimal levels of CCL2 or Tat (depending of the neutralizing antibody used) is used as the chemoattractant medium in the lower chamber for the migration experiments as described above in Section 3.5. It is important to include isotype matched control antibodies. The untreated medium is used as positive control (see **Fig. 20.3** for a schematic representation of the neutralization assay and **Fig. 20.4** for data from a typical experiment blocking migration induced by subtype B HIV-1_{ADA}).

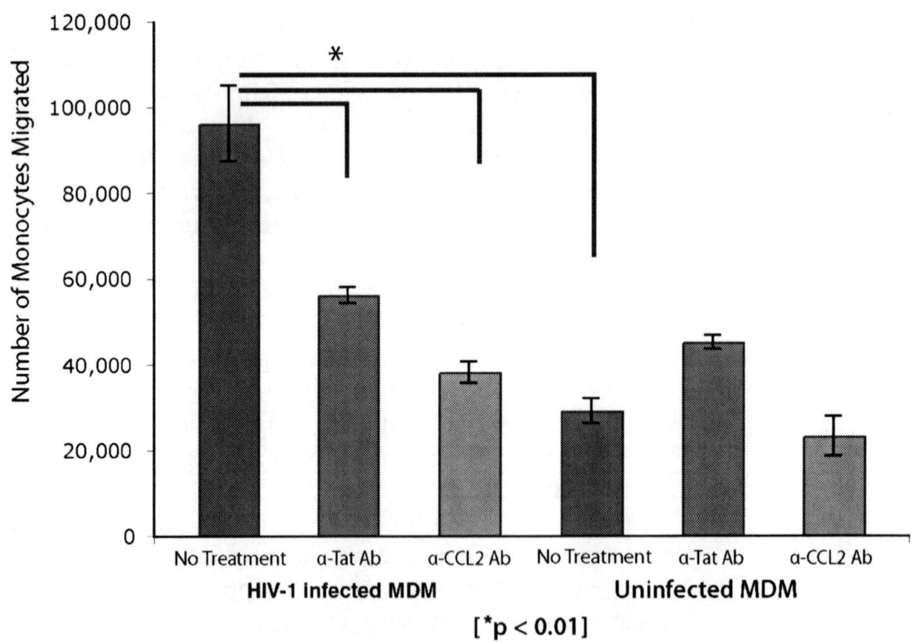

Fig. 20.3. Monocyte migration caused by spent medium from HIV-infected MDMs can be blocked by anti-Tat or anti-CCL2 antibodies. The numbers represent fluorescent cells detected in the lower chamber after 24 h incubation. Bars labeled 'anti-Tat Ab' and 'anti-CCL2' represent experiments in which media from HIV-infected MDMs were first treated with respective antibodies. This demonstrates monocyte migration is clearly mediated by HIV-1 Tat and HIV infection induced CCL2.

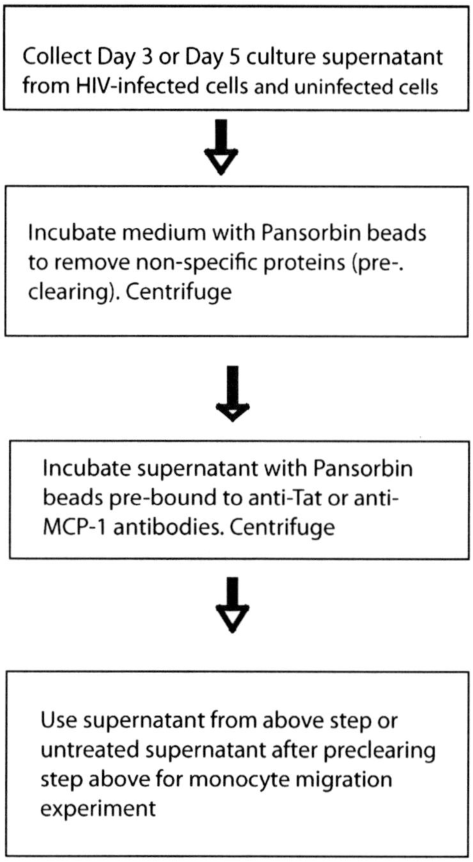

Collect Day 3 or Day 5 culture supernatant from HIV-infected cells and uninfected cells

Incubate medium with Pansorbin beads to remove non-specific proteins (pre-. clearing). Centrifuge

Incubate supernatant with Pansorbin beads pre-bound to anti-Tat or anti-MCP-1 antibodies. Centrifuge

Use supernatant from above step or untreated supernatant after preclearing step above for monocyte migration experiment

Fig. 20.4. A schematic representation of the key steps involved in immunodepletion of soluble factors that promote monocyte migration assay.

4. Notes

1. 48 h post-transfection time point is determined for HIV-1_{ADA} and HIV-1_{IndieC} to be ideal for maximum virus yield. When working with different viruses, be sure to collect supernatants at various time points 6 h apart from 36 to 72 h and measure p24 values to determine what time point yields the maximum p24 measured by ELISA.
2. Observe the GHOST cells under a fluorescent microscope to make sure there is not a high amount of background fluorescence prior to infection. Ghost cells can be passaged for a maximum of three cycles for use in infectivity experiments.
3. Different MOI's need to be titered to achieve equal viral loads with two or more different viruses. Also keep in mind that sometimes there might be cell death with higher MOIs.

4. If the recommended dilutions of the samples do not yield values which correspond to the standard curve depending on whether the values or too high or too low increase or decrease the dilution factor.
5. Monocytes are infused with fluorescent 200 nM CFSE (Sigma) by incubating in 5% FBS-PBS for 15 min at 37 °C in a volume of 1 mL. Spin down the cells at 1,800 rpm for 10 min and wash twice with PBS to obtain fluorescent monocytes to be used for migration assays.
6. Collected supernatant can be used for multiple experimental procedures including the determination of infection by HIV-p24 ELISA, endogenous levels of chemokines or cytokines, neutralization of viral and immune factors and to examine unknown factors released from cultures of uninfected and HIV-infected macrophages by many approaches such as proteomics or arrays. In general these supernatants are stable for 3 months.
7. Antibody concentrations are titered prior to doing the neutralization experiments. Also, amounts of CCL2 and other secretory chemokines can be determined prior to the neutralization experiments by ELISA to get an idea of how much antibody is needed for complete neutralization.

References

1. McArthur, J. C., Hoover, D. R., Bacellar, H., et al. (1993) Dementia in AIDS patients: incidence and risk factors. Multicenter AIDS Cohort Study. *Neurology* 43, 2245–52.
2. Sacktor, N., Lyles, R. H., Skolasky, R., et al. (2001) HIV-associated neurologic disease incidence changes:: Multicenter AIDS Cohort Study, 1990–1998. *Neurology* 56, 257–60.
3. Dore, G. J., McDonald, A., Li, Y., et al. (2003) Marked improvement in survival following AIDS dementia complex in the era of highly active antiretroviral therapy. *Aids* 17, 1539–45.
4. Langford, T. D., Letendre, S. L., Larrea, G. J., et al. (2003) Changing patterns in the neuropathogenesis of HIV during the HAART era. *Brain Pathol* 13, 195–210.
5. Sperber, K., Shao, L. (2003) Neurologic consequences of HIV infection in the era of HAART. *AIDS Patient Care STDS* 17, 509–18.
6. Gourdou, I., Mabrouk, K., Harkiss, G., et al. (1990) [Neurotoxicity in mice due to cysteine-rich parts of visna virus and HIV-1 Tat proteins]. *C R Acad Sci III* 311, 149–55.
7. Sabatier, J. M., Vives, E., Mabrouk, K., et al. (1991) Evidence for neurotoxic activity of tat from human immunodeficiency virus type 1. *J Virol* 65, 961–7.
8. Nath, A., Psooy, K., Martin, C., et al. (1996) Identification of a human immunodeficiency virus type 1 Tat epitope that is neuroexcitatory and neurotoxic. *J Virol* 70, 1475–80.
9. Eugenin, E. A., Osiecki, K., Lopez, L., et al. (2006) CCL2/monocyte chemoattractant protein-1 mediates enhanced transmigration of human immunodeficiency virus (HIV)-infected leukocytes across the blood-brain barrier: a potential mechanism of HIV-CNS invasion and NeuroAIDS. *J Neurosci* 26, 1098–106.
10. King, J. E., Eugenin, E. A., Buckner, C. M., et al. (2006) HIV tat and neurotoxicity. *Microbes Infect* 8, 1347–57.
11. Eugenin, E. A., King, J. E., Nath, A., et al. (2007) HIV-tat induces formation of an LRP-PSD-95- NMDAR-nNOS complex that promotes apoptosis in neurons and astrocytes. *Proc Natl Acad Sci U S A* 104, 3438–43.
12. Genis, P., Jett, M., Bernton, E. W., et al. (1992) Cytokines and arachidonic metabolites produced during human immunode-

ficiency virus (HIV)-infected macrophage-astroglia interactions: implications for the neuropathogenesis of HIV disease. *J Exp Med* 176, 1703–18.
13. Nottet, H. S., Gendelman, H. E. (1995) Unraveling the neuroimmune mechanisms for the HIV-1-associated cognitive/motor complex. *Immunol Today* 16, 441–8.
14. Persidsky, Y., Gendelman, H. E. (2003) Mononuclear phagocyte immunity and the neuropathogenesis of HIV-1 infection. *J Leukoc Biol* 74, 691–701.
15. Ghafouri, M., Amini, S., Khalili, K., et al. (2006) HIV-1 associated dementia: symptoms and causes. *Retrovirology* 3, 28.
16. Wiley, C. A., Baldwin, M., Achim, C. L. (1996) Expression of HIV regulatory and structural mRNA in the central nervous system. *Aids* 10, 843–7.
17. Sabatier, J. M., Vives, E., Mabrouk, K., et al. (1991) Evidence for neurotoxic activity of tat from human immunodeficiency virus type 1. *J Virol* 65, 961–7.
18. Prendergast, M. A., Rogers, D. T., Mulholland, P. J., et al. (2002) Neurotoxic effects of the human immunodeficiency virus type-1 transcription factor Tat require function of a polyamine sensitive-site on the N-methyl-D-aspartate receptor. *Brain Res* 954, 300–7.
19. Nath, A., Haughey, N. J., Jones, M., et al. (2000) Synergistic neurotoxicity by human immunodeficiency virus proteins Tat and gp120: protection by memantine. *Ann Neurol* 47, 186–94.
20. Haughey, N. J., Nath, A., Mattson, M. P., et al. (2001) HIV-1 Tat through phosphorylation of NMDA receptors potentiates glutamate excitotoxicity. *J Neurochem* 78, 457–67.
21. Jeang, K. T., Xiao, H., Rich, E. A. (1999) Multifaceted activities of the HIV-1 transactivator of transcription, Tat. *J Biol Chem* 274, 28837–40.
22. Ensoli, B., Buonaguro, L., Barillari, G., et al. (1993) Release, uptake, and effects of extracellular human immunodeficiency virus type 1 Tat protein on cell growth and viral transactivation. *J Virol* 67, 277–87.
23. Bonavia, R., Bajetto, A., Barbero, S., et al. (2001) HIV-1 Tat causes apoptotic death and calcium homeostasis alterations in rat neurons. *Biochem Biophys Res Commun* 288, 301–8.
24. Albini, A., Ferrini, S., Benelli, R., et al. (1998) HIV-1 Tat protein mimicry of chemokines. *Proc Natl Acad Sci U S A* 95, 13153–8.
25. Conant, K., Garzino-Demo, A., Nath, A., et al. (1998) Induction of monocyte chemoattractant protein-1 in HIV-1 Tat-stimulated astrocytes and elevation in AIDS dementia. *Proc Natl Acad Sci U S A* 95, 3117–21.
26. Weiss, J. M., Nath, A., Major, E. O., et al. (1999) HIV-1 Tat induces monocyte chemoattractant protein-1-mediated monocyte transmigration across a model of the human blood-brain barrier and up-regulates CCR5 expression on human monocytes. *J Immunol* 163, 2953–9.
27. D'Aversa, T. G., Yu, K. O., Berman, J. W. (2004) Expression of chemokines by human fetal microglia after treatment with the human immunodeficiency virus type 1 protein Tat. *J Neurovirol* 10, 86–97.
28. Lafrenie, R. M., Wahl, L. M., Epstein, J. S., et al. (1996) HIV-1-Tat protein promotes chemotaxis and invasive behavior by monocytes. *J Immunol* 157, 974–7.
29. Ranga, U., Shankarappa, R., Siddappa, N. B., et al. (2004) Tat protein of human immunodeficiency virus type 1 subtype C strains is a defective chemokine. *J Virol* 78, 2586–90.
30. Hetherington, H., Kuzniecky, R., Pan, J., et al. (1995) Proton nuclear magnetic resonance spectroscopic imaging of human temporal lobe epilepsy at 4.1 T. *Ann Neurol* 38, 396–404.

Chapter 21

Novel Mouse Models for Understanding HIV-1 Pathogenesis

Aviva Joseph, Kaori Sango, and Harris Goldstein

Abstract

Small animal models in which in vivo HIV-1 infection, pathogenesis, and immune responses can be studied would permit both basic research on the biology of the disease, as well as a system to rapidly screen developmental therapeutics and/or vaccines. To date, the most widely-used models have been the severe combined immunodeficient (SCID)-hu (also known as the thy/liv SCID-hu) and the huPBL-SCID mouse models. Recently three new models have emerged, i.e., the intrasplenic huPBL/SPL-SCID model, the NOD/SCID/IL2Rγ^{null} mouse model, and the Rag2$^{-/-}\gamma_c^{-/-}$ mouse model. Details on the construction, maintenance and HIV-1 infection of these models are discussed.

Key words: Mouse HIV-1 models, intrasplenic huPBL/SPL-SCID model, NOD/SCID/IL2R γ^{null} mouse model, Rag2$^{-/-}\gamma_c^{-/-}$ mouse model.

1. Introduction

Following the discovery of the HIV-1 virus, great efforts were made to develop rodent models that could mimic the course of the viral infection [reviewed in Ref. *(4)*]. To date, the most commonly used rodent models in HIV-1 investigation are variations of the severe combined immunodeficient (SCID) mouse, i.e., the SCID-hu (also known as the thy-liv SCID) model *(35)* and huPBL-SCID model *(38)*. In the SCID-hu mouse, fragments of human fetal thymus and liver are implanted under the kidney capsules of a SCID mouse. Two to 3 months later, a thymus-like conjoint human organ develops which supports long-term multi-lineage thymopoeisis leading to maturation of human thymocytes and circulation of human T-cells in the mouse peripheral blood *(4)*. When large quantities of fetal thymic and liver tissues are implanted under both kidney capsules, significant numbers of human T cells can be detected in the peripheral blood, spleens,

Vinayaka R. Prasad, Ganjam V. Kalpana (eds.), *HIV Protocols: Second Edition, vol. 485*

DOI 10.1007/978-1-59745-170-3_21 Springerprotocols.com

and lymph nodes *(28)*. Injection of HIV-1 into the implant or intraperitoneally (i.p.) results in the killing, activation and severe depletion of human CD4+ cells within a few weeks, associated with increased viral load *(34)*. After intraimplant injection of HIV-1, significant HIV-1 infection is detected by quantitative coculture, not only in the hu-thy/liv implant, but in the spleen and peripheral blood as well *(14,44)*. Over time, HIV-1 infection is associated with inversion of the CD4/CD8 ratio of peripheral human T cells or almost complete depletion of the CD4 cells *(27)*. A limitation of this model is that, the implanted tissues infected with HIV-1 are of fetal origin, and thus may not necessarily reflect the structure, function and response to HIV infection of their adult counterparts. Also, although circulating human T-cells persist in the peripheral blood for over a year, no mature B cells can be detected *(41,47)*, and no humoral or cellular immune responses to the viral infection, are induced in this model *(4,34)*. In addition, access to human fetal tissue is required to construct the model; proficient surgical skills are needed to be able to implant the tissue under the kidney capsule; and there is a two month delay following surgery, to allow the implanted tissue to grow to sufficient size, before the mouse can be used.

In the huPBL-SCID model, peripheral blood mononuclear cells (PBMCs) are injected into the peritoneum of a SCID mouse resulting in a population of human leukocytes, predominantly T cells, persisting in the peritoneal cavity *(38)*. These cells produce high levels of immunoglobulin (Ig) and are readily infected with HIV-1, with a typical infection persisting for up to 16 weeks associated with a rapid loss of CD4+ T cells *(4,39)*. Over time, however, the range of B and T cell receptors become oligoclonal, T cells become anergic to stimulation, and the ability to generate cytotoxic T lymphocytes (CTLs) against viral antigens is lost *(4,11,37,51,55)*. Although it has been reported by several groups, the capacity of these mice to produce primary immune responses is not definitive *(32,54,55)*. While the HIV-1 huPBL-SCID model is a valuable tool for studying primary infection with HIV-1 virus, its usefulness is limited by the short-term nature of infection and the absence of a robust primary cellular or humoral responses.

To overcome the restrictions of the aforementioned models, three novel mouse models have been developed over the past 2 years. All of these models have been implemented and utilized in our laboratory and so we present here detailed procedures we have adapted from the original reports for making and using them. These models are:

1. The intrasplenic huPBL/SPL-SCID model for the assessment of CTL activity in acute HIV-1 infection;
2. The HIV-1 Rag2$^{-/-}\gamma_c^{-/-}$ mouse model;
3. The HIV-1 NOD/SCID/IL2Rγ^{null} model.

1.1. The Intrasplenic huPBL/SPL-SCID Model for the Assessment of CTL Activity in Acute HIV-1 Infection

The initial control of HIV-1 replication during the acute phase of infection is associated with the generation of HIV-1 epitope-specific CD8+ cytotoxic T lymphocytes *(16,23,29,36,43)*, and the loss of this CTL activity and/or viral escape from CTL recognition is associated with increased viral loads and accelerated progression to AIDS *(17,20,45,59)*. A strong correlation exists between the potency and specificity of this response and: (a) viral load and clinical outcome *(40,42)*; (b) delay or arrest in disease progression in non-progressors *(10,18,46)*; and (c) protection from infection in some HIV-1-exposed individuals *(8,31,48,49)*. Understanding the in vivo correlates of protection conferred by HIV-1-specific CTLs would be greatly facilitated by the development of an in vivo model of acute infection in which novel therapies and basic research can be performed. Other requirements include that the model be cheap, relatively simple to make, capable of high-throughput studies and have a high degree of experimental reproducibility. A model that fulfills these criteria is the intrasplenic huPBL/SPL-SCID model for the assessment of CTL activity in acute HIV-1 infection *(24,25)*. This model can be used to evaluate in vivo anti-HIV CTL activity during acute HIV-1 infection. PBMCs, isolated from a leukopack, are acutely infected with HIV-1 and then injected into the spleens of SCID mice with or without HLA matched or mismatched HIV-1 specific CTLs. The procedure takes about 10 min per mouse and since so many PBMCs are isolated from one leukopack, it is possible to generate over 20 mice from the same donor resulting in very reproducible results. One week later the spleen is harvested and the number of HIV-1-infected PBMCs is determined by limiting dilution co-culture. The human PBMCs persist in the mouse for at least 2 weeks, are readily detectable by flow cytometry and constitute over 1% of the leukocytes in the SCID mouse spleen.

The model is very similar to the huPBL-SCID mouse, but while the human PBMCs are injected intraperitoneally in the huPBL-SCID model and are predominantly localized to the peritoneal cavity, the hu-PBL/SPL-SCID mouse model utilizes intrasplenic injections. An advantage of directly injecting the CTLs and PBMCs into the spleen is that it localizes both cells into one distinct lymphoid organ, which increases the probability of the two cell types interacting, as opposed to the disseminated and diffuse cell distribution of the cells among the organs when injecting i.p. *(24,25)*. Another advantage of this model is that the location where the CTL and infected cells interact is in a lymphoid organ which parallels what occurs in HIV-1 infected individuals.

1.2. The HIV-1 Rag2$^{-/-}\gamma_c^{-/-}$ Mouse Model

Since the initial description of the humanized Rag2$^{-/-}\gamma_c^{-/-}$ mouse model *(13,33,56,58)*, several reports have demonstrated that this mouse model can be used in the study of HIV-1 pathogenesis *(1–3,15,50,60)*. The interleukin-2

(IL-2) receptor γ chain, referred to as the "gamma common (γ_c) chain", plays a critical role in lymphoid development, due to its role as a constituent of the IL-2, IL-4, IL-7, IL-9, and IL-15 receptors *(6,9)*. Mouse knock-outs for the γ_c-chain gene lack natural killer (NK) cells and have diminished T and B cells. Recombinase-Activating-Gene-2 (Rag-2) plays a significant role in V(D)J recombination during B and T cell development, and loss of Rag-2 abrogates the maturation of these cells [reviewed in Ref. *(7)*]. When γ_c-chain-null mice are crossed onto a Rag-2 deficient background, residual T and B cells are eliminated. This double knockout lacks functional T, B and NK cells *(7,9,52)*, and is, therefore, permissive for human cell transplantation.

In the HIV-1 Rag2$^{-/-}\gamma_c^{-/-}$ mouse model, newborn mice (1–6 days old) are sublethally irradiated and then intrahepatically injected with human CD34+ hematopoietic stem cells (from human cord blood or human fetal liver). At about 1 month, mice develop de novo B, T and dendritic cells, form primary and secondary lymphoid organs, and are capable of generating functional immune responses *(12,56)*. At 2–8 months of age, mice can be injected either intraperitoneally [i.p. *(1–3,15)*] intravenously [i.v. *(60)*] or intrasplenically [i.s. *(50)*] with HIV-1 leading to productive viral replication, CD4+ T cell depletion, lymphadenopathy, and in some cases anti-HIV-1 immunoglobulin production.

The advantages of using Rag2$^{-/-}\gamma_c^{-/-}$ mice are: (a) unlike SCIDs, that are reported to be "leaky", Rag2$^{-/-}\gamma_c^{-/-}$ mice do not generate functional murine T and B cells *(5)*; (b) the engraftment with human lymphocytes promote formation of primary and secondary lymphoid tissues and development of functional immune responses *(12,56)*; and (c) Rag2$^{-/-}\gamma_c^{-/-}$ are less radiosensitive, and have longer life span, allowing long-term follow-up *(7)*.

1.3. The HIV-1 NOD/SCID/IL2Rγ^{null} Model

Another mouse model, very similar to the Rag2$^{-/-}\gamma_c^{-/-}$ mouse, is the nonobese-diabetic/severe combined immunodeficient (NOD/SCID) mouse line harboring a complete null mutation of the common cytokine receptor γ chain [NOD/SCID/interleukin 2 receptor (IL2r) γ^{null}]. Just like the Rag2$^{-/-}\gamma_c^{-/-}$ mouse, this mouse efficiently supports development of functional human hemato-lymphopoiesis *(19,21,22,53,57)*. An advantage however, over the Rag2$^{-/-}\gamma_c^{-/-}$ mouse, however, is that engraftment occurs even when the mice are injected with human CD34+ HSC at 4–6 weeks of age. When purified human CD34+ stem cells are transplanted intravenously into NOD/SCID/IL2rγ^{null} adult or newborn mice, 10–70% repopulation occurs. This model is of particular interest to HIV-1 investigators due to the fact that human IgA-secreting B cells have been found in the intestinal mucosa of these mice. These mice can also be infected with HIV-1 *(57)*.

2. Materials

2.1. The Intrasplenic huPBL/SPL-SCID Model for the Assessment of CTL Activity in Acute HIV-1 Infection

2.1.1. Mice

1. Specific pathogen-free male and female SCID mice, 8–12 weeks old, are bred and maintained in gnotobiotic conditions.

2.1.2. CTL Clones

1. HIV-1 specific clones can be either isolated from an HIV infected patients or derived from in vitro immunization. For example, we have seen similar efficiency of this model with the patient derived CTLs: 18030.D23.18, 161jXA14 (HLA-A2 restricted CTLs that recognize the $P17_{77-85}$ SLYNTVATL (SL9) peptide) and 68A62 [an HLA-A2 restricted CTL that recognizes the $POL_{464-472}$ ILKEPVHGV (IV9) peptide] clones, generously provided by Dr. Bruce Walker *(25)*, and in vitro immunization derived clones: CC2C and CC5A which are A*0201-restricted SL9-specific CTL clones, generously provided by Dr. June Kan-Mitchell *(26)*.
2. CTLs are expanded bi-weekly by adding 1×10^5 CTLs to 6×10^6 irradiated HLA-mismatched PBMCs to a well of a 12-well plate in complete media [RPMI (Gibco) +10% Fetal Bovine Serum (FBS, Mediatech Inc.) + penicillin/streptomycin (Gibco) + Hepes buffer (Gibco) + L-glutamate (Gibco)] containing 30 ng/mL of anti-CD3 (Orthoclone OKT3, Ortho-Biotech). The next day IL-2 (Roche) is added (25 units/mL) and re-added in fresh complete media two to three times a week at 50 units/mL *(26)*. *See* **Note 1**.

2.2. The HIV-1 $Rag2^{-/-}\gamma_c^{-/-}$ Mouse Model

2.2.1. Mice

$RAG2^{-/-}\gamma_c^{-/-}$ mice (*see* **Note 2**) are bred and maintained at gnotobiotic conditions. Mating pairs are continually set up since implantation of human stem cells can only be done in newborn mice younger than 7 days. *See* **Note 3**.

2.3. The HIV-1 NOD/SCID/IL2Rγ^{null} Model

2.3.1. Mice

NOD/SCID/IL2rγ^{null} newborns [within 48 h of birth *(21)*] or adults [6–10 weeks old *(19,22,53,57)*] are bred and maintained at gnotobiotic conditions.

3. Methods

3.1. The Intrasplenic huPBL/SPL-SCID Model for the Assessment of CTL Activity in Acute HIV-1 Infection

3.1.1. In Vivo Acute Killing Assay

For more details on this assay, please refer to Ref. *(24,25)*.

1. Human PBMCs are isolated from a leukopack (yields 500–800 × 10^6 PBMCs) by collecting the interface of a Ficoll–Paque gradient (typically, 5 mLs of leukopack blood is diluted in 25 mL of phosphate-buffered saline (PBS) and loaded over 15 mL of Ficoll-Paque PLUS (GE Healthcare) in 50 mL conical tubes and spun for 30 min (stopped without applying brakes) at 25 °C at 1,500 rpm).
2. Human Leukocyte Antigen (HLA)-type of the donor PBMCs is determined by flow cytometry, e.g., we typically determine HLA-A2 positivity or negativity with the fluorescein (FITC)-labeled BB7.2 anti HLA-A2 antibody (BD Pharmingen).
3. The HLA-matched and mismatched PBMCs are activated with phytohemagglutinin [PHA, 5 μg/mL (Sigma)] and IL-2 for 24 h in complete media.
4. To acutely infect cells, we incubate > 100 × 10^6 PBMCs with infectious HIV-1, e.g., $HIV-1_{JR-CSF}$ (800 $TCID_{50}$) at 37 °C in 1 mL of RPMI (no FBS). Four hours later 9 mL of complete media is added and the cells are incubated at 37 °C overnight.
5. The next day, the cells are washed and resuspended at 5–10 × 10^6 in 0.1 mL of sterile cold PBS in an insulin syringe with or without CTLs (the infected PBMCs:CTL ratio is 1:1).
6. SCID mice are anesthetized with ketamine/xylazine (100 mg/kg + 40 mg/kg) by i.p. injection, a small incision is made in the skin and peritoneum. The spleen is gently pulled

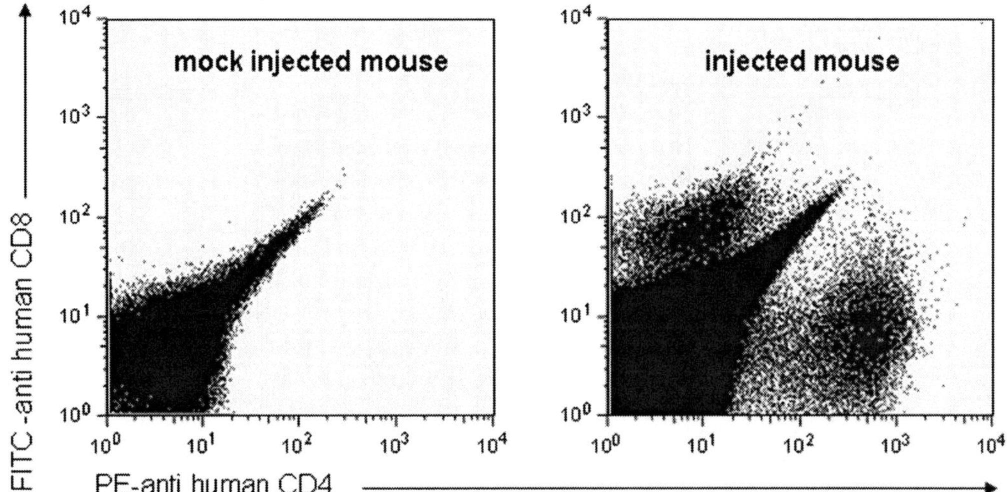

Fig. 21.1. **Persistence of PBMCs in the spleen of a SCID mouse.** 10×10^6 activated PBMCs in 0.1 mL of sterile cold PBS were injected (or mock injected with PBS only) into the spleens of male SCID mice (aged 8–12 weeks) – see Materials and Methods (**Sections 2 and 3**). Spleens were removed at 1 week and a single cell suspension was prepared. Red blood cells were lysed, and cells were stained for anti-human CD4 (FITC, BD Pharmingen) and CD8 (PE, BD Pharmigen) and DAPI (for viability) and collected on an LSR II.

through the hole and injected with the cells. The peritoneal incision is sutured and the skin is stapled closed.

7. The procedure takes about 10 min and since we isolate so many PBMCs from one leukopack, we are able to generate over 20 mice from the same donor that can be used immediately, resulting in very reproducible results.
8. The human PBMCs persist in the mouse for at least 2 weeks, are readily detectible by flow cytometry and constitute over 1% of the cells in the SCID mouse spleen (*see* **Fig. 21.1**).
9. Seven days later mice are killed and the spleens are removed.
10. A single cell preparation is made by disrupting the spleen and filtering the cells through a 70-μm cell strainer (BD Falcon).
11. The cells are then loaded over a Ficoll–Paque density gradient (*see above* **3.1.1.1**). The interface is collected, washed and cells are counted. *See* **Note 4**.
12. HIV-1-infected cell titer in the spleen is measured by limiting dilution co-culture as described below (*see* **3.1.2**).

3.1.2. Limiting Dilution Co-culture

1. 1×10^6 uninfected, PHA + IL-2 activated PBMCs (*see* **3.1.1.3**) are plated in 1 mL of complete media with IL-2 (50 units/ml) in 24-well plates.
2. Splenocytes are then added to the wells in fivefold limited dilutions, in quadruplicate, ranging from 1×10^6 cells/well to 320 cells/well.
3. One week later, p24 antigen levels in the supernatants of all the wells are determined by p24-antigen-specific ELISA (*see*

	average $TCID_{50}$
HLA-A*0201-positive HIV infected PBMCs	3125±0
HLA-A*0201-negative HIV infected PBMCs	3125±0
HLA-A*0201 restricted CTL 1 + HLA-A*0201-positive HIV infected PBMCs	0±0
HLA-A*0201 restricted CTL 1 + HLA-A*0201-negative HIV infected PBMCs	3125±0
HLA-A*0201 restricted CTL 2 + HLA-A*0201-positive HIV infected PBMCs	15±14.14
HLA-A*0201 restricted CTL 2 + HLA-A*0201-negative HIV infected PBMCs	3125±0
HLA-A*0201 restricted CTL 3 + HLA-A*0201-positive HIV infected PBMCs	0±0
HLA-A*0201 restricted CTL 3 + HLA-A*0201-negative HIV infected PBMCs	3125±0
HLA-A*0201 restricted CTL 4 + HLA-A*0201-positive HIV infected PBMCs	25±0
HLA-A*0201 restricted CTL 4 + HLA-A*0201-negative HIV infected PBMCs	3125±0
HLA-A*0201 restricted CTL 5 + HLA-A*0201-positive HIV infected PBMCs	1±0.71
HLA-A*0201 restricted CTL 5 + HLA-A*0201-negative HIV infected PBMCs	3125±0

Fig. 21.2. **Novel in vivo method for assessment of CTL activity against HIV-1 infected cells.** $5–10 \times 10^6$ HLA-A*0201 positive or negative cells were acutely infected with HIV-1 overnight in complete RPMI. The next day cells were harvested and injected with or without CTLS into the spleens of SCID mice. One week later serial ×5 dilutions, starting at 1×10^6 cells were added to 1×10^6 activated PBMCs in a total of 2 mL of complete media +IL-2. Seven days later supernatants were collected and p24 ELISA was done. The results presented are the means of three experiments done in duplicate.

3.1.3) and the results are expressed in $TCID_{50}$, i.e., the level of dilution of the cells at which half of the wells (i.e., two of wells of the quadruplicate) contain infectious virus *(30)*. *See* **Fig. 21.2** for an example of results from such a killing assay.

3.1.3. p24 ELISA

1. Typically, a commercial p24 kit can be used. In the past we have successfully used the Alliance HIV-1 P24 ANTIGEN ELISA Kit (Perkin Elmer) but it is also possible to use a lab-generated ELISA to measure p24 antigen concentrations without the kit *(24)*:
2. For the lab-generated ELISA, MaxiSorp ELISA plates (NUNC) are coated for 1 h at 37 °C, with 100 μL of anti-p24 HIV-1 monoclonal antibody (1 μg/mL; ImmunoDiagnostics) in 0.1 M $NaHCO_3$ buffer pH 8.5 and blocked for 2 h at 37 °C with 0.2 mL blocking buffer (0.1% casein in wash buffer (1.44 M NaCl, 0.5% Tween 20, 250 mM Trizma base, pH 7.5).
3. Plates are aspirated and samples (diluted in blocking buffer) are added and incubated for 1 h at 37 °C.
4. The plate is washed ×4 with wash buffer, and secondary antibody is added (biotynylated rabbit anti-p24 antibody in blocking buffer – made in-house) and incubated for 1 h at 37 °C.
5. The plate is washed again and Streptavidin–horse radish peroxidase (BD Pharmingen) is added and incubated for 1 h at 37 °C.

6. Plates are washed again and substrate is added (0.1 mL of Sigma *FAST* OPD) and incubated at room temperature for 30 min.
7. The reaction is stopped with 0.1 mL of 4 N sulfuric acid and read at 490 nm (above background of 650 nm) using an ELISA reader.

3.2. The HIV-1 Rag2$^{-/-}\gamma_c^{-/-}$ Mouse Model

3.2.1. Sources of Human Stem Cells

3.2.1.1. Isolation of CD34+ Cells from Human Cord Blood

Human CD34+ stem cells can be isolated from either cord blood or fetal liver.

1. Human cord blood can be obtained with written parental informed consent from healthy full-term newborns with the approval of the local Institutional Research Board.
2. The cord blood is loaded over a Ficoll–Paque gradient (typically 15 mL of cord blood is diluted with 15 mL of PBS and then gently loaded over 15 mL of Ficoll) and spun (with no brake) for 30 min at 1,500 rpm at 25 °C. The interface is collected and washed twice with PBS.
3. CD34+ cells are then enriched using immunomagnetic beads according to manufacturer's instructions (CD34+ Progenitor Cell Isolation Kit; Miltenyi Biotec). *See* **Note 5**.
4. Cells can either frozen [10% dimethyl sulfoxide (DMSO, Sigma), in complete media] or transplanted immediately.
5. Numbers and purity of CD34+ cells are typically evaluated by fluorescence-activated cell sorting (FACS). CD34+ cell purity is typically > 90% following density gradient and CD34+ enrichment.

3.2.1.2. Isolation of CD34+ Cells from Human Fetal Tissue

1. Human fetal liver can be obtained from elective abortion after obtaining written informed consent.
2. Human fetal liver cells are isolated by gentle disruption of the tissue in PBS – typically, we cut the liver up into 1-inch pieces and then disrupt the tissue with the plunger of a 10-mL syringe. We then collect the single cell suspension into the 10-mL syringe and filter it twice through a 70-μm cell strainer (BD Falcon).
3. The cells are then centrifuged and washed again with PBS. We typically load cells derived from one liver into 10 50-mL conical tubes.
4. These cells are then layered over Ficoll (as described above) and density gradient centrifuged.
5. The interface is collected and washed of any residual Ficoll.

6. CD34+ cells are then enriched by using immunomagnetic beads according to manufacturer's instructions (CD34+ Progenitor Cell Isolation Kit; Miltenyi Biotec).
7. Cells can either frozen (10% DMSO in complete media) or transplanted immediately.
8. Numbers and purity of CD34+ cells are typically evaluated by FACS. CD34+ cell purity is typically > 90% following density gradient and CD34+ enrichment – *see* **Fig. 21.3**.

3.2.2. Transplantation of Human CD34+ Cells into $Rag2^{-/-}\gamma_c^{-/-}$

1. Newborn mice are irradiated within 1 week of age (usually at day 1–3 following birth) with sub-lethal irradiation doses that range between 3.50 and 5 Gy *(1–3,15,50,60)*, others have also demonstrated that intrapartum busulfan administration to pregnant dams results in higher engraftment (15 mg/kg administered subcutaneously as a 20% DMSO solution at day 18 postcoitus) *(15)*.
2. At 1–5 h postirradiation, mice are transplanted with the CD34+ cells in 0.03–0.1 mL of PBS by intrahepatic injection. When CD34+ cells are isolated from cord blood, animals typically receive $0.01–0.6 \times 10^6$ cells *(12,15,60)*, while when CD34+ cells are isolated from fetal liver, animals typically

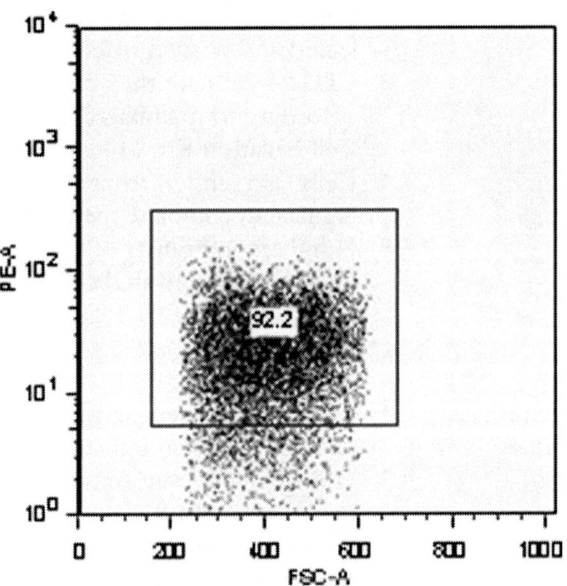

Fig. 21.3. **CD34 enrichment**. Single cell preparations are made from human fetal liver by gentle disruption of the tissue in PBS. The single cell preparation is then collected and loaded over Ficoll and density gradient centrifuged. The interface is collected and washed of any residual Ficoll. CD34+ cells are then enriched by using immunomagnetic beads according to manufacturer's instructions (CD34+ Progenitor Cell Isolation Kit; Miltenyi Biotec). Cells can either frozen (10% DMSO in complete media) or transplanted immediately. Numbers and purity of CD34+ cells are typically evaluated by fluorescence-activated cell sorting (FACS).

receive 1–5 × 10^6 cells *(1,3,50,60)*. The dark colored liver is easily discernible through the skin of the pup, making this injection technically easy. *See* **Note 6**.

3.2.3. Analysis of Engraftment

1. At 1–2 months of age, peripheral blood is collected by retro-orbital or tail vein bleeding.
2. Red blood cells are lysed by suspending the collected blood in Ammonium Chloride (8 g/mL) for 10 min. The cells are then washed twice with PBS, and stained with anti-human CD45, CD4, CD8, CD14 (or CD11b), CD19 (BD Bioscience) and any other marker of interest in FACs buffer (1 g/L NaN3). In order to see significant populations, at least 1,000,000 cells usually need to be collected.
3. Human CD45+ peripheral blood cell percentages are extremely variable, being as low as 1% and as high as 80% *(1–3,12,15,56,60)*. We typically consider a mouse that has at least 1% CD45+ cells of the lymphocyte gate to be engrafted – *see* **Fig. 21.4**.
4. The mice are then infected by injection with HIV-1.

3.2.4. HIV-1 Infection of Humanized RAG2$^{-/-}\gamma c^{-/-}$

1. The mice are injected either i.p. *(1–3,15)*, i.v. *(60)* or i.s. *(50)* with HIV-1 preparations typically in 0.1–0.2 mL of PBS at titers that range between 0.01 and 2 × 10^6 $TCID_{50}$ *(1–3, 12,15,60)*.

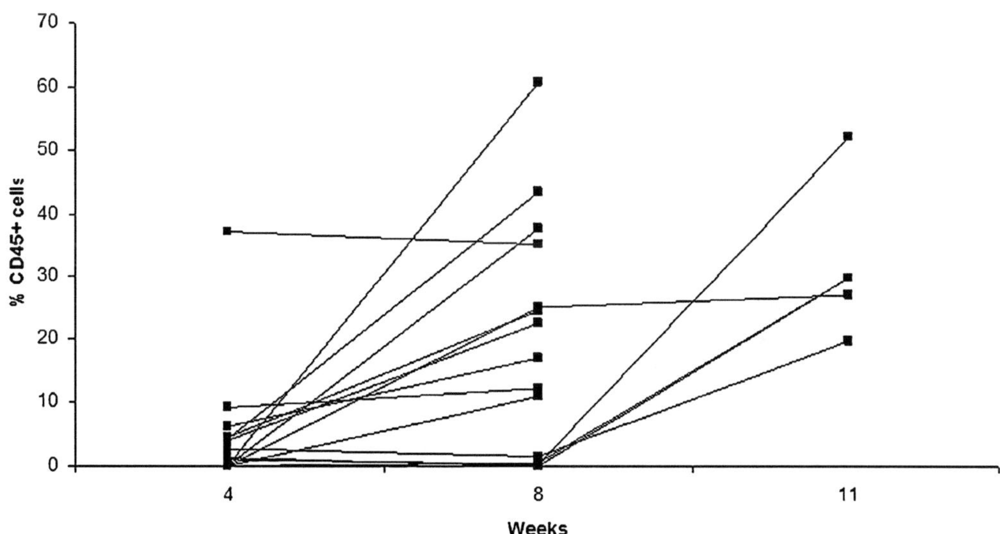

Fig. 21.4. **Human Cell Engraftment in Rag2$^{-/-}\gamma_c^{-/-}$**. CD34+ human hematopoietic stem cells were injected intrahepatically into 1- to 7-day-old new born Rag2$^{-/-}$ $\gamma_c^{-/-}$ mice. Mice were bled at 4, 8 and/or 11 weeks and PBMCs were stained for the expression of anti-human CD45. Presented is the percentage of anti-human CD45+ cells for 12 individual mice.

3.2.5. Analysis of HIV-1 Infection and Immune Response

Mice are killed at different time points following infection and peripheral blood, liver, spleen, bone marrow, thymus, lungs and lymph nodes are typically harvested.

3.2.5.1. HIV-1 Infection Analysis

1. Plasma viral load is typically monitored with the Roche Amplicor Monitor v.1.5. assay (Roche Diagnostics).
2. Infectious HIV-1 in all other organs is typically detected by limiting dilution co-culture of these cells with PHA/IL-2 activated PBMCs as described above (*see* **3.1.2**) – *see* **Fig. 21.5**.
3. In addition, several groups have performed immunohistochemistry with anti-p24 antibody on paraffin sections showing successful infections of nearly all organs.

3.2.5.2. CD4+ Cells Depletion

There are two ways to determine CD4+ cell depletion:

1. In order to determine CD4+ depletion, all tissues are usually stained for CD45, CD4, CD8, etc. and FACSed. A ratio is then made between the CD4+ population of the CD45 gated cells, and the CD8+ population of the CD45 cells. Before HIV-1 infection this is typically 2:1 to 4:1 (within normal range observed in healthy humans) – this however inverts following HIV-1 infection in most reports *(1–3,15)*.
2. Another way of determining CD4+ cell decline is to use the GUAVA Easycytes kit according to manufacturer's instructions for a readout of number of CD4+ cells/mL of blood *(60)*.

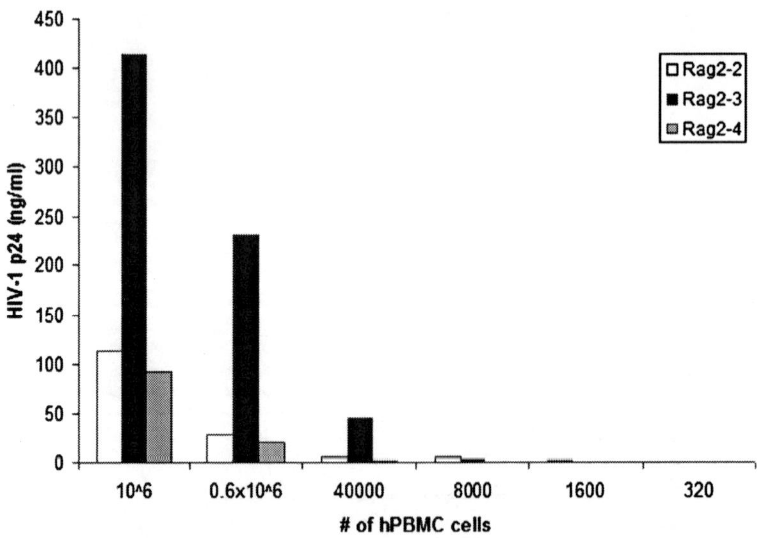

Fig. 21.5. **HIV-1 titer of splenocytes isolated from Rag2$^{-/-}$ $\gamma_c^{-/-}$ infected intrasplenically**. HIV-1$_{JR-CSF}$ (800 $TCID_{50}$) was injected intrasplenically into humanized Rag2$^{-/-}$ $\gamma_c^{-/-}$. For to 8 weeks following infection, splenocytes were harvested and cocultured by limiting dilution with 1×10^6 activated human PBMCs. One week later, supernatants were collected and assayed for productive HIV-1 infection by measuring p24 concentration by ELISA. Presented are the data of three separate mice.

3.2.5.3. Measuring Humoral Immune Responses to the HIV-1 Infection

1. The levels of human IgM and human IgG in peripheral blood has been determined by some groups. In one report, IgG against p34, gp41, p52, p58, and gp160 was found in one animal by Western blot *(2)*. In another, no antibodies were found against HIV-1 proteins *(1)*. Another group found no antibodies to HIV-1 proteins but did show that these mice could launch an immune response when vaccinated against *Haemophilus influenzae* type b with the Act HIB (Aventius Paster Inc.) conjugate vaccine *(15)*. *See* **Notes 7** and **8**.

3.3. The HIV-1 NOD/SCID/IL2Rγ^{null} Model

3.3.1. Sources of Human Stem Cells

See the description provided in **Section 3.2.1**.

3.3.2. Transplantation of Human CD34+ Cells into NOD/SCID/IL2Rγ^{null}

1. Mice are sub-lethally irradiated [1 Gy for newborns *(21)* and between 2.4 and 3 Gy for adults *(19,22,53,57)*].
2. Within 12 h of irradiation, mice are intravenously transplanted with $0.01–1.5 \times 10^6$ CD34+ cells. Newborns are injected via a facial vein *(21)*, while adults are injected via a tail vein *(19,22,53,57)*.

3.3.3. Analysis of engraftment

See the description provided in **Section 3.2.3**.

3.3.4. HIV-1 infection of humanized NOD/SCID/IL2rγ^{null}

1. In one report, mice were injected i.v. *(57)* with HIV-1 virus at titers that ranged between 200 and 65,000 $TCID_{50}$.

3.3.5. Analysis of HIV-1 infection and immune response

See the description provided in **Section 3.2.5**.

4. Notes

1. Fresh media should always be prepared when thawing CTLs – we find this is best for CTL viability.
2. Most report that is best to use Rag2$^{-/-}\gamma_c^{-/-}$ on a BALB/c background.
3. We find that the highest repopulation occurs when mice are injected within the first 3 days of life.
4. We typically layer the spleens in small volumes over ficoll, i.e., 1 mL of splenocytes (in PBS) over 1 mL of Ficoll.

5. It is very important to filter chord blood derived cells to ensure that they do not clog the filters.
6. One technical problem we have encountered in making these mice is that handling the mice following irradiation and injection may cause the mother to reject the pups and stop nursing them. We have circumvented this, by smearing the pups with urine collected from the mother before returning the pups to the mother.
7. Though at least two groups have attempted to look at HIV-specific cellular immune responses in both the HIV-1 Rag2$^{-/-}\gamma_c^{-/-}$ mouse model and the HIV-1 NOD/SCID/IL2Rγ^{null} model, to date, no-one has successfully shown the emergence or presence of such an immune response *(1,2)*.
8. To date no-one has shown mucosal immunity in either the HIV-1 Rag2$^{-/-}\gamma_c^{-/-}$ mouse model or the HIV-1 NOD/SCID/IL2Rγ^{null} model which could prove invaluable in an HIV-1 model.

Acknowledgments

This work was supported by the National Institutes of Health (National Institute of Allergy and Infectious Diseases AI67136, and the Einstein/MMC Center for AIDS Research AI51519).

References

1. An, D. S., Poon, B., Fang, R. H., et al. (2007) Use of a novel chimeric mouse model with a functionally active human immune system to study human immunodeficiency virus type 1 infection. *Clin Vaccine Immunol* 14, 391–396.
2. Baenziger, S., Tussiwand, R., Schlaepfer, E., et al. (2006) Disseminated and sustained HIV infection in CD34+ cord blood cell-transplanted Rag2-/-gamma c-/- mice. *Proc Natl Acad Sci U S A* 103, 15951–15956.
3. Berges, B. K., Wheat, W. H., Palmer, B. E., et al. (2006) HIV-1 infection and CD4 T cell depletion in the humanized Rag2-/-gamma c-/- (RAG-hu) mouse model. *Retrovirology* 3, 76.
4. Borkow, G. (2005) Mouse models for HIV-1 infection. *IUBMB Life* 57, 819–823.
5. Bosma, M. J., Carroll, A. M. (1991) The SCID mouse mutant: definition, characterization, and potential uses. *Annu Rev Immunol* 9, 323–350.
6. Cao, X., Shores, E. W., Hu-Li, J., et al. (1995) Defective lymphoid development in mice lacking expression of the common cytokine receptor gamma chain. *Immunity* 2, 223–238.
7. Chicha, L., Tussiwand, R., Traggiai, E., et al. (2005) Human adaptive immune system Rag2-/-gamma(c)-/- mice. *Ann N Y Acad Sci* 1044, 236–243.
8. De Maria, A., Cirillo, C., Moretta, L. (1994) Occurrence of human immunodeficiency virus type 1 (HIV-1)-specific cytolytic T cell activity in apparently uninfected children born to HIV-1-infected mothers. *J Infect Dis* 170, 1296–1299.
9. DiSanto, J. P., Muller, W., Guy-Grand, D., et al. (1995) Lymphoid development in mice with a targeted deletion of the interleukin 2 receptor gamma chain. *Proc Natl Acad Sci U S A* 92, 377–381.
10. Ferbas, J., Kaplan, A. H., Hausner, M. A., et al. (1995) Virus burden in long-term survivors of human immunodeficiency virus

(HIV) infection is a determinant of anti-HIV CD8+ lymphocyte activity. *J Infect Dis* 172, 329–339.
11. Garcia, S., Dadaglio, G., Gougeon, M. L. (1997) Limits of the human-PBL-SCID mice model: severe restriction of the V beta T-cell repertoire of engrafted human T cells. *Blood* 89, 329–336.
12. Gimeno, R., Weijer, K., Voordouw, A., et al. (2004) Monitoring the effect of gene silencing by RNA interference in human CD34+ cells injected into newborn RAG2-/- gammac-/- mice: functional inactivation of p53 in developing T cells. *Blood* 104, 3886–3893.
13. Goldman, J. P., Blundell, M. P., Lopes, L., et al. (1998) Enhanced human cell engraftment in mice deficient in RAG2 and the common cytokine receptor gamma chain. *Br J Haematol* 103, 335–342.
14. Goldstein, H., Pettoello-Mantovani, M., Katopodis, N. F., et al. (1996) SCID-hu mice: a model for studying disseminated HIV infection. *Semin Immunol* 8, 223–231.
15. Gorantla, S., Sneller, H., Walters, L., et al. (2007) Human immunodeficiency virus type 1 pathobiology studied in humanized BALB/c-Rag2-/-gammac-/- mice. *J Virol* 81, 2700–2712.
16. Gotch, F., Gallimore, A., McMichael, A. (1996) Cytotoxic T cells–protection from disease progression–protection from infection. *Immunol Lett* 51, 125–128.
17. Goulder, P. J., Phillips, R. E., Colbert, R. A., et al. (1997) Late escape from an immunodominant cytotoxic T-lymphocyte response associated with progression to AIDS. *Nat Med* 3, 212–217.
18. Harrer, T., Harrer, E., Kalams, S. A., et al. (1996) Cytotoxic T lymphocytes in asymptomatic long-term nonprogressing HIV-1 infection. Breadth and specificity of the response and relation to in vivo viral quasispecies in a person with prolonged infection and low viral load. *J Immunol* 156, 2616–2623.
19. Hiramatsu, H., Nishikomori, R., Heike, T., et al. (2003) Complete reconstitution of human lymphocytes from cord blood CD34+ cells using the NOD/SCID/gammacnull mice model. *Blood* 102, 873–880.
20. Huynen, M. A., Neumann, A. U. (1996) Rate of killing of HIV-infected T cells and disease progression. *Science* 272, 1962.
21. Ishikawa, F., Yasukawa, M., Lyons, B., et al. (2005) Development of functional human blood and immune systems in NOD/SCID/IL2 receptor {gamma} chain(null) mice. *Blood* 106, 1565–1573.
22. Ito, M., Hiramatsu, H., Kobayashi, K., et al. (2002) NOD/SCID/gamma(c)(null) mouse: an excellent recipient mouse model for engraftment of human cells. *Blood* 100, 3175–3182.
23. Jassoy, C., Walker, B. D. (1997) HIV-1-specific cytotoxic T lymphocytes and the control of HIV-1 replication. *Springer Semin Immunopathol* 18, 341–354.
24. Joseph, A., Zheng, J. H., Follenzi, A., et al. (2008) Lentiviral vectors encoding human immunodeficiency virus type 1 (HIV-1)-specific T-cell receptor genes efficiently convert peripheral blood CD8 T lymphocytes into cytotoxic T lymphocytes with potent in vitro and in vivo HIV-1-specific inhibitory activity. *J Virol* 82, 3078–89.
25. Joseph, A., Zheng, J., Patel, M., et al. (2007) An intrasplenic huPBL/SPL SCID model for the study of acute in vivo CTL activity against HIV-1 infected cells. *In preparation.*
26. Kan-Mitchell, J., Bisikirska, B., Wong-Staal, F., et al. (2004) The HIV-1 HLA-A2-SLYNTVATL is a help-independent CTL epitope. *J Immunol* 172, 5249–5261.
27. Kollmann, T. R., Kim, A., Pettoello-Mantovani, M., et al. (1995) Divergent effects of chronic HIV-1 infection on human thymocyte maturation in SCID-hu mice. *J Immunol* 154, 907–921.
28. Kollmann, T. R., Pettoello-Mantovani, M., Zhuang, X., et al. (1994) Disseminated human immunodeficiency virus 1 (HIV-1) infection in SCID- hu mice after peripheral inoculation with HIV-1. *J Exp Med* 179, 513–522.
29. Koup, R. A., Safrit, J. T., Cao, Y., et al. (1994) Temporal association of cellular immune responses with the initial control of viremia in primary human immunodeficiency virus type 1 syndrome. *J Virol* 68, 4650–4655.
30. LaBarre, D. D., Lowy, R. J. (2001) Improvements in methods for calculating virus titer estimates from TCID50 and plaque assays. *J Virol Methods* 96, 107–126.
31. Langlade-Demoyen, P., Ngo-Giang-Huong, N., Ferchal, F., et al. (1994) Human immunodeficiency virus (HIV) nef-specific cytotoxic T lymphocytes in noninfected heterosexual contact of HIV-infected patients. *J Clin Invest* 93, 1293–1297.
32. Markham, R. B., Donnenberg, A. D. (1992) Effect of donor and recipient immunization protocols on primary and secondary human antibody responses in SCID mice reconstituted with human peripheral blood mononuclear cells. *Infect Immun* 60, 2305–2308.
33. Mazurier, F., Fontanellas, A., Salesse, S., et al. (1999) A novel immunodeficient mouse

model–RAG2 x common cytokine receptor gamma chain double mutants–requiring exogenous cytokine administration for human hematopoietic stem cell engraftment. *J Interferon Cytokine Res* 19, 533–541.
34. McCune, J. M. (1991) SCID mice as immune system models. *Curr Opin Immunol* 3, 224–228.
35. McCune, J. M., Namikawa, R., Kaneshima, H., et al. (1988) The SCID-hu mouse: murine model for the analysis of human hematolymphoid differentiation and function. *Science* 241, 1632–1639.
36. McMichael, A. J., Rowland-Jones, S. L. (2001) Cellular immune responses to HIV. *Nature* 410, 980–987.
37. Mosier, D. E. (1996) Modeling AIDS in a mouse. *Hosp Pract (Minneap)*, 31, 41–48, 53–45, 59–60.
38. Mosier, D. E. (1996) Viral pathogenesis in hu-PBL-SCID mice. *Semin Immunol* 8, 255–262.
39. Mosier, D. E., Gulizia, R. J., Baird, S. M., et al. (1988) Transfer of a functional human immune system to mice with severe combined immunodeficiency. *Nature* 335, 256–259.
40. Mosier, D. E., Gulizia, R. J., Baird, S. M., et al. (1991) Human immunodeficiency virus infection of human-PBL-SCID mice. *Science* 251, 791–794.
41. Musey, L., Hughes, J., Schacker, T., et al. (1997) Cytotoxic-T-cell responses, viral load, and disease progression in early human immunodeficiency virus type 1 infection. *N Engl J Med* 337, 1267–1274.
42. Namikawa, R., Weilbaecher, K. N., Kaneshima, H., et al. (1990) Long-term human hematopoiesis in the SCID-hu mouse. *J Exp Med* 172, 1055–1063.
43. Ogg, G. S., Jin, X., Bonhoeffer, S., et al. (1998) Quantitation of HIV-1-specific cytotoxic T lymphocytes and plasma load of viral RNA. *Science* 279, 2103–2106.
44. Pantaleo, G., Demarest, J. F., Soudeyns, H., et al. (1994) Major expansion of CD8+ T cells with a predominant V beta usage during the primary immune response to HIV. *Nature* 370, 463–467.
45. Pettoello-Mantovani, M., Kollmann, T. R., Katopodis, N. F., et al. (1998) thy/liv-SCID-hu mice: a system for investigating the in vivo effects of multidrug therapy on plasma viremia and human immunodeficiency virus replication in lymphoid tissues. *J Infect Dis* 177, 337–346.
46. Phillips, R. E., Rowland-Jones, S., Nixon, D. F., et al. (1991) Human immunodeficiency virus genetic variation that can escape cytotoxic T cell recognition. *Nature* 354, 453–459.
47. Rinaldo, C., Huang, X. L., Fan, Z. F., et al. (1995) High levels of anti-human immunodeficiency virus type 1 (HIV-1) memory cytotoxic T-lymphocyte activity and low viral load are associated with lack of disease in HIV-1-infected long-term nonprogressors. *J Virol* 69, 5838–5842.
48. Roncarolo, M. G., Carballido, J. M., Rouleau, M., et al. (1996) Human T-and B-cell functions in SCID-hu mice. *Semin Immunol* 8, 207–213.
49. Rowland-Jones, S., Sutton, J., Ariyoshi, K., et al. (1995) HIV-specific cytotoxic T-cells in HIV-exposed but uninfected Gambian women. *Nat Med* 1, 59–64.
50. Rowland-Jones, S. L., Dong, T., Fowke, K. R., et al. (1998) Cytotoxic T cell responses to multiple conserved HIV epitopes in HIV-resistant prostitutes in Nairobi. *J Clin Invest* 102, 1758–1765.
51. Sango, K., Joseph, A., Buhl, S., et al. (2007) HIV-1 specific humoral responses induced by intrasplenic infection of HIV-1 in a humanized Rag2-/-gamma-common-chain-/- mouse model. *In preparation.*
52. Saxon, A., Macy, E., Denis, K., et al. (1991) Limited B cell repertoire in severe combined immunodeficient mice engrafted with peripheral blood mononuclear cells derived from immunodeficient or normal humans. *J Clin Invest* 87, 658–665.
53. Shinkai, Y., Rathbun, G., Lam, K. P., et al. (1992) RAG-2-deficient mice lack mature lymphocytes owing to inability to initiate V(D)J rearrangement. *Cell* 68, 855–867.
54. Shultz, L. D., Lyons, B. L., Burzenski, L. M., et al. (2005) Human lymphoid and myeloid cell development in NOD/LtSz-scid IL2R gamma null mice engrafted with mobilized human hemopoietic stem cells. *J Immunol* 174, 6477–6489.
55. Somasundaram, R., Jacob, L., Adachi, K., et al. (1995) Limitations of the severe combined immunodeficiency (SCID) mouse model for study of human B-cell responses. *Scand J Immunol* 41, 384–390.
56. Tary-Lehmann, M., Saxon, A., Lehmann, P. V. (1995) The human immune system in hu-PBL-SCID mice. *Immunol Today* 16, 529–533.
57. Traggiai, E., Chicha, L., Mazzucchelli, L., Bronz, L., Piffaretti, J. C., Lanzavecchia, A. and Manz, M. G. (2004) Development of a human adaptive immune system in cord blood cell-transplanted mice. *Science* 304, 104–107.
58. Watanabe, S., Terashima, K., Ohta, S., et al. (2007) Hematopoietic stem cell-engrafted NOD/SCID/IL2Rgamma null

mice develop human lymphoid systems and induce long-lasting HIV-1 infection with specific humoral immune responses. *Blood* 109, 212–218.

59. Weijer, K., Uittenbogaart, C. H., Voordouw, A., et al. (2002) Intrathymic and extrathymic development of human plasmacytoid dendritic cell precursors in vivo. *Blood* 99, 2752–2759.

60. Wolinsky, S. M., Korber, B. T., Neumann, A. U., et al. (1996) Adaptive evolution of human immunodeficiency virus-type 1 during the natural course of infection. *Science* 272, 537–542.

61. Zhang, L., Kovalev, G. I., Su, L. (2007) HIV-1 infection and pathogenesis in a novel humanized mouse model. *Blood* 109, 2978–2981.

Section IV

Immunological Studies of HIV

Subsection A

Mucosal Immunology

Chapter 22

Mucosal Antibody Responses to HIV

Zina Moldoveanu and Jiri Mestecky

Abstract

The measurement of antibodies specific for the majority of infectious agents in various external secretions is important in the evaluation of potentially protective immune responses at various sites of pathogen entry. Importantly, due to differences in the isotypes of antibodies in various body fluids, levels of total and antigen-specific antibodies in sera and secretions often display independent patterns. The measurement of mucosal antibodies to HIV presents several unique problems. Although controversial, recent results from several laboratories indicate that HIV-specific antibodies are mainly of the IgG and not IgA isotype, despite the pronounced dominance of total IgA in almost all external secretions. Due to the low levels of total IgG in such secretions, highly sensitive methods must be used, including chemiluminescence-enhanced Western blot analyses and ELISA. However, the results generated by ELISA must be interpreted with caution because of a relatively high frequency of false-positive results. Finally, due to the enormous variability of Ig levels not only in various secretions, but also in the same secretion collected at different times, determinations of total Ig levels must be performed to generate meaningful results.

Key words: HIV, mucosal antibodies, IgA, IgG, IgM, Western blot, ELISA, external secretions.

1. Introduction

Because of the simplicity of collection and the relevance for protection, external secretions are frequently used for the evaluation of humoral immune responses. Furthermore, antibodies present in sera and in the majority of external secretions display a considerable degree of independence, including the pronounced variations in absolute levels and dominance of immunoglobulin isotype distribution, and kinetic and perseverance of humoral immune responses.

Vinayaka R. Prasad, Ganjam V. Kalpana (eds.), *HIV Protocols: Second Edition, vol. 485*

DOI 10.1007/978-1-59745-170-3_22 Springerprotocols.com

Reliable measurements of mucosal antibodies are highly dependent on the use of appropriate and uniform collection methods. It is beyond the scope of this chapter to specify all the devices, required supplies, and processing and storage procedures. All these aspects and methodologies have been described and discussed in great detail with specifications characteristic of individual secretions including tears, parotid and whole saliva, colostrum and milk, urine, nasal, rectal, and vaginal washes, semen and pre-ejaculate, feces and fecal extracts, intestinal fluid, and uterine cervical secretion *(1)*.

There are several important factors that must be considered before the collection, processing, and subsequent analyses of various external secretions. Although it is easy to collect human urine, whole saliva or milk, the collection of, for example, tears, nasal, rectal, or vaginal washes may be difficult because of discomfort to the volunteer, esthetic objection, unavailability of trained and licensed personnel to collect intestinal or genital tract fluids, lack of equipment and supplies to process some secretions (e.g., intestinal lavages), and the possibility of a low volunteers' compliance to be subjected to repeated collections from many different mucosal sites.

In addition to these inevitable limitations, the levels of total as well as antigen-specific immunoglobulins are significantly influenced by a variety of factors, which must be always considered in the interpretation of experimental results. Thus, the dilution of the specimen due to the collection procedure (e.g., vaginal or rectal washes), marked variations in flow-rates of secretions stimulated by mechanical or chemical means (e.g., tears, saliva, or milk), and the hormonal status (e.g., variability of immunoglobulin levels in vaginal washes and cervical fluid during the menstrual cycle) must be considered. Furthermore, the presence of immunoglobulin-degrading proteolytic enzymes (e.g., in the intestinal fluid and semen), presence of immunoglobulins in complexes with humoral factors of innate immunity (e.g., mucin or lactoferrin), and contamination with indigenous bacteria present in abundance in most secretions, can singly, or in combination, influence immunoglobulin measurement.

Because of the enormous variability in the levels of total immunoglobulins of all major isotypes in different secretions ranging, for example, from $\sim 0.1\,\mu g/mL$ in urine to $\geq 12,300\,\mu g/mL$ in the colostrum (**Table 22.1**), it is imperative to express the amount of specific antibodies in the context of total immunoglobulin levels, or of levels of other less variable proteins, such as serum albumin or transferrin, which are not produced in mucosal tissues *(2)*. A comparison of the ratios of immunoglobulins to such serum proteins may provide insight into local versus circulation-derived immunoglobulins in external secretions. Furthermore, not only the total levels but also the molecular forms

Table 22.1
Levels of total immunoglobulins in selected human body fluids (μg/mL)

Fluid	IgA	IgG	IgM
Serum	500–3,500	7,000–12,000	500–1,500
Tears	80–400	0–16	0–18
Parotid saliva	15–319	0.4–5	0.4
Colostrum and milk	470–12,340	40–168	50–610
Intestinal fluid	166	4	8
Urine	0.1–1.0	0.06–0.6	
Ejaculate	12–23	16–33	0–8
Uterine cervical fluid	3–333	1–285	5–118

of IgA, such as proportions of monomeric (m), polymeric (p), and secretory IgA (S-IgA), and the distribution of IgA subclasses greatly varies in individual secretions *(1)*. This heterogeneity of molecular forms must be considered in the selection of standards necessary for assays of antibodies in various secretions. Finally, the level, isotype distribution, and molecular forms of IgA may be influenced by inflammatory conditions in a given tissue, and by the selection of an appropriate collection procedure. These explicit and intrinsic characteristics of individual external secretions reflect differences in the dominant source of immunoglobulins: either local production in mucosal tissues or production in the bone marrow, spleen, and lymph nodes. Furthermore, the expression of epithelial receptors responsible for immunoglobulin transport, and the differences in the regulation of immunoglobulin transport by hormones and cytokines may influence the properties of antibodies in various secretions.

The presence and immunoglobulin isotype distribution of HIV-1-specific antibodies in all external secretions described above, and collected from HIV-1-infected or HIV-1-exposed but uninfected individuals have been reported in many studies (for reviews *see 3–5*). A critical evaluation of these reports revealed that results generated in different laboratories may yield remarkable variances and discrepancies, even in cases in which the identical secretion was shared by seven different laboratories *(3)*. Obviously, the method used for the sample collections and analyses of HIV-1-specific antibodies in human external secretions have not been standardized. The present review is based on our previous studies, performed in collaboration with members of the AIDS Vaccine Evaluation Group, and validated in several participating laboratories *(1, 3, 4)*.

2. Materials

2.1. Collections of Human External Secretions

We refer to our earlier published detailed review *(1)* in which we provided all information pertaining to the procedures, supplies and equipment necessary for collection, processing, and safe storage of tears, whole and parotid saliva, nasal, rectal and vaginal washes, cervical uterine fluid, semen and pre-ejaculate, intestinal fluid, feces, colostrum and milk, and urine.

2.2. ELISA

1. Microtiter plates: high binding capacity 96-well polystyrene (Nunc, Rochester, NY or, Nunc, Roskilde, Denmark).
2. Coating diluent: phosphate-buffered saline (PBS) pH = 7.2 with 0.01% sodium azide (Sigma, St. Louis, MO).
3. Capture antibodies: unlabeled $F(ab')_2$ fragment of goat IgG specific for human IgA, IgG or IgM isotypes (Jackson ImmunoResearch Laboratories, West Grove, PA) or mouse monoclonal reagents specific for human IgA1, IgA2 and IgG1-4 subclasses, commercially available (Nordic Immunological Laboratories/Accurate Chemical, Westbury, NY; Southern Biotech, Birmingham AL; Binding Site, Birmingham, UK).
4. Capture antigens: include HIV gp160, gp120, gp 41 and Gag proteins, or HIV antigens included in various experimental vaccines. A limited amount of HIV antigens (~ 100 μg) can be obtained from the NIH (AIDS Research and Reference Reagent Program). Recombinant HIV-1 Env (gp160, gp120, gp41), Gag (p55, p66, p24) and Nef proteins are commercially available (e.g., Protein Sciences, Meriden, CT; RDI, Concord, MA, Advanced BioScience Laboratories, Kensington, MD).
5. Washing buffer: PBS-0.05% Tween-20 (PBST).
6. Blocking solution, consists of 5% fetal calf, or goat serum (Atlanta Biological, Atlanta, GA), or of 1% bovine serum albumin (Sigma) in PBST.
7. Standards: pool of human serum calibrated for total immunoglobulin isotype levels and colostral S-IgA can be purchased from Binding Site (Birmingham, UK) or from Microgenics corp (Fremont, CA).
8. Internal controls: serum and secretion samples with known levels of antibodies specific for the HIV-1 antigen to be analyzed. Both the internal controls and the standard sera should be preserved frozen in multiple, single use aliquots.
9. Developing antibodies: $F(ab')_2$ fragment of goat IgG specific for human IgA, IgG or IgM in the form of biotin-conjugates (e.g., BioSource, Camarillo, CA) or, the mouse monoclonal

antibodies specific for IgA or IgG subclasses as horseradish peroxidase (HRP)-conjugates (Binding Site).

10. HRP-labeled ExtrAvidin (Sigma).
11. Substrate buffer: 0.1M phosphate-citrate pH 5.0, prepared by dissolving citric acid (7.3 g) and disodium phosphate (23.87 g of $Na_2HPO_4 \cdot 12H_2O$) in 1 L distilled water. The pH needs to be checked, and adjusted to 5 and the buffer should be stored at 4 °C until use. All reagents can be purchased from Sigma.
12. Chromogenic substrate for HRP: the ortho-phenylene-diamine (OPD) and hydrogen peroxide (5 mg and 1.2 μL, respectively per 10 mL substrate buffer) can be purchased from Sigma. OPD needs to be stored at −20 °C in a desiccator and the substrate solution should be prepared just before use. Ready-made substrate solution of 3,3′,5,5′-tetramethylbenzidine (TMB, e.g., from BioFX Laboratories, Owings Mills, MD, InvitroGen, San Diego, CA, or KPL, Gaithersburg, MD) can be used as well. HRP substrates are light-sensitive. For reproducibility of color-developing time, the substrate should always have the same temperature (e.g., room temperature or 4 °C) when added to the plates.
13. Stop solution: 1 M sulfuric or hydrochloric acid (Sigma).

2.3. Chemiluminiscence-Enhanced Western Blotting (ECL-WB)

1. Commercially available Western blot nitrocellulose strips containing electrophoretically separated antigens of partially purified inactivated HIV-1. Strips are produced by Cambridge Biotech and can be purchased from Maxim Biomedical Inc. (Rockville, MD). Strips should be stored at 2–8 °C in a dark place. Maxim Biochemical Inc. also provides a Cambridge Biotech HIV-1 Western Blot Kit that includes the necessary reagents and a detailed protocol description.
2. Mini-incubation trays (Bio Rad, Hercules, CA).
3. Blocking buffer: Superblock (in Tris-buffered saline) from Pierce (Rockford, IL) supplemented with 0.05% Tween 20. Storage at 4 °C is required.
4. Washing buffer: 0.02M Tris-buffered saline pH 8.0, with 0.05% Tween 20 (TBST).
5. Biotin-conjugates of $F(ab')_2$ fragment of goat IgG specific for human IgA, IgG, or IgM (e.g., BioSource).
6. HRP-conjugated Neutravidin (Pierce).
7. ECL-SuperSignal West Pico chemiluminescent substrate (Pierce) should be stored at 4 °C. If the substrate shows color, it indicates that it is oxidized and should be discarded.
8. Whatman 1 chromatography paper (Fisher Scientific, Pittsburgh, PA).
9. High resolution BioMax Light film (Kodak Molecular Imaging Systems, Rochester, NY).

10. Film-developing cassette preferable with intensifying screen (Kodak BioMax, Rochester, NY). Clear plastic sheet protector for the strips placed into cassette (office supplies stores).
11. Rocker or rotary platform.
12. Forceps or tweezers.
13. Aspirator with disinfectant (e.g., bleach solution) trap.

3. Methods

Immunoglobulins in external secretions are present at much lower levels than in serum or plasma, with marked variations as to the individual isotypes (*see* **Table 22.1**). Furthermore, due to the presence of endogenous as well as exogenous (bacterial) proteases, the samples must be processed immediately after the collection and kept frozen at −70 °C. Sample processing specific for individual external secretions has been described *(1)*. To prevent disappointment with the measurement of HIV-specific antibodies, we strongly recommend to first determine the levels of total IgG, IgM, and IgA: if such levels are low (< 2 μg/mL), the probability of detecting HIV-specific antibodies is not high. Samples to be evaluated must be diluted so that a probable range of total Ig levels in a given secretion is reached (*see* Table 22.1). Therefore, the dilution of, for example, saliva will be different from that of milk or semen. Furthermore, in the ELISA, false-positive signals for the detection of HIV-specific antibodies of the IgA isotype in saliva and rectal washes are not uncommon *(3, 6, 7)*. Although the reason for such false-positivity is not known, it is likely that the presence of trace amounts of other contaminating proteins in the recombinant HIV antigen preparations, such as those derived from the baculovirus are recognized by naturally occurring IgA antibodies. This phenomenon can be avoided if lysates of HIV-infected cells are used for the ELISA plate coating. Therefore, it is recommended that sera and external secretions from HIV-infected and healthy individuals be used as positive and negative controls.

3.1. ELISA for Total Levels of IgA, IgG, and IgM Antibodies

1. Coat microtiter plates with 100 μL/well PBS-sodium azide solution containing unlabeled anti-human immunoglobulin isotype reagents and incubate overnight at 4 °C. For total levels of IgA, IgG or IgM, the Jackson ImmunoResearch reagents work well at 1 μg/mL concentration, although batch-to-batch variations might occur.
2. Wash plates ~ 3 times, each time filling the wells with ~ 300 μL PBS/well.

3. Block the coated plates for at least 1 h at room temperature with 200 μL/well of a suitable blocking reagent in PBST (*see* **Materials**).
4. Wash plates ~ 3 times, each time filling the wells with ~ 300 μL PBST/well.
5. Add 100 μL/well of serially diluted samples of external secretions to be analyzed, as well as standards in blocking solution, and incubate overnight at 4 °C. The standards, consisting of a pool of human serum with pre-determined amounts of each immunoglobulin isotype or of purified S-IgA, are necessary to determine the total levels of IgA, IgG or IgM (and IgA or IgG subclasses). The standards should be diluted with blocking buffer, so that the concentration for each isotype analyzed range between 50 and 0.195 ng/mL. The suggested starting dilutions for measurement of total immunoglobulin levels in samples to be analyzed are presented in **Table 22.2**. Vaginal secretions collected at the time of menstruation are contaminated with blood that can increase the immunoglobulin levels. Consequently the sample dilution should be higher.
6. Wash plates ~ 3 times, each time filling the wells with ~ 300 μL PBST/well.
7. For the detection of antibodies captured on the plate, add 100 μL/well of appropriate dilution of biotin-labeled $F(ab')_2$ of goat IgG anti-human immunoglogulins in blocking solution (e.g., 1:5,000 dilution) and incubate for 3 h at 37 °C. Bring plates to room temperature before washing.

Table 22.2
Suggested dilutions for the measurement of total IgG, or IgA or IgM antibodies in various external secretions by ELISA

External secretion	Starting dilution IgG	IgA	IgM
Tears	1:1,000	1:5,000	
Saliva – whole	1:200	1:8,000	1:100
Saliva – parotid	1:200	1:8,000	1:100
Nasal wash	1:500	1:1,000	1:50
Milk	1:10,000	1:100,000	
Intestinal fluid	1:20	1:100	1:20
Rectal wash	1:10	1:200	
Urine	1:100	1:50	
Vaginal wash	1:1,000	1:1,000	1:50
Semen	1:1,000	1:200	1:50

8. Wash plates ~ 3 times, each time filling the wells with ~ 300 μL PBST/well.
9. Add 100 μL/well HRP-conjugated ExtrAvidin 1:2,000 dilution, or Neutravidin 1:5,000 dilution in blocking solution and incubate for 1 h at 37 °C. Allow plates to reach room temperature.
10. Wash plates ~ 3 times, each time filling the wells with ~ 300 μL PBS/well. The washing buffer should not contain any detergent or sodium azide, which inhibits the peroxidase activity.
11. Add 100 μL/well solution of HRP substrate (OPD or TMB) and incubate at room temperature until color develops (~ 15–30 min).
12. Stop the color reaction with 1M sulfuric or hydrochloric acid (100 μL/well).
13. Read the absorbance at 490 nm (OPD) or 450 nm (TMB) in a microplate reader and store the data in the connected computer.

ELISA results: The immunoglobulin levels are calculated by interpolating the ODs of samples on calibration curves constructed from the ODs of known amounts (ng/ml) of immunoglobulin isotype from standardized serum or S-IgA, by using a computer program (e.g., Delta Soft, BioMettalics, Princeton, NJ). The sensitivity of the assay for total immunoglobulin is < 1 ng/mL.

3.2. ELISA for HIV-1 Antigen-Specific Antibodies

For indirect assessment of levels of HIV-1-specific antibodies in secretions, standard curves for total isotype should be constructed using the standardized serum or the purified colostral IgA *(8)*. The steps to be followed are similar to those described for total immunoglobulin levels. The differences are emphasized below:

1. Coat plates with 100 μL/well PBS-sodium azide solution containing the selected HIV-1 antigen and two rows, dedicated to standard, with unlabeled anti-human immunoglobulin reagent of the isotype pertinent to the analysis of HIV-1-specific antibodies (e.g., to measure IgG anti-HIV-1 in secretions, the rows for standardized serum should be coated with goat anti-human IgG). The optimal concentration for each HIV-1 antigen should be determined by checkerboard titration. Generally, concentrations of 0. 5–1 μg/mL are appropriate. For the standard, the coating concentration should be the same as for the level of total isotype (see above).

2, 3 and 4. are the same steps as for total immunoglobulin level ELISA.

5. Add 100 μL/well of serially diluted samples of external secretions to be analyzed, as well as standards and internal control samples diluted in blocking solution; incubate overnight

at 4 °C. Internal controls and standards should be included to assess the inter-assay variations and to generate quantitative data, expressed in ng/mL. The positive controls consist of serum or secretions from HIV-1-infected individuals, preferably corresponding to the samples to be assayed. The negative controls are samples obtained from healthy individuals and are necessary for assessing the nonspecific binding of immunoglobulins from sera and external secretions to the selected HIV-1 antigens. To determine levels of HIV-specific antibodies in secretions, several dilutions of samples should be examined according to the antigen and the secretion tested. The background "noise," determined with samples from healthy volunteers, can indicate the starting dilution for HIV-1 antibody-positive samples. Generally a range from 1:10 to 1: 320 is suitable for detection of HIV-1-specific antibodies in external secretions. The dilution for standards is described in 3.1.

6 to 13. are the same steps as for determination of total immunoglobulin levels.

The results for HIV-1-specific antibodies can be expressed as ng/ml or as arbitrary units. They are obtained from extrapolation of sample OD values against the standard curves generated with known levels of standardized serum or colostral S-IgA (as for total immunoglobulin levels), or with dilutions of positive control, respectively *(8)*. Samples can be designated as positive for HIV-1 antibodies if the values are > 2 SD above the mean of the values obtained for the negative control. The HIV-1-specific values can be related to total level of a particular immunoglobulin isotype and expressed as percentage of total isotype, or as ELISA units/μg of immunoglobulin isotype.

For both HIV-1-specific and total immunoglobulin levels ELISA, quality assurance should be incorporated into the analysis. All samples need to be assayed in duplicate and the coefficients of variation (CV) values should be assessed. Only CV values $\leq$ 10% should be accepted; if higher values are obtained, the samples should be rerun. For total immunoglobulin determinations, the standard curves should be monitored for consistent slope and r.

Three categories of quality control should be considered:

— intra-assay variation, represented by variation between replicates of the same sample;
— inter-assay variation, represented by the reproducibility of assay results obtained on different occasions;
— the consistency of results obtained from different dilutions of the sample and how different samples behave in this respect (the so-called "parallelism" problem).
— The precautions mentioned above will ensure consistent performance.

3.3. Chemiluminescence-Enhanced Western Blotting

Because of the low levels of the total as well as HIV-1-specific antibodies in external secretions, regular processing of WB strips containing HIV-1 antigens yields unsatisfactory results. Therefore, we adopted a highly sensitive ECL-WB assay *(9, 10)*, which allows a reliable determination of HIV-1-specific antibodies of all major isotypes *(4)*. Furthermore, ECL-WB permits the simultaneous detection of antibodies to up to nine HIV-1 antigens.

3.3.1. Development of Western Blot Strips Containing HIV-1 Antigens by ECL

1. Remove strips, with a forceps, from the sealed tube and place them on a clean aluminum foil. Make sure the number marked on the strip faces up. Each tube contains 27 numbered strips. If fewer strips are to be used, remove the needed strips and leave the unused strips in their original tube and store in the dark at 2–8 °C.
2. Cut off ~ 0. 5 cm from the pink colored end of each strip (opposite to the numbers).
3. Place the strips with a forceps in a clean mini incubation tray. Strips should be grasped only at the numbered edge.
4. Add 2 mL of wash buffer, brought at room temperature, to each well and incubate for 10 min on a rocker. Trays should be rocked at an angle of approximately 7° at 15–18 cycles per minute. Aspirate the liquid.
5. Add 1.0 mL Superblock buffer to each well and incubate on rocker for 60 min before adding the samples to be analyzed. This step is necessary in order to reduce background staining which may present difficulties in the interpretation of the results of samples with weak reactivity with HIV-1 antigens (see **Note 1, 2**). Place the tray cover.
6. Using separate pipette tips, add to each strip a sample of either positive, or negative control or the secretion to be analyzed (to the numbered end of the stips). Replace cover. For the detection of IgA or IgG antibodies to HIV-1, strips should be incubated with appropriately diluted samples according to the level of total immunoglobulin isotype in each secretion (determined by ELISA, or **Table 22.1**). The sample dilutions should range from 1:10 to 1:100. For each antibody isotype analyzed and for each batch of strips, one strip should be incubated only with the blocking buffer without sample in order to determine the background of developing reagents.
7. Incubate on rocker overnight at 4 °C. Bring tray to room temperature before step 8.
8. Using a separate pipet tip for each well, aspirate the fluid and then add 1.5 mL washing buffer to each well. Replace cover.
9. Incubate on rocker for 10 min at room temperature.
10. Repeat steps 8 and 9 three times.

11. Aspirate wells and add to each well 1.0 mL of biotin-labeled polyclonal anti-immunoglobulin isotype-specific antibodies diluted in Superblock buffer (e.g., 1:30,000–1:50,000 for anti-IgA and 1:50,000–1:70,000 for anti-IgG). Replace cover.
12. Incubate on rocker for 4 h at room temperature or, overnight at 4 °C.
13. Aspirate wells and add 1.5 mL washing buffer to each well. Replace cover.
14. Incubate on rocker for 10 min at room temperature.
15. Repeat steps 13 and 14 three times.
16. Add 1.0 mL of 1:20,000 dilution of HRP-labeled Neutravidin in Superblock buffer and cover the tray.
17. Incubate on rocker for 60 min at room temperature.
18. Aspirate wells and add 1.5 mL washing buffer to each well. Replace cover.
19. Incubate on rocker for 10 min at room temperature.
20. Repeat steps 18 and 19 four times.
21. Aspirate wells and add 1.0 mL of chemiluminescence substrate solution to each well. Make sure that strips are immersed in the substrate solution. Replace cover.
22. Incubate on rocker for 10 min at room temperature.
23. Aspirate wells.
24. Carefully place strips on Whatman paper and blot them dry.
25. Developed strips should be stored in the dark at room temperature. Extended exposure to light may cause bands to fade. It is recommended that strips be interpreted as soon as possible after the strips are dried (maximum 30 min).
26. Place the dry blots in a cassette, using a clear plastic sheet cover, and expose them to high resolution film (for 2 s up to 20 min).
27. Develop film and evaluate the number and intensity of positive bands on a light box.

    The results should be interpreted in relation to the strongly positive and negative controls included in each assay (**Fig. 22.1**). The interpretation is based on the number of positive bands and their intensity. According to an extensive evaluation of several thousand samples *(11)*, the serum antibody responses in early HIV infection are limited only to Env or Env and p24 antigens. However, at later stages, antibodies with reactivity to most, if not all, of the nine HIV antigens on WB strips are detectable. Although very rare (~ 0.0004%), false-positive WB bands, resembling the early sero-conversion, can occur due to cross-reactivity with an epitope on gp41 or p24. Samples can be considered positive when at least two of the p24, gp41, gp120/160 bands, with intensity comparable to the positive control, are present.

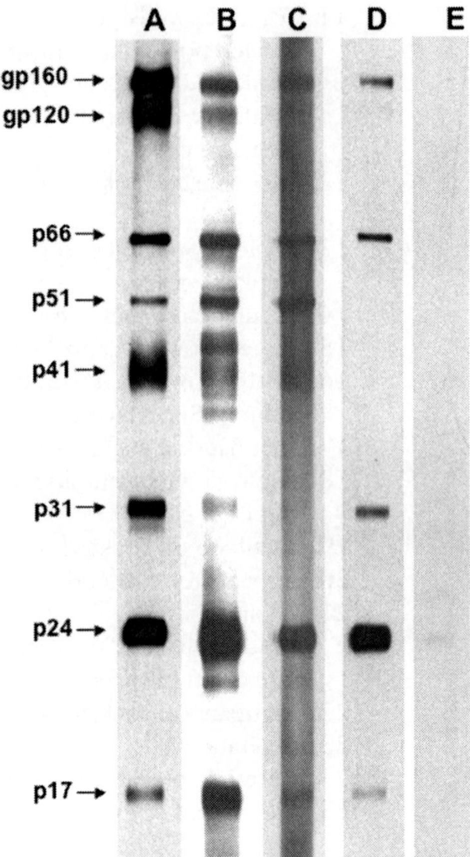

Fig. 22.1. Demonstration of HIV-1-specific antibodies in vaginal secretions by ECL-WB. **A**. Serum from an HIV-1-positive individual, strongly reacting with eight HIV-1 antigens on the blot. **B** and **C** are WB incubated with the vaginal secretion from an HIV-1-infected patient and developed with anti-human IgG (B) and with anti-human IgA (C). **D** shows the presence of IgG antibodies, in vaginal lavage of another subject, with specificity for only four of the HIV-1 antigens tested. **E**. IgG anti-HIV-1 antibodies in the vaginal secretion of a healthy volunteer.

4. Notes

1. For quantitative determinations of HIV-1-specific antibodies in ng/mL, we construct the standard curve, as described. Ideally, experimental values should be related to standard curves generated with the use of polyclonal mIgA1, pIgA1, and S-IgA (from, for example milk), IgG1-4, and IgM. This, however, is practically an impossible task. In addition to the fact that the levels of HIV-1-specific antibodies are generally low in all fluids and, when present, may occur in both mIgA and pIgA forms, acid-dissociation used for the release of specific

antibodies from affinity-chromatography columns may alter the binding affinity and specificity of such desorbed antibodies *(12)*. Moreover, IgA is prone to self-aggregation, particularly at acid pH. Therefore, the described procedure represents the best alternative for HIV-specific antibody quantitation.

2. In ECL-WB, the background staining may be especially strong with intestinal fluids and rectal washes, which are used at minimal dilution. The procedures used for the collection of these fluids result in a considerable dilution of samples and consequently, in low levels of total as well as HIV-1 antigen-specific antibodies.

References

1. Jackson, S., Mestecky, J., Moldoveanu, Z., et al. (2005) Appendix I: Collection and processing of human mucosal secretions, in (Mestecky, J., Bienenstock, J., Lamm, M. E., Mayer, L., McGhee, J. R., and Strober, W., eds.), *Mucosal Immunology*, 3rd ed., pp. 1829–1839 Elsevier/Academic Press, Amsterdam.
2. Delacroix, D. L., Hodgson, H. J. F., McPherson, A., et al. (1982) Selective transport of polymeric immunoglobulin A in bile. *J. Clin. Invest.* 70, 230–241.
3. Wright, P. F., Kozlowski, P. A., Rybczyk, G. K., et al. (2002) Detection of mucosal antibodies in HIV-1 Type 1-infected individuals. *AIDS Res. Human Retrovir.* 18, 1291–1300.
4. Mestecky, J., Jackson, S., Moldoveanu, Z., et al. (2004) Paucity of antigen-specific IgA responses in sera and external secretions of HIV-1-infected individuals. *AIDS Res. Human Retrovir.* 20, 972–988.
5. Mestecky, J. (2007) Humoral immune responses to the human immunodeficiency virus Type-1 (HIV-1) in the genital tract as compared to other mucosal sites. *J. Reprod. Immunol.* 73, 86–97.
6. Jackson, S., Prince, S., Kulhavy, R., et al. (2000) False positivity of enzyme-linked immunosorbent assay for measurement of secretory IgA antibodies directed at HIV type 1 antigens. *AIDS Res. Human Retrovir.* 16, 595–602.
7. Raux, M., Finkielsztejn, L., Salmon-Ceron, D., et al. (1999) Development and standardization of methods to evaluate the antibody response to an HIV-1 candidate vaccine in secretions and sera of seronegative vaccine recipients. *J. Immunol. Methods* 222, 111–124.
8. Russell, M.W., Mestecky, J. Julian, B.A., et al. (1986) IgA-associated renal diseases: antibodies to environmental antigens in sera and deposition of immunoglobulins and antigens in glomeruli. *J. Clin. Immunol.* 6, 74–86.
9. Mohamed, A.O., Ashley, R., Goldstein, A., et al. (1994) Detection of rectal antibodies to HIV-1 by a sensitive chemiluminiscent Western blot immunodetection method. *J. Acquir. Immune Defic. Syndr.* 7, 375–380.
10. Constantine, N. T., Bansal, J., Zhang, X., et al. (1994) Enhanced chemiluminescence as a means of increasing the sensitivity of Western blot assays for HIV antibody. *J. Virol. Methods* 47, 153–164.
11. Kleinman, S., Busch, M.P., Hall, L., et al.. (1998) False-positive HIV-1 test results in a low-risk screening setting of voluntary blood donation. Retrovirus epidemiology donor study. *J. Am. Med. Assoc.* 280, 1080–1085.
12. McMahon, M.J., O'Kennedy, R. (2000) Polyreactivity as an acquired artefact, rather than a physiologic property, of antibodies: evidence that monoreactive antibodies may gain the ability to bind to multiple antigens after exposure to low pH. *J. Immunol. Methods* 241, 1–10.

Chapter 23

Isolating Mucosal Lymphocytes from Biopsy Tissue for Cellular Immunology Assays

Barbara L. Shacklett, J. William Critchfield, Donna Lemongello

Abstract

Mucosal tissues of the gastrointestinal and genitourinary tracts serve as major portals of HIV-1 transmission, and recent literature has highlighted the important role of these tissues in pathogenesis. However, our understanding of human mucosal T-cell responses remains limited. We have previously reported methods for isolating, culturing and analyzing mucosal T-lymphocytes obtained from gastrointestinal biopsy tissue. This method of acquiring tissue is minimally invasive and well accepted by patients, and allows sampling of sites that would not otherwise be accessible without surgical intervention. This chapter summarizes the approach currently in use in our laboratory to isolate and study CD4+ and CD8+ T-cells from rectal biopsies obtained through flexible sigmoidoscopy. These methods are also applicable, with minor modifications, to small tissue samples obtained from other lymphoid tissues.

Key words: Mucosa, lymphocyte, CTL (cytotoxic T-cell).

1. Introduction

The gastrointestinal tract is considered to be the largest lymphoid organ in the body, and contains the majority of the body's lymphocytes. The intestinal mucosa plays an important role in immune defense, and must strike a balance between tolerance to food antigens and commensal microbes, and responsiveness to mucosal pathogens. In keeping with this dual role in tolerance and responsiveness, gastrointestinal T cells have been characterized as adopting a unique, partially activated memory phenotype that distinguishes them from blood T cells *(1–3)*.

Vinayaka R. Prasad, Ganjam V. Kalpana (eds.), *HIV Protocols: Second Edition, vol. 485*

DOI 10.1007/978-1-59745-170-3_23 Springerprotocols.com

The gut contains immunological inductive sites, such as Peyer's patches and organized lymphoid aggregates within the intestine, and adjacent mesenteric lymph nodes *(4)*. Antigen-specific T cells that are primed in mucosal inductive sites are evidently preprogrammed to return to the mucosa, due to expression of mucosal homing integrins and chemokine receptors. Intestinal effector cells include two distinct populations: intraepithelial lymphocytes (IEL), which have a cytotoxic phenotype and are primarily CD8+, and lamina propria lymphocytes (LPL), which are mainly CD4+ in healthy individuals.

The gastrointestinal lamina propria is a site of profound CD4+ T-cell depletion during acute HIV-1 infection *(5–8)*, and restoration of these cells following initiation of antiretroviral therapy is delayed and incomplete *(8, 9)*. Despite the importance of this site in HIV-1 infection, little is known of the immune response to HIV-1 in mucosal tissues. To better assess the phenotypic characteristics and effector functions of mucosal T cells, several groups have developed protocols for isolating intestinal lymphocytes, either from surgical resection specimens or from biopsies obtained during endoscopy, colonoscopy or sigmoidoscopy. This chapter presents a simple method for processing biopsy tissue that has been adapted from earlier protocols, which used primarily surgical resection tissue *(10–12)*.

In optimizing this protocol, we compared the efficiency of four different techniques for isolating mucosal mononuclear cells from rectal biopsy tissue: manual disruption with a stainless steel mesh screen, mechanical disruption in a commercial tissue homogenizer, enzymatic digestion using a mixture of collagenase and dispase enzymes, and digestion with collagenase type II alone *(13)*. While the mechanical disruption approach was the most rapid and technically straightforward, collagenase type II digestion yielded the greatest number of viable lymphocytes, generally four- to fivefold greater than that obtained by other methods *(13)*. Both yield and cell viability were enhanced using this approach.

The protocol consists of two steps: first, fresh tissue biopsies are incubated with collagenase, type II; second, viable lymphocytes are separated from enterocytes and debris using density gradient centrifugation. The resulting lymphocyte suspension contains a mixture of intraepithelial and lamina propria T cells, as well as B cells. The suspension may be processed in parallel with peripheral blood mononuclear cells (PBMC), and is suitable for most phenotypic and functional assays, including MHC class I tetramer staining, cell surface phenotyping, and cytokine flow cytometry (CFC) *(13–15)*. If desired, cells also may be polyclonally expanded and cultured, then analyzed by ELISPOT or used as effector cells in bulk ^{51}Cr release assays *(16, 17)*.

2. Materials

2.1. Biopsy Collection

1. Flexible sigmoidoscope equipped with single-use disposable biopsy forceps (*see* **Note 1**).
2. Collection medium: RPMI-1640, supplemented with 15% fetal bovine serum, penicillin/streptomycin and L-glutamine (also referred to as "complete medium").
3. Additional antibiotics: piperacillin-tazobactam (Zosyn®, 500 μg/mL) and amphotericin B (optional) (1.25 μg/mL) (*see* **Note 2**).

2.2. Lymphocyte Isolation from Biopsy Tissue

1. Collagenase type II-S (clostridiopeptidase A from Clostridium histolyticum)
2. Shaking incubator, 37 °C.
3. 30-cm^3 disposable syringes attached to a blunt-ended 16-gauge needle.
4. Sterile plastic strainers, 70 μm mesh.

2.3. Discontinuous Percoll Gradient Centrifugation

1. Isotonic Percoll solution: Mix 1 part 10× PBS (or Hank's Basic Salt Solution, HBSS) with nine parts Percoll. Store at 4 °C.
2. 35% and 60% isotonic Percoll in PBS. For 100 mL of 60% Percoll, mix 60 mL of isotonic Percoll with 40 mL PBS. To prepare 100 mL of 35% Percoll, mix 35 mL of isotonic Percoll with 65 mL PBS. Add a few drops of 0.5% phenol red dye (in PBS) to the 60% Percoll solution. This will permit easy visual differentiation of layers within the discontinuous gradient. Store both solutions at 4 °C.

2.4. Polyclonal Expansion

1. For stimulation, select one of the following:
   a. Phorbol myristate acetate (PMA) and ionomycin, 5 ng/mL and 0.5 μg/mL, respectively, in complete medium *(18)*.
   b. Anti-CD3 monoclonal antibody 12F6, 0.1 μg/mL in complete medium (antibody provided by Dr. Johnson T. Wong, Massachusetts General Hospital). Bispecific antibodies, recognizing either CD3:CD4 or CD3:CD8, are available from the same source *(19, 20)*. The CD3:CD4 antibody stimulates depletion of CD4+ T-cells and expansion of CD8+ T-cells, while the CD3:CD8 antibody has the opposite effect. The bispecific MAbs are also used at 0.1 μg/mL in complete medium.
2. Human recombinant interleukin-2 (hrIL-2), 25,000 U/mL (500 ×), stored in aliquots at −80 °C.
3. Piperacillin-tazobactam (Zosyn®, 500 μg/mL) and amphotericin B (optional) (1.25 μg/mL).

4. Autologous PBMC, irradiated with 3000 cGy.
5. 24-well sterile polystyrene tissue culture plates.

3. Methods

3.1. Sample Collection

1. Biopsy samples are collected using sterile, single-use disposable biopsy forceps and are immediately placed in a sterile 50 mL conical centrifuge tube containing 10 mL of complete medium (*see* **Note 3**).
2. After releasing each tissue piece in the collection medium, rinse biopsy forceps with sterile PBS or distilled water before re-introducing into the patient.
3. At the conclusion of the biopsy procedure, close the collection tube and package according to local and federal shipping requirements. Transport the sample to the laboratory as quickly as possible, maintaining at room temperature (25 °C) during transport (*see* **Note 4**).

3.2. Lymphocyte Isolation from Biopsy Tissue

1. Immediately before each procedure, prepare a fresh solution of collagenase type II consisting of 0.5 mg collagenase/mL in RPMI with 5% FCS. Filter the solution (0. 22 μm) to sterilize. One hundred milliliters will be required for processing up to 20–25 individual tissue biopsies (*see* **Note 3**).
2. In a laminar flow biosafety cabinet, transfer the biopsies from the collection tube to a 70-μm nylon mesh screen. Gently wash the biopsies twice with complete medium. This is easily done by pipetting medium over the biopsies while holding the mesh screen over a waste receptacle or a 50-mL conical centrifuge tube.
3. *First incubation*: After washing, transfer the biopsies to a fresh 50 mL conical tube containing 30 mL collagenase solution. Incubate at 37 °C in a shaking incubator for 30 min.
4. After 30 min, break down tissue pieces by repeatedly (10–12 times) passing them through a 12-cm^3 syringe topped with a blunt-ended 16 gauge needle. This should greatly reduce the size of individual tissue fragments.
5. Strain the resulting cell suspension through a disposable nylon 70 μm strainer to separate free cells from undigested tissue. Collect the cell suspension in a fresh 50 mL conical tube. Label this tube "First Digestion".
6. *Second incubation*: Return the undigested tissue pieces to a 50-mL conical tube, add fresh collagenase solution (30 mL), and incubate in the shaking 37 °C incubator for 30 min.
7. Immediately remove collagenase from the cells that passed through the strainer ("First Digestion") by washing twice with PBS or culture medium. This is done by centrifuging

for 5 min in a tabletop centrifuge at 700 × *g*, 25 °C, to pellet the cells. Discard the supernatant. Immediate removal of collagenase helps preserve cell integrity and viability.

8. Resuspend the cell pellet in 40–50 mL PBS. Centrifuge again as in Step 7 and discard the supernatant.
9. Repeat Step 8.
10. After the second wash, resuspend the cell pellet in 6 mL complete medium. Place the tube containing the washed cells, loosely capped to allow gas exchange, in a tissue culture incubator at 37 °C, 5% CO_2.
11. Now return to the tissue pieces incubating in the presence of collagenase. Repeat steps 4 thru 6 for a total of three collagenase treatments. By the end of the third treatment, few tissue fragments will remain and most lymphocytes will have been released (*see* **Note 5**).
12. After the second and third collagenase treatments, follow steps 7–10 in order to remove excess collagenase. After washing, the cells that were isolated during the first, second and third treatments should be pooled such that the final total volume is 18 mL.

3.3. Percoll Gradient Enrichment

1. In 15 mL conical centrifuge tubes, prepare the Percoll gradients by underlaying 4 mL of 35% Percoll with 4 mL of 60% Percoll. Prepare as many tubes as needed for the number of biopsies collected. Each tube will accommodate 6 mL of cell suspension; thus, three tubes will be required for 18 mL of cell suspension. The 60% Percoll solution, which will be the lower solution in the tube, is tinted red in order to readily distinguish it from the 35% Percoll.
2. Refrigerate the gradients at 4 °C for one hour before using.
3. Gently layer 6 mL of cell suspension on top of each gradient. Final volume in each 15 mL conical tube will be 14 mL. Centrifuge at 700 × g for 20 min, 4 °C, *without brake.*
4. Enterocytes are located primarily at the top interface (between media and 35% Percoll). Lymphocytes are located primarily at the lower interface (between 35% and 60% Percoll) (*see* **Note 6**).
5. Place the harvested layers in a sterile, 50-mL conical centrifuge tube. Add PBS to a final volume of 50 mL. Invert the tube several times to mix well, then centrifuge for 10 min, 700 × g, 25 °C (*see* **Note 7**).
6. Discard supernatant and repeat the washing step: resuspend the cell pellet in a small volume of PBS, then add PBS to a final volume of 50 mL. Centrifuge as in step 5.
7. After the second wash, discard supernatant. Resuspend the cell pellet in 5 mL of complete medium. Count viable cells in a hemocytometer using trypan blue (*see* **Note 8**).

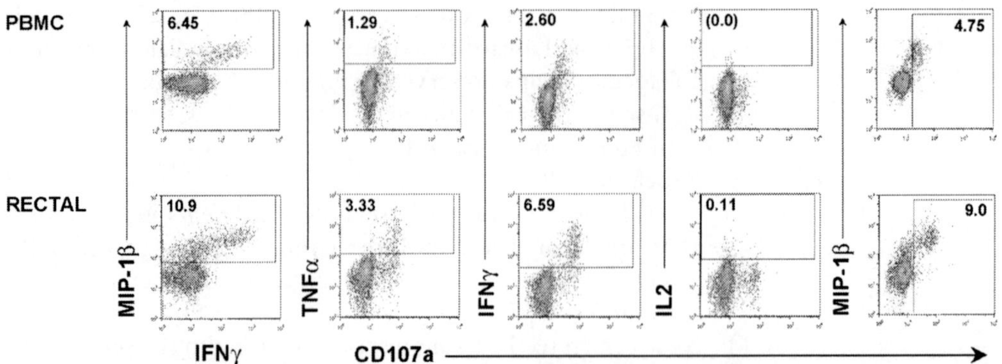

Fig. 23.1. HIV-1 gag-specific T-cell responses in rectal mononuclear cells and PBMC. This figure illustrates a standard cytokine flow cytometry (CFC) experiment performed in parallel on freshly isolated PBMC and rectal cells from an HIV-1-positive individual. Cells were stimulated with overlapping 15-mer peptides spanning the entire HIV gag protein and analyzed by nine-color flow cytometry for production of IFNγ, TNFα, IL2, MIP-1β. Degranulation was assessed by staining for CD107a. PBMC are shown on the *top row*; rectal mononuclear cells are shown on the *bottom row*. Numbers indicate percent CD8+ T-cells positive for each marker after subtracting background responses.

8. Adjust the volume as needed to give a final density of 1–2 × 10^6 cells/mL. Add piperacillin–tazobactam to a final concentration of 0.5 mg/mL (*see* **Note 2**).
9. Place the cell suspension in the tissue culture incubator at 37 °C, 5% CO_2. Cells may be used immediately or "rested" overnight in the incubator (*see* **Note 9**).
10. **Applications**. Mucosal lymphocytes may be used directly without further manipulation for most flow cytometry staining protocols and cellular immunology assays, including MHC class I tetramer staining, CFC, cell surface phenotyping, magnetic bead sorting, and flow cytometry-based cell sorting (*see* **Notes 10, 11**) (*see* **Fig**. 23.1). For ELISPOT assays, high levels of cytokine secretion preclude analysis of fresh mucosal cells. However, background becomes acceptably low in most cases after polyclonal expansion *(13, 16)* (*see* **Note 12**). Polyclonal expansion may also be used to generate bulk cultures that may be tested for cytotoxic function in bulk ^{51}Cr assays *(17)*.

3.4. Polyclonal Expansion

1. Start with 1–2.0 × 10^6 cells. Wash and resuspend at a density of 1–2 × 10^6/mL in complete medium supplemented with piperacillin-tazobactam (Zosyn®, 500 μg/mL), amphotericin B (optional) (1.25 μg/mL), and hrIL-2 (50 U/mL) (*see* **Note 2**).
2. Add **either** PMA/ionomycin (5 ng/mL and 0.5 μg/mL, respectively) **or** 0.1 μg/mL CD3:4 or CD3:8 bispecific Abs, or anti-CD3 MAb.
3. Three days later, add 5–10 × 10^6 autologous irradiated PBMC.

4. Twice weekly, change medium by removing as much supernatant as possible without disturbing the cells. Replace with fresh complete medium supplemented with IL-2 (50 U/mL) and antibiotics. Do not add more PMA or antibody.
5. Every 21–28 days, restimulate cells by adding fresh PMA/ionomycin or antibody.
6. Typically, $0.5–1.0 \times 10^6$ cells are expanded by 10- to 20-fold in 14–21 days using these approaches *(13, 16, 19, 21, 22)* (*see* **Note 12**).

4. Notes

1. Biopsies may be acquired from other sites, including jejunum, ileum, duodenum, and colon. These biopsies are typically 1–3 mm in diameter.
2. Piperacillin-tazobactam (Zosyn®) is particularly efficient at reducing contamination due to intestinal flora *(13)*. Amphotericin B is useful to prevent overgrowth of yeast, but may also retard growth of human cells.
3. The number of biopsies obtained will vary depending upon the site being biopsied and the experimental protocol. Individual research studies will have different protocols for collection of biopsy tissue. Details of any research protocols involving human subjects are subject to review by Institutional Review Boards of participating institutions. Research protocols involving human subjects should always be planned in consultation with collaborating clinicians and local regulatory authorities. Similarly, the details of any protocols involving nonhuman primates are subject to approval of Institutional Animal Care and Use Committees and should be worked out in consultation with veterinary clinicians.
4. It should be stressed that specimen collection and transport issues can dramatically affect cell yield. For best results, biopsy specimens should be transported to the laboratory immediately upon collection, and processed within 2–3 h. Delayed sample processing may greatly reduce cell yield, viability, and assay results. Overnight shipping of specimens is not recommended.
5. Particularly small biopsies may require only two collagenase incubations.
6. Depending on cell yield, the layers may be difficult to see; when in doubt, harvest the interface even if a distinct cell population is not visible. Although the upper Percoll interface should contain mostly epithelial cells, lymphocytes are frequently present as well. Yield can therefore be increased by combining the two layers. A triple gradient, consisting of

20–44–67% Percoll layers, may provide improved separation of epithelial cells and intestinal lymphocytes; in this configuration, enterocytes are found at the interface between 20% and 44% Percoll, and lymphocytes are enriched between the 44% and 67% Percoll layers *(18, 23)*. The presence of epithelial cells is generally not problematic for flow cytometry-based assays, since gating of T-cells is based on size, granularity and expression of lineage-specific surface antigens. However, if desired, epithelial cells may be selectively removed using magnetic beads coated with monoclonal antibody BerEP4 *(18)*.

7. It is important to wash in a large volume of PBS (45–50 mL) in order to dilute the Percoll, allowing cells to pellet.
8. Based upon our experience with human rectal tissue, the yield may vary from approximately 0.2 to 1×10^6 viable lymphocytes per biopsy piece (1–3 mm each). Variation may be caused by a number of factors, including the size of the biopsy forceps, patient-to-patient variation, and details of the isolation procedure.
9. Due to the length of time required to obtain and process primary tissue specimens, it is often convenient to “rest” cells in the incubator overnight before proceeding with experiments. We have found this procedure to be satisfactory in most cases; however, some loss of viability is to be expected.
10. Care should be used when designing flow cytometry experiments using cells that have been subjected to enzymatic digestion protocols. Some surface antigens are adversely affected, resulting in loss or modification of critical epitopes that allow detection using monoclonal antibodies. Collagenase is generally less detrimental to surface antigens than other enzymes such as dispase and pancreatin. Nevertheless, before undertaking these experiments, it is prudent to test all flow cytometry panels in parallel with untreated PBMC and PBMC that have been incubated with collagenase.
11. It is not practical to isolate IEL and LPL separately from small biopsies; however, if purified populations are desired, larger tissue pieces may be obtained from intestinal resection or gastric bypass surgery. IEL are then isolated by rapid shaking of tissue fragments at 37 °C in a solution of 0.75 mM ethylenediamine tetraacetic acid in HBSS. LPL are subsequently isolated by collagenase treatment (*10–12*).
12. Polyclonal expansion can modify activation status and expression of certain phenotypic markers *(13, 17)*. For this reason, freshly isolated cells are preferred for most cellular immunology assays. However, polyclonal expansion provides the advantage of increased cell numbers. In addition, for assays requiring a qualitative or semiquantitative readout, these differences may not be significant. Thus, this approach may

provide an appealing option in cases where cell number is limiting *(21)*. Before using polyclonally expanded cells for a particular application, the performance of fresh and expanded populations should be directly compared.

Acknowledgements

The authors thank Drs. Beth Jamieson, Otto Yang, Peter Anton and colleagues at UCLA, and Delandy Young and Timothy Hayes at UC Davis for assistance with protocol development. We also thank Dr. Johnson T. Wong, Harvard University for providing bispecific monoclonal antibodies and Dr. Ron Veazey, Tulane University, for sharing protocols and advice. This work was supported by NIH-NIAID R01 AI-057020 and grant CH05-D-606 from the California Universitywide AIDS Research Program (UARP).

References

1. Masopust, D., Vezys, V., Wherry, E. J., et al. (2006). Cutting edge: gut microenvironment promotes differentiation of a unique memory CD8 T cell population. *J Immunol* 176, 2079–2083.
2. Wang, H. C., Zhou, Q., Dragoo, J., et al. (2002). Most murine CD8(+) intestinal intraepithelial lymphocytes are partially but not fully activated T cells. *J Immunol* 169, 4717–4722.
3. Hayday, A., Theodoridis, E., Ramsburg, E., et al. (2001). Intraepithelial lymphocytes: exploring the Third Way in immunology. *Nat Immunol* 2, 997–1003.
4. Mowat, A., Viney, J. (1997). The anatomical basis of intestinal immunity. *Immunol Rev* 156, 145–166.
5. Mehandru, S., Poles, M. A., Tenner-Racz, K., et al. (2004). Primary HIV-1 infection is associated with preferential depletion of CD4+ T lymphocytes from effector sites in the gastrointestinal tract. *J Exp Med* 200, 761–770.
6. Mattapallil, J. J., Douek, D. C., Hill, B., et al. (2005). Massive infection and loss of memory CD4+ T cells in multiple tissues during acute SIV infection. *Nature* 434, 1093–1097.
7. Li, Q., Duan, L., Estes, J. D., et al. (2005). Peak SIV replication in resting memory CD4+ T cells depletes gut lamina propria CD4+ T cells. *Nature* 434, 1148–1152.
8. Guadalupe, M., Reay, E., Sankaran, S., et al. (2003). Severe CD4+ T-cell depletion in gut lymphoid tissue during primary human immunodeficiency virus type 1 infection and substantial delay in restoration following highly active antiretroviral therapy. *J Virol* 77, 11708–11717.
9. Guadalupe, M., Sankaran, S., George, M. D., et al. (2006). Viral suppression and immune restoration in the gastrointestinal mucosa of human immunodeficiency virus type 1-infected patients initiating therapy during primary or chronic infection. *J Virol* 80, 8236–8247.
10. Veazey, R. S., Rosenzweig, M., Shvetz, D. E., et al. (1997). Characterization of gut-associated lymphoid tissue (GALT) of normal rhesus macaques. *Clin Immunol Immunopathol* 82, 230–242.
11. Bull, D. M., Bookman, M. A. (1977). Isolation and functional characterization of human intestinal mucosal lymphoid cells. *J Clin Invest* 59, 966–974.
12. Davies, M. D., Parrott, D. M. (1981). Preparation and purification of lymphocytes from the epithelium and lamina propria of murine small intestine. *Gut* 22, 481–488.
13. Shacklett, B. L., Yang, O. O., Hausner, M. A., et al. (2003). Optimization of methods to assess human mucosal T-cell responses to HIV infection and vaccination. *J Immunol Methods* 279, 17–31.
14. Shacklett, B. L., Cox, C. A., Quigley, M. F., et al. (2004). Abundant expression of granzyme A, but not perforin, in granules of CD8+ T cells in GALT: implications

for immune control of HIV-1 infection. *J Immunol* 173, 641–648.

15. Shacklett, B. L., Cox, C. A., Sandberg, J. K., et al. (2003). Trafficking of HIV-1-specific CD8+ T-cells to gut-associated lymphoid tissue (GALT) during chronic infection. *J. Virol.* 77, 5621–5631.
16. Ibarrondo, F. J., Anton, P. A., Fuerst, M., et al. (2005). Parallel human immunodeficiency virus type 1-specific CD8+ T-lymphocyte responses in blood and mucosa during chronic infection. *J Virol* 79, 4289–4297.
17. Shacklett, B. L., Beadle, T. J., Pacheco, P. A., et al. (2000). Characterization of HIV-1-specific cytotoxic T lymphocytes expressing the mucosal lymphocyte integrin CD103 in rectal and duodenal lymphoid tissue of HIV-1-infected subjects. *Virology* 270, 317–327.
18. Lundqvist, C., Hammarstrom, M. L., Athlin, L., et al. (1992). Isolation of functionally active intraepithelial lymphocytes and enterocytes from human small and large intestine. *J Immunol Methods* 152, 253–263.
19. Wong, J. T., Colvin, R. B. (1987). Bi-specific monoclonal antibodies: selective binding and complement fixation to cells that express two different surface antigens. *J Immunol* 139, 1369–1374.
20. Yang, O. O., Kalams, S. A., Trocha, A., et al. (1997). Suppression of human immunodeficiency virus type 1 replication by CD8+ cells: evidence for HLA class I-restricted triggering of cytolytic and noncytolytic mechanisms. *J Virol* 71, 3120–3128.
21. Jones, N., Agrawal, D., Elrefaei, M., et al. (2003). Evaluation of antigen-specific responses using in vitro enriched T cells. *J Immunol Methods* 274, 139–147.
22. Bihl, F. K., Loggi, E., Chisholm, J. V., et al. (2005). Simultaneous assessment of cytotoxic T lymphocyte responses against multiple viral infections by combined usage of optimal epitope matrices, anti- CD3 mAb T-cell expansion and "RecycleSpot". *J Transl Med* 3, 20.
23. Leon, F., Roy, G. (2004). Isolation of human small bowel intraepithelial lymphocytes by annexin V-coated magnetic beads. *Lab Invest* 84, 804–809.

Subsection B

Measuring T Cell Responses via Flow Cytometry

Chapter 24

Quantifying HIV-1-Specific *CD8+* T-Cell Responses Using ELISPOT and Cytokine Flow Cytometry

Barbara L. Shacklett, J. William Critchfield, and Donna Lemongello

Abstract

Since the initial description and characterization of the agent that causes AIDS, human immunodeficiency virus (HIV-1), numerous research groups have characterized immune responses to this virus. Much effort has been directed towards identifying potential correlates of protection that may be useful for the development of vaccines and immunotherapies. In addition, several investigations have focused on comparing patients with rapid vs. slow disease progression profiles in an attempt to identify the characteristics of a "successful" immune response. Although many gaps remain in our understanding of the host–pathogen relationship, great progress has been made during the past 20 years in elucidating the adaptive, cell-mediated response to HIV-1. These investigations have benefited in recent years from the development of new approaches to the analysis of antigen-specific CD8+ T-cell function, notably the ELISPOT assay and cytokine flow cytometry. This chapter provides simple protocols for these two methods.

Key words: Cytokine, ELISPOT, flow cytometry, T-cell.

1. Introduction

CD8+ T-lymphocytes provide the body's major adaptive cellular immune response to viral infection. Antigen-specific CD8+ T-cells recognize viral antigen through the antigen-binding site of the T-cell receptor (TCR), encoded by the variable regions of the TCR α and β chains. Antigen is presented as a short peptide bound to an MHC class I molecule on the surface of an antigen-presenting cell. Antigen-specific CD8+ T-cells may be identified *in vitro* using reagents that mimic the MHC class I-peptide complexes encountered *in vivo*. This principle has been used to develop MHC class I dimers and multimers (tetramers, pentamers) that are folded around peptides corresponding to

Vinayaka R. Prasad, Ganjam V. Kalpana (eds.), *HIV Protocols: Second Edition, vol. 485*

DOI 10.1007/978-1-59745-170-3_24 Springerprotocols.com

known immunodominant viral epitopes *(1, 2)*. These complexes are typically labeled with fluorochromes, and can be used to identify T-cells bearing the cognate TCR by flow cytometry. While this elegant approach has proven to be both highly specific and sensitive, it does not provide information about antiviral effector functions.

The traditional method for measuring CD8+ T-cell effector function has been the ^{51}Cr release assay *(3, 4)*. In this assay, MHC class I matched or mismatched target cells expressing antigen are labeled with radioactive ^{51}Cr and incubated with potential effector cells for 4–6 h. Cytotoxicity is measured indirectly by quantifying ^{51}Cr released into the culture supernatant by dying target cells. A variation on this approach, known as limiting dilution analysis (LDA), or cytotoxic T-cell precursor frequency analysis, provides an estimate of the frequency of antigen-specific T-cells present in blood *(3)*.

The major advantage of the ^{51}Cr release assay is that it provides direct evidence of the ability of CD8+ T-cells to kill target cells expressing relevant antigen. However, disadvantages of this approach include technical complexity, reliance on radioactivity, and a high degree of inter-assay variability *(5)*. Over the past several years, these disadvantages have led to a wide acceptance of newer methods for assessing antigen-specific CD8+ T-cell function, notably the ELISPOT and cytokine flow cytometry (CFC) assays. Comparative studies in several viral systems have demonstrated that the older LDA approach systematically underestimated the frequency of circulating antigen-specific T-cells *(5–7)*. These findings, coupled with the rapid development of digital image analysis and multiparameter flow cytometry, have led to widespread acceptance of the ELISPOT and CFC assays.

CD8+ T-cells are most often associated with MHC class I-restricted cytotoxicity directed towards infected target cells, and the terms "antigen-specific CD8+ T-cell" and "cytotoxic T-cell (CTL)" are frequently used interchangeably. However, another important effector function of CD8+ T-cells is the release of immunomodulatory and antiviral cytokines. CD8+ T-cells can express a range of cytokines and chemokines, including but not limited to IFN-γ, TNF-α, IL-2, IL-10, TGF-β, MIP-1α, MIP-1β and RANTES. Both the ELISPOT and CFC assays are designed to detect the production of these factors upon TCR stimulation.

The enzyme-linked immuno-spot, or ELISPOT assay, uses a 96-well, membrane-bottom plate and an enzymatic reporter system to detect cytokines released by T-cells (**Fig. 24.1**). The assay may be conveniently spread out over three days: on the afternoon of the first day, plates are coated with primary antibody; on the second day, cells are added to the plate along with stimulating peptides, and incubated overnight; on the third day, the cells are washed away and the assay is developed, revealing cytokine

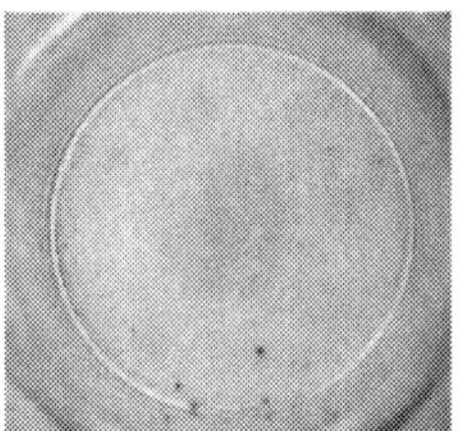 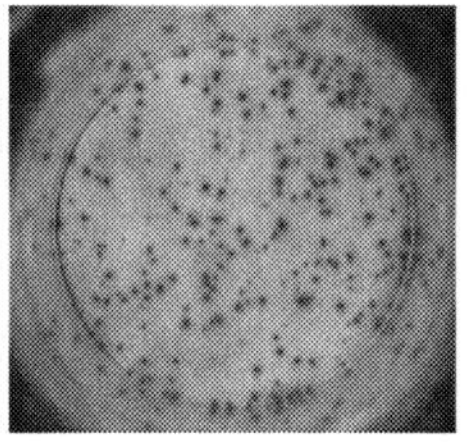 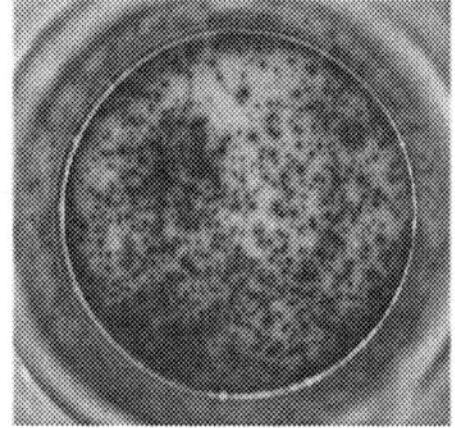

Fig. 24.1. **Elispot assay**. The photographs show a negative control well (*left*), positive control well (*right*), and a test well stimulated with HIV-1 peptides (*center*).

"spots" bound to the membrane. The ELISPOT can detect 50 or fewer antigen-specific T-cells per million lymphocytes, and is at least 1 $\log_{10}$ more sensitive than traditional LDA or bulk ^{51}Cr release assay *(8)*. ELISPOT may be used for rapid mapping of T-cell responses using peptides organized in a matrix *(9)*; the convenient assay format also lends itself to comprehensive mapping of human immunodeficiency virus (HIV-1)-specific responses using pooled peptides corresponding to all viral proteins *(10)*. The protocol outlined below is a basic ELISPOT protocol for detecting IFNγ production by antigen-specific T-cells. With appropriate primary and secondary antibodies, the same protocol may be adapted to detect numerous other cytokines and soluble factors.

Like ELISPOT, the CFC assay, also referred to as intracellular cytokine staining, measures cytokines produced by antigen-specific T-cells in response to TCR stimulation (**Fig. 24.2**). In this approach, cells are stimulated with antigen for several hours (usually 4–6 h) in the presence of compounds (brefeldin A, monensin) that inhibit cytokine release. At the end of the incubation period, the cells are stained with fluorochrome-labeled monoclonal antibodies recognizing surface antigens. The cells are then fixed, permeabilized and stained with fluorescent antibodies recognizing cytokines of interest and intracellular antigens.

The CFC assay provides several clear advantages over ELISPOT. First, flow cytometry allows visualization and phenotyping of cytokine-producing cells. Dead and dying dells, as well as non-T-cells, may be excluded from analysis, limiting false-positive results. Specific populations of T-cells (i.e., CD8+, CD4+, various memory/effector subsets) may be identified with confidence. Second, rapid new developments in the field have provided the capability of measuring as many as 19 parameters simultaneously *(11)*. This provides the potential for assessing multiple cytokines in a single assay, and for identifying cells expressing specific combinations of cytokines. An example of the power of this approach is the recent work suggesting that the "quality" of the HIV-1-specific CD8+ T-cell response, in terms of cytokines and effector molecules, may be an important determinant of disease progression *(12)*. The protocol presented here is a concise, basic four-color CFC assay that may be

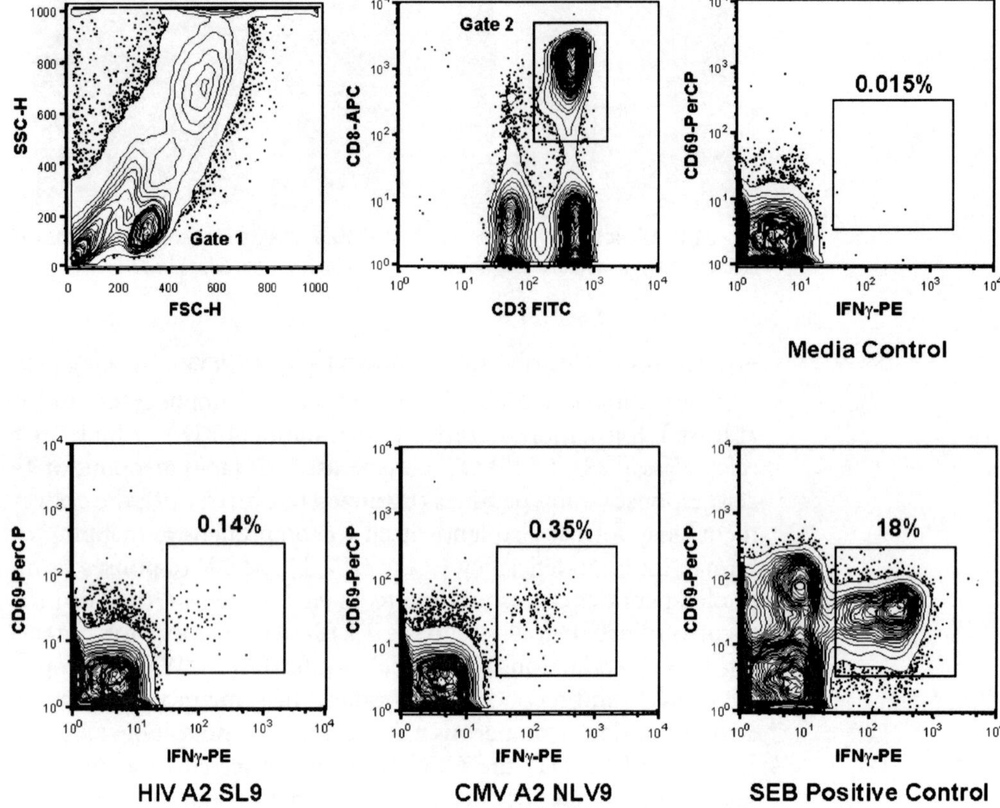

Fig. 24.2. **Four-color cytokine flow cytometry**. Results of a four-color assay using immunodominant peptides from HIV-1 gag and CMV pp65. *Top row*: negative control sample. Gate 1 (*left panel*) was drawn based on forward and side scatter. Gate 2 (*center panel*) was drawn based on CD3 and CD8. The final output (*right panel*) shows cells that have been selected sequentially by gates 1 and 2. The percentage refers to CD3 + /CD8+ cells producing IFNγ. *Bottom row*: PBMC from an HLA-A*0201 positive, HIV-1 infected individual respond to both the HIV-1 gag SL9 epitope (*left panel*) and the CMV pp65 NLV9 epitope (*center panel*). The *right panel* shows cells from the same patient stimulated with SEB. Percentages indicate CD3 + /CD8+ cells producing IFNγ.

readily adopted by users with simple dual-laser flow cytometers (e.g., FACS Calibur; Becton-Dickinson). It is based upon protocols by Picker *(13)* and Komanduri *(14)*. With relatively minor modifications and appropriate instrumentation, this protocol may be adapted to detect additional cytokines and cytolytic granule constituents *(15)*.

2. Materials

2.1. ELISPOT Assay

1. Primary antibody: anti-IFNγ mAb clone 1-DIK, 1 mg/mL (Mabtech Cat. No. 3420-3, Stockholm, Sweden). Store at 4 °C.

2. Coating buffer: sterile phosphate-buffered saline without Ca^{2+} and Mg^{2+} (PBS). Immediately prior to coating plates, prepare a solution of primary antibody (5 μg/mL) by adding 25 μL antibody to 5 mL PBS. 5 mL will be required for each ELISPOT plate. ELISPOT plates: sterile MultiScreen-HA 96-well filtration plate (Millipore MAHA S4510, Molsheim, France) (*see* **Note 1**).
3. Complete medium: RPMI 1640 supplemented with 10% heat-inactivated fetal calf serum, penicillin-streptomycin, and L-glutamine. Store at 4 °C.
4. HIV peptides: Suitable peptides are available from a number of sources. The NIH AIDS Research and Reference Reagent Program (Rockville, MD), supplies individual peptides as well as peptide sets corresponding to all HIV-1 proteins and multiple isolates. These may be used individually or prepared as pools. Peptides should be stored at −80 °C, and multiple freeze/thaw cycles should be avoided. Both CD8+ and CD4+ T-cells are adequately stimulated by 15-mer peptides *(16–18)* (*see* **Note 2**).
5. Positive control: Phytohemagglutinin (PHA). PHA is resuspended in PBS at 1 mg/mL and aliquoted, then stored at −80 °C. It is used at a final concentration of 5–10 μg/mL in the ELISPOT and CFC assays (*see* **Note 3**).
6. Wash solution: 0.05% Tween-20 (Sigma). Store at 4 °C.
7. Secondary antibody: anti-IFNγ mAb, clone 7-B6-1-Biotin, 1 mg/mL (Mabtech Cat. No. 3420–6). This will be diluted in PBS to a final concentration of 1 μg/mL immediately prior to use (*see* **Note 1**).
8. Detection reagent: Streptavidin-Horseradish Peroxidase (SA-HRP). This will be diluted 1:100 (or as appropriate based upon titration experiments) immediately prior to use in PBS containing 10% FCS.
9. Substrate: Stable diaminobenzidine (DAB), (Invitrogen Cat. No. 750118). Stable DAB must not be subjected to repeated freeze-thaw cycles. For best results, dispense in 5- to 10-mL aliquots upon receipt, store at −20 °C and thaw only once, immediately prior to use. The solution should be reddish in color when thawed; if it turns brown or a precipitate forms, it should be discarded.
10. Trypan Blue viability dye and hemocytometer for cell counting.
11. Disposable plastic reagent reservoirs.
12. Liquid waste containers.

2.2. Cytokine Flow Cytometry

1. Monoclonal antibodies (*see* **Note 4**):
Costimulatory antibodies (unlabeled): CD28, CD49d; Fluorochrome-conjugated antibodies to CD3, CD8, IFNγ, and CD69. A suitable antibody combination for a typical

four-color experiment would include: CD3-FITC, IFNγ-PE, CD69-PerCP (or PE-Cy5, CyChrome, Tricolor), CD8-APC (or Alexa 647). This assumes a dual-laser cytometer equipped with a 488-nm blue laser and a 635-nm red laser. Appropriate fluorochrome-labeled antibodies are available from a variety of sources.

2. Brefeldin A. Resuspend and aliquot in DMSO at 5 mg/mL, and store at −80 °C. Immediately prior to use, thaw and dilute 1:5 in medium; 2 μL will be needed for each test well (*see* **Note 5**).
3. Assay Plates: 96-well V-bottom polystyrene plates and lids.
4. HIV Peptides: Suitable peptides are available from a number of sources. The NIH AIDS Research and Reference Reagent Program (Rockville, MD) supplies individual peptides and peptide sets corresponding to all HIV-1 proteins. Peptide stocks should be stored at −80 °C; repeated freeze/thaw cycles should be avoided. Both CD4+ and CD8+ T cells are adequately stimulated by 15-mer peptides *(16–18)*.
5. Positive Control: Staphylococcal enterotoxin B. SEB should be resuspended in PBS at 500 μg/mL and stored in aliquots at −80 °C. Alternatively, phorbol myristate acetate (PMA) and ionomycin may be used as a positive control. PMA is reconstituted at 1 mg/mL in DMSO. A 10 μg/mL stock can then be made by diluting 1:100 in DMSO. Both stock solutions should be aliquoted and stored at −80 °C. Ionomycin is reconstituted at 1 mM in 100% EtOH, then aliquoted and stored at −80 °C (*see* **Note 3**).
6. FACS buffer: PBS with 2% heat-inactivated FCS.
7. FACS buffer (as above) with 0.5 mM EDTA.
8. Formaldehyde: Dilute 20% ultrapure aqueous formaldehyde (Tousimis, Rockville, MD) in PBS to prepare 4% and 1% solutions (*see* **Note 6**).
9. Permeabilization reagent: FACS-Perm (10× concentrate, BD Biosciences). Prepare by diluting 1 mL of the commercial reagent in 9 mL of dH_2O. Important: do not substitute PBS for dH_2O.

3. Methods

3.1. ELISPOT Assay

Note: All work should be performed in a laminar flow tissue culture hood at the appropriate biosafety level.

3.1.1. Day 1: Coating of ELISPOT Plates

1. Prepare the primary antibody solution (*see* Materials) (*see* **Note 1**).

2. Add 50 μL of the primary antibody solution to each well of a sterile ELISPOT plate, ensuring that the bottom of each well is entirely covered; refrigerate at 4 °C overnight (*see* **Note 7**).
3. Plan the layout of the ELISPOT plate and determine the number of cells required for the experiment. For best results, 1×10^5 or 2×10^5 cells, depending upon cell availability, should be added to each well. Sensitivity has been shown to be optimal when 2×10^5 cells are used *(8)*. Each condition should be assayed in duplicate or triplicate. Allow sufficient cells to include negative and positive control wells in duplicate or triplicate.
4. If using cryopreserved PBMC, these should be thawed, washed twice in medium, counted and resuspended at $1–2 \times 10^6$/mL. They should then be "rested" in a tissue culture incubator (37 °C, 5% CO_2) until the following day.

3.1.2. Day 2: Addition of Cells and Stimulating Antigen

1. Invert the plate briskly over a liquid waste container. Discard unbound primary antibody and pat down firmly onto paper towels to remove excess liquid from wells. Using a multichannel pipettor, add 200 μL PBS to each well. Carefully avoid touching the bottom of wells with pipet tips. Invert the plate briskly to discard liquid, and again pat down firmly onto paper towels. Repeat the above washing step with PBS three more times (*see* **Note 8**).
2. Block the plate by adding 50 μL of complete medium to each well. Incubate the plate at 37 °C, 5% CO_2 for 1 h.
3. While the plate is blocking, count and wash the cells, and determine the viability using Trypan blue. Resuspend the cells in complete medium at a concentration of 1×10^5 or 2×10^5 cells per 100 μL (*see* **Note 9**).
4. After the 1-h blocking is complete, do not remove the blocking solution. Add 100 μL cell suspension to each well. The final volume in each well should then be 150 μL. It is critical at this step that cells be thoroughly resuspended. Cell clumps can interfere with spot quantitation and lead to inaccurate results (*see* **Note 10**).
5. Add antigen and controls. If using synthetic peptides to stimulate cells, add peptide such that the final concentration is 5–10 μg/mL. For each patient sample, allow two negative control wells containing peptide vehicle or complete medium in place of peptide. Also allow two positive control wells for each patient sample. Add PHA to these wells such that the final concentration is 10 μg/mL. Return the ELISPOT plate to the incubator at 37 °C, 5% CO_2 and incubate overnight (16–20 h). Do not move or jar the plate during this time (*see* **Notes 11, 12**).

3.1.3. Day 3: Color Development

1. Prepare a solution of biotinylated secondary antibody at 1 μg/mL in PBS. 5 mL of this solution will be required for each ELISPOT plate (*see* **Note 1**).
2. Remove the ELISPOT plate from the incubator. As in step 5 above, invert the plate briskly over a liquid waste container and discard the cell suspension (*see* **Note 13**). Immediately add 200 μL washing solution to each well. Invert the plate again to discard washing solution and pat firmly on paper towels to remove excess liquid from wells. Repeat washing three more times.
3. Add 50 μL of the secondary antibody solution (step 10) to each well, and incubate the plate at 37 °C, 5% CO_2 for 2 h (*see* **Note 12**).
4. Before the end of the 2-h incubation, prepare the SA-HRP solution. Dilute SA-HRP 1:100 in PBS containing 10% FCS. Five milliliters of this solution will be required for each ELISPOT plate (*see* **Note 1**).
5. Invert plate to discard unbound secondary antibody and pat firmly on paper towels to remove excess liquid from wells. Add 200 μL 0.1% Tween solution to each well (500 μL Tween 20, 500 mL PBS). Discard wash solution as above and repeat wash three more times.
6. Add 50 μL of the SA-HRP solution to each well and incubate the plate at room temperature on a flat surface for 1 h (*see* **Note 12**).
7. Wash wells four times as in step 14. To reduce background, the plastic backing may be carefully removed and the underside of the plate rinsed with wash buffer using a plastic squeeze bottle. Take care not to damage the membrane at this stage.
8. Add 50 μL stable DAB to each well. Incubate at room temperature for 5 min on a flat surface (*See* **Note 12**).
9. Invert plate to discard DAB. Rinse wells thoroughly 3× with dH_2O. As before, remove the plastic backing and carefully rinse the back side of wells with water to help to reduce background staining. Allow the plate to air-dry.
10. Count spots under an inverted microscope or with the aid of an automated ELISPOT reader. Automated readers are believed to be superior to manual counting in terms of reproducibility, consistency and ease of data management *(19, 20)*.
11. Analyze ELISPOT data by subtracting the average spot-forming cells (SFC) in replicate background wells (stimulated by media alone) from the average SFC in test wells. Convert the result to SFC per 10^6 cells by multiplying by 5 (in the case of 2×10^5 cells/well) or 10 (in the case of 1×10^5 cells/well).

12. Determine positive responses. A useful cutoff is 50 SFC per 1×10^6 cells *(8)*; however, investigators may wish to determine their own criteria for positive responses based upon results in low-risk negative controls (*see* **Note 14**).

3.2. Basic 4-Color Cytokine Flow Cytometry

Note: All work should be performed in a laminar flow tissue culture hood at the appropriate biosafety level.

1. Either fresh or cryopreserved cells may be used. Cryopreserved PBMC should be thawed the day before the assay and "rested" overnight prior to assaying. Cells should be washed once in medium, resuspended at $1\text{–}2 \times 10^6$/mL, and maintained in a tissue culture incubator (37 °C, 5% CO_2) until use. For best results, $5 \times 10^5 - 1 \times 10^6$ cells should be tested per assay condition (i.e., per well). Lower numbers may be used, but sensitivity will be compromised since results are reported as a percentage of CD8+ T-cells expressing cytokine (*see* **Notes 15, 16**).
2. Label a 96-well V-bottom plate for the assay. For each patient sample, prepare a plate map showing each of the conditions to be tested as listed below. To limit the possibility of cross-contamination, skip every second column and every second row of the 96-well plate (*see* **Note 17**).
   i. Unstimulated cells (media and costimulatory antibodies only).
   ii. HIV peptides or peptide pools.
   iii. Positive control (SEB or PMA/ionomycin) (*see* **Note 3**).
   iv. Unstained cells (1 well) and compensation controls (4 wells) (*see* **Note 18**).
3. Create a Master Mix to deliver costimulatory antibodies and Brefeldin A by combining the following reagents: 1 μL CD28, 1 μL CD49d, 2 μL Brefeldin A (final concentration 10 μg/mL). The volumes shown are per sample and should be multiplied by the total number of patient samples (*see* **Notes 19 and 20**).
4. Distribute 4 μL of the Master Mix to the bottom of wells i thru iv for each patient sample.
5. **Antigen Stimulation**. Add appropriate stimuli to individual wells as follows. For best results, peptides, (including commercial preparations) should be titrated prior to the assay to determine the optimal volume. Numbers in parentheses are final concentrations.
   i. Unstimulated cells: no additions.
   ii. HIV peptides or peptide pools (5–10 μg/mL).
   iii. 5 μL SEB (5 μg/mL) or PMA/ionomycin (25 ng/mL and 1 μM, respectively).
   iv. Unstained cells and compensation controls: no additions.
6. To each of the test wells, add 1×10^6 cells in 200 μL media. Mix well by pipetting up and down at least 5 to 10 times. For

unstained cells and compensation controls, only 50–100,000 cells are required.

7. Cover the 96-well plate and incubate for 5 h in a tissue culture incubator, 37 °C, 5% CO_2.
8. Remove the plate from the incubator and centrifuge at 4 °C for 5 min at 700 × g. Remove supernatant by quickly inverting the plate over a liquid waste container and giving it one brisk shake. Gently blot the plate on absorbent towels.
9. Using a multichannel pipettor, resuspend cells thoroughly in 200 μL FACS buffer containing 0.5 mM EDTA. Mix 5–10 times manually using the pipettor. Incubate the plate for 10 min at 37 °C, 5% CO_2.
10. Remove the plate from the incubator and centrifuge at 4 °C for 5 min at 700 × g. Remove supernatant by quickly inverting the plate over a liquid waste container and giving it one brisk shake. Gently blot the plate on absorbent towels.
11. Resuspend cells thoroughly in 50 μL FACS buffer without EDTA. Mix 5 to 10 times manually using the pipettor.
12. **Surface Staining**. Add antibodies corresponding to surface stains (i.e., CD8) to all wells except the unstained control. Mix gently by pipetting 5–10 times. Antibodies should be carefully titrated; typical volumes range from 1 to 15 μL per stain for most commercial antibodies (*see* **Note 21**).
13. Incubate the plate for 20 min in the dark at 4 °C (*see* **Note 22**).
14. To each well, add 150 μL FACS buffer. Mix 5 to 10 times with pipettor. Then centrifuge the plate at 4 °C for 5 min at 700 × g. Remove supernatant as described above.
15. **Fixation**. Resuspend cell pellets in 200 μL of 4% formaldehyde. Mix 5–10 times with pipettor. Incubate for 5 min at 37 °C, 5% CO_2. Centrifuge as described above and decant supernatant (*see* **Note 23**).
16. **Permeabilization**. Resuspend cell pellets in 200 μL of 1X FACS Perm. Mix 5 to 10 times with pipettor. Adequate mixing is particularly important at the permeabilization stage. Incubate at room temperature for 10 min. Centrifuge and decant (*see* **Note 24**).
17. Resuspend cell pellets in 200 μL of FACS buffer. Mix 5 to 10 times with pipettor. Centrifuge and decant.
18. **Intracellular Staining**. Add appropriately titred intracellular antibodies in a total volume of 50 μL of FACS buffer to all wells except the unstained control. Intracellular antibodies will include CD3, IFNγ, CD69, and any other cytokines being tested (*See* **Notes 20 and 21**). Mix gently by pipetting 5 to 10 times. All antibodies should be carefully titrated; typical volumes range from 1 to 15 μL per stain for most commercial antibodies.
19. Incubate at 4 °C for 20 min in the dark.

20. Add 150 μL of FACS buffer. Mix gently by pipetting 5 to 10 times. Centrifuge and decant.
21. Repeat step 20, this time adding 200 μL of FACS buffer (*see* **Note 25**).
22. Transfer cells from the 96-well plate to 12 × 75 mm FACS tubes using 2 × 150 μL 1% Paraformaldehyde (PFA) in PBS. The final volume in each FACS tube will be 300 μL. Store samples at 4 °C protected from light for a minimum of 1 h and a maximum of 24 h prior to data acquisition.
23. Analyze by flow cytometry, acquiring 100,000 lymphocyte events for each sample. For unstained controls and compensation beads, collect 20,000 events.
24. Calculate antigen-specific responses by subtracting the percent positive events in the media (unstimulated) control from the percent positive events in the peptide-stimulated sample. Response magnitude will vary by donor (*see* **Note 26**).

4. Notes

1. ELISPOT plates are available with either mixed cellulose ester or polyvinylidene fluoride membranes. This protocol was developed for cellulose ester plates. Optimal antibody concentrations should always be verified by titration.
2. ELISPOT protocols utilizing recombinant vaccinia viruses (r-VV) have also been described *(21)*. r-VV expressing HIV-1 proteins are available from the NIH AIDS Research and Reference Reagent Program.
3. It is critically important to run both positive and negative control wells for each individual patient sample. Positive control wells serve not only to validate the experimental conditions, but also as controls for cell responsiveness. A poor response to stimulation (i.e., fewer than 200 SFC/10^6) may indicate poor viability or improper cryopreservation. Many compounds can serve as appropriate positive controls. PHA, frequently used in ELISPOT assays, is a plant lectin believed to nonspecifically cross-link TCRs. PMA, a protein kinase activator, along with a Ca^{2+} ionophore such as ionomycin, is commonly used in CFC assays due to its broad induction of cytokines *(13)*. Staphylococcal enterotoxins A and B (SEA, SEB), act as superantigens, triggering only cells expressing certain Vβ chains *(13)*. A useful antigen-specific positive control for both assays is the recently described CEF peptide pool available from the NIH AIDS Research and Reference Reagent Program and several commercial suppliers. This reagent contains immunodominant peptides from three common viruses (CMV, EBV, Influenza A) and is recognized by 85% of the general population *(22)*.

4. The CFC assay was initially described with the inclusion of CD69 as a marker for early activation *(13)*. However, this marker is not absolutely essential. In addition to IFNγ, TNFα and IL2 have been particularly useful for delineating functional CD8+ T-cell subsets *(12,23,24)*. The recently described degranulation assay includes antibodies directed towards CD107a, a component of cytolytic granules *(15, 25)*. With appropriate instrumentation, such as 3- or 4-laser cytometers, more elaborate panels may be constructed, incorporating 10 or more colors *(11, 12, 26)*.
5. Brefeldin A is a macrolide antibiotic that specifically blocks protein transport from the endoplasmic reticulum to the Golgi complex. Its inclusion in the assay is necessary to prevent cytokine release. Some protocols also include monensin *(15)*. Monensin neutralizes the pH of endosomal and lysosomal compartments, reducing loss of fluorescence signal by antibodies in these compartments. Monensin also inhibits secretion of certain cytokines, including IFNγ, IL2, and MIP-1β. It is also absolutely required for assessing degranulation by staining for CD107a *(15)*. However, unlike brefeldin A, monensin does not prevent release of TNFα.
6. In aqueous solution, PFA becomes formaldehyde. PFA in solid form is quite stable, but inconvenient to manipulate; aqueous formaldehyde is easier to use. Once diluted in PBS, the working solution is stable for only 1–2 weeks at 4 °C. Concentrated aliquots may be kept for several months if frozen at −20 °C or below.
7. Coated plates should be left in the refrigerator for a minimum of 5 h prior to use, or can remain at 4 °C for up to 5 days before proceeding to the next step. Plates should be protected with plastic film to reduce evaporation and should be discarded if wells have dried out.
8. Throughout the protocol, ELISPOT plates may be washed using a handheld squeeze bottle dispensing wash solution. It is critical that the ELISPOT plate not be allowed to dry out at any time during the assay.
9. Poor cell viability will inevitably lead to poor results. Improper cryopreservation and/or thawing may decrease viability *(8)*; controlled stepwise cryopreservation is optimal for preserving cell integrity *(27)*.
10. To assess CD4+ or CD8+ T-cells separately, magnetic bead depletion may be used to remove either population prior to the ELISPOT assay.
11. For best results, the final concentration of DMSO in the reaction should be $< 1\%$ *(8, 27)*. Optimal peptide concentration is in the range of 5–10 μg/mL; 2 μg/mL may be sufficient to obtain reasonable stimulation, and concentrations at or above 20 μg/mL may be inhibitory *(8)*. Recent evidence suggests

that the cytokine response profile is dictated in part by peptide concentration *(28)*.

12. Incubation times should be respected carefully. The overnight incubation should last 16–20 h *(8)*. Exceeding the recommended incubation times for secondary antibody, enzyme, and/or substrate may result in high background.
13. When working with biohazardous material, it is imperative to follow standard procedures for inactivating infectious agents. Infectious material may be discarded by briskly inverting the ELISPOT plate over a liquid waste container containing bleach (sodium hypochlorite). Alternatively, infectious material may be removed from the ELISPOT plate using a multichannel pipettor, then immediately discarded in a liquid waste container. To inactivate HIV, a final concentration of 0.5 percent sodium hypochlorite, equivalent to a 1:10 dilution of household bleach, is recommended *(29)*.
14. Different research groups have reported different cutoffs, based on empirical *(10, 22)* and statistical approaches *(8, 19, 30)*. ELISPOT sensitivity may be enhanced by adding exogenous cytokines *(31)*, costimulatory antibodies (CD28, CD49d) *(32)*, or antigen-presenting cells (B-cells, dendritic cells) *(33, 34)*. However, these approaches may also increase nonspecific background.
15. Treatment with RNAse-free DNAse I (final concentration 2.5 U/mL) for 5 min at 37 °C may reduce clumping of thawed cryopreserved cells *(35)*. Cells should be washed following DNAse treatment.
16. CFC protocols using whole blood are also available; these require lysis of red blood cells using commercially available lysis buffers *(36)*.
17. Individual samples may also be processed in 12 × 75 mm polystyrene tubes rather than 96-well plates.
18. Compensation beads may be used in place of cells. These are polystyrene microparticles that bind mouse, rat or hamster immunoglobulins bearing κ light chains (BD Pharmingen).
19. Costimulatory antibodies should be titrated to determine optimal signal/noise ratios.
20. When measuring degranulation using antibodies to CD107a, these must be added along with the Master Mix *(15)*. To detect CD107a, monensin must also be added at this time (1 μM final concentration), and the brefeldin concentration should be reduced to 5 μg/mL. Master Mix should be adjusted accordingly.
21. In CFC protocols, cell surface phenotyping is performed before cell fixation to prevent destruction of relevant antigens by the fixative. However, some cell surface antigens, such as CD3, are internalized upon stimulation. CD4 is internalized upon stimulation with PMA and ionomycin *(13)*.

22. It is critically important that cells and monoclonal antibodies be maintained at 4 °C or on ice unless otherwise indicated. Fluorochrome-labeled antibodies (and cells that have been stained) should be shielded from light.
23. It is critical to fix the cells after the 5-h incubation and prior to permeabilization. Fixation serves two purposes: it prevents loss of intracellular cytokines once the cells are subsequently permeabilized, and also serves to crosslink and retain monoclonal antibodies bound to cell surface antigens. However, cells must not be "over-fixed", as this may prevent antibodies from binding to intracellular antigens.
24. Permeabilization is critical for allowing access of antibodies to intracellular antigens. Inadequate permeabilization may result in poor cytokine staining.
25. Adequate washing is critical following intracellular staining; inadequate washing may lead to high background staining.
26. DNA intercalating dyes such as ethidium monoazide (EMA) and amine reactive dyes may be used to exclude dead cells from subsequent gating *(35, 37)*. Also, a lymphocyte gate based on forward vs. side scatter cannot exclude small monocytes and B cells. To exclude these populations, antibodies to CD14 and CD19 labeled with the *same* fluorochrome may be added to define a "dump" channel *(35)*. These two approaches are rather impractical for four-color experiments, but should be considered seriously for experiments using six or more colors.

Acknowledgments

This work is supported by NIH/NIAID AI057020 and by the California Universitywide AIDS Research Program (UARP), grant CH05-D-606. The authors thank their colleagues in the Department of Medical Microbiology and Immunology, UC Davis, for assistance with protocol optimization. We also thank Drs. Douglas Nixon (University of California, San Francisco) and Michael Betts (University of Pennsylvania) for sharing protocols and technical information.

References

1. Altman, J. D., Moss, P. A., Goulder, P. J., et al. (1996) Phenotypic analysis of antigen-specific T lymphocytes. *Science* 274, 94–96.
2. Ogg, G. S., Jin, X., Bonhoeffer, S., et al. (1998) Quantitation of HIV-1-specific cytotoxic T lymphocytes and plasma load of viral RNA. *Science* 279, 2103–2106.
3. Coligan, J., Kruisbeek, A., Margulies, D., et al. (1999), Current Protocols in Immunology, Wiley Interscience, New York.
4. Nixon, D. F., Townsend, A. R., Elvin, J. G., et al. (1988) HIV-1 gag-specific cytotoxic T lymphocytes defined with recombinant vaccinia virus and synthetic peptides. *Nature* 336, 484–487.

5. Doherty, P. C. (1998) The new numbers game for virus-specific CD8+ T cells. *Science* 280, 227.
6. Tan, L. C., Gudgeon, N., Annels, N. E., et al. (1999) A re-evaluation of the frequency of CD8+ T cells specific for EBV in healthy virus carriers. *J Immunol* 162, 1827–1835.
7. Lalvani, A., Brookes, R., Hambleton, S., et al. (1997) Rapid effector function in CD8+ memory T cells. *J Exp Med* 186, 859–865.
8. Russell, N. D., Hudgens, M. G., Ha, R., et al. (2003) Moving to human immunodeficiency virus type 1 vaccine efficacy trials: defining T cell responses as potential correlates of immunity. *J Infect Dis* 187, 226–242.
9. Anthony, D. D., Lehmann, P. V. (2003) T-cell epitope mapping using the ELISPOT approach. *Methods* 29, 260–269.
10. Addo, M. M., Yu, X. G., Rathod, A., et al. (2003) Comprehensive epitope analysis of human immunodeficiency virus type 1 (HIV-1)-specific T-cell responses directed against the entire expressed HIV-1 genome demonstrate broadly directed responses, but no correlation to viral load. *J Virol* 77, 2081–2092.
11. Perfetto, S. P., Chattopadhyay, P. K., Roederer, M. (2004) Seventeen-colour flow cytometry: unravelling the immune system. *Nat Rev Immunol* 4, 648–655.
12. Betts, M. R., Nason, M. C., West, S. M., et al. (2006) HIV nonprogressors preferentially maintain highly functional HIV-specific CD8+ T cells. *Blood* 107, 4781–4789.
13. Picker, L. J., Singh, M. K., Zdraveski, Z., et al. (1995) Direct demonstration of cytokine synthesis heterogeneity among human memory/effector T cells by flow cytometry. *Blood* 86, 1408–1419.
14. Komanduri, K. V., Viswanathan, M. N., Wieder, E. D., et al. (1998) Restoration of cytomegalovirus-specific CD4+ T-lymphocyte responses after ganciclovir and highly active antiretroviral therapy in individuals infected with HIV-1. *Nat Med* 4, 953–956.
15. Betts, M. R., Koup, R. A. (2004) Detection of T-Cell Degranulation: CD107a and b, in: Cytometry, New Developments, Vol. 75, pp. 497–512 (Darzynkiewicz, Z., Roederer, M. and Tanke, H. J., Eds.) Elsevier Press, San Diego, CA.
16. Kern, F., Surel, I. P., Brock, C., et al. (1998) T-cell epitope mapping by flow cytometry. *Nat Med* 4, 975–978.
17. Betts, M. R., Ambrozak, D. R., Douek, D. C., et al. (2001) Analysis of Total Human Immunodeficiency Virus (HIV)-Specific CD4(+) and CD8(+) T-Cell Responses: Relationship to Viral Load in Untreated HIV Infection. *J Virol* 75, 11983–11991.
18. Cherepnev, G., Volk, H. D., Kern, F. (2004) Use of Peptides and Peptide Libraries as T-Cell Stimulants in Flow Cytometric Studies, in (Darzynkiewicz, Z., Roederer, M. and Tanke, H. J., Eds.), *Methods in Cell Biology*, Vol. 75, pp. 453–479 Elsevier, San Diego.
19. Herr, W., Protzer, U., Lohse, A. W., et al. (1998) Quantification of CD8+ T lymphocytes responsive to human immunodeficiency virus (HIV) peptide antigens in HIV-infected patients and seronegative persons at high risk for recent HIV exposure. *J Infect Dis* 178, 260–265.
20. Janetzki, S., Schaed, S., Blachere, N. E., et al. (2004) Evaluation of Elispot assays: influence of method and operator on variability of results. *J Immunol Methods* 291, 175–183.
21. Larsson, M., Jin, X., Ramratnam, B., et al. (1999) A recombinant vaccinia virus based ELISPOT assay detects high frequencies of Pol-specific CD8 T cells in HIV-1-positive individuals. *AIDS* 13, 767–777.
22. Currier, J. R., Kuta, E. G., Turk, E., et al. (2002) A panel of MHC class I restricted viral peptides for use as a quality control for vaccine trial ELISPOT assays. *J Immunol Methods* 260, 157–172.
23. Harari, A., Rizzardi, G. P., Ellefsen, K., et al. (2002) Analysis of HIV-1- and CMV-specific memory CD4 T-cell responses during primary and chronic infection. *Blood* 100, 1381–1387.
24. Sandberg, J. K., Fast, N. M., Nixon, D. F. (2001) Functional heterogeneity of cytokines and cytolytic effector molecules in human cd8(+) t lymphocytes. *J Immunol* 167, 181–187.
25. Betts, M. R., Brenchley, J. M., Price, D. A., et al. (2003) Sensitive and viable identification of antigen-specific CD8+ T cells by a flow cytometric assay for degranulation. *J Immunol Methods* 281, 65–78.
26. Roederer, M., De Rosa, S., Gerstein, R., et al. (1997) 8 color, 10-parameter flow cytometry to elucidate complex leukocyte heterogeneity. *Cytometry* 29, 328–339.
27. Mwau, M., McMichael, A. J., Hanke, T. (2002) Design and validation of an enzyme-linked immunospot assay for use in clinical trials of candidate HIV vaccines. *AIDS Res Hum Retroviruses* 18, 611–618.
28. Betts, M. R., Price, D. A., Brenchley, J. M., et al. (2004) The functional profile of primary human antiviral CD8+ T cell effector activity is dictated by cognate peptide concentration. *J Immunol* 172, 6407–6417.
29. CDC (1987) Recommendations for Prevention of HIV Transmission in Health Care Settings. *MMWR* 36, 3S–18S.

30. Hudgens, M. G., Self, S. G., Chiu, Y. L., et al. (2004) Statistical considerations for the design and analysis of the ELISpot assay in HIV-1 vaccine trials. *J Immunol Methods* 288, 19–34.
31. Jennes, W., Kestens, L., Nixon, D., et al. (2002) Enhanced ELISPOT detection of antigen-specific T cell responses from cryopreserved specimens with addition of both IL-7 and IL-15-the Amplispot assay. *J Immunol Methods* 270, 99–108.
32. Ott, P. A., Berner, B. R., Herzog, B. A., et al. (2004) CD28 costimulation enhances the sensitivity of the ELISPOT assay for detection of antigen-specific memory effector CD4 and CD8 cell populations in human diseases. *J Immunol Methods* 285, 223–235.
33. Altfeld, M. A., Trocha, A., Eldridge, R. L., et al. (2000) Identification of dominant optimal HLA-B60- and HLA-B61-restricted cytotoxic T-lymphocyte (CTL) epitopes: rapid characterization of CTL responses by enzyme-linked immunospot assay. *J Virol* 74, 8541–8549.
34. Larsson, M., Wilkens, D. T., Fonteneau, J. F., et al. (2002) Amplification of low-frequency antiviral CD8+ T-cell responses using autologous dendritic cells. *AIDS* 16, 171–180.
35. Lamoreaux, L., Roederer, M., Koup, R. (2006) Intracellular cytokine optimization and standard operating procedure. *Nature Protocols* 1, 1507–1516.
36. Maecker, H. T., Rinfret, A., D'Souza, P., Darden, J., Roig, E., Landry, C. et al. (2005) Standardization of cytokine flow cytometry assays. *BMC Immunol* 6, 13.
37. Perfetto, S. P., Chattopadhyay, P. K., Lamoreaux, L., et al. (2006) Amine reactive dyes: an effective tool to discriminate live and dead cells in polychromatic flow cytometry. *J Immunol Methods* 313, 199–208.

Chapter 25

Multiparameter Flow Cytometry Monitoring of T Cell Responses

Holden T. Maecker

Abstract

HIV vaccine research increasingly uses polychromatic flow cytometry as a tool to monitor T cell responses. The use of this technology allows for the analysis of highly defined subsets of cells with unique phenotypes and functions. Ultimately, such studies may identify surrogate markers of protection from disease progression. However, this powerful technology comes with a number of technical hurdles, and there is a need to standardize the assays and protocols used in clinical trial monitoring. Here an optimized protocol, with variations for specific circumstances, is presented. This protocol covers the analysis of multiple cytokines, cell surface markers, and other functional markers such as perforin, CD107, and CD154. While the protocol can be adapted to various numbers of fluorescence parameters, optimized panels of 8–10 colors are presented.

Key words: Antigen-specific, intracellular staining, multicolor, polychromatic, fixation, permeabilization, AIDS vaccine research.

1. Introduction

HIV vaccine researchers have been among the first to adopt polychromatic flow cytometry (more than four colors) as a tool to dissect T cell responses to HIV infection and to HIV vaccines. There is little doubt that a successful HIV vaccine will need to induce a strong cellular immune response, as well as neutralizing antibodies *(1)*. However, the nature of a protective cellular immune response to HIV is only beginning to be elucidated, largely by studies in the SIV model *(2–8)*, and by examination of HIV+ long-term nonprogressors *(9 –11)*. The ability of polychromatic flow cytometry to interrogate multiple subsets of immune cells for their functional capacities makes it a powerful

Vinayaka R. Prasad, Ganjam V. Kalpana (eds.), *HIV Protocols: Second Edition, vol. 485*

DOI 10.1007/978-1-59745-170-3_25 Springerprotocols.com

tool for discovering potential surrogates of protection. In fact, it could be argued that this technology provides the most in-depth view currently possible into the workings of the human immune system at a cellular level.

What are the potential surrogates of T cell-based protection from HIV progression? Obviously, no definitive answers are yet available, as the only information comes from animal models or correlative studies in humans. However, current evidence suggests that HIV+ nonprogressors maintain proliferative capacity of their HIV-specific CD4+ and CD8+ T cells *(9, 10)*, at least in part by maintenance of the ability to produce IL-2 *(10–12)*. In fact, recent studies using polychromatic flow cytometry suggest that a greater proportion of HIV-specific CD8+ T cells in nonprogressors are "multifunctional" *(11)*, being able to produce several cytokines (e.g., IFNγ, TNFα, IL-2, and MIP-1β) and to degranulate (as evidenced by cell-surface CD107 expression).

Phenotypic studies in SIV models suggest that protected animals maintain T cells of a "central memory" phenotype (CD28+, CD95 +) in contrast to nonprotected animals *(7, 8)*. HIV+ progressors are known to show altered differentiation of HIV-specific T cells, as seen by staining for markers such as CCR7, CD62L, CD27, CD28, CD45RA, and CD127 *(13–19)*. Very recent work indicates that upregulation of a death receptor, PD-1, on HIV-specific T cells may lead to dysfunction of those cells and consequent disease progression *(20, 21)*. And extensive literature has correlated an elevated expression of activation markers such as CD38 and HLA-DR on CD8+ T cells with poor prognosis [reviewed in *(22)*].

Some researchers may believe that functions (e.g., cytokines and degranulation capacity) are more important than phenotypes (e.g., memory/effector markers) in categorizing antigen-specific T cells. But the emergence of markers such as PD-1 suggests that cell-surface proteins may hold important prognostic value as well. Thus, researchers find themselves in the position of wanting to monitor an ever-growing number of phenotypic and functional markers; and polychromatic flow cytometry is the best available tool to meet that goal.

1.1. Instrumentation Considerations

Polychromatic flow cytometry has been made practical only recently by the availability of commercial instruments with digital signal processing and detectors for up to 18 colors. Digital processing is important in that it allows for more precise calculation of optical spillover between detectors, and thus more precise compensation than was possible with analog systems *(23, 24)*. Equally important, software routines are now available that will automate the compensation process when presented with a set of single-color fluorescent controls. This makes it possible to collect data in eight or more colors with nearly the same ease as traditional two to four color experiments, despite an exponentially more complex spillover matrix.

Much of the instrumentation for polychromatic flow cytometry is customized, which means that each user's system can have different lasers and optical filters. This in turn can alter the efficiency of detection of particular fluorochromes, such that the same antibody panel on one instrument will not yield identical results on another instrument. Some degree of standardization, or at least awareness of these variables, is beginning to occur, with the emergence of groups sharing their experiences (see for example http://maeckerlab.typepad.com). A minimum level of instrument standardization will need to be defined in order to achieve comparable results with polychromatic antibody panels on different instruments. Such standardization is a prerequisite to doing multicenter trials in which polychromatic flow cytometry will be performed at more than one site; and it is certainly a prerequisite to comparing results from different vaccine trials.

Even with two identical cytometers, setup of the instrument is still a variable that can vastly alter results. There has been recent progress in automated software routines using standardized particles to create optimal instrument setups (e.g., CST, BD Biosciences). For users not equipped with such automated setup paradigms, manual procedures need to be defined that optimize photomultiplier tube (PMT) voltage gains *(25)*, as well as optimize instrument performance in general *(26)*. Finally, monitoring of certain cytometer parameters over time is an important quality control tool to ensure consistent data and to anticipate and/or identify potential problems that might require service.

1.2. Reagent Considerations

The mere possession of an instrument capable of detecting, for example, 12 fluorescence parameters, does not guarantee success in 12-color flow cytometry. Choice of fluorochromes, antibody specificities, and the combination of these into an optimized reagent panel, are important considerations. In many cases, increasing a panel by one additional fluorochrome detracts so severely from the resolution sensitivity in other detectors, that it is not warranted. This occurs because of fluorescence spillover, the fact that each fluorochrome contributes some signal to neighboring detectors as well as to its primary detector. Because the optical spectrum is limited, the addition of new fluorescent reagents becomes more and more difficult without creating severe spillover problems, as the number of fluorochromes in an experiment increases. This is clearly a case where more is not always better.

Selection of fluorochromes and antibody conjugates based on brightness and minimal spillover has been recently reviewed *(24)*. In general, one should start by selecting a set of fluorochromes that offer the greatest brightness, within the constraints of the

Table 25.1
Suggested fluorochrome configurations

Laser	6-color	8-color	10-color	12-color
488 nm	FITC	FITC	FITC	FITC
	PE	PE	PE	PE
488 nm or 532 nm			PE-Texas Red, PE-Alexa 594 or PE-Alexa 610	PE-Texas Red, PE-Alexa 594 or PE-Alexa 610
	PerCP-Cy5.5	PerCP-Cy5.5	PerCP-Cy5.5	PerCP-Cy5.5
	PE-Cy7	PE-Cy7	PE-Cy7	PE-Cy7
	APC	APC	APC	APC
633 nm			Alexa 680 or Alexa 700	Alexa 680 or Alexa 700
	APC-Cy7	APC-Cy7	APC-Cy7	APC-Cy7
		Pacific Blue	Pacific Blue	Pacific Blue
405 nm		AmCyan	AmCyan	AmCyan
				Qdot 655
				Qdot 705

user's instrument, while minimizing spectral overlaps between detectors. Suggestions of fluorochrome sets to use for common instrument configurations are given in **Table 25.1**.

Next, one should assign antibody specificities to particular fluorochromes by matching the dimmest specificities with the brightest fluorochromes. Further adjustments should then be made to minimize potential spillover issues. For example, two fluorochromes with significant spectral overlap might be used to identify non-overlapping cell populations, thereby negating their spillover. Conversely, one should avoid compromising a reagent for which high sensitivity is required by having a reagent in a neighboring detector that brightly stains the same cell population. For example, use of CD8 APC-Cy7 with anti-IL-2 APC is a potential problem, if IL-2 is to be detected on CD8+ cells. In this case, not only does APC-Cy7 cause spillover into the APC detector, but the tandem dye can also degrade, resulting in false positive signals in APC *(24)*. Among tandem dyes, APC-Cy7, followed by PE-Cy7, are most sensitive to such degradation, which is catalyzed by light, increased temperature, and exposure to fixative *(24)*. Finally, all of the above considerations need to be tempered by what antibody conjugates are commercially available or can be made by the investigator.

Once an antibody panel is selected, titration of certain reagents is often required to achieve an optimal signal:noise ratio. While many reagents are sold pre-titered, this does not always mean that the specified titer is optimal in a given application. Since polychromatic experiments already compromise sensitivity due to spillover between detectors, the need for optimal signal:noise is critical. It should also be noted that the optimal titer for cell-surface staining with an antibody is often higher than that for intracellular staining (after fixation and permeabilization). For many antibody specificities, resolution is compromised so drastically by fixation and permeabilization that these antibodies need to be used prior to application of a fixative.

Given the many considerations in optimizing a reagent panel for polychromatic flow cytometry, it makes sense to take advantage, wherever possible, of panels already validated by others. **Table 25.2** shows some staining panels successfully used in the

Table 25.2
Some suggested multicolor antibody panels

8-color[1]	10-color[2]	10-color[3]
Anti-IFNγ FITC	CD27 FITC	Anti-IFNγ FITC
Anti-IL-2 PE	CD154 PE	Anti-IL-2 PE
	CD107 PE-Alexa 610	Anti-TNFα PE-Alexa 610
CD28 PerCP-Cy5.5	CD4 PerCP-Cy5.5	CD28 PerCP-Cy5.5
CD45RA PE-Cy7	Anti-IFNγ PE-Cy7	CD45RA PE-Cy7
CD27 APC	Anti-IL-2 APC	CD27 APC
	Anti-TNFα Alexa 700	CD3 Alexa 700
CD8 APC-Cy7	CD8 APC-Cy7	CD8 APC-Cy7
CD3 Pacific Blue	CD3 Pacific Blue	CD4 Pacific Blue
CD4 AmCyan	CD14 AmCyan	CD14 AmCyan

[1]Used in ref. *(19)*.
[2]Used for International Flow Cytometry School (IFCS) 2006, Florence, Italy. Note that readout of IL-2 on CD8+ cells could be compromised by the use of CD8 APC-Cy7 with anti-IL-2 APC (*see* **Section 1.2**).
[3]Used for Multicolor ICS Users Group standardization studies (see also http://maeckerlab.typepad.com for additional panel suggestions and details).

author's laboratory. These can be taken as a starting point, given that small variations are still likely to be successful, as opposed to starting from "scratch" in the design of a new panel.

The basics of intracellular cytokine staining have been reviewed elsewhere *(27, 28)*, and tips for optimizing protocols have also been recently published *(29)*. The main focus of the protocol presented here is to show how intracellular cytokine staining can be adapted to a polychromatic format, which may include readout not only of phenotypic markers, but also of functional markers such as CD107 or CD154, in addition to cytokines. Variables in the stimulation and processing steps that apply to these markers and others are summarized in **Table 25.3**.

Table 25.3
Procedural variables for different functional markers

Variable	Covered by this protocol IL-2, IL-4, IL-5, IL-10[2], IL-13, IFNγ, MIP-1β, TNFα	CD107, CD154	Not covered[1] TGFβ
Stimulation conditions	6–12 h	5–6 h in the presence of staining antibodies for these markers	16–24 h in serum-free medium[3]
Secretion inhibitor	brefeldin A	monensin[4]	monensin
Fixation/ permeabilization system	FACS Lysing Solution, FACS Permeabilizing Solution 2[5]	FACS Lysing Solution, FACS Permeabilizing Solution 2[5]	Cytofix, Cytoperm[6]

[1]This marker would be difficult to combine with those in the first two columns, mainly due to the longer optimal stimulation time.
[2]We have performed IL-10 staining with some positive results under the listed conditions, but have not attempted to optimize conditions for IL-10 detection.
[3]Serum-free medium (e.g., AIM V, Invitrogen, Grand Island, NY) produces much stronger TGFβ responses, presumably because serum contains free TGFβ that blocks staining for this marker.
[4]CD107 and CD154 are taken up by endocytic vescicles and degraded. Monensin blocks this degradation by preventing acidification of these vescicles. When doing combined assays with cytokines, a monensin+brefeldin A combination is recommended (see **Note 5**).
[5]BD Biosciences, San Jose, CA. Note that the Cytofix/Cytoperm system (BD Biosciences, San Diego, CA) is also used successfully for these markers by many investigators.
[6]BD Biosciences, San Diego, CA.

2. Materials

2.1. Reagents

1. Freshly isolated or cryopreserved PBMC, isolated by Ficoll gradient centrifugation or via Cell Preparation Tubes (CPT; BD Vacutainer, Franklin Lakes, NJ) or equivalent.
2. RPMI-1640 medium with 20 m*M* HEPES, 10% fetal bovine serum, and antibiotic/antimycotic solution (cRPMI-10, components from Sigma Chemical Co., St. Louis, MO).
3. Stimulation antigens, e.g., peptide mixes [*see* **ref.** *(30)*], or SEB as a positive control. Optional: Preconfigured plates containing lyophilized stimulation reagents and secretion inhibitor(s) can be purchased [BD Lyoplate, BD Biosciences, San Jose, CA *(31)*].
4. Recommended: costimulatory antibodies to CD28 and CD49d, 0.1 mg/mL each in sterile PBS (FastImmune, BD Biosciences).
5. Brefeldin A, 5 mg/mL in DMSO (Fast Immune, BD Biosciences); or brefeldin A+monensin, 2.5 mg/mL each in 50% DMSO+50% methanol. For the latter stock, combine brefeldin A and monensin (Golgistop, BD Biosciences) 1:1. Aliquot and store both stocks at −20°C.
6. EDTA, 20 m*M* in PBS (pH 7.4) (FastImmune, BD Biosciences)
7. Cell fixation reagent, e.g., BD FACS Lysing Solution (BD Biosciences) or equivalent.
8. Cell permeabilizing reagent, e.g., BD FACS Permeabilizing Solution 2 (BD Biosciences) or equivalent.
9. Fluorescent-labeled antibodies (*see for example* **Table 25.2**). Optional: Preconfigured plates containing lyophilized antibody cocktails [BD Lyoplate, BD Biosciences *(31)*].
10. Wash buffer: 0.5% bovine serum albumin +0. 1% NaN_3 in PBS.
11. Recommended: BD CompBeads [anti-mouse Ig κ, anti-rat Ig κ, or anti-rat/hamster Ig κ (BD Biosciences)], for creating single-color compensation controls.
12. Optional: To reduce biohazard potential, or if samples will be stored > 24 h prior to acquisition: 1% paraformaldehyde in PBS [dilute 10% paraformaldehyde (EM Science, Gibbstown, NJ) 1:10 in PBS]; or BD Stabilizing Fixative (BD Biosciences).

2.2. Equipment

1. 96-well conical bottom polypropylene plates with lids [e.g., BD Falcon (Bedford, MA) or equivalent] (*see* **Note 1**).
2. 12-channel aspiration manifold with 7-mm prongs (V&P Scientific, San Diego, CA).
3. Plate holders for table-top centrifuge [e.g., Sorvall Instruments (Newtown, CT)].

4. Polychromatic flow cytometer with digital signal processing, e.g., BD LSR II (BD Biosciences) or Dako Cyan ADP (Dako Corporation, Fort Collins, CO).
5. Optional: 96-well plate loader for flow cytometer.

3. Methods

3.1. Sample Collection

1. For fresh PBMC (*see* **Note 2**): Resuspend at 5 × 10^6 to 1 × 10^7 viable lymphocytes/mL in warm (37 °C) cRPMI-10 (*see* **Note 3**).
2. For cryopreserved PBMC (*see* **Note 4**): Thaw briefly in a 37 °C water bath, then slowly dilute up to 10 mL with warm (37 °C) cRPMI-10 and centrifuge for approximately 7 min at 250 × *g*. Resuspend in a small volume of warm cRPMI-10, perform a viable cell count, and dilute to a final concentration of 5 × 10^6 to 1 × 10^7 viable lymphocytes/mL (*see* **Note 3**).
3. Add 200 μL of cell suspension per well to a 96-well plate (*see* **Section 2.2.1** for appropriate plates). For cryopreserved PBMC, incubate at 37 °C for 6–18 h prior to stimulation (*see* **Note 4**).

3.2. Cell Activation

1. For assays not involving CD107 or CD154: Thaw an aliquot of 5 mg/mL brefeldin A stock (*see* **Note 5**). Dilute 1:10 in sterile PBS to make a 50× working stock.
2. For assays measuring CD107 and/or CD154: Thaw an aliquot of 2.5 mg/mL brefeldin A+2.5 mg/mL monensin stock (*see* **Note 5**). Dilute 1:10 in sterile PBS to make a 50× working stock.
3. For assays using preconfigured lyophilized stimulation reagents in plates: Add 200 μL of cell suspension directly to the appropriate wells, let sit for a few minutes, then pipet up and down thoroughly to mix. Skip to **step 3.2.7**.
4. Resuspend peptides or peptide mixes in DMSO at a concentration of 500 μg/mL/peptide or greater (*see* **Note 6**). Store resuspended peptides in aliquots at −80 °C. Dilute peptide stocks in sterile PBS, if necessary, to achieve a 50× working stock that is between 50 and 100 μg/mL/peptide (when diluted 1:50, this will yield a final concentration of 1–2 μg/mL/peptide).
5. Prepare a 50× SEB stock of 50 μg/mL in sterile PBS. Store this stock at 4 °C (aliquoting is not necessary).
6. For each stimulation condition, prepare a "master mix" of the 50× working stocks and costimulatory antibodies as follows:
   4 μL/well peptides, SEB (positive control), or PBS (negative control).

4 μL/well brefeldin A or brefeldin A+monensin.
4 μL/well CD28 + CD49d Ab stock (*see* **Note 7**).

7. Pipet 12 μL of the appropriate master mix into each well containing cells. Mix by gently pipetting.
8. For assays involving CD107 and/or CD154, also add the recommended titer of the antibody conjugate(s) to each well. Minimize exposure to light, particularly for tandem dye conjugates (*see* **Note 8**).
9. Incubate covered plate for 6–12 h at 37 °C (*see* **Notes 9** and **10**).

3.3. Sample Processing

1. To halt activation and detach adherent cells, add 20 μL per well of 20 mM EDTA in PBS and mix by pipetting.
2. Incubate 15 min at room temperature, then mix again by vigorous pipetting to fully resuspend adhered cells.
3. Centrifuge plate at 250 × *g* for 5 min. Aspirate supernatant with 7 mm vacuum manifold (*see* **Note 11**).
4. For assays using amine-reactive dye for staining non-viable cells: Resuspend the amine dye at optimum concentration in PBS (usually around 2.5 μg/mL, but this should be determined for individual lots of dye). Resuspend each well with 100 μL of this solution, incubate 20 min at room temperature, then add 100 μL wash buffer, and wash as in **step 3.3.3** above.
   a. For assays using liquid reagents and cell-surface markers other than CD3, CD4, and CD8: Resuspend each well in 100 μL wash buffer and add optimal titers of all Abs to cell-surface markers (*see* **Note 12**), incubate 30–60 min at room temperature, then add 100 μL wash buffer, and wash as in **step 3.3.3** above.
   b. For assays using preconfigured lyophilized staining reagents and cell-surface staining Abs: Resuspend the appropriate wells of the surface Ab plate with 50 μL of wash buffer. Let sit for a few minutes, then pipet up and down thoroughly to mix. Transfer the solution to appropriate wells of the cell plate, incubate 30–60 min at room temperature in the dark, then add 100 μL wash buffer, and wash as in **step 3.3.3** above.
5. Resuspend cell pellets with 100 μL of 1× BD FACS Lysing Solution per well. Incubate at room temperature for 10 min (*see* **Notes 13** and **14**).
6. Add 100 μL wash buffer to each well, then centrifuge plate at 500 × *g* for 5 min (*see* **Note 15**). Aspirate supernatant with 7 mm vacuum manifold.
7. Resuspend cell pellets with 200 μL of 1× BD FACS Permeabilizing Solution 2 per well. Incubate at room temperature for 10 min (*see* **Note 14**).

8. Centrifuge plate at 500 × *g* for 5 min (*see* **Note 15**). Aspirate supernatant with 7 mm vacuum manifold.
9. Add 200 μL wash buffer to each well, and wash as in **step 3.3.8** above.
10. Again add 200 μL wash buffer to each well, and wash as in **step 3.3.8** above.
11. a. For assays using liquid reagents: Resuspend pellet in 100 μL wash buffer and add optimal titers of all Abs to intracellular markers. Incubate in the dark at room temperature for 60 min, mixing by pipetting or gentle agitation every 15–20 min.
    b. For assays using preconfigured lyophilized intracellular staining reagents: Resuspend the appropriate wells of the intracellular Ab plate with 50 μL of wash buffer. Let sit for a few minutes, then pipet up and down thoroughly to mix. Transfer the solution to the appropriate wells of the cell plate, and incubate at room temperature in the dark for 60 min, mixing by pipetting or gentle agitation every 15–20 min.
12. Add 200 μL of wash buffer to each well, and wash as described in Sample Processing step 8 above.
13. Again add 200 μL wash buffer to each well, and wash as described in Sample Processing step 8 above.
14. Resuspend pellets with 150 μL wash buffer. Store at 4 °C in the dark until ready for data acquisition, which should be performed within 24 h. Optional: resuspend pellets with 150 μL of 1% paraformaldehyde in PBS or BD Stabilizing Fixative (*see* **Note 16**).

3.4. Data Acquisition and Analysis

1. First determine optimal PMT settings for the instrument and reagent panel in question. If automated software for this purpose is not available, follow guidelines as described in reference *(24)*. In brief:
    a. Establish minimum baseline PMT settings for the instrument by acquiring a set of dim particles at various voltages, and choosing the lowest voltage for each PMT for which the CV of these particles is minimized.
    b. Run a sample stained with the full reagent cocktail in question, and adjust baseline PMT voltages as needed so that events are mostly above zero but do not register in the highest fluorescence channel.
2. Create a set of compensation controls consisting of single-stained cells or beads (*see* **Note 17**). Acquire these controls and use the software's automated algorithm to calculate compensation (*see* **Note 18**).
3. Create a template for acquisition that displays the relevant parameters in the test samples in the form of dot plots. This template need not be the same as that used for analysis, i.e., it

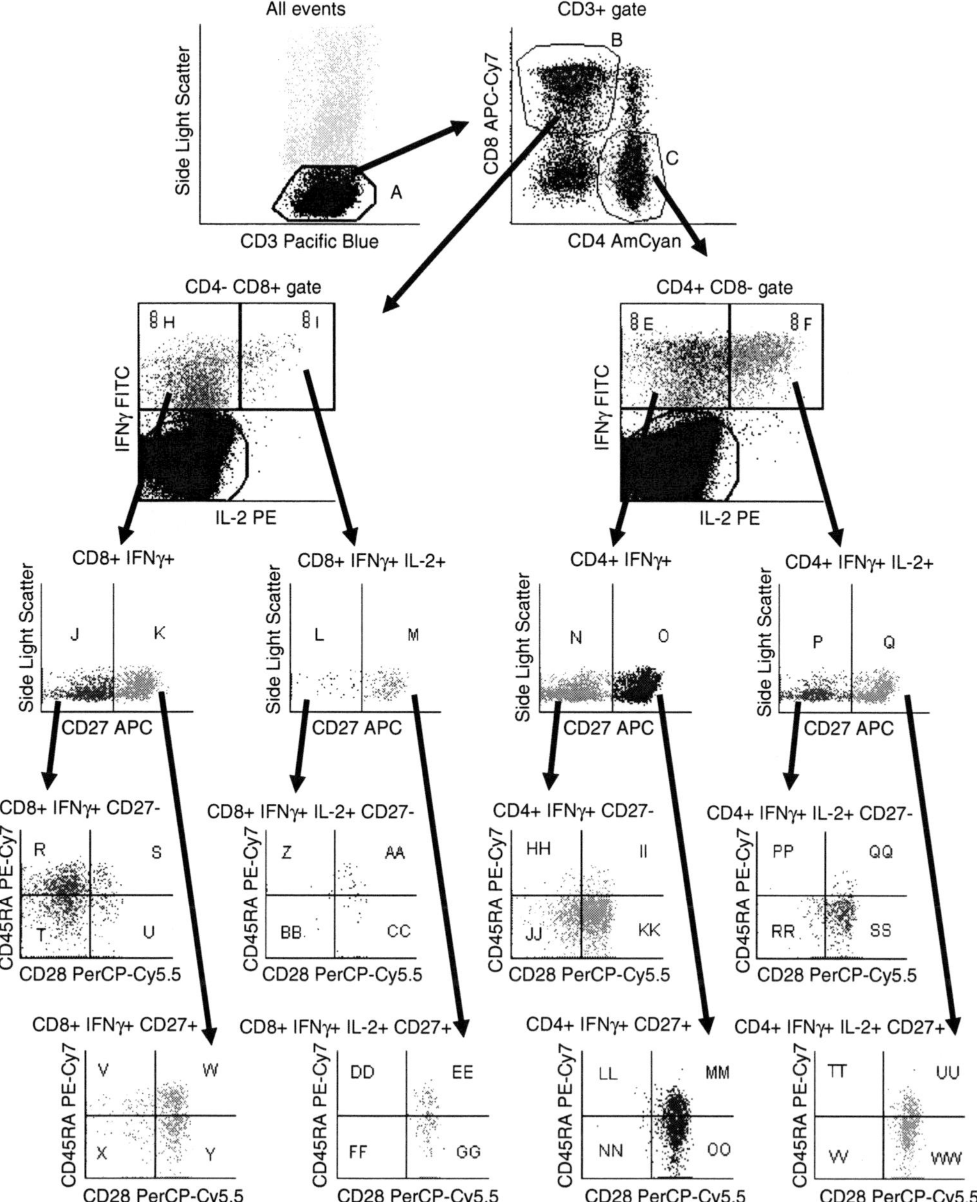

Fig. 25.1. Gating scheme for eight-color ICS study [reproduced from reference *(19)*]. An acquisition threshold was used to exclude CD3-negative cells. Dynamic gates were used for defining CD3+, CD4+, and CD8+ lymphocytes, tethered regions were used to define the cytokine-positive cells, and conventional (static) regions or quadrants were used to define all other (rare) subsets. Note that all possible combinatorial subsets of CD27, CD28, and CD45RA were reported for each subset of cytokine-positive cells, resulting in a set of 32 populations whose frequencies describe a response "fingerprint" for that sample.

does not need to specify all gates or regions of interest. In fact, a simplified acquisition template will allow faster processing of data. However, the template should show any gates used to define the saved population of cells or the stopping criteria (e.g., CD3+ cells).

4. Set an appropriate threshold, usually on FSC, to eliminate small debris, and set the stopping and storage criteria. It is usually safest to store all events (rather than a gated subset) to allow exploration of all data on analysis. However, sometimes a threshold or gate on CD3+ cells may be employed in order to reduce file sizes (*see* **Note 19**). When using a plate loader, be sure to set a stopping criterion based on time, so that samples with insufficient cells will not run dry.
5. Acquire data from a fully stained sample to verify that the settings chosen are appropriate. If any changes to PMT voltages are made, be sure to re-run compensation samples and re-calculate compensation based on the new voltages.
6. Record data from samples.
7. Analyze data using the acquisition software or compatible third-party software. Be sure to define all regions of interest and report the desired statistics on these (*see* **Note 20** and **Fig. 25.1**). Where possible, use a batch analysis function to analyze all samples from a given experiment or study and export the statistical data to a spreadsheet (*see* **Note 21**).
8. For large studies, it is helpful to create a database to accept the statistical output files from batch analysis. This database can then be queried to create data tables from subsets of the data, allowing rapid graphing, statistical analysis, background subtraction, conversion to absolute counts, etc.

4. Notes

1. Plates vs. tubes: Cells can also be stimulated in 15 mL conical polypropylene tubes, with staining in 12 × 75 mm polystyrene tubes (BD Falcon). However, plates are preferred for ease of handling multiple samples, and results for human PBMC are equivalent to those in tubes *(32)*.
2. Fresh PBMC: If PBMC are not to be cryopreserved, they should ideally be prepared on the day of blood draw, then either stimulated the same day, or rested at 37 °C in cRPMI-10 overnight and stimulated the following day. Overnight resting at 37 °C increases the staining intensity of cytokines, but the effect is more pronounced with cryopreserved samples. Overnight shipping of whole blood or PBMC at ambient temperatures can cause a variable decrease in cell function

and should be avoided if possible, though shipping PBMC is preferable to shipping whole blood.

3. Higher cell concentrations (1×10^7/mL, 2×10^6/well) should be used when possible, especially when response levels are low and/or there are many cell subsets to enumerate.
4. Cryopreserved PBMC: If cells cannot be stimulated within 24 h of blood draw, they should be cryopreserved by a validated protocol *(33)*. Upon thawing, recoveries of > 60% and viabilities of > 80% should be obtained to minimize loss of functional responses. The method of thawing is equally as important as that of cryopreservation *(33)*. Thawed cells should be rested in cRPMI-10 for 6–18 h at 37 °C to maximize cytokine staining intensity *(32)*. Some cell loss may occur during this period, so plating a slight excess of cells is desirable. Alternately, the cells can be rested in bulk (e.g., in a slanted 15 mL conical polypropylene tube), then recounted and resuspended at the desired concentration after resting.
5. Brefeldin A vs. monensin: Secretion of most cytokines of interest (IFNγ, IL-2, etc.) is best inhibited by brefeldin A at 10 μg/mL cells. However, CD107 and CD154 are transiently expressed on the cell surface. Therefore, staining Abs to CD107 and/or CD154 are added to the stimulation culture to bind the antigen(s) as soon as they are expressed. Monensin increases the intensity of staining under these conditions by preventing the acidification and degradation of lysosomal vesicles that contain the recycled CD107 and CD154. Thus, for combined cytokine and CD107 or CD154 detection, 5 μg/mL each of brefeldin A and monensin is recommended.
6. Peptide mixes: Peptide mixes can be prepared and lyophilized as premixed pools of up to several hundred peptides *(30)*. These can then be resuspended in DMSO at high concentration per peptide, avoiding DMSO toxicity. The total concentration of DMSO in the assay should be kept at < 0. 5%.
7. Costimulatory antibodies: Antibodies to CD28 and CD49d can increase the cytokine response to protein antigens, peptides, and SEB by amplifying the signal for low-affinity T cells *(34)*. In occasional donors, they increase cytokine production in the absence of antigen (TNFα is usually most affected).
8. Adding staining Abs during stimulation: As described in **Note 5**, staining Abs to CD107 and CD154 are best added during stimulation, to capture the transiently expressed antigen. Fluorochrome conjugated Abs are sensitive to light exposure, so they should be handled in low light and, once added, the samples should be incubated in the dark. Certain tandem dyes such as APC-Cy7 and PE-Cy7 are particularly sensitive to light and temperature *(24)* and are not optimal choices for use in stimulation cultures.

9. Stimulation time: A minimum of 5–6 h allows adequate detection of most proinflammatory cytokines like IFNγ, TNFα, and IL-2 *(35)*. Increasing the time of incubation (in the presence of brefeldin A) increases cytokine staining intensity, but is not recommended for CD107 or CD154. For whole proteins requiring intracellular processing, a preincubation of 2 h prior to adding brefeldin A and/or monensin is recommended *(35)*. CD8 responses to whole protein antigens can sometimes be detected, and are increased with longer incubation in antigen alone, but not in all donors *(36)*.
10. Automating incubation times: A programmable heat block, incubator, or water bath can be used to time activation, cooling the samples to 4–18 °C at the end of a specified period at 37 °C, and holding them for later processing.
11. A fixed-length vacuum manifold helps achieve consistent washing without undue cell loss in microtiter plates. Because of the small wash volume, a sufficient number of washes and efficient removal of supernatant are essential.
12. CD3, CD4, and CD8 can be stained either before or after fixation and permeabilization. Down-modulation of these antigens occurs to a variable degree depending upon the stimulus. Cells that have down-modulated these antigens can be better detected by intracellular staining (postfixation and permeabilization) *(30)*, although the overall staining intensity is usually decreased. Most other cell-surface antigens are optimally stained before fixation.
13. Freezing of activated samples: Samples can be frozen at −80 °C directly in FACS Lysing Solution *(35, 37)*. This allows for samples to be sent to another laboratory for processing, or for longitudinal samples to be accumulated for batch processing. Lysed whole blood should be washed once prior to freezing.
14. Fixation and permeabilization steps: Solutions for these steps should be stored and used at 22–25 °C. FACS Lysing Solution simultaneously lyses erythrocytes and fixes leukocytes. While erythrocyte lysis is not required for PBMC samples, fixation is still helpful to prevent cell loss prior to permeabilization.
15. Centrifugation speed: All centrifugation post-fixation should be done at higher *g* force (500 × *g*) due to increased cell buoyancy.
16. Use of paraformaldehyde is only helpful when samples are stored for more than 24 h prior to acquisition, or to ensure neutralization of potentially biohazardous samples. In addition to subtle effects on cell scatter and fluorescence, storage in paraformaldehyde can cause degradation of tandem dyes such as APC-Cy7 and PE-Cy7. An alternative fixative is available that protects these tandems from degradation (BD Stabilizing Fixative, BD Biosciences), but it is not compatible with AmCyan staining.

17. Compensation controls: Where possible, anti-immunoglobulin coated capture beads (BD Biosciences) are preferred as compensation controls, because they provide a bright and homogeneous population of events stained with the antibody conjugate of interest. Ideally, the same lot of antibody should be used for compensation as is used in the experiment. In practice, however, this is only important for certain tandem conjugates, such as APC-Cy7 and PE-Cy7. The compensation controls should ideally be treated identically to the experimental samples in terms of fixation, etc., although this too is only important for the above tandem dyes.
18. When to apply compensation: While compensation can be calculated and changed at any time by software packages such as FloJo (TreeStar, Ashland, OR) or FACSDiva (BD Biosciences), it is helpful to perform compensation before sample acquisition, so that any setup problems can be more readily detected.
19. Number of events to collect: Because multiparameter ICS assays tend to divide responding populations of cells into ever-smaller subsets, it is important to process and collect enough cells per sample to allow statistically significant differences between samples to be detected. The number of events required will depend upon the anticipated levels of responses and background, as well as the number of subsets of responding cells being identified. Statistical tools for sample size calculation can be found at http://maeckerlab.typepad.com.
20. Gating of down-modulated cells: Be sure that gates set on CD3, CD4, and CD8 parameters include dim-positive cells, since down-modulation of these markers occurs with activation. When using dynamic gating (*see* **Note 21**), set the region size to the maximum value possible without causing inclusion of neighboring populations. Some donors have a significant population of CD4 + CD8dim T cells. This population contains a disproportionate number of cells specific for chronic antigens such as CMV and HIV, and should be included in the CD4+ T cell gate to avoid under-reporting of responses.
21. Batch analysis: Dynamic gating tools such "Snap-To" gates in FACS Diva (BD Biosciences) can be used to accommodate staining differences between samples for populations such as CD3+, CD4+, and CD8+ cells (*see* **Fig. 25.1**). This in turn allows use of a single analysis template and batch analysis across multiple samples in an experiment or study. However, dynamic gates are not always useful for rare populations, and their specifications (size and movement) need to be adjusted for the data set being analyzed. Batch analysis and dynamic gating thus do not replace the need for visual inspection of all data.

Acknowledgements

Details of this protocol were optimized by Laurel Nomura and Maria Suni (BD Biosciences).

References

1. Pantaleo, G., Koup, R. A. (2004).Correlates of immune protection in HIV-1 infection: what we know, what we don't know, what we should know. *Nat Med* 10, 806.
2. Hel, Z., Nacsa, J., Tryniszewska, E., et al. (2002).Containment of simian immunodeficiency virus infection in vaccinated macaques: correlation with the magnitude of virus-specific pre- and postchallenge CD4(+) and CD8(+) T cell responses. *J Immunol* 169, 4778.
3. Mooij, P., Nieuwenhuis, I. G., Knoop, C. J., et al. (2004).Qualitative T-helper responses to multiple viral antigens correlate with vaccine-induced immunity to simian/human immunodeficiency virus infection. *J Virol* 78, 3333.
4. Boyer, J. D., Maciag, P. C., Parkinson, R., et al. (2005).Rhesus macaques with high levels of vaccine induced IFN-gamma producing cells better control viral set-point following challenge with SIV239. *Vaccine.*
5. Sun, Y., Schmitz, J. E., Acierno, P. M., et al. (2005).Dysfunction of simian immunodeficiency virus/simian human immunodeficiency virus-induced IL-2 expression by central memory CD4+ T lymphocytes. *J Immunol* 174, 4753.
6. Acierno, P. M., Schmitz, J. E., Gorgone, D. A., et al. (2006).Preservation of Functional Virus-Specific Memory CD8+ T Lymphocytes in Vaccinated, Simian Human Immunodeficiency Virus-Infected Rhesus Monkeys. *J Immunol* 176, 5338.
7. Sun, Y., Schmitz, J. E., Buzby, A. P., et al. (2006).Virus-specific cellular immune correlates of survival in vaccinated monkeys after simian immunodeficiency virus challenge. *J Virol.*
8. Letvin, N. L., Mascola, J. R., Sun, Y., et al. (2006).Preserved CD4+ central memory T cells and survival in vaccinated SIV-challenged monkeys. *Science* 312, 1530.
9. Migueles, S. A., Laborico, A. C., Shupert, W. L., et al. (2002).HIV-specific CD8(+) T cell proliferation is coupled to perforin expression and is maintained in nonprogressors. *Nat Immunol* 3, 1061.
10. Boaz, M. J., Waters, A., Murad, S., et al. (2002).Presence of HIV-1 Gag-specific IFN-gamma + IL-2+ and CD28 + IL-2 + CD4 T cell responses is associated with nonprogression in HIV-1 infection. *J Immunol* 169, 6376.
11. Betts, M. R., Nason, M. C., West, S. M., et al. (2006).HIV nonprogressors preferentially maintain highly functional HIV-specific CD8+ T-cells. *Blood* 107, 4781.
12. Younes, S. A., Yassine-Diab, B., Dumont, A. R., et al. (2003).HIV-1 viremia prevents the establishment of interleukin 2-producing HIV-specific memory CD4+ T cells endowed with proliferative capacity. *J Exp Med* 198, 1909.
13. Chen, G., Shankar, P., Lange, C., et al. (2001).CD8 T cells specific for human immunodeficiency virus, Epstein-Barr virus, and cytomegalovirus lack molecules for homing to lymphoid sites of infection. *Blood* 98, 156.
14. Champagne, P., Ogg, G. S., King, A. S., et al. (2001).Skewed maturation of memory HIV-specific CD8 T lymphocytes. *Nature* 410, 106.
15. Harari, A., Rizzardi, G. P., Ellefsen, K., et al. (2002).Analysis of HIV-1- and CMV-specific memory CD4 T-cell responses during primary and chronic infection. *Blood* 100, 1381.
16. Ellefsen, K., Harari, A., Champagne, P., et al. (2002).Distribution and functional analysis of memory antiviral CD8 T cell responses in HIV-1 and cytomegalovirus infections. *Eur J Immunol* 32, 3756.
17. Appay, V., Dunbar, P. R., Callan, M., et al. (2002).Memory CD8+ T cells vary in differentiation phenotype in different persistent virus infections. *Nat Med* 8, 379.
18. Paiardini, M., Cervasi, B., Albrecht, H., et al. (2005).Loss of CD127 expression defines an expansion of effector CD8+ T cells in HIV-infected individuals. *J Immunol* 174, 2900.
19. Nomura, L. E., Emu, B., Hoh, R., et al. (2006).IL-2 production correlates with effector cell differentiation in HIV-specific CD8+ T cells. *AIDS Res Ther* 3, 18.

20. Day, C. L., Kaufmann, D. E., Kiepiela, P., et al. (2006).PD-1 expression on HIV-specific T cells is associated with T-cell exhaustion and disease progression. *Nature* 443, 350.
21. Trautmann, L., Janbazian, L., Chomont, N., et al. (2006).Upregulation of PD-1 expression on HIV-specific CD8(+) T cells leads to reversible immune dysfunction. *Nat Med* 12, 1198.
22. Savarino, A., Bottarel, F., Malavasi, F., et al. (2000).Role of CD38 in HIV-1 infection: an epiphenomenon of T-cell activation or an active player in virus/host interactions? *Aids* 14, 1079.
23. Roederer, M. (2001).Spectral compensation for flow cytometry: visualization artifacts, limitations, and caveats. *Cytometry* 45, 194.
24. Maecker, H. T., Frey, T., Nomura, L. E., et al. (2004).Selecting fluorochrome conjugates for maximum sensitivity. *Cytometry A* 62, 169.
25. Maecker, H. T., Trotter, J. (2006).Flow cytometry controls, instrument setup, and the determination of positivity. *Cytometry A*.
26. Perfetto, S. P., Ambrozak, D., Nguyen, R., et al. (2006).Quality assurance for polychromatic flow cytometry. *Nat Protocols* 1, 1522.
27. Maecker, H. T. (2004). in (T. S. Hawley, and R. G. Hawley, eds.), *Flow Cytometry Protocols*, p. 95. Humana Press, Totowa, NJ.
28. Maecker, H. T. (2007). in (T. Kieber-Emmons, ed.), *Cancer Vaccine Protocols*. Humana Press, Totowa, NJ.
29. Lamoreaux, L. L., Roederer, M., Koup, R. (2006).Intracellular cytokine optimization and standard operating procedure. *Nat Protocols* 1, 1507.
30. Maecker, H. T., Dunn, H. S., Suni, M. A., et al. (2001).Use of overlapping peptide mixtures as antigens for cytokine flow cytometry. *J Immunol Methods* 255, 27.
31. Maecker, H. T., Rinfret, A., D'Souza, P., et al. (2005).Standardization of cytokine flow cytometry assays. *BMC Immunol* 6, 13.
32. Suni, M. A., Dunn, H. S., Orr, P. L., et al. (2003).Performance of plate-based cytokine flow cytometry with automated data analysis. *BMC Immunology* 4, 9.
33. Disis, M. L., dela Rosa, C., Goodell, V., et al. (2006).Maximizing the retention of antigen specific lymphocyte function after cryopreservation. *J Immunol Methods* 308, 13.
34. Waldrop, S. L., Davis, K. A., Maino, V. C., et al. (1998).Normal human CD4+ memory T cells display broad heterogeneity in their activation threshold for cytokine synthesis. *J Immunol* 161, 5284.
35. Nomura, L. E., Walker, J. M., Maecker, H. T. (2000).Optimization of whole blood antigen-specific cytokine assays for CD4(+) T cells. *Cytometry* 40, 60.
36. Maecker, H. T., Ghanekar, S. A., Suni, M. A., et al. (2001).Factors affecting the efficiency of CD8+ T cell cross-priming with exogenous antigens. *J Immunol* 166, 7268.
37. Nomura, L. E., DeHaro, E. D., Martin, L. N., et al. (2003).Optimal preparation of rhesus macaque blood for cytokine flow cytometric analysis. *Cytometry* 53A, 28.

Subsection C

Antiviral Responses

Chapter 26

Measuring HIV Neutralization in a Luciferase Reporter Gene Assay

David C. Montefiori

Abstract

Neutralizing antibody (NAb) assays for human immunodeficiency virus (HIV) are used to study the immune response in infected individuals, to examine monoclonal antibodies and viral diversity, and to judge the potential value of candidate vaccine immunogens in preclinical and clinical trials. An important aspect of these efforts is an ability to achieve and document equivalent assay performance across multiple laboratories. Recent advances in assay technology have led to major improvements in how HIV NAbs are measured. Stable cell lines containing HIV Tat-regulated reporter genes are now available that permit rapid, sensitive and reproducible measurements of virus neutralization after a single round of infection in a high throughput format.Moreover, these assays may be used with molecularly cloned Env-pseudotyped viruses for greater reagent stability and traceability.A luciferase (Luc) reporter gene assay performed in TZM-bl (JC53bl-13) cells was recently optimized and many of its performance parameters have been validated. This assay has become the main endpoint neutralization assay used by the NIH-sponsored HIV Vaccine Trials Network and by a growing number of laboratories worldwide.

Key words: Neutralizing antibodies, HIV vaccines, reporter gene, pseudovirus, assay standardization.

1. Introduction

Infection with human immunodeficiency virus type (HIV), the causative agent of Acquired Immunodeficiency Syndrome (AIDS), gives rise to B- and T-cell immune responses of various magnitudes against multiple viral antigens. The most important antigens for neutralizing antibodies (NAbs) are the surface gp120 and transmembrane gp41 envelope glycoproteins. These glycoproteins arise after proteolytic cleavage of a gp160 precursor molecule and form a trimolecular complex of heterodimers

Vinayaka R. Prasad, Ganjam V. Kalpana (eds.), *HIV Protocols: Second Edition, vol. 485*

DOI 10.1007/978 1 59745 170 3_26 Springerprotocols.com

that mediate virus entry into cells *(1)*. Entry is a multi-step process that begins when gp120 engages its cellular receptor, CD4 *(2)*. Subsequent interactions between gp120 and a coreceptor molecule, usually CCR5 or CXCR4 *(3, 4)*, expose the N-terminal peptide region of gp41 that then inserts into the cell membrane and draws the viral and cellular membranes into close proximity for fusion to take place *(5–7)*. NAbs can block entry at each of these discrete steps *(8–10)*, thereby disabling the virus.

An assay is described here that measures HIV neutralization as a function of reductions in Tat-regulated Luc reporter gene expression after a single round of infection in TZM-bl cells. This assay was first developed by Dr. George Shaw and colleagues at the University of Alabama, Birmingham *(11)* and was later optimized and validated at Duke University. TZM-bl (also know as JC53BL-13) is a CXCR4-positive HeLa cell clone that was engineered to express CD4 and CCR5 *(12)*. The cells were further engineered to contain integrated reporter genes for firefly luciferase and *Escherichia coli* β-galactosidase under control of an HIV long-terminal repeat sequence *(13)*. TZM-bl cells are permissive to infection by a wide variety of HIV, SIV and SHIV strains, including primary HIV isolates and molecularly cloned Env-pseudotyped viruses. Assay stocks of Env-pseudotyped viruses are produced in 293T/17 cells by co-transfection with an Env expression plasmid and a second plasmid expressing the entire HIV genome except Env. Only the latter *env*-minus plasmid is transcribed into viral genomic RNA that is packaged by the pseudovirions for delivery of the *tat* gene to TZM-bl cells. Thus, co-transfection generates pseudovirus particles that are infectious but are unable to produce infectious progeny virions for subsequent rounds of infection. Reporter gene expression is induced in trans by viral Tat protein soon after single cycle infection. DEAE dextran is added to the medium to enhance infection and has been found to have no obvious effects on NAb activity. Luciferase activity is quantified as relative luminescence units (RLU) and is directly proportional to the number of infectious virus particles present in the initial inoculum over a wide range of values. The assay is performed in 96-well plates for high throughput capacity.

2. Materials

1. Dulbecco's Modified Eagle's Medium (DMEM) (Gibco/BRL, Bethesda, MD) supplemented with 10% heat-inactivated fetal bovine serum (PBS, HyClone, Ogden, UT), 25 mM HEPES (Gibco/BRL, Bethesda, MD) and 50 μg

gentamicin/ml (Sigma Chemical Company, St. Louis, MO). Complete medium should be stored at 4 °C.

2. DEAE dextran, average mol. wt. 500,000 (Sigma Chemical Company). Prepare a 7.5 mg/mL stock in sterile water. Store in aliquots at −20 to −80 °C. Protect from light.
3. Trypsin-EDTA (0.25% trypsin, 1 mM EDTA), sterile (Invitrogen Corp., Carlsbad, CA). Store at 4 °C.
4. Britelite Luminescence Reporter Gene Assay System (PerkinElmer Life & Analytical Sciences, Wellesley, MA). Reconstitute one vial of lyophylized Britelite Substrate Solution with 250 mL of Britelite Substrate Buffer Solution. After the substrate has dissolved completely (about 1 min), mix gently and distribute to 10.5–15 mL conical polypropylene tubes and store at −70 °C. Thaw in a room temperature water bath in the dark immediately before each use. Mix gently and use within 60 min of thawing. Excess reagent may be stored at −70 °C and used once more. *Caution: The lyophylized Britelite substrate is classified as hazardous. Latex gloves, surgical gown and eye protection are required when working with this reagent.*
5. FuGENE-6 HD® Transfection Reagent (Roche Applied Sciences, Indianapolis, IN).
6. TZM-bl cells (NIH AIDS Research and Reference Reagent Program, Rockville, MD).
7. 293T/17 cells (American Type Culture Collection, Manassas, VA).
8. pSG3Δenv backbone plasmid DNA (NIH AIDS Research and Reference Reagent Program).
9. HIV-1 gp160 reference clones for production of infectious Env-pseudotyped viruses (NIH AIDS Research and Reference Reagent Program).

3. Methods

The assay is designed to test five samples at eight dilutions in duplicate in a single 96-well plate. Typical neutralization curves for samples that test positive are linear in the range of 20–85% reductions in RLU. Thus, greatest precision is achieved by defining the NAb titer as the 50% inhibitory dose (ID50), which lies midway in the linear portion of the neutralization curve. Using the plate layouts as shown in **Fig. 26.1**, automated data reduction is possible by using an Excel macro that may be downloaded from a public website. A positive control sample with a known neutralization titer should be included on at least one plate in each series to assure the assays were performed properly. In addition, suitable negative control samples should be included to provide essential

Titers

	1	2	3	4	5	6	7	8	9	10	11	12
A	CC	VC	Dil 8	Dil 8	Dil 8	Dil 8	Dil 8	Dil 8	Dil 8	Dil 8	Dil 8	Dil 8
B	CC	VC	Dil 7	Dil 7	Dil 7	Dil 7	Dil 7	Dil 7	Dil 7	Dil 7	Dil 7	Dil 7
C	CC	VC	Dil 6	Dil 6	Dil 6	Dil 6	Dil 6	Dil 6	Dil 6	Dil 6	Dil 6	Dil 6
D	CC	VC	Dil 5	Dil 5	Dil 5	Dil 5	Dil 5	Dil 5	Dil 5	Dil 5	Dil 5	Dil 5
E	CC	VC	Dil 4	Dil 4	Dil 4	Dil 4	Dil 4	Dil 4	Dil 4	Dil 4	Dil 4	Dil 4
F	CC	VC	Dil 3	Dil 3	Dil 3	Dil 3	Dil 3	Dil 3	Dil 3	Dil 3	Dil 3	Dil 3
G	CC	VC	Dil 2	Dil 2	Dil 2	Dil 2	Dil 2	Dil 2	Dil 2	Dil 2	Dil 2	Dil 2
H	CC	VC	Dil 1	Dil 1	Dil 1	Dil 1	Dil 1	Dil 1	Dil 1	Dil 1	Dil 1	Dil 1
			Sample 1		*Sample 2*		*Sample 3*		*Sample 4*		*Sample 5*	

Screening

	1	2	3	4	5	6	7	8	9	10	11	12
A	CC	VC	S#4 Post	S#4 Post	S#4 Post	S#8 Post	S#8 Post	S#8 Post	S#12 Post	S#12 Post	S#12 Post	BLK
B	CC	VC	S#4 Pre	S#4 Pre	S#4 Pre	S#8 Pre	S#8 Pre	S#8 Pre	S#12 Pre	S#12 Pre	S#12 Pre	BLK
C	CC	VC	S#3 Post	S#3 Post	S#3 Post	S#7 Post	S#7 Post	S#7 Post	S#11 Post	S#11 Post	S#11 Post	BLK
D	CC	VC	S#3 Pre	S#3 Pre	S#3 Pre	S#7 Pre	S#7 Pre	S#7 Pre	S#11 Pre	S#11 Pre	S#11 Pre	BLK
E	CC	VC	S#2 Post	S#2 Post	S#2 Post	S#6 Post	S#6 Post	S#6 Post	S#10 Post	S#10 Post	S#10 Post	BLK
F	CC	VC	S#2 Pre	S#2 Pre	S#2 Pre	S#6 Pre	S#6 Pre	S#6 Pre	S#10 Pre	S#10 Pre	S#10 Pre	BLK
G	CC	VC	S#1 Post	S#1 Post	S#1 Post	S#5 Post	S#5 Post	S#5 Post	S#9 Post	S#9 Post	S#9 Post	BLK
H	CC	VC	S#1 Pre	S#1 Pre	S#1 Pre	S#5 Pre	S#5 Pre	S#5 Pre	S#9 Pre	S#9 Pre	S#9 Pre	BLK

Fig. 26.1. Standard assay templates. *Top*: Template for measuring the titer of NAbs in five samples per 96-well plate. *Bottom*: Template for screening preimmune and postimmune samples from 20 study subjects. CC, cell control wells (cells only); VC, virus control wells (virus and cells but no test sample); Dil, dilution; Pre, preimmune; Pst, postimmune.

information about nonspecific background activity. The activity of negative control samples may be used as a baseline to decide a true positive response.

An important decision needs to be made about the choice of HIV strain(s) to use in the assay. This decision is dependent on the particular study design and the scientific questions being addressed. For example, the potential impact of the NAb response in infected individuals is best evaluated by using virus isolates and matched (autologous) serum samples from the study subjects. Because HIV exhibits an extraordinary degree of genetic diversity *(14)* and is armed with multiple neutralization-escape mechanisms *(1, 11, 15)*, assays with mismatched (heterologous) serum samples and virus strains do not provide accurate information about the autologous NAb response. Another important consideration is the timing of virus isolation and serum collection as they relate to the dynamics of neutralization-escape and an evolving NAb antibody response. For example, many primary patient isolates are resistant to neutralization by contemporaneous autologous serum but are sensitive to neutralization by autologous serum that is collected weeks or months after the virus was isolated *(11, 16)*.

A much different approach is needed when assessing vaccine-elicited NAbs, where the goal for HIV is to generate a broadly cross-reactive NAb response against circulating strains of the virus *(17)*. Antibodies produced in response to vaccination are most likely to neutralize the matched vaccine strain(s) and one or more of a small subset of strains that are unusually sensitive to neutralization. Assays with these strains (Tier 1) are recommended as a means to determine whether any NAbs were produced and to provide a measure of titer *(18)*. However, it should not be assumed that an ability to neutralize Tier 1 viruses will predict an effective NAb response. If positive results are obtained in this first round of testing, additional assays are needed to determine the extent of cross-neutralization against circulating strains of the virus (Tier 2). To facilitate this latter effort, initial panels of HIV reference strains have been created as Env-pseudotyped viruses for Tier 2 assessments of neutralization *(19, 20)*. Additional panels of Tier 2 reference strains are under development that will permit a greater depth of analysis by covering a broader range of viral diversity *(21, 22)*.

3.1. Splitting Cells

1. Decant the culture medium from a confluent monolayer of cells in a T75 culture flask. Remove residual serum by rinsing the monolayer with 6 mL of sterile PBS.
2. Slowly add 2.5 mL of 0.25% trypsin-EDTA solution to cover the cell monolayer. Incubate at room temperature for 30–45 s.
3. Decant the trypsin-EDTA solution. For TZM-bl cells, incubate for 4 min at 37 °C in a 5% CO_2/95% air environment. For 293T/17 cells, incubate for 1 min at room temperature.
4. Add 10 mL of complete DMEM medium and suspend the cells by gentle pipet action. Count cells.
5. Seed new T75 culture flasks with approximately 10^6 cells in 15 mL of complete DMEM. Cells should be split approximately every 3 days.

3.2. Preparation of Env-Pseudotyped Viruses by Transfection in 293T/17 Cells

1. Seed 3×10^6 293T/17 cells in a T75 culture flask containing 15 mL complete DMEM. Incubate overnight at 37 °C in a 5% CO_2/95% air environment. Monolayers should be 50–80% confluent at the time of transfection.
2. Mix 4 μg of HIV Env plasmid DNA and 8 μg pSG3Δenv backbone plasmid DNA in a 1.5-mL sterile conical

polypropylene tube. Adjust the final volume to 100 μL with DMEM (without serum and antibiotic).

3. In a second 1.5-mL sterile conical polypropylene tube add 652 μL DMEM (without serum and antibiotic). Pipet 48 μL of FuGENE-6 reagent directly into the DMEM without contacting the sides of the tube. Mix gently.
4. Transfer the entire contents of the plasmid DNA mixture from the first tube to the second tube containing the FuGENE-6 solution. Mix by pipet action or by brief vortexing. Incubate at room temperature for 30 min to allow complexes to form.
5. Transfer the entire contents of the reaction mixture to a monolayer of 293T/17 cells that is 50–80% confluent in a T75 culture flask containing 15 mL of complete DMEM. Swirl the contents gently to ensure uniform distribution across the cell monolayer. Incubate at 37 °C in a 5% CO_2/95% air environment for 3–8 h to permit plasmid DNA to enter cells.
6. Replace the DNA-FuGENE-6 medium with 15 mL fresh complete DMEM and continue the incubation for 2 days.
7. Collect the pseudovirion-containing culture fluid and adjust the concentration of FBS to 20%. Filter the fluid using a 0.45-micron nitrocellulose filter. Store the filtered pseudovirion preparations in 1-mL aliquots at −70 °C.
8. Complete DMEM (15 mL) may be added to the flask for a second harvest of pseudovirus after an additional overnight incubation.

3.3. Titration of Infectious Pseudovirions

1. Place 100 μL complete DMEM per well in all wells of a 96-well flat-bottom culture plate. Transfer 25 μL of virus to the first four wells of a dilution series (column 1, rows A–D for one virus and rows E–H for a second virus), mix, do serial fivefold dilutions (i.e., transfer 25 μL, mixing each time) for a total of 11 dilutions. Discard 25 μL from the 11th dilution. Wells in column 12 serve as controls for background luminescence caused by low-level constitutive Luc expression in uninfected cells.
2. Add 100 μL of TZM-bl cells at a density of 10,000 cells/mL in complete DMEM containing DEAE dextran (30 μg/mL). Rinse pipet tips in a reservoir of PBS between each step to minimize carry-over. Incubate at 37 °C in a 5% CO_2/95% air environment for 48 h (*see* **Notes 1** and **2**).
3. Remove 100 μL of culture medium from each well, leaving approximately 100 μL. Dispense 100 μL Britelite Substrate Solution to each well. Incubate at room temperature for 2 min to allow complete cell lysis.
4. Mix by pipet action (two strokes) and transfer 150 μL to a corresponding 96-well black plate. Read the plate immediately in a luminometer.

5. Calculate the TCID50 using the Reed-Muench formula as described *(23)*. Wells with RLU > 2.5 times background are considered positive for the calculation. An Excel macro for automated calculation of TCID50 may be found at: http://www.hiv.lanl.gov/content/nab-reference-strains/html/home.htm.

3.4. Neutralization Assay

1. Using the 96-well template shown in **Fig. 26.1** (Top), place 150 μL of complete DMEM in all wells of column 1 (cell control). Place 100 μL in all wells of columns 2–12 (column 2 is the virus control). Place an additional 40 μL in all wells of columns 3–12, row H (to receive test samples) (*see* **Note 3**).
2. Add 11 μL of test sample to each well in columns 3 & 4, row H. Add 11 μL of a second test sample to each well in columns 5 & 6, row H. Add 11 μL of a third test sample to each well in columns 7 & 8, row H. Add 11 μL of a fourth test sample to each well in columns 9 & 10, row H. Add 11 μL of a fifth test sample to each well in columns 11 & 12, row H. Mix the samples in row H and transfer 50 μL to row G. Mix and repeat the transfer of samples through row A (these are serial threefold dilutions). After final transfer and mixing is complete, discard 50 μL from the wells in columns 3–12, row A into a waste container of disinfectant (*see* **Notes 4 , 5 and 6**).
3. Thaw assay stocks of Env-pseudotyped virus in an ambient temperature water bath. When completely thawed, dilute the virus in complete DMEM to achieve a concentration of 4,000 TCID50/mL.
4. Dispense 50 μL of Env-pseudovirions (200 TCID50) to all wells in columns 2–12, rows A through H. Cover plates and incubate at 37 °C in a 5% CO_2/95% air environment for 1 h (*see* **Notes** 7 and **8**).
5. Prepare a suspension of TZM-bl cells (trypsinized approximately 10–15 min prior to use) at a density of 1×10^5 cells/mL in complete DMEM containing DEAE dextran (37.5 μg/mL). Dispense 100 μL of this cell suspension (10,000 cells) to each well in columns 1–12, rows A though H (*see* **Notes 1 and 2**).
6. Cover plates and incubate for 48 h.
7. Remove 150 μL of culture medium from each well, leaving approximately 100 μL. Dispense 100 μL Britelite Substrate Solution to each well. Incubate at room temperature for 2 min to allow complete cell lysis. Mix by pipet action (two strokes) and transfer 150 μL to a corresponding 96-well black plate (*see* **Note 9**). Read the plate immediately in a luminometer.
8. Calculate the percent neutralization at each sample dilution by taking the difference in average RLU between test wells (cells + serum sample + virus) and cell control wells (cells only, column 1), dividing this result by the difference in

average RLU between virus control (cell + virus, column 2) and cell control wells (column 1), subtracting from 1 and multiplying by 100. NAb titers are expressed as the serum dilution required to reduce RLU by 50% (*see* **Note 10**). An Excel macro and user guide for automated data reduction may be found at: http://www.hiv.lanl.gov/content/nab-reference-strains/html/home.htm.

9. Assay performance may be considered acceptable when: (1) virus-induced syncytium formation is absent in virus control wells, (2) the mean RLU for the virus control is ≥ 10x the mean RLU for the cell control (in a typical assay, the mean for the cell control is between 500 and 2500 RLU), and (3) the neutralization titer of the positive control sample is within 3-fold of the expected titer.

4. Notes

1. TZM-bl cells are stable in culture for approximately 6 months (~ 60 passages). After this time, the cells exhibit an abrupt decline in CD4 and CCR5 surface expression and become less susceptible to infection.
2. DEAE dextran from different sources and different lots may exhibit substantial variability in potency and cell toxicity. For this reason, each new batch of DEAE dextran should be titrated to determine the optimal concentration for use in the assay. New solutions of DEAE dextran may be titrated by performing serial dilutions in a 96-well plate and adding a representative Env-pseudotyped virus and TZM-bl cells as described for the neutralization assay. The optimal concentration of DEAE dextran is determined from the dilution that yields the highest RLU and has no detrimental effects on the cells as observed by light microscopy after a 48-h incubation.
3. The protocol as written here uses a plate layout as shown in **Fig. 26.1** (Top) to measure NAb titers that are in the range of 1:20 to 1:43,740. A different range of titers may be examined by changing the sample dilutions in this same layout. For example, a starting dilution of 1:10 or lower may be used to detect very low levels of NAbs. In cases where samples contain very high titers of NAbs, either a higher starting dilution or an increased magnitude of dilution at each transfer step (e.g., 1:4) may be necessary. Alternatively, samples may be screened at a single dilution for higher throughput **(Fig. 26.1**, Bottom). This latter option is especially attractive when performing large-scale assessments of cross-reactive neutralizing activity against many different strains of the virus. For greatest specificity in the screening assay, it is best to measure neutralization against the corresponding preimmune sample,

as this provides a more precise adjustment for background activity at a single dilution.

4. All serum and plasma samples should be heat-inactivated at 56 °C for 1 h prior to assay. This heat-inactivation step is required to destroy complement. HIV is highly sensitive to complement-mediated lysis and inactivation by serum and plasma from mice, guinea pigs, rabbits and other small animals. Although the virus generally resists the lytic pathway of human and macaque complement *(24)*, heat-inactivation is still advised.
5. Serum and plasma may be placed at −70 °C for long-term storage either before or after heat-inactivation. Thawed samples may be stored at 4 °C for up to 4 weeks. Prolonged storage at 4 °C might compromise the integrity of the sample. In general, neutralizing activity of serum and plasma samples remains stable after a limited number of freeze-thaw cycles (approximately four cycles). Monoclonal antibodies require additional attention because their stability can be sensitive to different storage media and to a small number freeze–thaw cycles.
6. A number of factors can introduce false positive results in the assay. For example, serum and plasma samples may contain substances that either possess nonspecific antiviral activity or are toxic to the cells. Maximum sensitivity in the assay is achieved by testing the lowest sample dilution that does not produce these unwanted effects.

   Anticoagulants in plasma are a common interfering substance that can be toxic to cells at plasma dilutions $< 1:30$. This toxicity limits the sensitivity of the assay by precluding valid measurements of NAbs at low sample dilutions. An additional concern is that some forms of the anticoagulant heparin possess anti-HIV activity that may be mistaken for the presence of NAbs. When designing studies it is always best to plan for the collection of serum rather than plasma. If plasma is collected, heparin is the least desirable choice of anticoagulant for studies of HIV NAbs.

   Another concern is that serum and plasma from HIV-infected individuals who are on antiretroviral therapy may contain adequate concentrations of the drugs to inhibit the virus and be mistaken for the presence of NAbs. The magnitude of antiviral activity can vary depending on drug type and the time when the last dose was taken prior to blood draw. The assay is particularly sensitive to inhibitors of virus fusion and of reverse transcription (RT) – two functions that are needed for transactivation of the Tat-regulated Luc reporter gene. It may be possible to circumvent the activity of RT inhibitors by using Env-pseudotyped viruses that carry a drug-resistant RT in the *gag-pol* gene of the helper plasmid. In this case, drug-resistant variants can display lower replication fitness that should be taken into account. The assay is not affected by drugs that target later stages of virus replication, such as viral protease inhibitors.

7. High concentrations of Env-pseudotyped viruses may induce syncytium formation and cause cytopathic effects that will invalidate the assay. Cytopathic effects are usually avoided by using a standard inoculum size of 200 TCID50. It may be necessary to use a lower virus doses for highly cytopathic strains of the virus but it is important to keep in mind that as the dose of input virus is decreased, RLU in virus control wells may be reduced to unacceptable levels ($< 10\times$ background). In general, a direct linear relationship exists between RLU and the number of infectious pseudovirions over a range of approximately 5–500 TCID50/well (providing no cell-killing is observed at the highest virus doses). NAb titers are affected very little ($\leq$ 3-fold) over this 100-fold range of input virus doses.
8. This protocol may be adapted for use with uncloned, replication-competent stocks of virus produced in human lymphocytes; however, this often results in a lack of sensitivity compared to other assays *(19, 25)*. Sensitivity in this assay equals or surpasses that of other assays when 293T/17-derived Env-pseudotyped viruses are used.
9. Either white or black 96-well plates may be used when measuring luminescence. Overall luminescence values will be approximately $10\times$ lower with black plates. Black plates are preferred because, in cases of very high luminescence, some bleeding of light to adjacent wells may occur and cause artificially high readings. This artifact is avoided by using back plates.

Acknowledgments

The author thanks Dr. Marcella Sarzotti-Kelsoe for guidance in optimizing and validating the assay, and Kelli Greene for her critical review of the document. He also thanks Drs. George Shaw and John Kappes for sharing their reagents. This work was supported by NIH grants AI30034 and AI46705.

References

1. Wyatt, R., Sodroski, J. (1998) The HIV-1 envelope glycoproteins: fusogens, antigens, and immunogens. *Science* 280, 1884–1888.
2. Lasky, L. A., Nakamura, G., Smith, D. H., et al. (1987) Delineation of a region of the human immunodeficiency virus gp120 glycoprotein critical for interaction with the CD4 receptor. *Cell* 50, 975–985.
3. Berger, E. A. (1997) HIV entry and tropism: the chemokine receptor connection. *AIDS* 11(suppl. A), S3–S16.
4. Berger, E. A., Doms, R. W., Fenyö, E.-M., et al. (1998) A new classification for HIV-1. *Nature* 391, 240.
5. Weissenhorn, W., Dessen, A., Harrison, S. C., et al. (1997) Atomic structure of the

ectodomain from HIV-1 gp41. *Nature* 387, 426–430.
6. Chan, D. C., Fass, D., Berger, J. M., et al. (1997) Core structure of gp41 from the HIV envelope glycoprotein. *Cell* 89, 263–273.
7. Chan, D. C., Kim, P. S. (1998) HIV entry and its inhibition. *Cell* 93, 681–684.
8. Salzwedel, K., Smith, E. D., Dey, B., et al. (2000) Sequential CD4-coreceptor interactions in human immunodeficiency virus type 1 Env function: soluble CD4 activates Env for coreceptor-dependent fusion and reveals blocking activities of antibodies against cryptic conserved epitopes on gp120. *J. Virol.* 74, 326–333.
9. McDougal, J. S., Kennedy, M. S., Orloff, S. L., et al. (1996) Mechanism of human immunodeficiency virus type 1 (HIV-1) neutralization: irreversible inactivation of infectivity by anti-HIV-1 antibody. *J. Virol.* 70, 5236–5245.
10. Sullivan, N., Sun, Y., Sattentau, Q., et al. (1998) CD4-induced conformational changes in the human immunodeficiency virus type 1 gp120 glycoprotein: consequences for virus entry and neutralization. *J. Virol.* 72, 4694–4703.
11. Wei, X., Decker, J. M., Wang, S., et al. (2003) Antibody neutralization and escape. *Nature* 422, 307–312.
12. Platt, E. J., Wehrly, K., Kuhmann, S. E., et al. (1998) Effects of CCR5 and CD4 cell surface concentrations on infection by macrophage tropic isolates of human immunodeficiency virus type 1. *J. Virol.* 72, 2855–2864.
13. Wei, X., Decker, J. M., Liu, H., et al. (2002) Emergence of resistant human immunodeficiency virus type 1 in patients receiving fusion inhibitor (T-20) monotherapy. *Antimicrob Agents Chemother.* 46, 1896–1905.
14. Leitner, T., Korber, B., Daniels, M., et al. (2005) HIV-1 subtype and circulating recombinant form (CRF) reference sequences, 2005. in (T. Leitner, B. Foley, B. Hahn, P. Marx, F. McCutchan, J. W. Mellors, S. Wolinski, and B. Korber eds.), *HIV Sequence Compendium 2005*, pp. 41–48. Theoretical Biology and Biophysics Group, Los Alamos National Laboratory, Los Alamos, N. Mex. (http://hiv.lanl.gov).
15. Kwong, P. D., Doyle, M. L., Casper, D. J., et al. (2002) HIV-1 evades antibody-mediated neutralization through conformational masking of receptor-binding sites. *Nature* 420, 678–682.
16. Richman, D. D., Wrin, T., Little, S. J. et al. (2003) Rapid evolution of the neutralizing antibody response to HIV type 1 infection. *Proc. Natl. Acad. Sci. (USA)* 100, 4144–4149.
17. Haynes, B. F, Montefiori, D. C. (2006) Aiming to induce broadly reactive neutralizing antibody responses with HIV-1 vaccine candidates. *Expert Rev. Vaccines* 5, 347–363.
18. Mascola, J. R., D'Souza, P., Gilbert, P., et al. (2005) Recommendations for the design and use of standard virus panels to assess the neutralizing antibody response elicited by candidate human immunodeficiency virus type 1 vaccines. *J. Virol.* 79, 10103–10107.
19. Li, M., Gao, F., Mascola, J. R., et al. (2005) Human immunodeficiency virus type 1 env clones from acute and early subtype B infections for standardized assessments of vaccine-elicited neutralizing antibodies. *J. Virol.* 79, 10108–10125.
20. Li, M., Salazar-Gonzalez, J. F., Derdeyn, C. A., et al. (2006) Genetic and neutralization properties subtype C human immunodeficiency virus type 1 molecular *env* clones from acute and early heterosexually acquired infections in southern Africa.. *J. Virol.*, in press.
21. Brown, B. K., Darden, J. M., Tovanabutra, S., et al. (2005) Biologic and genetic characterization of a panel of 60 human immunodeficiency virus type 1 (HIV-1) isolates, representing clades A, B, C, D, CRF01_AE, and CRF02_AG, for the development and assessment of candidate vaccines. *J. Virol.* 79, 6089–6101.
22. Esparza, J. and the Coordinating Committee of the Global HIV/AIDS Vaccine Enterprise. (2005) The Global HIV/AIDS Vaccine Enterprise: Scientific strategic plan. *PLoS Medicine* 2(2), e25.
23. Johnson, V. A., Byington, R. E. (1990) Infectivity assay (virus yield assay), in (Aldovani, A., and Walker, B. D., eds.), Techniques in HIV Research, pp71–76. Stockton Press, New York, N.Y.
24. Montefiori, D. C., Cornell, R. J., Zhou, J. Y., et al. (1994) Complement control proteins, CD46, CD55 and CD59, as common surface constituents of human and simian immunodeficiency viruses and possible targets for vaccine protection. *Virology* 205, 82–92.
25. Louder, M. K., Sambor, A., Chertova, E., et al. (2005) HIV-1 envelope pseudotyped viral vectors and infectious molecular clones expressing the same envelope glycoprotein have a similar neutralization phenotype, but culture in peripheral blood mononuclear cells is associated with decreased neutralization sensitivity. *Virology* **339**, 226–238.

Chapter 27

Assessing the Antiviral Activity of HIV-1-Specific Cytotoxic T Lymphocytes

Otto O. Yang

Abstract

Currently available assays for detecting HIV-1-specific cytotoxic T lymphocytes (CTL) are remarkable for their technical ease, sensitivity, and precision of measurement. However, it is becoming increasingly apparent that CTL responses vary in their antiviral activity at the clonal level. These assays do not reveal the antiviral efficacies of the CTL they detect. Thus, an experimental approach that has been developed for this purpose has been the coculture of CTL with HIV-1-infected cells, to measure the impact of the CTL on viral replication. Data from such experiments can be useful to elucidate the contribution of viral and CTL factors to the antiviral efficiency of CTL.

Key words: cytotoxic T lymphocytes, CD8+ T cells, antiviral activity, assays.

1. Introduction

The MHC class I-restricted, $CD8^+$ cytotoxic T lymphocyte (CTL) response against HIV-1 has been shown to be an important arm of immunity that helps contain HIV-1 during chronic infection. However, it is notable that standard assays of the CTL response are indirect with regard to antigen presentation (utilizing synthetic peptides or recombinant vaccinia virus to express HIV-1 epitopes) and/or readout of CTL function (measuring cytokine release or T cell receptor, TCR, binding), pertaining to the actual antiviral activity of these cells *(1)*. To date, none of these commonly utilized assays have been shown to directly correlate to the function of CTL to suppress HIV-1 replication, despite the great sensitivity and quantitative accuracy of several assays (such as ELISpot and MHC-peptide tetramer staining) to

Vinayaka R. Prasad, Ganjam V. Kalpana (eds.), *HIV Protocols: Second Edition, vol. 485*

DOI 10.1007/978-1-59745-170-3_27 Springerprotocols.com

detect HIV-1-specific CTL. This is underscored by the observation that the quantitative measurements yielded by such assays do not correlate to the immune control of HIV-1 in vivo *(2, 3)*.

This inability to predict CTL efficacy with these measurements is likely because such assays do not directly measure the interaction of CTL with HIV-1-infected cells *(1)*. The suppression of HIV-1 replication in an acutely infected cell depends on the completeness and timing of CTL activity against the cell, which in turn depends on the efficiency and timing of TCR recognition of the epitope-MHC complex on the surface of the cell, which in turn depends on the efficiency and timing of protein expression and epitope processing/transport/MHC binding. For example, a CTL targeting an epitope that is expressed at low levels late in cell infection would be much less antiviral than those targeting an epitope that is expressed at high levels early in infection, but both would be equally detectable and indistinguishable by an assay such as ELISpot for interferon-γ production in response to excess exogenous synthetic epitope. Thus the indirect nature of measuring CTL responses using either uninfected cells labeled with peptides or infected with recombinant vaccinia virus (ELISpot, intracellular cytokine staining, chromium release assay) or peptide-MHC tetramers as a surrogate for actual HIV-1-infected cells bypasses the physiologic processes of epitope generation that are crucial factors affecting the antiviral potential of the CTL.

Given this barrier to using standard assays to predict the antiviral activity of HIV-1-specific CTL, several groups have developed assays using HIV-1-infected target cells *(4–10)*. These assays have been used to more directly measure the antiviral activity of CTL, and to allow controlled comparisons to evaluate the impact of various factors in their ability to suppress HIV-1 replication. The cleanest evaluations of these factors have utilized highly controlled settings using clonal cell populations, avoiding confounding variables such as variability of virus replication in primary $CD4^+$ T cells, mixed CTL populations, and mixed lymphocyte reactions due to MHC mismatching. Here I present methods to perform functional assessment of the antiviral activity of HIV-1-specific CTL by direct challenge with HIV-1-infected target cells. This system allows direct semi-controlled comparisons of factors such as CTL epitope specificity or viral sequence for the ability of CTL to suppress HIV-1 replication.

2. Materials

2.1. Cell Culture and CTL Cloning/Maintenance

1. R10: RPMI-1640 medium (Sigma, St. Louis, MO) supplemented with 10% heat-inactivated fetal calf serum (Sigma), 10 nM *N*-2-hydroxyethylpiperazine-*N*′-2-ethanesulfonic acid

(HEPES), 2 mM L-glutamine, 50U penicillin/mL, and 50 μg streptomycin/mL. R10-50: R10 with 50 U/mL recombinant human IL-2.
2. 96-well U-bottom and 24-well flat bottom plates, and T25 flasks with vented caps for tissue culture.
3. Anti-CD3 stimulatory antibody (OKT-3 or similar antibody) used at the manufacturer's recommended dose for T cell proliferative activity (1 μg/mL for OKT-3).
4. HIV-1 peptides: custom synthesized epitopes of known HLA restriction, or 15-mer peptides (NIH AIDS Reference and Reagent Repository).
5. Ficoll-purified healthy donor PBMC, irradiated with 3000 rads, for use as feeder cells.

2.2. Chromium Release Assay

1. 96-well U-bottom plates for tissue culture.
2. R10 (*see* **Section 2.1**).
3. $Na_2{}^{51}CrO_4$ (New England Nuclear, Boston, MA).
4. Triton X-100 (Sigma), 5% solution diluted in water.
5. Scintillation counter.

2.3. HIV-1-Permissive Cells and HIV-1

1. HIV-1-permissive $CD4^+$ lymphocytes (ATCC) of defined HLA type appropriate for CTL clones to be tested.
2. Titered HIV-1 stock, aliquotted and stored at −80 °C.

2.4. Quantitative p24 ELISA

1. Quantitative HIV-1 p24 antigen ELISA kit (Dupont, Boston, MA, or other commercial kit).
2. ELISA plate reader.

3. Methods

The basic principle of measuring inhibition of HIV-1 replication by CTL is quite simple; the technical difficulty is deriving HIV-1-specific CTL clones, and the complexities are in the experimental design to harness this assay to yield meaningful conclusions (*see* **Note 1**). HIV-1-permissive target cells are infected with HIV-1 at a low multiplicity (M.O.I. of 10^{-2} to allow viral spread), and cultured with or without HIV-1-specific CTL. Viral replication is then monitored by quantitative p24 antigen ELISA. The degree of suppression is determined by comparing p24 production in the absence and presence of CTL.

3.1. Peptide-Stimulation and Enrichment of HIV-1-Specific CTL Within PBMC

1. Use Ficoll-purified PBMC (fresh or cryopreserved) from an HIV-1-infected individual, for whom the HLA type is known and/or a peptide-specific response has been identified (via ELISpot or other mapping assays using overlapping 15-mer peptides).

2. Take 10^6 PBMC and suspend in 2 ml R10-50 with the peptide of interest at 2 μg/mL for a 15-mer or 100 ng/mL for a minimal epitope, in a well of a 24-well plate.
3. An alternative to step 2 is immunomagnetic enrichment after short-term peptide stimulation (*see* **Note 2**).
4. Feed by removing 1 mL medium and replacing it with 1 mL fresh R10-50 every 3–4 days.
5. After 14 days, test the cells by chromium release assay using HLA matched target cells (for the HLA type of the minimal epitope) or autologous EBV-transformed B cells if HLA restriction is unknown.

3.2. Screening by Chromium Release Assay

1. Centrifuge target cells bearing the appropriate HLA molecule (defined HLA or autologous EBV-transformed B cells) in two 15-mL tissue culture centrifuge tubes (approximately 10^6 cells per tube).
2. Remove supernatant leaving approximately 200 μL of the medium behind, and resuspend pellet in remaining medium.
3. Add 100 μCi $Na_2{}^{51}CrO_4$ to each tube, and add the appropriate peptide to a final concentration of 10 μg/mL to one tube.
4. Incubate for 1 h at 37 °C with loose caps in a humidified 5% CO_2 environment.
5. Wash twice with cold (4 °C) R10 in a 4 °C refrigerated centrifuge; count cells and resuspend at 10^5/mL in R10.
6. Suspend effector cells to be tested at 5×10^5/mL and 10^6/mL for testing at effector:target ratios of 5:1 and 10:1, respectively.
7. In duplicate wells of a 96-well U-bottom plate, add 100 μL of target cells (with or without peptide) and no effector cells (spontaneous release control), 100 μL of effector cells, or 100 μL of 5% Triton X-100 solution (maximum release control).
8. Incubate plates at 37 °C in a humidified 5% CO_2 environment for 4 h.
9. Spin plates to pellet cells.
10. Harvest supernatant for scintillation counting of released ^{51}Cr.
11. Using averages of the duplicate values, calculate % specific lysis by each effector as: %specific lysis = 100× (Experimental Release - Spontaneous Release)/(Maximal Release - Spontaneous Release).

3.3. Limiting Dilution Cloning

1. If the prior chromium release screening showed activity from the peptide-stimulated PBMC, the enriched HIV-1-specific CTL can be cloned at limiting dilution from the mixed population.

2. Prepare fresh irradiated PBMC feeder cells and resuspend these cells at 10^6/mL in R10-50. Prepare 10 mL of this feeder solution per cloning plate.
3. Add anti-CD3 stimulatory antibody to the feeders to a final concentration of 0.2 μg/mL.
4. Add 100-μL feeder solution to each well of each cloning plate.
5. Take the peptide-stimulated PBMC and dilute them to 100, 30, and 10 cells/mL in R10-50. Make 10 mL dilution for each cloning plate.
6. Add 100 μL of these cells to be cloned to each well of the cloning plate, yielding 10 cells/well, 3 cells/well, and 1 cell/well for each respective dilution.
7. Maintain the plates in a cell culture incubator.
8. Approximately every 7 days, remove 100 μL of the supernatant, being careful not to disturb the cell pellet, and feed with 100 μL fresh R10-50.
9. Observe for growing cell pellets by visual inspection between 14 and 21 days.
10. When pellets are visible, transfer them to a 24-well plate containing anti-CD3 stimulatory antibody and 10^6 fresh irradiated feeders in 2 mL R10-50. Feed every 3–4 days by replacing 1 mL of supernatant (leaving the cells undisturbed) with 1 mL of fresh R10-50. Try to pick pellets from plates with 12 or fewer pellets; those from cloning plates with more than 12 pellets are less likely to be clonal.
11. After 10–14 days, test the expanded cells by chromium release assay as described in 3.2 above, to screen for clones that are HIV-1-specific.

3.4. Maintenance of HIV-1-Specific CTL Clones

1. CTL clones require periodic stimulation every 1 or 2 weeks to expand and to maintain viability. Start with 5×10^5 to 10^6 CTL for expansion; these should be washed and cultured overnight in R10 (without IL-2).
2. Prepare a T25 flask with 15×10^6 feeders in 10 mL R10 with OKT-3. Add the CTLs.
3. The next day, add 10 mL R10-50.
4. Every 3–4 days thereafter, feed by removing 10-mL supernatant (leaving cells undisturbed) and replacing it with 10 mL fresh R10-50.
5. At 7–14 days after stimulation, test the expanded cells by chromium release assay as described in 3.2 above.
6. At this time, CTL can be viably cryopreserved. When thawed, they should be immediately stimulated and expanded.

3.5. Testing HIV-1 Inhibition by CTL Clones

1. Choose an appropriate HIV-1-permissive cell line that is HLA matched to the restricting HLA allele(s) for the CTL clone(s) of interest (*see* **Note 3**), such as T1 *(11)*, H9 *(12)*, Jurkat

Table 27.1
HIV-1-permissive cell lines with defined HLA types. These cell lines have been successfully used in assays of CTL suppression of HIV-1 in our laboratory

Cell line	A Alleles	B Alleles	C Alleles
T1	*02(*0201)/*19	*05/*40	*01/*03
H9	*01	*15	*03/*06
Jurkat	*09/*10	*07/*14	N.D.
C8166	*01/*09	*08/*12	N.D.
CEM	*01/*19	*08/*40	N.D.
PB-1	*02	*07/*57	*06/*15

N.D.=not determined.

(13), C8166 *(14)*, CEM *(15)*, PB-1 *(16)* (**Table 27.1**). This cell line should be just sub-confluent in growth. Alternatively, primary autologous cells can be utilized as target cells (*see* **Note 4**).

2. Centrifuge approximately 10^6 cells in a 15-mL tissue culture centrifuge tube, and remove the supernatant leaving about 100 μL. Resuspend the cell pellet in the remaining medium.
3. Add 10^4 infectious units of the HIV-1 strain to be tested. Vortex well, and incubate with the cap loose in a tissue culture incubator for 4 h. Agitate the tube about every 30 min.
4. Wash the cells twice with 10 mL R10. Resuspend the cells at a concentration of 5×10^5 cells/mL in R10-50. Add 100 μL to each test well of a 96-well U-bottom plate.
5. Resuspend the CTL clone(s) to be tested at 1.25×10^5 cells/mL in R10-50. Add 100 μL to triplicate wells. Include control wells with only R10-50 (no CTL) added. Addition of CTL to HLA mismatched target cells is another useful control. Alternatively, nonclonal cells such as purified $CD8^+$ T lymphocytes or enriched CTL cell lines can be utilized as the effector cells (*see* **Note 5**).
6. At 2- to 4-day intervals, remove 100-μL supernatant (without disturbing the cells), inactivate by adding 10 μL of 5% Triton X-100 in water, and keep for p24 ELISA. After inactivation, the supernatants can be stored at −20 °C. Replace with 100 μL fresh R10-50 to feed the cells. Repeat until p24 in the no-CTL control wells begins to plateau, usually in 7–10 days. Inhibition should be calculated at a time point just before plateau of viral production, usually 5–7 days..

7. Calculate $\log_{10}$ units suppression as: $\log_{10}$(p24 antigen without CTL) – $\log_{10}$(p24 antigen with CTL) for each set of triplicates (means). For comparisons where different viruses are used, the $\log_{10}$ units of inhibition can be normalized as a fraction of total $\log_{10}$ units in the control without CTL (e.g. one culture with 3 $\log_{10}$ units of inhibition and 6 $\log_{10}$ units of replication yields 50% $\log_{10}$ units of suppression, compared to another culture with 2 $\log_{10}$ units of inhibition and 4 $\log_{10}$ units of replication also yielding 50% $\log_{10}$ units of suppression). Suppression should be calculated at a time point just before HIV-1 growth reaches a plateau (usually day 6 or 7).

4. Notes

1. The most variable factor in inhibition measurements is the activity of the CTL. HIV-1-specific CTLs are prone to senescence and loss of activity, and this introduces variability into the assay for comparisons of different CTL clones. The best-controlled comparisons are between two viruses using the same CTL clone (**Fig. 27.1**). For example, the impact of a mutation in Nef affecting HLA class I downregulation can be tested in this manner by comparing viruses differing only in

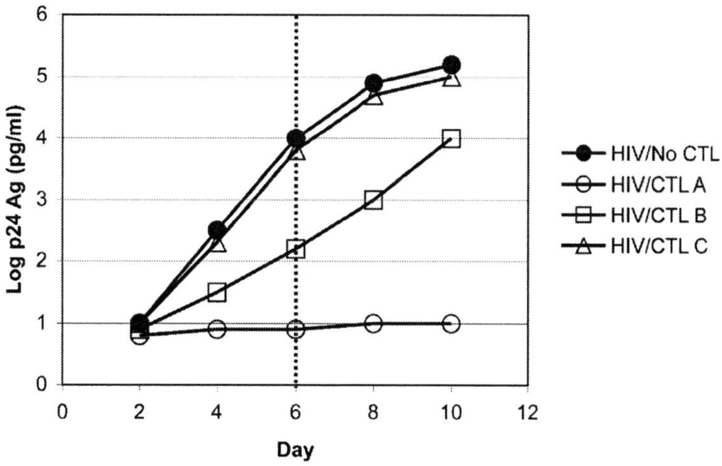

Fig. 27.1. **Example of comparing inhibition of a strain of HIV-1 by three different CTL clones.** Inhibition of HIV-1 by three hypothetical CTL clones (clones A–C) is evaluated. On day 6, clone A demonstrates $3.1\log_{10}$ units of inhibition, while clone B demonstrates $1.8\log_{10}$ units of inhibition, and clone C demonstrates $0.2\log_{10}$ units of inhibition.

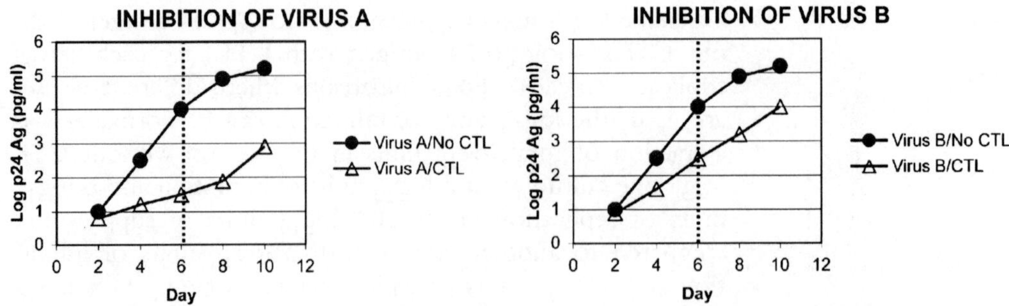

Fig. 27.2. **Example of comparing inhibition of two strains of HIV-1 by a CTL clone.** Two hypothetical strains of HIV-1 (strain A and B) are compared for their suppression by a CTL clone in parallel inhibition assays. On day 6, Virus A demonstrates $2.5 \log_{10}$ units of inhibition, while Virus B demonstrates $1.5 \log_{10}$ units of inhibition.

Nef *(17)*. Comparisons of different CTL clones to each other (**Fig. 27.2**) must be interpreted with caution, because differences can be due to clone activity in vitro rather than true differences in other properties such as epitope targeting. In general, a highly active clone should demonstrate about 50% specific lysis when tested by chromium release at a 1:1 effector:target cell ratio.

2. CTL enrichment before cloning can also be accomplished by interferon-γ live cell capture after short-term peptide-stimulation using an immunomagnetic purification kit (IFN-γ Secretion Assay - Cell Enrichment and Detection Kit, Miltenyi Biotech, Auburn, CA). These purified cells can then be expanded using anti-CD3 antibody and irradiated feeder PBMC to generate a peptide-specific line for cloning.
3. Target cells of appropriate HLA type are a limiting factor. Autologous EBV-immortalized B cells transduced with CD4 are infectable with X4 HIV-1 *(6)* and can be utilized. Utilization of autologous immortalized $CD4^+$ T lymphocytes (HTLV *tax*-transduced or fused with immortalized cell lines) is another option that our group has recently been exploring.
4. Primary $CD4^+$ T lymphoblasts *(10)* or other HIV-1-permissive cell types *(18)* can be utilized as target cells in the inhibition assay. For detailed quantitative comparisons, this is not preferable, because viral replication is less reproducible in primary cells, and primary lymphocytes can react in mixed lymphocyte reactions against the CTL, variably affecting viral replication and effector cell function.
5. Primary polyclonal $CD8^+$ T lymphocytes or MHC-peptide tetramer sorted cells can be utilized as effector cells in the inhibition assay. Again, a major caveat is that there can be mixed lymphocyte reactions that could affect the results of the assay in unpredictable ways.

References

1. Yang, O. O. (2003) Will we be able to 'spot' an effective HIV-1 vaccine? *Trends Immunol* 24, 67–72.
2. Addo, M. M., Yu, X. G., Rathod, A., et al. (2003) Comprehensive epitope analysis of human immunodeficiency virus type 1 (HIV-1)-specific T-cell responses directed against the entire expressed HIV-1 genome demonstrate broadly directed responses, but no correlation to viral load. *J Virol* 77, 2081–2092.
3. Betts, M. R., Ambrozak, D. R., Douek, D. C., et al. (2001) Analysis of total human immunodeficiency virus (HIV)-specific CD4(+) and CD8(+) T-cell responses: relationship to viral load in untreated HIV infection. *J Virol* 75, 11983–11991.
4. Ahearne, P. M., Morgan, R. A., Sebastian, M. W., et al. (1995) Multiple CTL specificities against autologous HIV-1-infected BLCLs. *Cell Immunol* 161, 34–41.
5. Buseyne, F., Fevrier, M., Garcia, S., et al. (1996) Dual function of a human immunodeficiency virus (HIV)-specific cytotoxic T-lymphocyte clone: inhibition of HIV replication by noncytolytic mechanisms and lysis of HIV-infected CD4+ cells. *Virology* 225, 248–253.
6. McElrath, M. J., Rabin, M., Hoffman, M., et al. (1994) Evaluation of human immunodeficiency virus type 1 (HIV-1)-specific cytotoxic T-lymphocyte responses utilizing B-lymphoblastoid cell lines transduced with the CD4 gene and infected with HIV-1. *J Virol* 68, 5074–5083.
7. Shankar, P., Sprang, H., Lieberman, J. (1998) Effective lysis of HIV-1-infected primary CD4+ T cells by a cytotoxic T-lymphocyte clone directed against a novel A2-restricted reverse-transcriptase epitope. *J Acquir Immune Defic Syndr Hum Retrovirol* 19,111–120.
8. Tomiyama, H., Akari, H., Adachi, A., et al. (2002) Different effects of Nef-mediated HLA class I down-regulation on human immunodeficiency virus type 1-specific CD8(+) T-cell cytolytic activity and cytokine production. *J Virol* 76, 7535–7543.
9. Yang, O. O., Kalams, S. A., Rosenzweig, M., et al. (1996) Efficient lysis of human immunodeficiency virus type 1-infected cells by cytotoxic T lymphocytes. *J Virol* 70, 5799–5806.
10. Yang, O. O., Kalams, S. A., Trocha, A., et al. (1997) Suppression of human immunodeficiency virus type 1 replication by CD8+ cells: evidence for HLA class I-restricted triggering of cytolytic and noncytolytic mechanisms. *J Virol* 71,3120–3128.
11. Salter, R. D., Howell, D. N., Cresswell, P. (1985) Genes regulating HLA class I antigen expression in T-B lymphoblast hybrids. *Immunogenetics* 21, 235–246.
12. Popovic, M., Sarngadharan, M. G., Read, E., et al. (1984) Detection, isolation, and continuous production of cytopathic retroviruses (HTLV-III) from patients with AIDS and pre-AIDS. *Science*. 224, 497–500.
13. Weiss, A., Wiskocil, R. L., Stobo, J. D. (1984) The role of T3 surface molecules in the activation of human T cells: a two-stimulus requirement for IL 2 production reflects events occurring at a pre-translational level. *J Immunol* 133, 123–128.
14. Salahuddin, S. Z., Markham, P. D., Wong-Staal, F., et al. (1983) Restricted expression of human T-cell leukemia–lymphoma virus (HTLV) in transformed human umbilical cord blood lymphocytes. *Virology* 129, 51–64.
15. Foley, G. E., Lazarus, H., Farber, S., et al. (1965) Continuous Culture of Human Lymphoblasts from Peripheral Blood of a Child with Acute Leukemia. *Cancer* 18, 522–529.
16. Rozmyslowicz, T., Kijowski, J., Conover, D. O., et al. (2001) New T-lymphocytic cell lines for studying cell infectability by human immunodeficiency virus. *Eur J Haematol* 67, 142–151.
17. Yang, O. O., Nguyen, P. T., Kalams, S. A., et al. (2002) Nef-mediated resistance of human immunodeficiency virus type 1 to antiviral cytotoxic T lymphocytes. *J Virol* 76, 1626–1631.
18. Severino, M. E., Sipsas, N. V., Nguyen, P. T., et al. (2000) Inhibition of human immunodeficiency virus type 1 replication in primary CD4(+) T lymphocytes, monocytes, and dendritic cells by cytotoxic T lymphocytes. *J Virol* 74, 6695–6699.

Chapter 28

Methods for Quantitating Antigen-Specific T Cell Responses Using Functional Assays in Rhesus Macaques

Rama Rao Amara

Abstract

Ex vivo enumeration of the absolute frequencies of antigen-specific T cells is key for evaluating the immunogenicity of T cell-based vaccines. Currently there are three methods that are widely used to quantify cellular immune responses: Enzyme-Linked Immuno Spot assay (ELISpot), Intracellular Cytokine Staining assay (ICS) and Tetramer assay. These three different assays offer different information. In this chapter, I discuss the two functional assays, ELISpot and ICS. The ELISpot and ICS assays use short term in vitro stimulation to assay the frequency and cytokine expression profiles of responding cells. The ELISpot assay scores spots of captured cytokine produced by individual cells whereas, ICS uses flow cytometry to profile individual cells for surface markers and the production of cytokines.

Key words: T cells; ELISpot, intracellular cytokine staining; cytokines, rhesus macaques.

1. Introduction

Ex vivo enumeration of the absolute frequencies of antigen-specific T cells is key for evaluating the immunogenicity of T cell-based vaccines. Currently there are three methods that are widely used to quantify cellular immune responses: Enzyme Linked Immuno Spot assay (ELISpot), Intracellular Cytokine Staining assay (ICS) and Tetramer assay. All these three methods are highly sensitive. ELISpot and ICS assays measure the frequency of antigen-specific cells based on their ability to produce different cytokines whereas, tetramer assay measures the frequency of antigen-specific cells independent of their function. Each of these techniques has advantages and disadvantages that are summarized in **Table 28.1**.

Vinayaka R. Prasad, Ganjam V. Kalpana (eds.), *HIV Protocols: Second Edition, vol. 485*

DOI 10.1007/978-1-59745-170-3_28 Springerprotocols.com

Table 28.1
Comparison of current quantitative T cell assays

	Tetramer staining	ELISPOT	Intracellular Cytokine Staining (ICS)
Sensitivity:	Sensitive	Sensitive	Sensitive
Ease of performance:	Easy to perform. Requires a flow cytometer	Requires careful technique and an Immunospot reader	Requires careful technique and a flow cytometer
Substrate for assay:	Can be done on whole blood or PBMC. Requires very little blood. Can be done on frozen cells	Must be done on PBMC. Can be done on frozen cells	Can be done on whole blood or PBMC. Can be done on frozen cells
Time for assay:	Very short (2 h) assay	About 36 h	About 10 h.
Stimulation:	Stimulation is not required – can detect anergic cells	Stimulation is required – failure to detect anergic cells	Stimulation is required – failure to detect anergic cells
Information on functionality of cells:	No information about functionality of cells	Information about functionality of cells	Information about functionality of cells
Need for HLA typing and HLA-specific reagents:	Prior knowledge about the epitope and its HLA restriction is required	Do not need prior knowledge about epitopes and their HLA restriction	Do not need prior knowledge about epitopes and their HLA restriction
Read out for assay:	Detects epitope specific CD8 T cells	Detects total T cell response (CD4 and CD8) unless specific subsets are depleted	Detects and delineates both CD4 and CD8 T cell responses
Potential for phenotypic analyses:	Phenotypic analysis of responding T cells is possible	No phenotypic analysis of responding T cells	Phenotypic analysis of responding T cells is possible
Cost:	Expensive	Most cost effective	Expensive

These three different assays offer different information. The tetramer assay directly tests samples for the frequency of CD8 T cells with a specific T cell receptor *(1)*; and, when combined with multicolor flow cytometry, offers the opportunity to determine the phenotypes of cells without in vitro stimulation. However, tetramer assay is limited by the availability of reagents and the need to tailor tests to the histocompatibility type of each individual. The ELISpot and ICS assays use short term in vitro stimulations to assay the frequency and cytokine profiles of responding cells *(2–6)*. The ELISpot assay scores spots of captured cytokine produced by individual cells whereas, ICS uses flow cytometry to profile individual cells for surface markers and the production of cytokines.

ELISpot assay has been used as the standard assay for macaque *(7, 8)* and human trials because it is easy to perform, does not require major equipment like flow cytometer, less labor intensive and cost effective. One major limitation of this assay is its failure to distinguish a CD4 response from a CD8 response unless depletion of specific subsets is performed. It is always beneficial to perform ICS assay that will allow phenotyping of responding T cells into CD4 and CD8, and studying cytokine co-expression profile, at least on a subset of samples at critical time points. In this chapter, I will discuss the two functional assays, ELISpot and ICS. **Table 28.2** describes the combination of antibodies that can be used to detect macaque IFN-γ, IL-2, IL-4 and IL-13.

Table 28.2
Antibody pairs for detection of cytokines IL-2, IL-4 and IL-13 using ELISpot assay

Cytokine	Coating Ab suggested concentration	Detection Ab suggested concentration
IFN-γ	Pharmingen (Cat#554698) 4 μg/mL	MABTECH (Cat# 3420-6-250) 2 μg/mL
IL-2	R&D Systems (Cat#Mab 202) 10 μg/mL	R&D Systems (Cat#BAF 202) 2 μg/mL
IL-4	Pharmingen (Cat#554482) 5 μg/Lml	Bio-source International (Cat#NON0059) 1:5000 dil
IL-13	R&D Systems (Cat#AF 213) 5 μg/mL	R&D Systems (Cat#BAF 213) 2 μg/mL

2. Materials

2.1. ELISpot

1. Millipore IP Opaque 96 well plate (e.g., Millipore MAIP S 4510).
2. PBS (e.g., Cellgro #21-040-CV).
3. Tween 20 (e.g., Sigma #S-7949).
4. Wash buffer: PBS with 0.05% Tween 20 (PBST).
5. RPMI-1640 (e.g., Fisher #MT-10-040-CVRF).
6. Fetal Bovine Serum – heat inactivated at 56 °C for 30 min (e.g., HyClone # SH30070.03).
7. Antibiotics (100 ×) – Penicillin (10,000 iu/mL) and streptomycin (10, 000 μg/mL) (e.g., Cellgro # 30-002-Cl).
8. Complete medium: RPMI-1640 with 10% FBS and 1× antibiotics.
9. Staphylococcal enterotoxin B (SEB) (e.g., Sigma #S-4881).
10. Strepavidin-AP (e.g., Mabtech # 3310-8).
11. NBT-BCIP (e.g., Pierce #34042).
12. Anti- human CD28 (e.g., B.D. Pharmagin #555725).
13. Anti- human CD49d (e.g., B.D. Pharmingin #314701).

2.2. Intracellular Cytokine Staining

1. 5.0 mL polypropylene tube (e.g., BD Falcon #352063).
2. Brefeldin A (e.g., Sigma #B-7651) or Golgi plug (e.g., BD Pharmingen).
3. EDTA (e.g., Sigma # ED255).
4. Cytofix/Cytoperm Plus – Contains Cytofix/Cytoperm, Perm wash, Golgi plug (BD Pharmingen).
5. Formalin 10% (e.g., Polysciences Inc. #04018).
6. Staining Buffer – 2% heat-inactivated FBS and 0.05% (*w/v*) NaN3 in PBS.
7. Anti-Human CD3-FITC conjugate (e.g., BD # PN IM1281).
8. Anti-human CD8-PerCP conjugate (e.g., BD #347314).
9. Anti-human CD3-APC conjugate (e.g., BD # PN IM2467).
10. Anti-human IL2-PE conjugate (e.g., BD Pharmingen # 554566).
11. Anti-human IFN-γ-APC conjugate (e.g., BD Pharmingen # 554702).

3. Methods

3.1. ELISpot

3.1.1. Day Before the Assay

1. Establish format for the plate. A typical plate includes quadruplicate wells for no antigen (nostim), duplicate wells for

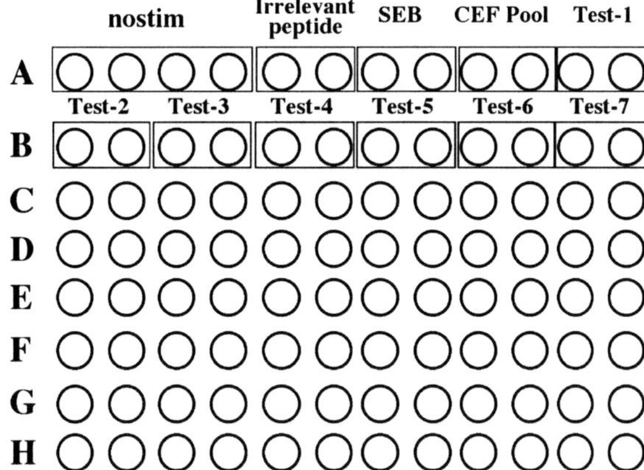

Fig. 28.1. A typical format for the organization of samples for ELISpot assay.

the negative control such as an irrelevant peptide or peptide pool, duplicate wells for positive controls such as SEB and the CEF pool (only for human samples) and duplicate wells for the test samples. The example in **Fig. 28.1** represents a typical format.

2. Pre-wet the membrane of each well of the plate by spraying 70% ethanol. Remove ethanol thoroughly by flicking and dabbing. Air-dry briefly.
3. Coat each well with 100 μL of mouse anti-human IFN-γ (capture antibody) at a concentration of 4 μg/mL in PBS. Gently tap the plate to make sure that the antibody solution is at the bottom of the wells.
4. Seal the plate with plastic cling wrap (Saran Wrap) and incubate overnight at 4 °C (also *see* **Note 1**).

3.1.2. Day 1

1. Prepare fresh PBMC or thaw frozen PBMC. Add co-stimulatory antibodies (anti-CD28 and anti CD49d) each at a final concentration of 2 μg/mL. If frozen PBMC are used, rest them over night at 37 °C in complete medium under 5% CO_2 at a concentration of 2×10^6 cells/mL in 15 mL conical polypropylene tubes.
2. Flick the plate to remove capture antibody solution.
3. Add 200 μL complete medium to each well and incubate at 37 °C for a minimum of 60 min.
4. Wash plate four times with plain RPMI and once with complete medium (200 μL each time).
5. Add the desired concentration of PBMCs in complete medium to each well in a volume of 100 μL. Cell concentrations can range from 5×10^5 to 1×10^4 PBMC depending on the expected frequencies of ELISpots. In the absence of any prior information about the expected frequencies, seed 2×10^5 and

5×10^4 PBMC in duplicate wells. It is important to note that the ELISpot assay is not linear when cells are seeded below 2×10^5 per well.

6. Add 100 μL of antigens of interest to appropriate wells. If peptides are used for stimulation, a final concentration of 1–2 μg/mL for each peptide in the pool is best. SEB is used at 1 μg/mL. Keep the final concentration of DMSO $\leq$ 1%.
7. Culture for 36 ± 6 h at 37 °C in a CO_2 incubator (do not disturb the plate during incubation).

3.1.3. Day 3

1. Manually wash plate three times with PBS to remove cells, and then 3 times with PBST to lyse and remove any remaining cells (200 μL each time).
2. Add biotinyated anti-IFN-γ antibody (diluted in PBST plus 2% FBS) to a concentration of 1 μg/mL, 100 μL per well.
3. Incubate at 37 °C for 2 h.
4. Wash plate six times with wash buffer (PBST, 200 μL each time).
5. Add 100 μL per well of a 1:1,000 dilution of streptavidin-AP diluted in PBST plus 2% FBS.
6. Incubate at 37 °C for 1 h.
7. Wash plate six times with wash buffer, 200 μL each time.
8. Add 100 μL of NBT-BCIP substrate. Incubate at room temperature for 20–40 min (*see* **Note 2**).
9. Watch for spot formation. When the desired intensity of spots has appeared, wash the plate thoroughly under a gentle stream of tap water. Make sure not to over-develop the plate. Remove the bottom plastic base of the plate and rinse the bottom of the plate thoroughly under the tap water. In general, spots appear much darker and brighter after the plate has dried.
10. Dry and count using the automated ImmunoSpot reader. If a reference standard is used, make sure that the frequencies in the standard are in the expected range.

3.2. Intracellular Cytokine Staining

1. For each sample, set up 5 mL polypropylene tubes for stimulation and compensation controls. Set up a separate compensation control tube for each fluorochrome that is used on the test samples (*see* **Note 3**). Aliquot one million test PBMCs in 100 μL of complete medium containing 2 μg/mL of anti-CD28 antibody and 2 μg/mL of anti-CD49d antibody and antibiotics (Penicillin (100 iu/mL) and streptomycin (100 μg/mL) into each tube. The following is an example for stimulation and compensation controls:

Tube Name	Stimulant	Staining antibody
Comp-FITC	None	CD3-FITC
Comp-PE	None	CD4-PE
Comp-PerCP	None	CD8-PerCP
Comp-APC	None	CD3-APC
Positive stimulation	SEB	Same stains as on test samples
No stimulation	None	Same stains as on test samples

2. For stimulations, add the desired amount of antigen (peptide should be in the range of 1–2 μg/mL) in a volume of 100 μL. Keep the final concentration of DMSO ≤ 1%. Vortex gently and incubate for 2 h at 37 °C, 5% CO_2.
3. At 2 h, vortex gently and add Golgi-Plug (0. 2 μL of stock) or Brefeldin A (BFA, at 10 μg/mL) in a volume of 10 μL in complete medium. Vortex gently. Incubate for an additional 4 h (for a total of 6 h incubation) at 37 °C, 5% CO_2 (*see* **Note 4**). At this point, cells can be stored at 4 °C for up to 12–16 h without loss of signal.
4. Vortex gently. Add 2 mL of staining buffer. Centrifuge at 1500 rpm for 5 min. Aspirate all but ~ 100 μL of the supernatant.
5. Vortex gently. Stain for the desired surface antigens in the 100 μL volume by adding cell surface MAb staining reagents for 30 min at 4 °C in the dark using the concentrations recommended by the manufacturer. It is always important to titrate the concentration of antibody for the best signal to noise ratio.
6. Wash cells one time in 2 ml of staining buffer. Centrifuge at 1500 rpm for 5 min. Aspirate supernatant.
7. Fix and permeabilize cells by adding 250 μL of Cytofix/Cytoperm Solution. Vortex and incubate for 20 min at 4 °C in the dark.
8. Add 1 mL perm/wash solution (dilute stock 1:10 in sterile H_2O). Centrifuge at 1500 rpm for 5 min. Decant supernatant. Wash once more with 1 mL perm/wash solution. Aspirate all but ~ 100 μL of supernatant.
9. Vortex gently. Stain for the desired intracellular antigens in the ~ 100-μL volume by adding MAb staining reagents for

30 min at 4 °C in the dark using the concentrations recommended by the manufacturer.

10. Wash cells once with 2 mL of perm/wash solution and once with 2 mL of PBS.
11. Re suspend cell pellet in 200 μL of 1% formalin in PBS. At this point cells can be stored for up to 24 h in the dark at 4 °C.
12. Set up compensations using Comp tubes and acquire within 24 h on FACSCalibur using CellQuest. Analyze the data using Flowjo software (Treestar, Inc., San Carlos, California).

4. Notes

1. Coating can also be done at room temperature for a minimum of 6 h.
2. Five to ten minutes may be sufficient for detection of IFN-γ.
3. Compensation is performed to correct the spillage of signal from one channel to the other. A clear positive and negative population is required to set up compensation. So, it is important to use antibodies that identify a clear positive and negative population. In general, it is advisable to use antibodies to abundant CD antigens such as CD3, CD4 and CD8.
4. Cells can be cultured up to 12 h in the presence of Brefeldin A.

References

1. Altman, J. D., Moss, P. A., Goulder, P. J., et al. (1996). Phenotypic analysis of antigen-specific T lymphocytes. *Science* 274, 94–96.
2. Rininsland, F. H., Helms, T., Asaad R. J., et al. (2000). Granzyme B ELISPOT assay for ex vivo measurements of T cell immunity. *J Immunol Methods* 240, 143–155.
3. Kumar, A., Weiss, W., Tine, J. A., et al. (2001). ELISPOT assay for detection of peptide specific interferon-gamma secreting cells in rhesus macaques. *J Immunol Methods* 247, 49–60.
4. Kalyuzhny, A., Stark, S. (2001). A simple method to reduce the background and improve well-to-well reproducibility of staining in ELISPOT assays. *J Immunol Methods* 257, 93–97.
5. Pahar, B., Li, J., Rourke T., et al. (2003). Detection of antigen-specific T cell interferon gamma expression by ELISPOT and cytokine flow cytometry assays in rhesus macaques. *J Immunol Methods* 282, 103–115.
6. Karlsson, A. C., Martin, J.N., Younger, S. R., et al. (2003). Comparison of the ELISPOT and cytokine flow cytometry assays for the enumeration of antigen-specific T cells. *J Immunol Methods* 283, 141–153.
7. Amara, R. R., Villinger, F., Altman, J. D., et al. (2001). Control of a Mucosal Challenge and Prevention of AIDS by a Multiprotein DNA/MVA Vaccine. *Science* 292, 69–74.
8. Barouch, D. H., Santra, S., Schmitz, J. E., et al. (2000). Control of viremia and prevention of clinical AIDS in rhesus monkeys by cytokine-augmented DNA vaccination. *Science* 290, 486–492.

Section V

Drug Resistant Viruses and Viral Evolution

Chapter 29

Isolation of Drug-Resistant Mutant HIV Variants Using Tissue Culture Drug Selection

Maureen Oliveira, Bluma G. Brenner, and Mark A. Wainberg

Abstract

The life cycle of HIV-1 can be affected in different manners by the various classes of antiviral agents. Genetic heterogeneity is a characteristic of this virus, which contributes significantly to the ability of the virus to generate mutations which overcome the efficacy of the drug therapy. The selection of drug resistant mutants in vitro can be readily accomplished by maintaining the virus in a state of sub-optimal growth, regulated by slowly increasing the amount of drug pressure applied. This technique is thought to mimic the consequences of drug therapy in patients. Therefore, in this way, novel compounds can be assessed for their selection profile in order to preview the likelihood of emergence of HIV-1 drug resistance in future clinical trials. In addition, combinations of drugs can be investigated in the same manner.

Key words: HIV-1, drug selection, drug resistance, resistance mutations.

1. Introduction

The application of increasing drug pressure on viral growth in vitro can select for variants bearing mutations which affect their sensitivity to the anti-viral agent used *(1, 2)*. By stringent control of the drug dose increments, it is possible to generate resistant species to any and all classes of anti-viral drugs. In addition, it is possible to select for mutant species with viruses from any HIV-1 strain, recombinant or clinical isolate from all sub-types, as well as HIV-2 and SIV viruses *(3)*.

The order of appearance of mutations, the rate of emergence of mutations as well as the pattern of mutations can be

Vinayaka R. Prasad, Ganjam V. Kalpana (eds.), *HIV Protocols: Second Edition, vol. 485*

DOI 10.1007/978-1-59745-170-3_29 Springerprotocols.com

ascertained by genotyping the viral culture supernatant at different time points.

In most cases, the mutational profiles arising through in vitro selection experiments with specific drugs reflect the clinical outcome that arises when these same compounds are employed in HIV therapy *(4)*.

2. Materials

2.1. Selection in Cell Lines

1. Cell lines: MT-2 or MT-4 from NIH AIDS Research and Reference Reagent Program.
2. 10% RPMI 1640 Medium: RPMI 1640 Medium supplemented with 10% fetal bovine serum (qualified FBS) 2 mM L-glutamine per milliliter, 100 U of penicillin per milliliter and 100 μg of streptomycin per milliliter.
3. 24-well tissue culture plates

2.2. Selection in Primary Cells

1. IL-2 Medium: RPMI 1640 Medium, supplemented with 10% qualified FBS, 20 U of Human Interleukin-2 (IL-2) per milliliter, 5 μg of Hydrocortisone per milliliter, 2 mM L-glutamine per milliliter, 100 U of penicillin per milliliter, and 100 μg of streptomycin per milliliter.
2. Primary cells, Cord Blood Mononuclear Cells (CBMC) or Peripheral Blood Mononuclear Cells (PBMC) stimulated for 3 days with 10 μg/mL Phytohemagglutinin A in RPMI 1640 Medium, supplemented with 10% qualified FBS, 2 mM L-glutamine per milliliter, 100 U of penicillin per milliliter and 100 μg of streptomycin per milliliter.
3. 24-well tissue culture plates
4. P24 antigen assay or RT activity assay

P24 antigen assay: Vironostika HIV-1 Antigen kit, bioMérieux, Boxtel, The Netherlands

RT Assay: 50 mM Tris-HCl (pH 7.9), 5 mM MgCl2, 150 mM KCl, 0.5 mM EGTA, 0.05% Triton X-100, 2% ethylene glycol, 5 mM dithiothreitol, 0.3 mM reduced glutathione, 50 μg/mL poly(rA)-oligo(dT), 5μCi [3H]dTTP (specific activity 50–80Ci/mmol).

3. Methods

Drug resistant variants can be obtained in both T-lymphocyte cell lines and primary cells. Generally, it is simpler and more easily accessible to utilize T cell lines, such as MT-2 and MT-4 *(5)*.

Recombinant clones such as the pNL4-3 or HXB-2D can be used to select drug resistant variants and the same clone can be adapted by site-directed mutagenesis to confirm the significance of the mutational profiles. The limitation of using cell lines is that the viruses used must be Syncytium Inducing (SI) in order to help monitor virus growth by visual inspection.

Employing primary cells, such as PBMC or CBMC will allow for both SI and NSI (Non-SI) variants to be successfully selected for resistance, but a quantitative assay such as the p24 antigen assay or the measurement of RT activity is required to assess the progress of the viral growth.

3.1. Selection of Drug Resistant Variants in a Cell Line

1. Infect MT-2 cells with HIV-1 strain (MOI 0.01) for 2 h at 37 °C. Wash 1X to eliminate any unbound virus. Seed 250,000 infected MT-2 cells per well in a 24-well plate.
2. Add drugs to be tested, 10–50 μL of a concentrated stock, beginning with a concentration below the IC_{50} level. Total volume in the well should be 2–2.5 mL of 10% RPMI 1640 Medium. Set up one well without drug pressure as a control. In the case of a novel compound, it is useful to include a known antiviral agent of same the class in parallel (*see* **Note 1,6**).
   When a culture well reaches the peak of infection, as observed visually by the presence of syncytia or CPE (cytopathic effect) in all cells (100%), collect 2 × 500 μL aliquots of culture supernatant and store at −80 °C.
3. The maximal cytopathic effect is usually reached by day 3 to day 7. If there are few CPE evident (< 40%) at day 3–4, split the well by resuspending the culture and transferring 250 μL to a new well, replenishing with the appropriate drug concentration and adding fresh media 1–1.5 mL. Collect when the peak CPE is achieved.
4. To start the next and subsequent rounds of infection, seed 250,000 fresh cells in two new wells and 250 μL of the previously frozen supernatant. Incubate at 37° for 2 h, and add drug stock to give the previous dose of the drug in one well, and a 2- to 2.5-fold higher concentration in the other. For example, if the first passage was with 0. 1 μM, the next round will be 0.1 and 0.25, the next 0.25 and 0.5, then 0.5 and 1.0, etc. Maintain the control well without drug treatment for each passage.
5. If the infection does not peak within 7 days, split the cells in the well again if there are a good number of CPE. If there are too few CPE, abort the well and continue with the lower concentration (*see* **Notes 4, 5**).
6. The selection is complete when in the repeated passage with high doses of the drug; the culture consistently peaks at the same interval as the control well, usually 3–5 days.

7. If required, amplify the selected virus supernatant, without drug present, into a large number of wells, pool, clarify by centrifugation and collect as a viral stock.
8. Perform genotypic analysis on the viral stock to identify the mutational profile of the resistant virus. If necessary, analyze previous time points using the aliquots frozen at each passage.
9. Perform a phenotypic analysis of the selected viral stock to determine the fold-change in sensitivity to the compound selected (*see* **Note 2**).

3.2. Selection of Drug Resistant Variants in Primary Cells

1. Primary cells (CBMC or PBMC) are infected with viral supernatant (MOI 0.1–1.0), for 2 h at 37 °C. Wash 1X to remove unbound virus. Resuspend cells in IL-2 Medium and plate 2–4 million cells per well in a 24-well plate.
2. Add 10–50 μL volumes of concentrated drug stocks, beginning with a concentration below the IC_{50} level. Bring total volume in each well to 2–2.5 mL. in IL-2 Medium. Include one untreated well as a control. When testing an untried compound, set up a parallel experiment with a known ant-viral agent of the same class (*see* **Note 1, 6**).
3. After 7 days in culture, test the viral activity of each well by RT or p24 assay. Compare all values to that of the untreated control well (100%). If the value of the drug-treated well is between 50–100% of the control, then raise the drug concentration by 2- to 2.5-fold. If it is between 10% and 50%, maintain the same concentration. If the value drops to < 10%, drop the dose to a minimal drug pressure or release the drug pressure entirely. When the value returns to 100%, resume the application of drug pressure (*see* **Notes 3, 4**).
4. Collect 1 × 500 μL aliquot and store at −80 °C. To initiate the next and succeeding passages, re-suspend the remaining contents of the well and transfer 500 μL to a new well. Add 2–4 million fresh cells, concentrated drug stock to the chosen concentration(s) and bring to a total volume of 2–2.5 mL with IL-2 Medium.
5. Conduct the succeeding passages in the same manner until the selected variant is under maximum drug pressure and the viral activity is similar to that of the untreated control well (*see* **Notes 7, 8**).
6. Amplify the selected virus without drug pressure to generate a stock, if required.
7. Genotype the selected viral stock to ascertain the occurrence of resistance mutations.
8. Phenotype the stock to determine the level of resistance to the drug tested as well as to other compounds of the same drug class.

4. Notes

1. In order to accurately determine the starting concentration for the test compound the IC_{50} for the virus should be screened prior to the start of the study on the cell type being used for the selection protocol.
2. After the selection is completed and a stock is collected, phenotyping of the resistant variant can be expanded to check for cross-resistance to other compounds from the same class.
3. Viruses, especially clinical isolates, can exhibit varied tolerances to the selection process. If, at every passage you can have more than one alternative drug concentration, you have a greater likelihood to have chosen a correct strategy.
4. It is possible to lose detectable viral growth when exerting effective drug pressure. Re-start the selection from previously frozen aliquots.
5. When using cell lines, since the peak day for collection varies, it is easiest to store the aliquots for the next passage until all are ready for the next round.
6. Combinations of drugs can be selected in the same way, although in many cases, a synergistic effect may require lower concentrations of the compounds being tested together.
7. The emergence of mutations can have an impact on viral fitness, thus lowering the replicative capacity of the selecting variant. In this case, although the RT activity may be lower than the untreated control, it should persist at a fairly constant level.
8. The accumulation of multiple mutations can result in an increase in viral fitness.

5. Summary

The selection of HIV variants that display resistance to antiviral drugs is of clinical importance and also provides insights into HIV pathogenesis. First, the rapidity with which resistance can be selected in regard to a given drug may be predictive of the rapidity with which resistance may develop when that drug is used in the clinic. In most cases, rapid selection of drug resistance is indicative of a low genetic barrier for diminished drug susceptibility. This means that only one or perhaps only two mutations may be required in order for resistance to develop. This is the situation for most inhibitors of HIV reverse transcriptase. In contrast, most protease inhibitors usually require an accumulation of at least six or seven mutations in order for resistance to develop, indicating that these compounds are ones that possess a high genetic barrier for resistance.

Understandably, resistance mutations occur primarily within the genes that encode the proteins that are the target of specific drugs. In this context, resistance mutations associated with protease inhibitors most commonly occur within the HIV protease gene. In contrast, mutations associated with nucleoside and non-nucleoside reverse transcriptase inhibitors occur within the reverse transcriptase gene.

The development of drug resistance has created a need for new drugs that can retain activity against isolates of HIV that are resistant to currently available drugs. This means that there is a need to continuously develop new compounds that retain activity against HIV-1 in the face of drug resistance. In fact, manufacturers and scientists understand the importance of possessing banks of viruses that display resistance against HIV-1 in order to demonstrate that new compounds that are undergoing development possess activity against drug resistance strains *(6)*.

An understanding of HIV drug resistance has also provided key insights in regard to HIV pathogenesis. Among other considerations, it is now evident that drug resistant variants of HIV-1 can be sexually transmitted. This has unfortunate consequences in regard to public health, as the possibility exists that a newly infected individual may not necessarily be responsive to currently approved drugs, if that person was infected with a drug resistant virus. This has created a need to perform HIV genotyping on all newly diagnosed individuals prior to therapy, in order to assure that they are treated with effective antiviral drugs. At the same time, it has become evident that certain drug resistance mutations may have negative consequences on the ability of HIV-1 to replicate. This has underscored that wild-type variants of HIV-1 have become predominant due to the fact that they have superior replication competence.

References

1. Gao, Q., Gu, Z., Parniak, M.A., et al. (1992) In Vitro Selection of Human Immunodeficiency Virus Type I Resistant to 3′-Azido-3′-Deoxythymidine and 2′,3′-Dideoxynosine. *J Virol* 66, 12–19
2. Gao., Q., Gu, Z., Salomon, H., et al. (1994) Generation of multiple drug resistance by sequential in vitro passage of the human immunodeficiency virus type 1. *Arch Virol* 136, 111–122
3. Richman, D., Shih, C-K., Lowy, I., et al. (1991) Human immunodeficiency virus type 1 mutants resistant to nonnucleoside inhibitors of reverse transcriptase arise in tissue culture. *Proc Natl Acad Sci USA* 88, 11241–11245
4. Larder, B., Kellam, P., Kemp, S. (1993) Convergent combination therapy can select viable multidrug- resistant HIV-1 in vitro. *Nature* 365, 451–453
5. Diallo, K., Brenner B., Oliveira, M., et al.. (2003) The M184V substitution in human immunodeficiency virus type 1 reverse transcriptase delays the development of resistance to amprenavir and efavirenz in subtype B and C clinical isolates. *Antimicrob Agents Chemother* July, 2376–2379

6. Wainberg MA, Brenner BG, Turner D. (2005)Changing patterns in the selection of viral mutations among patients receiving nucleoside and nucleotide drug combinations directed against human immunodeficiency virus type 1 reverse transcriptase. *Antimicrob Agents Chemother* 49(5), 1671–1678,.

Chapter 30

Virus Evolution as a Tool to Study HIV-1 Biology

Ben Berkhout and Atze T. Das

Abstract

Mutational analysis of the viral genome is frequently used to study the role of sequence or structural elements in HIV-1 replication. Many laboratories that use this approach have occasionally come across revertant viruses that overcome an introduced defect either by restoration of the original sequence or by the introduction of additional mutations in the viral genome. Similarly, replication of a wild type virus under selective pressure, due to the presence of inhibitors or due to specific culture settings, may result in the appearance of evolved variants that replicate more efficiently under the applied conditions. We have developed in vitro HIV-1 evolution from an anecdotal event to a systematic research tool to study different aspects of the viral replication cycle. In this manuscript, we will briefly review the method of forced virus evolution to study HIV-1 biology and provide several examples that illustrate the power of this method, as it frequently yielded interesting and unexpected information about the mechanism of virus replication.

Key words: HIV-1, virus evolution, revertant, forced evolution.

1. Introduction

Darwinian evolution is a two-step process. The first step consists of the generation of genetic variants that can serve as the material that will be exposed to phenotypic selection, which forms the second step of evolution. In HIV-1 biology, the error-prone reverse transcription process is responsible for the generation of variation, resulting in a viral quasispecies. The variant with the best fitness will subsequently outcompete the other variants. The first step of evolution is completely independent of the actual selection process. Even though phenotypic selection criteria will eventually shape the virus population, nucleotide substitution rates

Vinayaka R. Prasad, Ganjam V. Kalpana (eds.), *HIV Protocols: Second Edition, vol. 485*

DOI 10.1007/978-1-59745-170-3_30 Springerprotocols.com

will largely determine the set of mutants available within the viral quasispecies for the subsequent selection process. As we will present different evolutionary strategies, the importance of both steps will become apparent. In some protocols, we were able to stress the importance on the mutational event, minimizing or even avoiding the effect of phenotypic selection.

The most basic evolution protocol is described below. We prefer to start evolution experiments with a plasmid carrying the complete viral genome. Knowing the complete genome sequence of such an infectious molecular clone allows one to unequivocally score newly acquired mutations, although the mere selection of a mutation does not necessarily mean that this mutation is responsible for the improved replication capacity. The molecular clone can be used to mutate a sequence or structural element under investigation. If this element is important for replication, the introduced mutation will affect the replication capacity. We usually start testing the replication potential of the mutated virus in the SupT1 T cell line. We prefer this T cell line because it allows the formation of syncytia when a CXCR4-using HIV-1 strain is used, which makes it easy to visually detect virus spreading. In addition, virus replication is monitored by following the CA-p24 level in the culture supernatant.

We usually start an evolution experiment by introduction of 5–10 μg of the molecular clone by means of electroporation into 5×10^6 SupT1 cells (5 mL culture). Whereas the efficient replication of the original virus (wild type; wt) will result in the rapid appearance of syncytia and a prompt increase in the CA-p24 level (**Fig. 30.1**), the mutated virus may show a reduced level of replication or may not show any sign of replication (no syncytia, no CA-p24 production). If so, we usually split and maintain the cells when needed. Once signs of virus replication are observed (syncytia and/or CA-p24 values), we start to passage the culture medium onto fresh cells. Initially, we will use a large inoculum (0.5 mL) that in addition to the virus also contains virus-producing cells. In subsequent passages, the size of the inoculum is gradually decreased and cells are removed from the inoculum by centrifugation prior to infection. Fast-replicating variants usually produce a full infection within 1 week upon passage of only a few microliters of supernatant.

When we passage the virus, we also isolate the massively infected cells for subsequent sequence analysis of the integrated proviral genome. To do so, the cells are washed with phosphate-buffered saline and subsequently lysed in 10 mM Tris-HCl (pH 8.0)-1 mM EDTA-0.5% Tween20, followed by incubation with 200 μg/mL proteinase K at 56 °C for 30 min and at 95 °C for 10 min. The proviral DNA is amplified with virus-specific primers surrounding the viral sequences of interest. This PCR product is either directly sequenced, resulting in a sequence reflecting the

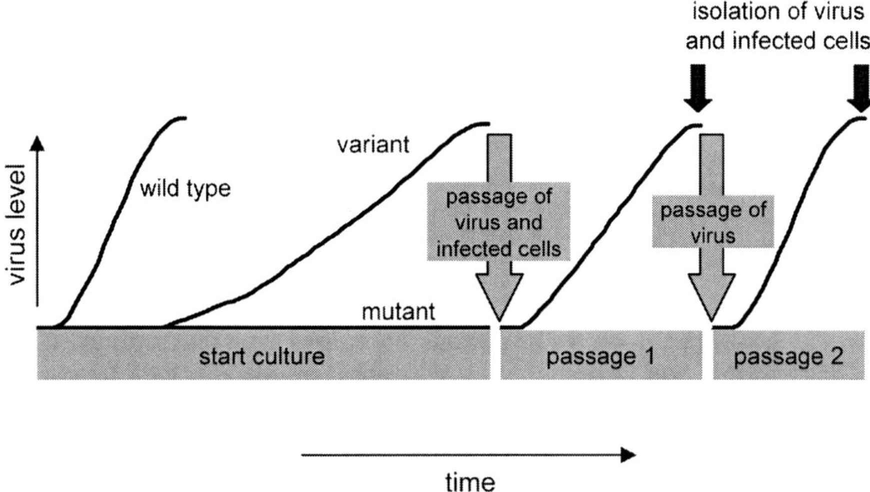

Fig. 30.1. The evolution protocol. Virus cultures are started by the introduction of plasmids encoding the complete wild-type or mutated HIV-1 genome into T cells. Virus replication can be monitored by measuring the CA-p24 level in the culture supernatant. Whereas the wild-type virus replicates efficiently, resulting in a rapid increase in the virus load, a mutant virus in which an essential element has been modified may replicate poorly, and the virus load may remain below the level of detection. A low level of ongoing replication, in combination with the error-prone viral replication machinery will drive evolution of the mutant, which may result in the emergence of variants with improved replication potential and a concomitant increase in the virus load. To allow further evolution of the virus, the culture medium is sequentially passaged onto fresh cells. Initially, we use a large inoculum that contains virus and infected cells. When virus replication has improved, indicated by the shorter period needed to reach a high virus load, the size of the inoculum is reduced and cells are removed prior to infection. At each passage, infected cells are isolated for subsequent sequence analysis of the integrated proviral genome. In most experiments, the start cultures are maintained for up to 12 weeks to allow the emergence of evolved variants.

complete virus population, or cloned into a vector (for example a TA-cloning vector), followed by the sequence analysis of multiple clones. The latter, more elaborate method allows the detection of subsets within the viral sequences present in the culture and this method is more suitable if the culture contains multiple virus variants. However, one should keep in mind that sequence changes identified in this way may have resulted from PCR errors, and it may be advisable to initially focus on mutations observed in multiple clones.

To demonstrate that an identified sequence change is indeed responsible for the increased replication capacity, the evolved sequence is re-cloned into the viral genome and the replication potential of this variant is compared to that of the ancestral mutated virus. A similar protocol can be used to select virus variants that are resistant to specific virus inhibitors [e.g., Reverse Transcriptase (RT) or Protease inhibitors] or that have adapted to specific conditions (e.g., a specific cell type) by long-term culturing of the virus under the selective pressure. It is obvious that not every evolution experiment succeeds, and we usually stop nonproductive experiments after 3 months of culturing, although we may continue in exceptional cases. To increase the chance of obtaining variants, we usually start multiple evolution cultures and we use more DNA (up to 40 μg per 5×10^6 SupT1 cells) in the initial transfection for very crippled viruses.

2. Mutation Frequency Drives Virus Evolution

In the previous section, we discussed that both mutation and selection are involved in most virus evolution settings. The mutation frequency in HIV-1 varies considerably for different kinds of mutations. For example, a G-to-A mutation is observed very frequently and is considered to be easy, transversions are seen less frequently than transitions, and double mutations are more difficult than single mutations and therefore much less frequently observed *(1, 2)*. Given the high HIV-1 population size, the rapid replication kinetics and the high error rate during viral replication, it can be calculated that all variant genotypes with single nucleotide substitutions will be available in an infected individual *(3)*, but it is less likely that complex substitution patterns are available in the initial virus pool. The virus populations that are handled in in vitro experiments will usually be smaller, which will limit the available genotypic variation.

Since mutation is also a chance process, one cannot put too much value on the mutational pattern seen in an individual culture, but should preferably analyze variants emerging in multiple cultures. Independent cultures may provide different answers to the same problem. Even the repeated selection of the same variant is informative, as it indicates that the problem can be solved very effectively (and perhaps exclusively) by that specific mutation. In most cases, multiple evolution routes are observed that provide a glimpse of the available sequence space that supports virus replication.

We hypothesized that if "difficult" drug resistance mutations are consistently observed at certain positions in the viral genome, it is likely that "easier" nucleotide substitutions at that codon did not pass the fitness selection criteria *(4)*. A notable example is provided by changes at codon 215 in the RT gene upon treatment with the nucleotide inhibitor AZT (zidovudine). Variants Thr215Phe or Thr215Tyr are selected that require two substitutions each (ACC → TTC and ACC → TAC, respectively). There are five alternative amino acid substitutions (Ile, Ala, Ser, Pro, Asn) that are easier to generate than Thr215Phe, and ten substitutions (Ile, Ala, Ser, Pro, Asn, Val, Met, Leu, Asp, Gly) that are easier to generate than Thr215Tyr. Yet, none of these alternative substitutions have been detected in vivo and in vitro. These combined data strongly suggest that only an aromatic residue at position 215 can provide the AZT-resistance phenotype. The results of this theoretical exercise are consistent with mutational analyses *(5)*.

More dramatic sequence changes can occur in virus evolution experiments in special cases, e.g., through recombination events or through duplication or deletion of repeated sequence elements. We reported another very special case in studies with HIV-1 variants with a mutated primer-binding site (PBS). This motif binds the $tRNA^{lys3}$ primer for reverse transcription, and the 18-nucleotide PBS sequence can revert to the wt sequence in a single step in which the $tRNA^{lys3}$ sequence is copied in the genome of the viral progeny *(6)*.

3. Clonal Selection of Individual Virus Variants

During virus evolution, several variants will emerge in parallel and competition will result in the outgrowth of the fittest variant. We described a limiting dilution evolution approach that excludes this strong competition force, and that thereby allows the outgrowth of suboptimal virus variants. We applied this protocol for the selection of previously unidentified HIV-1 variants that are resistant to the nucleoside RT inhibitor 3TC (lamivudine) *(7)*. In this approach, cells infected with HIV-1 were cultured in the presence of 3TC. Instead of maintaining a large cell-virus culture, the cells were serially diluted shortly after infection (**Fig. 30.2**). The

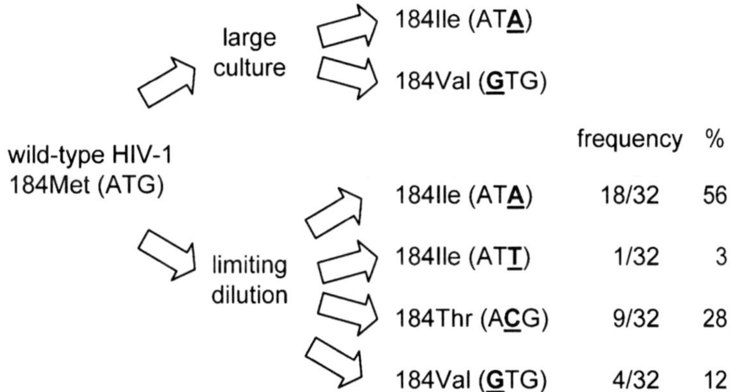

Fig. 30.2. Selection of 3TC resistant HIV-1 variants. Cells infected with HIV-1 were either maintained in a large culture volume or serially diluted shortly after infection, and cultured in the presence of 3TC. In all cultures in which HIV-1 escaped from 3TC inhibition, the virus acquired a single nucleotide substitution (underlined) at codon 184 of the RT gene. The large culture yielded only two types of 3TC-resistant variants, whereas the genotypic analysis of the 32 3TC-resistant viruses resulting from the limiting dilution protocol revealed four variants (the frequency of each variant is indicated).

large-culture studies and 3TC-therapy of HIV-1 infected persons had yielded only two types of 3TC-resistant variants, in which the wt methionine at position 184 within the catalytic domain of the RT enzyme is replaced by either isoleucine or valine. Our limiting dilution protocol yielded a third variant in which a threonine was present at position 184. Indeed, this 184Thr variant does not replicate as efficiently as the 184Ile or 184Val variants, and will easily be outcompeted by these variants in a larger population setting.

By avoiding the forces of competition, evolution is determined solely by the likelihood of generating a particular mutant. The results of the 3TC-selection experiment are fully consistent with this idea, as simple transitions (purine to purine or pyrimidine to pyrimidine mutations) are observed more frequently than more difficult transversions (purine to pyrimidine or pyrimidine to purine), and no double mutations were scored. When a large number of evolution cultures are followed, one will obviously also start to see the less likely events. For instance, we observed eighteen times the expected Ile variant due to a transition (ATG to ATA), but we also detected an alternative Ile codon due to a single transversion (ATG to ATT) in a single culture. In larger populations, all possible variants are present or within reach, and the top fit viruses will be selected.

The limiting dilution protocol allows one to identify suboptimally fit virus variants that otherwise would be lost due to competition. In real life, evolution can be a complex mixture of the mutation and selection steps. For instance, 3TC-treated patients may initially show the 184Ile variant, which is made by the most frequently occurring G-to-A transition *(1)*. This variant is subsequently outcompeted by the 184Val variant, which is more difficult to make via an A-to-G transition, but providing a subtly improved RT activity and viral fitness *(8)*. Thus, both mutation and selection participate in the outcome of most regular evolutionary scenarios.

4. In Vivo SELEX with Viral Genomes Containing Randomized Sequences

The Systematic Evolution of Ligands by EXponential enrichment (SELEX) technique has over the last 10 years proven to be a useful method for the selection of molecules with unique properties *(9)*. Whereas Darwinian evolution usually applies to living organisms over long periods of time, SELEX allows for the rapid in vitro evolution of functionally active nucleic acids from a pool of randomized sequences. We described an in vivo version of this nucleic acid evolution protocol in which selection and amplification take place inside living cells by means of HIV-1 replication *(10)*. In brief, we generated a library of HIV-1 DNA genomes with random sequences in a particular genetic domain. This mixture of HIV-1 genomes was transfected in human T cells and the outgrowth of the fittest viruses was observed within two weeks of viral replication. The flow diagram of a typical in vitro versus in vivo SELEX experiment is shown in **Fig. 30.3**. Compared to the laborious in vitro selection and amplification steps, much time and effort is saved in the in vivo approach. Furthermore, adaptive changes may arise in or outside the randomized genome segment during virus replication as the viral replication machinery is error-prone (**Fig. 30.3**: fine-tuning). We used this approach to probe the sequence requirements for short 3-nucleotide sequence segments in the TAR hairpin structure, which is the binding site for the viral Tat trans-activator protein *(10)*. The method has been successfully applied to study the replication of other retroviruses *(11, 12)*, hepatitis B virus *(13)*, plant viruses *(14, 15)*, and herpes simplex virus type 1 *(16)*. Although many RNA viruses exist as a quasispecies of closely related, but genetically distinct genotypes, their evolutionary potential is restricted because they probe only a limited area of sequence space around the qua-

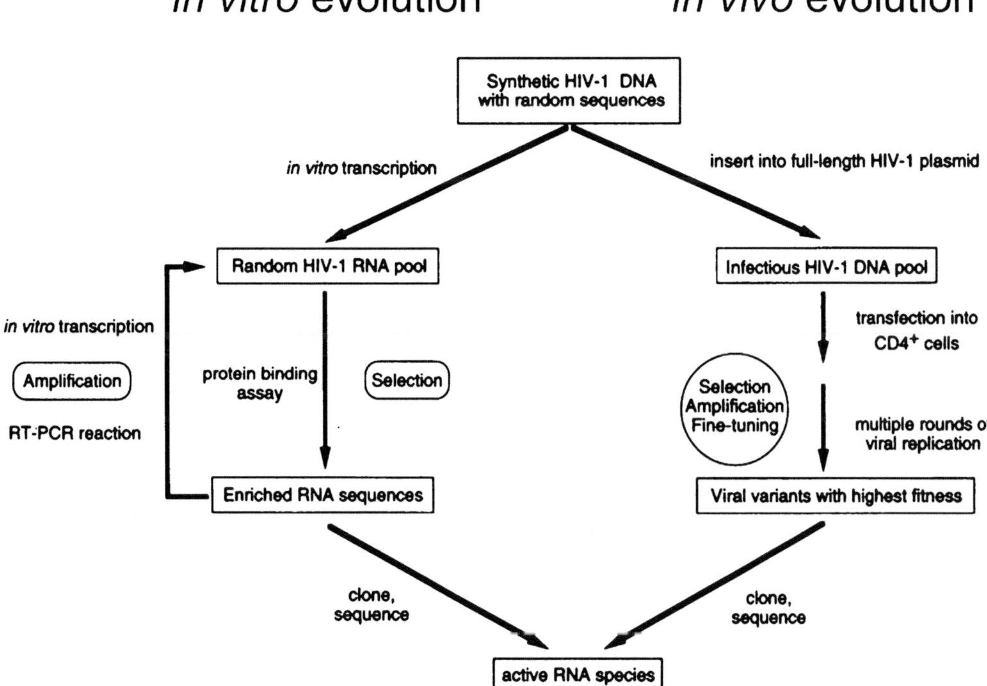

Fig. 30.3. Schematic diagram of the in vitro and HIV-based in vivo SELEX protocols. Reprinted with permission from Berkhout and Klaver (1993) *Nucleic Acids Research* 21 (22) 5020–5024. © 1993 Oxford University Press.

sispecies. In vivo SELEX provides the opportunity to sample more sequence space by randomization of multiple nucleotides, yielding valuable information and new molecules with interesting properties.

5. Second-Site Mutations

RNA interference (RNAi) is an evolutionarily conserved process that may provide the cell with an additional tool to regulate gene expression and to control infecting viruses. RNAi may also be a powerful method for intracellular immunization against HIV-1. Recently, we and others demonstrated long-term inhibition of virus replication in human T cells that stably express small interfering RNAs (siRNAs) directed against the viral Nef gene. However, viral escape variants emerged that were no longer inhibited *(17)*. An overview of nine escape routes/viruses is presented in **Fig. 30.4**. First of all, the escape routes demonstrate the exquisite sequence-specificity, as all – except one escape variant (R8) – carried a mutation within the 19-nucleotide target sequence. Second, this escape overview highlights the mutational variation that can trigger escape: we observed point mutations at different positions, but also deletions that partially or completely delete the target sequence. Third, the exceptional R8 mutation was studied in detail, and turned out to affect the local RNA structure, such that the target sequence is masked and not accessible for the RNAi machinery *(18)*. This result highlights why we usually perform so many evolution experiments in parallel. Evolution is a chance process, and one sometimes will only hit upon more exotic paths by analyzing many independent experiments. Such extensive analyses provide an opportunity to have a look at the sequence space that is available for a virus for rescue of replication.

The exceptional R8 revertant virus provides an example of a second-site mutation. In this case, its effect is understood at the level of RNA structure. In evolution studies with a Tat-inactivated HIV-1 variant, we observed a second-site mutation in the viral Tat protein. This mutation partially restored Tat activity and viral replication, although a structural explanation is currently lacking *(19)*. Sometimes second-site mutations occur in another genome segment. For instance, we selected an up-mutation in the Envelope protein in studies with an HIV-1 variant that had been translationally crippled through modification of the 5′ untranslated leader region *(20)*. Although it is attractive to think about a functional coupling of these two issues, further analysis revealed that the second-site mutation presented a general improvement of viral replication fitness that is not

A

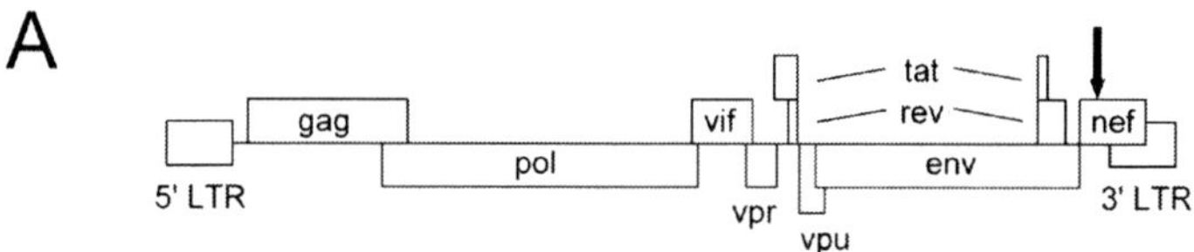

B

siRNA-Nef target (164–182)

	day	sequence	
LAI		GCTGCTTGTGCCTGGCTAGAAGCACAAG	
R1	43	----------------------------	106-nt deletion
R2	46	GCTGCTTGT-----------AGCACAAG	11-nt deletion
R3	27	GCTGCTTGTGCCTGGC**<u>G</u>**AGAAGCACAAG	1-nt substitution
R3'	62	GCTGCTTGT**<u>A</u>**CCTGGC**<u>G</u>**AGAAGCACAAG	2-nt substitution
R4	62	GCTGCT------TGGCTAGAAGCACAAG	6-nt deletion
R5	62	GCTGCTTGTGCCTGGCTAGAAG------	63-nt deletion
R6	80	GCTGCTTGTGCCTGGCTAGA**<u>G</u>**GCACAAG	1-nt substitution
R7	77	----------------------------	225-nt deletion
R8	74	**<u>A</u>**CTGCTTGTGCCTGGCTAGAAGCACAAG	1-nt substitution
R9	61	GCTGCTTGTGCCTGG**<u>A</u>**TAGAAGCACAAG	1-nt substitution

Fig. 30.4. HIV-1 escape variants that resist siRNA-Nef inhibition. (**A**) Schematic of the HIV-1 LAI proviral genome. The position of the siRNA-Nef target sequence is indicated with an arrow. (**B**) HIV-1 LAI variants resistant to siRNA-Nef were selected in nine independent cultures. The Nef target sequence (*gray box* indicates nucleotides 164–182 of the Nef gene) and flanking sequences are shown for the wt (LAI) and the evolved RNAi-resistant viruses (R1–R9). The day at which the escape variants were sequenced is indicated. Deletions are shown as dashes, substitutions are underlined and in bold. In the R1 virus, nucleotides 125–230 of the Nef gene are deleted. In the R5 virus, nucleotides 179–241 are deleted. In the R7 virus, we observed deletion of nucleotides 44–268 and a T269A substitution. Reprinted with permission from Westerhout et al. (2005) *Nucleic Acids Research* 33 (2) 796–804. © 2005 The Author.

directly linked to the leader modifications *(21)*. Indeed, this same Envelope mutation appeared in a completely unrelated evolution study *(22)*.

6. Analysis of RNA Structure

We have successfully used evolutionary techniques to study important structured RNA motifs in the HIV-1 genome. For instance, we were the first to describe the polyA hairpin structure that partially occludes the AAUAAA polyadenylation signal (*see* **Fig. 30.5**, the wt hairpin is shown on top in the middle with the polyadenylation signal in bold). Mutations that weaken this structure produce a viral replication defect, but mutations

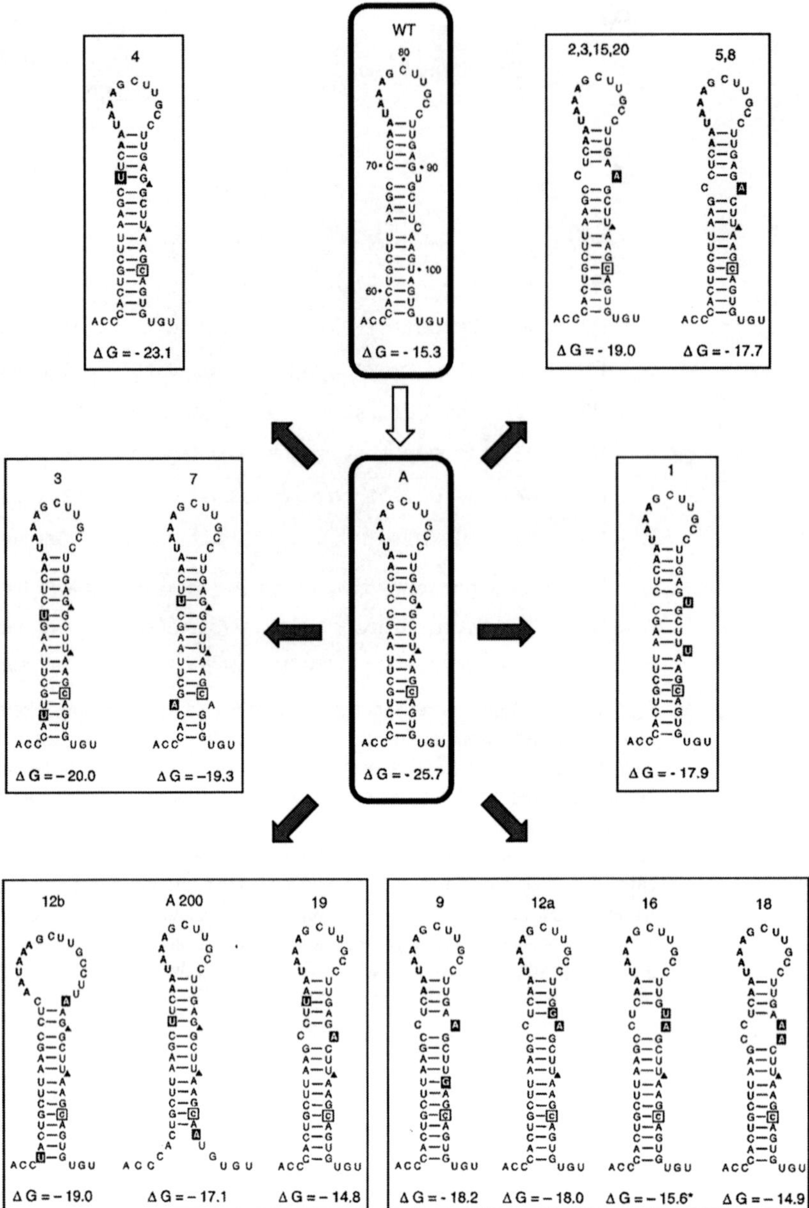

Fig. 30.5. Evolutionary pathways of the stabilized polyA hairpin. The polyA hairpin structure of wt HIV-1 and the stabilized mutant A are presented for reference (nucleotide substitutions are in open boxes, base deletions are indicated by ▲). Evolution of the poorly replicating mutant A in multiple independent long-term cultures resulted in virus variants that acquired additional mutations in the polyA hairpin region. The predicted structure for all evolved variants is shown, with the calculated helix stability indicated below the stem (ΔG in kcal/mol). *Black boxes* mark the differences with the starting sequence of mutant A. The structures are grouped according to the number and position of the acquired mutations. Reprinted with permission from Berkhout et al. (1997) *Nucleic Acids Research* 25 (5) 940–947. © 1997 Oxford University Press.

that further stabilize the hairpin are also notoriously detrimental. For example, in mutant A the base paired stem had been stabilized by removal of two bulging nucleotides (indicated as Δ in **Fig. 30.5**) and replacement of a weak G-U base pair by a strong G-C base pair (boxed). The thermodynamic stability of the hairpin changed from -15.3 to -25.7 kcal/mol due to these mutations. We started multiple independent cultures of this poorly replicating mutant A and we could select multiple variants with improved replication capacity. Sequencing of these evolved variants revealed a pleiotropy of additional mutations in the polyA hairpin and we identified variants with either a single or double mutation on the right and/or left hand side of the stem (**Fig. 30.5**). Contrasting with this enormous variation in repair strategies, we observed an invariable drift towards a hairpin with reduced stability ($\Delta G = -14.8/-23.1$ kcal/mol), and it is likely that some variants did not yet attain the most optimal configuration. Evolution of an HIV-1 mutant in which the polyA hairpin had been destabilized showed the reverse pattern, and resulted in the stabilization of the hairpin.

Very similar evolutionary strategies have been used for the RNA bacteriophage MS2 by the group of Van Duin, one of the few laboratories that have used evolutionary techniques for basic RNA studies *(23–25)*. For us, the results of evolutionary repair studies has led to the description of an alternative HIV-1 RNA leader structure in which the polyA hairpin opens up *(26)*. This riboswitch mechanism may be important in controlling RNA dimerization, and the exquisite sensitivity of the hairpin stability is likely due to the fine-tuning of the riboswitch.

7. Unusual Recombination Events

The systematic use of forced virus evolution techniques will also reveal unusual mutation and recombination events. We came across a novel recombination mechanism by studying the evolution of an HIV-1 mutant with an excessively stable hairpin introduced in the viral genome, which did not show any sign of replication in virus/cell cultures. Moderately stable hairpins can usually be destabilized by one or a few precise nucleotide mutations or deletions *(27)*, but more gross changes are needed to rescue a genome with a 300-bp hairpin. We selected four escape viruses in which the major part of the hairpin was deleted *(28)*. Surprisingly, the hairpin deletion was linked to patch insertion of tRNA sequences (**Fig. 30.6**). The tRNAs were inserted in the viral genome in the antisense orientation, indicating that tRNA-mediated recombination occurred during minus-strand DNA synthesis, and we proposed a detailed model for this hairpin-induced

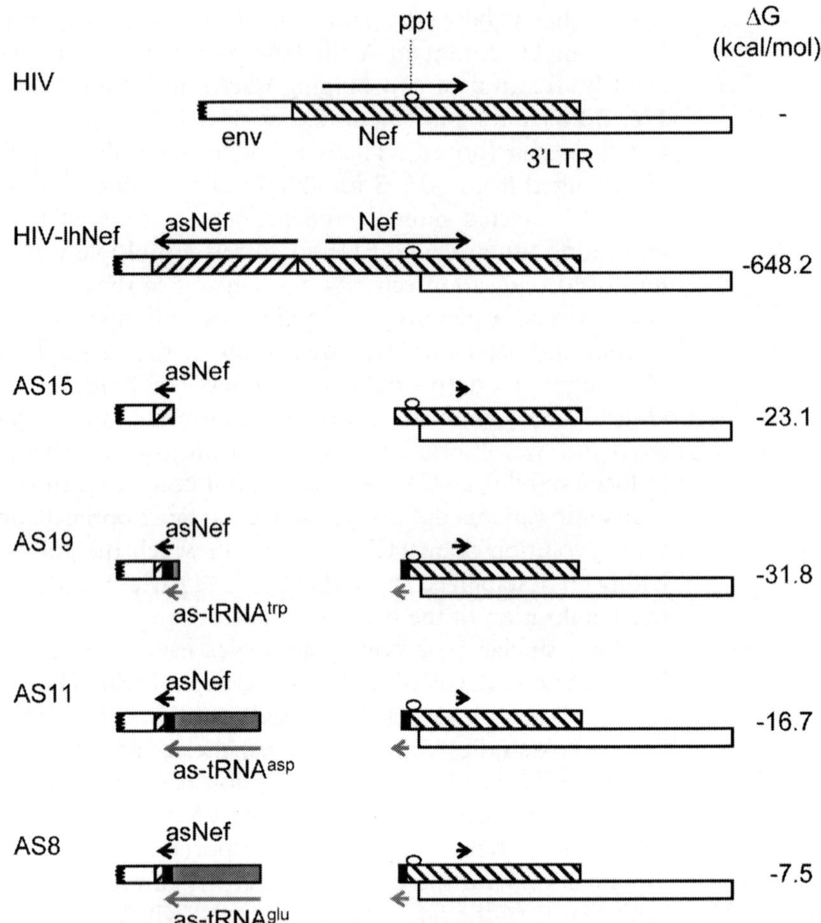

Fig. 30.6. Hairpin-induced tRNA-mediated (HITME) recombination in HIV-1. HIV-lhNef carries a 300-bp long hairpin structure in the Nef gene and does not replicate. In four independent cultures, replication-competent escape variants emerged. The major part of the hairpin was deleted in all escape viruses. In three cases, the hairpin deletion was linked to patch insertion of tRNAasp, tRNAglu or tRNAtrp sequences. The name of the escape viruses reflects the number of base pairs in the remaining RNA hairpin. The ΔG column shows the thermodynamic stability of the perfectly basepaired stem segment. Upstream Env gene and downstream 3′LTR sequences are indicated with *white boxes*. Nef and asNef sequences are indicated with *hatched boxes* and *black arrows* above these boxes indicate the complementarity between Nef and asNef. Human tRNA sequences are inserted in the antisense orientation at the recombination junctions in AS19, AS11 and AS8. The *grey arrows* indicate sequences complementary to the tRNAs, with the *black boxes* indicating the sequence homology with the HIV-lhNef genome and the *grey boxes* indicating the inserted tRNA sequence. Reprinted with permission from Konstantinova et al. (2006) *Nucleic Acids Research* 34 (8) 2206–2218. © 2006 The Author.

tRNA-mediated (HITME) recombination. The transient role of the cellular tRNA molecule as enhancer of retroviral recombination is illustrated by the eventual removal of the inserted tRNA sequences by a subsequent deletion event. The acquisition of new genetic material can also be seen in HIV-infected individuals. For instance, some drug-resistant RT variants acquire an insert in the fingers domain. This insert is usually one or two amino acids, and the corresponding three or six-nucleotide sequence insertion in the viral genome is generated by duplication of the neighboring sequences. However, some unusually long inserts have been described *(29)*. In one case, the insert sequence was shown to be derived from human genome sequences *(30)*.

8. Evolution of HIV-1 Vaccines

Evolutionary research strategies play an important role in the testing of HIV-1 vaccines. For instance, whereas AIDS vaccines based on a live-attenuated virus have shown promise in the SIV-macaque model, virus evolution studies revealed that such vaccines can be unsafe for use in humans *(31)*. The major safety concern is that chronic low level replication of the attenuated virus may eventually lead to selection of fitter and pathogenic virus variants *(32)*. The Δ3 HIV-1 variant, which contains deletions in three non-essential genome segments, was considered a vaccine candidate because it seemed impossible that the virus could evolve to a pathogenic variant through repair of the deletions. However, we demonstrated that this virus can regain replication fitness in prolonged in vitro cell culture infections *(33)*. Indeed, the deletions were not repaired, but an interesting sequence duplication was selected in the LTR promoter, which doubled the number of Sp1 binding sites from three to six and greatly improved virus replication. In general, multiplication or deletion of repeated sequence motifs seems a popular viral evolution strategy *(33, 34)*.

Evolution may also help to improve HIV-1 vaccine candidates. As a novel approach toward a safe live attenuated virus vaccine, we designed a doxycycline-dependent HIV-1 variant with multiple deletions, inactivating mutations and insertions in the viral genome (**Fig. 30.7**) *(35)*. The original construct replicated poorly, but greatly improved variants were subsequently selected in which the introduced components of the Tet-On system for inducible gene expression, the rtTA trans-activator protein and the tetO binding sites, had been optimized. The number of tetO binding sites was quickly reduced from eight to two or three, and the spacing between these motifs was reduced (**Fig. 30.7B,C**) *(34, 36)*. Intriguingly, the altered spacing resembles that of the E. coli Tn10 operon from which these genetic elements were originally derived (**Fig. 30.7D**). We also identified mutations in the rtTA protein that improved its activity more than sevenfold, and its sensitivity for doxycycline more than 100-fold (**Fig. 30.7A**) *(37–39)*. These novel rtTAs do not only improve replication of our conditional-live HIV-1 variant but also the performance of the Tet-On system, which is beneficial for other applications of this regulatory switch. The improvement of an

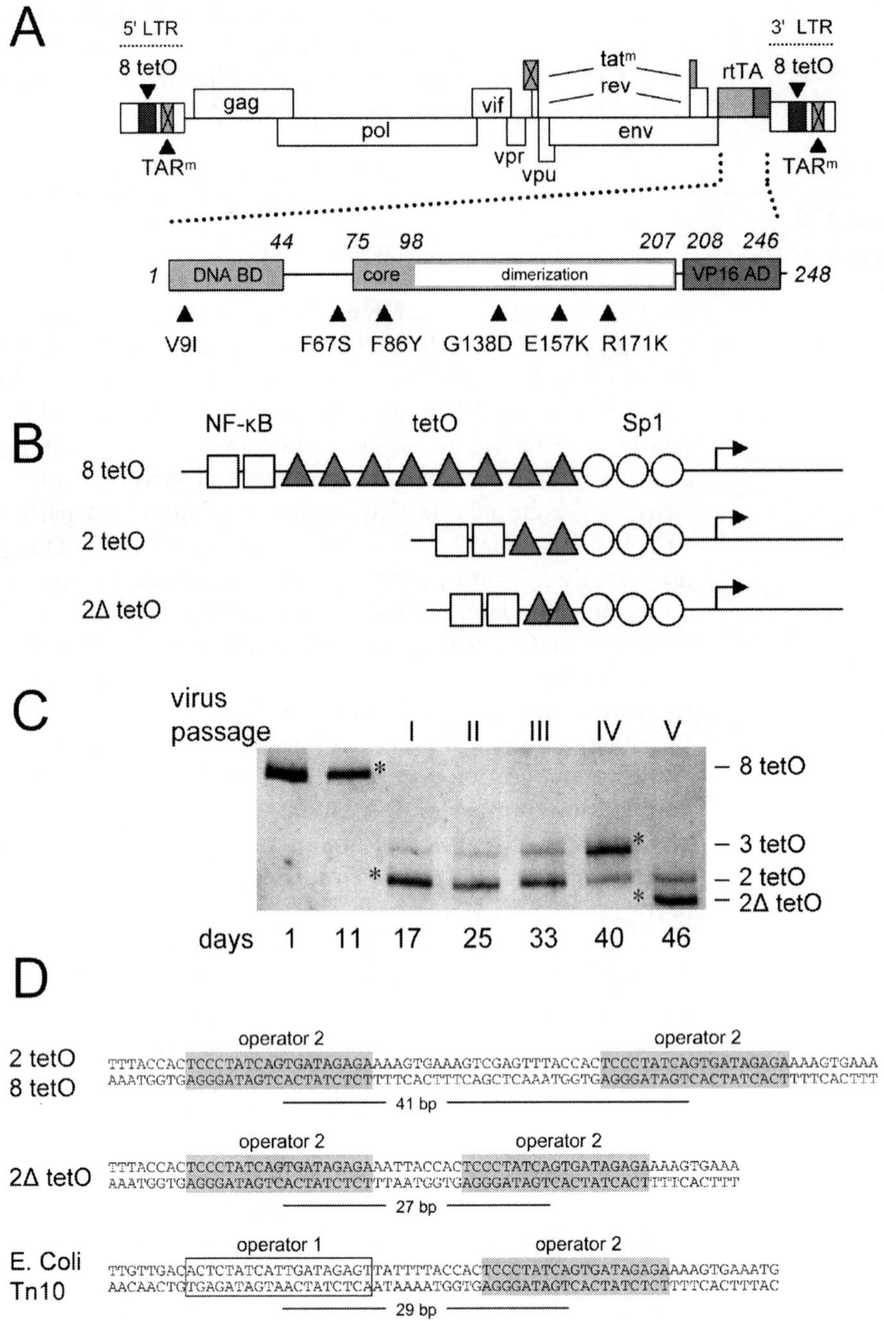

Fig. 30.7. In vitro evolution of the doxycycline-dependent HIV-rtTA virus. (**A**) In the HIV-rtTA virus, the Tat-TAR axis of transcription regulation has been inactivated by mutation of both Tat and TAR (tat^m and TAR^m, crossed boxes). Transcription and replication of the virus were made doxycycline-dependent by introduction of eight tetO elements in the LTR promoter region and replacing the nef gene by the rtTA gene. This 248-amino acid protein is a fusion of the E. coli Tet repressor (TetR, which can be subdivided into a DNA-binding domain [BD] and a regulatory core domain with a dimerization surface) and the VP16 activation domain (AD) of herpes simplex virus. Administration of the doxycycline effector induces a conformational switch in the rtTA protein, which subsequently can bind to the tetO-LTR promoter region and activate transcription of the proviral genome. Thus, transcription and replication of HIV-rtTA are critically dependent

exogenous gene function by means of virus evolution is rather unique, as inserts are usually removed from the viral genome because they do not positively contribute to virus replication. We also anticipated that evolution may not only move into the beneficial direction. Indeed, we also scored loss of doxycycline-control in some extended cultures, and therefore designed rtTA variants that block such unwanted evolution routes *(40, 41)*. We will soon test a similar drug-dependent SIV variant in the macaque model, which should provide relevant information on the safety and efficacy of this vaccine candidate.

9. Discussion

We presented some of the HIV-1 research in which we used virus evolution techniques. The examples provided underscore the enormous potential of forced virus evolution experiments. Although virus evolution in cell culture experiments can accurately mimic several in vivo evolution scenarios, e.g., during antiviral therapy of an HIV-1 infected individual, one selection pressure that is obviously missing in vitro is that of the humoral and cellular immune responses of the adaptive immune system. However, there is accumulating evidence that the innate immune system, including antiviral factors like interferon, APOBEC3G and TRIM5α, plays a major role in the control of invading retroviruses, and components of this system will be active in certain cell culture systems. Although the forced evolution strategy that we describe in this review may sound like a typical man-made research tool, it is also operating in vivo when an attenuated HIV-1 variant is transmitted to an individual *(42, 43)*.

Only a few laboratories have used the virus evolution method in a systematic manner to study certain virus replication steps. The major advantage compared to other assay systems is that forced evolution will disclose new phenomena in a physiological relevant setting. For those researchers that keep their eyes open, there will be surprises along the way. We have trained our students to foster unexpected escape routes because they can form the basis for truly new discoveries. This sometimes requires that the scientist

Fig. 30.7. (continued) on doxycycline. The black triangles indicate amino acid substitutions in rtTA that were observed upon evolution of HIV-rtTA in multiple, independent long-term cultures, and that were found to improve the transcriptional activity and doxycycline-sensitivity of rtTA *(37, 38)*. (**B**) Modified tetO configurations observed upon evolution of HIV-rtTA. (**C**) Representative example of an evolution experiment. The LTR promoter region was PCR-amplified at different times and resolved on gel. Asterisks indicate bands that were identified by sequencing. (**D**) Spacing of tetO elements in the E. coli Tn10 operon, the original HIV-rtTA virus (8tetO) and in evolved 2tetO and 2ΔtetO configurations.

has to ignore the traditional way of doing science, in which a pre-existing model is tested in specific experiments. We prefer to start with an unbiased view, and carefully listen to what the virus tries to tell us. This can provide an unexpected view of the versatility of biology and underscore the enormous potential of evolution. At the same time, this evolution approach can provide new insight into molecular mechanisms and ways to improve certain applications, e.g., gene regulation switches or virus vaccines.

References

1. Berkhout, B., Das, A. T., Beerens, N. (2001) HIV-1 RNA editing, hypermutation and error-prone reverse transcription. *Science* 292, 7.
2. Berkhout, B., de Ronde, A. (2004) APOBEC3G versus reverse transcriptase in the generation of HIV-1 drug-resistance mutations. *AIDS* 18, 1861–1863.
3. Coffin, J. M. (1995) HIV population dynamics in vivo: implications for genetic variation, pathogenesis, and therapy. *Science* 267, 483–489.
4. Keulen, W., Boucher, C., Berkhout, B. (1996) Nucleotide substitution patterns can predict the requirements for drug-resistance of HIV-1 proteins. *Antiviral Res* 31, 45–57.
5. Lacey, S. F., Larder, B. A. (1994) Mutagenic study of codons 74 and 215 of the human immunodeficiency virus type 1 reverse transcriptase, which are significant in nucleoside analog resistance. *J Virol* 5, 3421–3424.
6. Das, A. T., Klaver, B., Berkhout, B. (1995) Reduced replication of human immunodeficiency virus type 1 mutants that use reverse transcription primers other than the natural tRNA(3Lys). *J Virol* 69, 3090–3097.
7. Keulen, W., Back, N. K. T., van Wijk, A., et al. (1997) Initial appearance of the 184Ile variant in lamivudine-treated patients is caused by the mutational bias of the human immunodeficiency virus type 1 Reverse Transcriptase. *J Virol* 71, 3346–3350.
8. Back, N. K. T., Nijhuis, M., Keulen, W., et al. (1996) Reduced replication of 3TC-resistant HIV-1 variants in primary cells due to a processivity defect of the reverse transcriptase enzyme. *EMBO J* 15, 4040–4049.
9. Breaker, R. R. (2004) Natural and engineered nucleic acids as tools to explore biology. *Nature* 432, 838–845.
10. Berkhout, B., Klaver, B. (1993) In vivo selection of randomly mutated retroviral genomes. *Nucleic Acids Res* 21, 5020–5024.
11. Morris, S., Johnson, M., Stavnezer, E., et al. (2002) Replication of avian sarcoma virus in vivo requires an interaction between the viral RNA and the TpsiC loop of the tRNA(Trp) primer. *J Virol* 76, 7571–7577.
12. Doria-Rose, N. A., Vogt, V. M. (1998) In vivo selection of Rous sarcoma virus mutants with randomized sequences in the packaging signal. *J Virol* 72, 8073–8082.
13. Rieger, A., Nassal, M. (1995) Distinct requirements for primary sequence in the 5′- and 3′-part of a bulge in the hepatitis B virus RNA encapsidation signal revealed by a combination in vivo selection/in vitro amplification system. *Nucleic Acids Res* 23, 3909–3915.
14. Sun, X., Zhang, G., Simon, A. E. (2005) Short internal sequences involved in replication and virion accumulation in a subviral RNA of turnip crinkle virus. *J Virol* 79, 512–524.
15. Zhang, J., Simon, A. E. (2005) Importance of sequence and structural elements within a viral replication repressor. *Virology* 333, 301–315.
16. Yoon, M., Spear, P. G. (2004) Random mutagenesis of the gene encoding a viral ligand for multiple cell entry receptors to obtain viral mutants altered for receptor usage. *Proc Natl Acad Sci USA* 101, 17252–17257.
17. Das, A. T., Brummelkamp, T. R., Westerhout, E. M., et al. (2004) Human immunodeficiency virus type 1 escapes from RNA interference-mediated inhibition. *J Virol* 78, 2601–2605.
18. Westerhout, E. M., Ooms, M., Vink, M., et al. (2005) HIV-1 can escape from RNA interference by evolving an alternative structure in its RNA genome. *Nucleic Acids Res* 33, 796–804.
19. Verhoef, K., Berkhout, B. (1999) A second-site mutation that restores replication of a Tat-defective human immunodeficiency virus. *J Virol* 73, 2781–2789.
20. Das, A. T., van Dam, A. P., Klaver, B., et al. (1998) Improved envelope function selected by long-term cultivation of a translation-impaired HIV-1 mutant. *Virology* 244, 552–562.
21. Das, A. T., Land, A., Braakman, I., et al. (1999) HIV-1 evolves into a non-syncytium-

inducing virus upon prolonged culture in vitro. *Virology* 263, 55–69.
22. Baldwin, C. E., Berkhout, B. (2006) Second site escape of a T20-dependent HIV-1 variant by a single amino acid change in the CD4 binding region of the envelope glycoprotein. *Retrovirology* 3, 84.
23. Licis, N., Balklava, Z., van Duin, J. (2000) Forced retroevolution of an RNA bacteriophage. *Virology* 271, 298–306.
24. Klovins, J., Tsareva, N. A., de Smit, M. H., et al. (1997) Rapid evolution of translational control mechanisms in RNA genomes. *J Mol Biol* 265, 372–384.
25. Olsthoorn, R. C. L., van Duin, J. (1996) Evolutionary reconstruction of a hairpin deleted from the genome of an RNA virus. *Proc Natl Acad Sci USA* 93, 12256–12261.
26. Huthoff, H., Berkhout, B. (2001) Two alternating structures for the HIV-1 leader RNA. *RNA* 7, 143–157.
27. Berkhout, B., Klaver, B., Das, A. T. (1997) Forced evolution of a regulatory RNA helix in the HIV-1 genome. *Nucleic Acids Res* 25, 940–947.
28. Konstantinova, P., de Haan, P., Das, A. T., et al. (2006) Hairpin-induced tRNA-mediated (HITME) recombination in HIV-1. *Nucleic Acids Res* 34, 2206–2218.
29. van der Hoek L., Back, N., Jebbink, M. F., et al. (2005) Increased multinucleoside drug resistance and decreased replicative capacity of a human immunodeficiency virus type 1 variant with an 8-amino-Acid insert in the reverse transcriptase. *J Virol* 79, 3536–3543.
30. Takebe, Y., Telesnitsky, A. (2006) Evidence for the acquisition of multi-drug resistance in an HIV-1 clinical isolate via human sequence transduction. *Virology* 351, 1–6.
31. Mills, J., Desrosiers, R., Rud, E., et al. (2000) Live attenuated HIV vaccines: proposal for further research and development. *AIDS Res Hum Retroviruses* 16, 1453–1461.
32. Baba, T. W., Liska, V., Khimani, A. H., et al. (1999) Live attenuated, multiply deleted simian immunodeficiency virus causes AIDS in infant and adult macaques. *Nat Med* 5, 194–203.
33. Berkhout, B., Verhoef, K., van Wamel, J. L., et al. (1999) Genetic instability of live, attenuated human immunodeficiency virus type 1 vaccine strains. *J Virol* 73, 1138–1145.
34. Marzio, G., Verhoef, K., Vink, M., et al. (2001) In vitro evolution of a highly replicating, doxycycline-dependent HIV for applications in vaccine studies. *Proc Natl Acad Sci USA* 98, 6342–6347.
35. Verhoef, K., Marzio, G., Hillen, W., et al. (2001) Strict control of human immunodeficiency virus type 1 replication by a genetic switch: Tet for Tat. *J Virol* 75, 979–987.
36. Marzio, G., Vink, M., Verhoef, K., et al. (2002) Efficient human immunodeficiency virus replication requires a fine-tuned level of transcription. *J Virol* 76, 3084–3088.
37. Das, A. T., Zhou, X., Vink, M., et al. (2004) Viral evolution as a tool to improve the tetracycline-regulated gene expression system. *J Biol Chem* 279, 18776–18782.
38. Zhou, X., Vink, M., Klaver, B., et al. (2006) Optimization of the Tet-On system for regulated gene expression through viral evolution. *Gene Ther* 13, 1382–1390.
39. Zhou, X., Symons, J., Hoppes, R., et al. (2007) Improved single-chain transactivators of the Tet-On gene expression system. *BMC Biotechnol* 7, 6.
40. Zhou, X., Vink, M., Klaver, B., et al. (2006) The genetic stability of a conditional-live HIV-1 variant can be improved by mutations in the Tet-On regulatory system that restrain evolution. *J Biol Chem* 281, 17084–17091.
41. Zhou, X., Vink, M., Berkhout, B., et al. (2006) Modification of the Tet-On regulatory system prevents the conditional-live HIV-1 variant from losing doxycycline-control. *Retrovirology* 3, 82.
42. Churchill, M. J., Rhodes, D. I., Learmont, J. C., et al. (2006) Longitudinal analysis of human immunodeficiency virus type 1 nef/long terminal repeat sequences in a cohort of long-term survivors infected from a single source. *J Virol* 80, 1047–1052.
43. Huthoff, H., Das, A. T., Vink, M., et al. (2004) A human immunodeficiency virus type 1-infected individual with low viral load harbors a virus variant that exhibits an in vitro RNA dimerization defect. *J Virol* 78, 4907–4913.

INDEX

U

V

W

X

Y

Z

Lightning Source UK Ltd.
Milton Keynes UK
UKOW021944231212

204058UK00001B/6/P

UK1TE